Nano-Optics and Nanophotonics

The Springer Series in Nano-Optics and Nanophotonics provides an expanding selection of research monographs in the area of nano-optics and nanophotonics, science- and technology-based on optical interactions of matter in the nanoscale and related topics of contemporary interest. With this broad coverage of topics, the series is of use to all research scientists, engineers and graduate students who need up-to-date reference books. The editors encourage prospective authors to correspond with them in advance of submitting a manuscript. Submission of manuscripts should be made to the editor-in-chief, one of the editors or to Springer.

More information about this series at http://www.springer.com/series/8765

Peter Michler
Editor

Quantum Dots for Quantum Information Technologies

Springer

Editor
Peter Michler
Institut für Halbleiteroptik und Funktionelle Grenzflächen
Universität Stuttgart
Stuttgart
Germany

ISSN 2192-1970 ISSN 2192-1989 (electronic)
Nano-Optics and Nanophotonics
ISBN 978-3-319-56377-0 ISBN 978-3-319-56378-7 (eBook)
DOI 10.1007/978-3-319-56378-7

Library of Congress Control Number: 2017936032

Printed on acid-free paper

This Springer imprint is published by Springer Nature
The registered company is Springer International Publishing AG
The registered company address is: Gewerbestrasse 11, 6330 Cham, Switzerland

Preface

Many people worldwide are fascinated from quantum information science and from the prospects of novel quantum technologies, such as quantum computing, quantum communication, quantum metrology, and quantum sensing. Semiconductor quantum dots (QDs) have been identified as a promising hardware for implementing the basic building blocks, e.g., stationary and flying qubits in the solid state. This is because individual charge carriers in QDs can be generated, manipulated, and coherently controlled and can be strongly decoupled from their environment, so that processes destroying the coherence of the qubits can be largely suppressed. Moreover, miniaturized and integrated solutions with existing semiconductor technology are foreseeable.

Through a huge common effort during the last nearly two decades, the semiconductor quantum optics community has made important progress in spin manipulation, the generation of indistinguishable single and entangled photon states, controlling the light–matter interaction, and spin–photon entanglement. For example, 16 years after the first demonstration of a QD-based triggered single-photon source near-optimal QD single-photon sources are nowadays available. They clearly outperform the up to now most used spontaneous parametric down-conversion sources with respect to brightness for comparable photon indistinguishability. This breakthrough was possible by the continuous and common research efforts of many research groups. Important milestones are the development of optimized QDs and microcavity structures, control of charge fluctuations, the introduction of a fully deterministic fabrication processes, and truly resonant optical excitation techniques of QDs.

Our book aims to provide an overview of recent exciting developments in the field of semiconductor quantum optics with quantum dots. The topics addressed include the theory of cavity QED and phonon-dressed light–matter interactions, resonantly excited quantum dots for indistinguishable single-photon emission, polarization, time-bin entangled photon generation, and superradiance. Spin properties with a special emphasis on noise properties, on ultrafast manipulation of exciton spins, on nanophotonic spin–photon interface and spin–photon entanglement are also discussed. The last part is devoted to photonic integrated circuits with quantum dots.

Finally, I would like to thank all my colleagues for writing the various chapters and the very good cooperation in the course of outlining the book and editing their chapters.

Stuttgart, Germany Peter Michler

Contents

Contributors

Alistair J. Brash Department of Physics and Astronomy, University of Sheffield, Sheffield, UK

D. Cogan Department of Physics and the Solid State Institute, Technion Israel Institute of Technology, Haifa, Israel

Aymeric Delteil Institute of Quantum Electronics, ETH Zurich, Zurich, Switzerland

Fei Ding Institute for Integrative Nanosciences, IFW Dresden, Dresden, Germany; Institute for Solid State Physics, Leibniz University of Hannover, Hannover, Germany

Y. Don Department of Physics and the Solid State Institute, Technion Israel Institute of Technology, Haifa, Israel

Matthias Florian Institut für Theoretische Physik, Universität Bremen, Bremen, Germany

A. Mark Fox Department of Physics and Astronomy, University of Sheffield, Sheffield, UK

L. Gantz Department of Physics and the Solid State Institute, Technion Israel Institute of Technology, Haifa, Israel

Wei-bo Gao Division of Physics and Applied Physics, Nanyang Technological University, Singapore, Singapore

D. Gershoni Department of Physics and the Solid State Institute, Technion Israel Institute of Technology, Haifa, Israel

Christopher Gies Institut für Theoretische Physik, Universität Bremen, Bremen, Germany

T. Heindel Institut für Festkörperphysik, Technische Universität Berlin, Berlin, Germany

Tobias Huber Institut für Experimentalphysik, Universität Innbruck, Innsbruck, Austria; Joint Quantum Institute, National Institute of Standards and Technology and University of Maryland, Gaithersburg, MD, USA

Stephen Hughes Department of Physics, Engineering Physics and Astronomy, Queen's University, Kingston, ON, Canada

Ataç Imamoğlu Institute of Quantum Electronics, ETH Zurich, Zurich, Switzerland

Frank Jahnke Institut für Theoretische Physik, Universität Bremen, Bremen, Germany

Michael Jetter Research Center SCoPE and IQST, Institut für Halbleiteroptik und Funktionelle Grenzflächen (IHFG), University of Stuttgart, Stuttgart, Germany

Feng Liu Department of Physics and Astronomy, University of Sheffield, Sheffield, S3 7RH, UK

P. Lodahl The Niels Bohr Institute, University of Copenhagen, Copenhagen, Denmark

Peter Michler Research Center SCoPE and IQST, Institut für Halbleiteroptik und Funktionelle Grenzflächen (IHFG), University of Stuttgart, Stuttgart, Germany

Simone Luca Portalupi Institut für Halbleiteroptik und Funktionelle Grenzflächen, University of Stuttgart, Stuttgart, Germany

Ana Predojević Institute for Quantum Optics, University Ulm, Ulm, Germany

S. Reitzenstein Institut für Festkörperphysik, Technische Universität Berlin, Berlin, Germany

Ulrich Rengstl Research Center SCoPE and IQST, Institut für Halbleiteroptik und Funktionelle Grenzflächen (IHFG), University of Stuttgart, Stuttgart, Germany

S. Rodt Institut für Festkörperphysik, Technische Universität Berlin, Berlin, Germany

Kaushik Roy-Choudhury Department of Physics, Engineering Physics and Astronomy, Queen's University, Kingston, ON, Canada

E.R. Schmidgall Department of Physics and the Solid State Institute, Technion Israel Institute of Technology, Haifa, Israel; Department of Physics, University of Washington, Seattle, WA, USA

Oliver G. Schmidt Institute for Integrative Nanosciences, IFW Dresden, Dresden, Germany

I. Schwartz Department of Physics and the Solid State Institute, Technion Israel Institute of Technology, Haifa, Israel

A.S. Sørensen The Niels Bohr Institute, University of Copenhagen, Copenhagen, Denmark

Alexander Steinhoff Institut für Theoretische Physik, Universität Bremen, Bremen, Germany

S. Stobbe The Niels Bohr Institute, University of Copenhagen, Copenhagen, Denmark

Shuo Sun Department of Electrical and Computer Engineering, Institute for Research in Electronics and Applied Physics (IREAP), and Joint Quantum Institute (JQI), University of Maryland, College Park, MD, USA

Zhe Sun Institute of Quantum Electronics, ETH Zurich, Zurich, Switzerland

P. Tighineanu The Niels Bohr Institute, University of Copenhagen, Copenhagen, Denmark

Edo Waks Department of Electrical and Computer Engineering, Institute for Research in Electronics and Applied Physics (IREAP), and Joint Quantum Institute (JQI), University of Maryland, College Park, MD, USA

Richard J. Warburton Department of Physics, University of Basel, Basel, Switzerland

Gregor Weihs Institut für Experimentalphysik, Universität Innbruck, Innsbruck, Austria

Part I
Theory of Light-Matter Interaction

Chapter 1
Theory of Quantum Light Sources and Cavity-QED Emitters Based on Semiconductor Quantum Dots

Christopher Gies, Matthias Florian, Alexander Steinhoff and Frank Jahnke

Abstract The first chapter presents from a theoretical perspective fundamentals and advances made in the field of quantum light sources and cavity-QED devices that are based on self-organized semiconductor quantum dots (QDs) as active material. We summarize key physical properties of QDs as embedded solid-state emitters and how to account for their semiconductor properties, such as carrier scattering, dephasing, and non-resonant coupling in microscopic theories. In combination with a quantization of the electromagnetic field, these models allow for a quantitative description of device properties and non-classical effects that render few-emitter microcavity systems so useful for applications in the quantum-information technologies.

1.1 Introduction

Research on quantum-optical light sources is to a large extent driven by applications in the so-called *new quantum technologies* [1], such as quantum computing, sensing, metrology, and cryptography, that rely on the preparation, use, and control of quantum-mechanical properties of matter or light. Photons are the smallest units of energy of the quantized electromagnetic field and play a central role, as they can be manipulated individually. At the same time they propagate at the speed of light, enabling the transfer of information over large distances ("flying qubits"). While laser physics has largely advanced to an applied and technological stage, applications that make use of the quantum-mechanical properties of photons are one fundamental research topic of this decade. The development of new devices demands bright, efficient, and integrable quantum light sources that easily couple into optical fibres. Neutral atoms and ions in traps, solid-state emitters such as quantum dots and defects, molecules, and even nanomechanical systems are being investigated for

All authors have contributed in equal amounts to this chapter.

C. Gies (✉) · M. Florian · A. Steinhoff · F. Jahnke
Institut für Theoretische Physik, Universität Bremen, Bremen, Germany
e-mail: gies@itp.uni-bremen.de

P. Michler (ed.), *Quantum Dots for Quantum Information Technologies*, Nano-Optics and Nanophotonics, DOI 10.1007/978-3-319-56378-7_1

their potential, and interfacing these different approaches in terms of wavelength and bandwidth into "hybrid systems" is a key challenge of current research.

In this chapter, we present from a theoretical perspective fundamentals and advances made in the field of quantum light sources and cavity-QED devices that are based on self-organized semiconductor quantum dots (QDs) as active material. Semiconductor QDs are often regarded as artificial atoms due to the three-dimensional confinement potential leading to discrete electronic states. At a closer look, however, even QDs with only few confined electronic levels possess a large number of multi-exciton configurations due to many possibilities to accommodate excited carriers in these states. The Coulomb interaction separates many of these configurations energetically, thereby creating a large manifold of transition energies. This and the coupling of the QDs to their environment via phonons and excited carriers outside the QD result in two important phenomena, namely dephasing, and the possibility to enable the emission from a QD into a cavity mode even if its ground state transition is detuned by several meV. Both phenomena are an integral part of QD physics and distinguish their behavior from that of atoms. For the quantum-optical applications considered in this chapter that mostly rely on few or even a single emitter, they play an important role. In Chap. 2, nonresonant QD-cavity coupling is a central ingredient of the theory on resonance fluorescence.

After giving a short overview of theoretical approaches to solid-state cavity-QED and the characterization of light, this chapter contains four main topics. The wide range of emission properties that can be realized from single or few QDs in microcavities is discussed in Sect. 1.2. This includes cavity-QED lasing with few emitters and lasing in the presence of strong coupling, single-photon emission from few-QDs in a cavity, the direct generation of entangled photon pairs by cavity-enhanced two-photon emission, and single-photon generation with long electrical excitation pulses. In Sect. 1.3 we analyze in detail carrier scattering and dephasing processes. Section 1.4 addresses different physical mechanisms of non-resonant QD-cavity coupling. In Sect. 1.5 the role of radiative inter-emitter coupling giving rise to superradiance in QD nanolasers is discussed. It is complementary but related to the single-photon superradiance of Sect. 5.3, and both effects can strongly enhance the light-matter interaction in QD-based systems.

1.1.1 QDs Coupled to the Quantized Light Field

A central element of the theoretical description is the Jaynes–Cummings interaction of the electronic QD excitations with the quantized electromagnetic field. In atom-like systems, one typically considers a few-level model, where the electronic transitions between two selected levels is resonantly coupled to the cavity field. In QDs, the light field interacts with the multi-exciton states. These are formed from the discrete single particle states (which are the result of the three-dimensional confinement potential) and the many-body Coulomb configuration interaction.

For a small number of discrete electronic transitions, which are coupled to the Fock states of a single-mode cavity field, the Hilbert space of this system is small enough that it is possible to describe the full quantum dynamics in terms of matrix elements of the statistical operator $\rho(t)$, which follows the von Neumann-Lindblad equation,

$$\frac{\partial}{\partial t}\rho = -\frac{i}{\hbar}[H, \rho] + \sum_X \mathscr{L}_X(\rho)\,. \tag{1.1}$$

In H enters the Jaynes–Cummings Hamiltonian,

$$H_{JC} = \sum_i g_i \left(b^\dagger \sigma_i^- + \sigma_i^+ b\right)\,, \tag{1.2}$$

where $b^\dagger$ and b are creation and annihilation operators for photons in the cavity mode, $\sigma_i^\pm$ are raising and lowering operators for the electronic excitations, and g_i is the corresponding light-matter coupling strength.

When the QD is weakly excited, it is usually sufficient to consider only optical transitions between the ground state and the energetically lowest exciton state, or the biexciton to exciton to ground-state cascade. In this case, the QD-basis consists of a small set of multi-exciton states, and in the Hamiltonian $H = H_0 + H_{JC}$, the free part H_0 contains the multi-exciton energies while in H_{JC} the operators $\sigma_i^\pm$ describe transitions between different multi-exciton configurations i. Such an approach has been used, e.g., in [2–6].

At stronger pumping, several excited carriers can be present in the QD confinement potential, and their Coulomb configuration interaction leads to a large number of possible multi-exciton states. Furthermore, at higher excitation of the system, typically additional excited carriers reside in the delocalized QD barrier states. This environment leads to two important effects. The many-body Coulomb interaction between the QD carriers and the excited charge carriers outside the QD results in dephasing processes. Dephasing is also present due to the interaction of the QD carriers with phonons. Secondly, excited charge carriers outside the QD are responsible for screening of the Coulomb interaction, thus changing the configuration interaction of the QD carriers.

To describe such a situation, it is more suitable to use as basis states the many-particle configurations formed from products (Slater determinants) of single-particle states. Then the multi-exciton states follow from explicitly including the Coulomb interaction H_{Coul} in the Hamiltonian $H = H_0 + H_{Coul} + H_{JC}$ of (1.1). In this case, $\sigma_i^\pm$ describe transitions between different product states. Screening of the Coulomb interaction and the resulting dynamical changes of the multi-exciton configurations can be treated via the time dependent solution of (1.1). When including the Hamiltonian for the full QD-carrier Coulomb interaction, the Jaynes–Cummings interaction among multi-exciton states is obtained. The corresponding approach has been used in [7–9] and is the basis for various investigations discussed in this contribution.

The last term in (1.1) describes dissipation due to the coupling of the system of QD excitations and cavity photons to their environments. Here, X labels all possible transitions taking place in the system assisted by such interaction processes. Losses of cavity photons, in this chapter denoted by the rate κ, can be described via a coupling of the cavity mode to a bath of external modes [10]. Similarly, the spontaneous recombination of QD excitations into a quasi-continuum of lossy modes can be considered [10]. More challenging is the inclusion of semiconductor models for QD-carrier scattering and the resulting dephasing. How this can be realized and under which assumptions the carrier-phonon interaction and the Coulomb interaction with carriers in delocalized states can be cast into the form of Lindblad terms is the topic of Sect. 1.3. Also, the non-resonant QD-cavity coupling via many-body interaction can be described in this way, as discussed in Sect. 1.4.

1.1.2 Characterization of Light

In many experiments with semiconductor QDs in optical resonators it became customary to characterize the cavity field in terms of correlation properties of photons. Here one can distinguish between classical emission properties, like the intensity $\langle n \rangle$ (zero-order correlations) or coherence properties of the emission (first-order correlations). True quantum properties can be revealed in the intensity-autocorrelation function (second-order correlations) and in higher-order correlation functions [11].

The normalized first-order correlation function,

$$g^{(1)}(t,\tau) = \frac{\langle b^{\dagger}(t)\, b(t+\tau)\rangle}{\langle b^{\dagger}(t)\, b(t)\rangle}, \tag{1.3}$$

describes field-amplitude correlations as measured with a Michelson interferometer. In a stationary situation, when $g^{(1)}$ is independent of t, the coherence time of the radiation is given by

$$\tau_c = \int_{-\infty}^{\infty} d\tau \, \left|g^{(1)}(\tau)\right|^2 \tag{1.4}$$

and the emission spectrum follows from a Fourier transform of $g^{(1)}(\tau)$:

$$S(\omega) = \frac{1}{2\pi} \int_{-\infty}^{\infty} d\tau \, e^{-i\omega\tau} g^{(1)}(\tau). \tag{1.5}$$

Information on statistical properties of the radiation are contained in the second-order, or intensity autocorrelation function,

$$g^{(2)}(t,\tau) = \frac{\langle b^{\dagger}(t)\, b^{\dagger}(t+\tau)\, b(t+\tau)\, b(t)\rangle}{\langle b^{\dagger}(t)\, b(t)\rangle \langle b^{\dagger}(t+\tau)\, b(t+\tau)\rangle}. \tag{1.6}$$

In a stationary situation, $g^{(2)}(\tau = 0)$ can be used to identify thermal radiation, for which $g^{(2)}(0) = 2$, indicating high likelihood of finding two photons at the same time (photon bunching). This can be distinguished from coherent radiation obeying Poisson statistics, where $g^{(2)}(0) = 1$. This situation is realized in an ideal laser, where the only noise arises from the randomness of spontaneous emission. For nonclassical light the intensity correlation is $g^{(2)}(0) < 1$ (photon antibunching).

If the system is driven by an excitation pulse, no stationary state exists, and the correlation functions are explicitly t-dependent. It is straightforward to calculate $g^{(2)}(t, \tau = 0)$ e.g. by evaluating the expectation value from the time-dependent density operator, see Sect. 1.5. Delay-time τ dependent results can be obtained with the help of the quantum regression theorem [12]. In a density-matrix approach, this procedure is explained in detail in [13]. The spirit of the quantum-regression theorem can be transfered to hierarchies of equations that are obtained by cluster-, or correlation-expansion methods [14–19] to obtain $g^{(1)}(t, \tau)$ [20, 21] and $g^{(2)}(t, \tau)$ [20]. The calculation of $g^{(2)}(\tau)$ under pulsed excitation conditions is more intricate and requires averaging over a multitude of two-time calculations [22].

Finally, we point out the link between the second-order photon correlation function and the photon-number probability distribution function (a.k.a. photon statistics) p_n, which refers to the diagonal elements of the photon density matrix after the electronic degrees of freedom have been traced out. For equal-time operators, (1.6) can be rewritten as

$$g^{(2)}(t, \tau = 0) = \frac{\langle n^2(t)\rangle - \langle n(t)\rangle}{\langle n(t)\rangle^2}, \tag{1.7}$$

where $\langle n(t)\rangle = \langle b^\dagger(t)b(t)\rangle$. As the photon operators act only on the photonic degrees of freedom, in this representation it is straightforward to see that $\langle n\rangle$ and $\langle n^2\rangle$ are the first and second moments of the photon statistics, respectively. While a distribution function can be accurately represented by *all* of its moments, the autocorrelation function $g^{(2)}(t, \tau = 0)$ therefore contains only partial information on the full photon statistics. As such, an interpretation of $g^{(2)}(t, \tau = 0)$ requires at least some intuition about the underlying distribution function, otherwise results can be misleading. We give an example in Sect. 1.2.1.

1.2 Emission of Single and Few-QD Microcavity Systems

The discrete level structure and tunability of the electronic properties of semiconductor QDs can be exploited to design deterministic quantum-light sources. A single QD driven with short optical excitation pulses produces antibunched single photons on demand [23], the demonstration [24] of which has played a great part to promote solid-state systems into the quantum-information research field. The fabrication of nanostructures, where a single QD is embedded in a microresonator, offers particular advantages for applications both in quantum-information technologies and lasers. The confinement of the light field can be used to enhance the emission from

Table 1.1 Different operational regimes of single or few QDs in optical cavities and the quantities used to characterize the quantum-optical properties of the emission. The last point in the list refers to [27]

Operational regime:	***Characterisation:***
Lasing in weak coupling regime *(1-many QDs)* medium to strong excitation	$g^{(2)}(0) = 1$, $\langle n\rangle > 1$, increase in coherence time
Lasing in strong coupling regime *(1 QD)* high ratio of g/κ, low to medium excitation	$g^{(2)}(0) = 1$, footprints of higher excited states in emission spectrum
Sub- and superradiant emission *(2-1000 QDs)* QD-QD coupling via light field	super-thermal $g^{(2)}(0) \gg 2$, faster spontaneous emission
Single-photon emission *(1 QD)* no cavity or low-Q cavity, weak excitation	$g^{(2)}(0) \cong 0$, Hong-Ou-Mandel
Cavity-enhanced single-photon emission *(1-few QDs)* right balance between cavity losses, strong excitation	error/purity η of single-photon emission
Cavity-selected entangled-photon-pair emission *(1 QD)* high-Q cavity tuned to two-photon resonance	concurrence, fidelity, $g^{(2)}(0) > g^{(2)}(\tau > 0)$
N-photon Fock-state emission *(1 QD)* strong light-matter coupling, detuned excitation	time-ordered $g^{(2)}$

transitions between QD many-particle configurations, promising higher repetition rates for single-photon sources and smaller losses for high-efficiency lasers. At the same time, the cavity acts as a spectral filter and introduces the Jaynes–Cummings nonlinearity, enabling the use of effects like the photon blockade [25, 26]. The surprisingly diverse range of regimes in which QD-microcavity systems can operate may be fascinating and confusing at the same time. Table 1.1 is an attempt to provide an overview.

1.2.1 Single Photons from a Single and Few QDs in a Cavity

A typical single-photon source is realized without a cavity, or with a weakly-reflecting mirror to enhance the emission directionality. More recent technological advances have enabled the fabrication of microresonators that contain exactly one QD emitter in the field maximum of a single mode [28], allowing to use the benefits of cavity-QED effects for the design of single-photon sources [29–31].

The emission properties of a single QD in a cavity can vary from single-photon emission to lasing [7, 8, 32, 33]: In the low-Q cavity regime antibunched emission ($g^{(2)}(0) = 0$) of the single QD dominates. On the other hand, high-Q cavities with photon losses on ps timescales enable the accumulation of photons such that the single emitter can be driven close to or even into the regime of lasing, where coherent emission results from a Poissonian cavity light field. In the left panel of Fig. 1.1

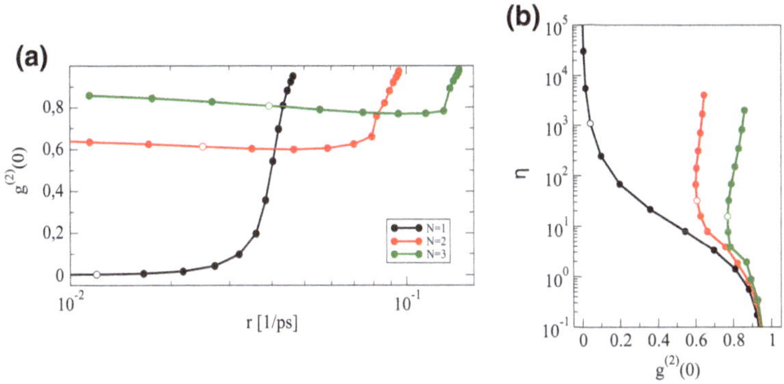

Fig. 1.1 **a** Autocorrelation function $g^{(2)}(0)$ as a function of the photon output rate for N = 1, 2, 3 emitters. Each *curve* is obtained from a series of calculations where the cavity-Q is varied. From *left* to *right*, $\kappa = 10, 6.3, 4, 2.5, 1.6, 1, 0.63, 0.4, 0.25, 0.16, 0.1, 0.063, 0.04, 0.025, 0.016, 0.01$/ps, *open circles* indicate $\kappa = 2.5$/ps on each *curve*. High excitation is used to drive the system into saturation. **b** Single-photon purity η versus $g^{(2)}(0)$. Starting from high η, the same values for κ are used, *open circles* indicate $\kappa = 0.63$/ps on each *curve*

we classify these two regimes as a function of the maximally attainable emission rate $r = \kappa\langle n\rangle$. Using strong incoherent excitation, each symbol on the black curve represents a single-QD-microcavity system with a unique cavity loss rate κ. To the left side one approaches the limit of a free QD emitter (large κ). To the right side, the cavity losses become smaller than the spontaneous emission rate, so that photons accumulate in the cavity and lead to coherent emission with $g^{(2)}(0) = 1$. In between, there is a wide regime with cavity-enhanced single-photon emission, where $0 < g^{(2)}(0) < 0.5$. In the solid-state community, this is a frequently used criterion to identify the source of the emission to be a single emitter, since the smallest possible $g^{(2)}(0)$ value from a two-emitter system is that of a two-photon Fock state yielding 0.5.

How do additional emitters influence this result? Non-resonant coupling of spectrally detuned QD transitions to the cavity mode distinguishes solid-state emitters from atoms (cf. Sect. 1.4). Most current realizations of SQD nanostructures are not free of residual emitters [32, 33]. When comparing to experiments, background effects from detuned emitters can play an important role and significantly modify the properties of a true single-emitter system. While it may seem intuitively clear that a single-photon source can only be realized using a single-emitter, this is not entirely true if the cavity acts as a filter for the statistical properties of photons originating from *several* emitters, or if these emitters are correlated in an entangled state. In the left panel of Fig. 1.1 results are also shown for two and three emitters in the cavity. Surprisingly, non-classical emission is also possible, even if perfect antibunching in terms of $g^{(2)}(0)$ values being as close as possible to zero is compromised in comparison to the single-emitter case. We take a closer look at the interpretation of these results, first reminding ourselves that according to (1.7), $g^{(2)}(0)$ contains information from the photon statistics in an averaged form via the first and second *moments* of the probability distribution function p_n. The performance of a single-photon source is more accurately characterized in terms of the rate r (how long on average must one wait for an emission event) and the error η^{-1} to have an unusable packet with more than one photon. This characterization can be realized if access to the full photon statistics is available, such as from density-matrix calculations or photon-number resolved measurements. We define the single-photon purity (inverse to the error) as

$$\eta = \frac{p_1}{\sum_{i \geq 2} p_i} \tag{1.8}$$

that relates the probability of the emission of a single photon to that of the emission of bundles of two or more photons [34, 35]. Single-photon purity is one of the criteria listed in Chap. 3 Sect. 2, to which we refer for further information on single-photon sources. In the right panel of Fig. 1.1 we show the attainable η and the corresponding $g^{(2)}(0)$ for the κ values used in the left panel. While the error can only be arbitrarily minimized in the single-emitter system, high η can also be obtained with two and three emitters in the cavity. Most interestingly, the autocorrelation function is unable to capture this behavior, as for the same purity of single-photon emission, $g^{(2)}(0)$ values can be very different. Moreover, the same high η may be achieved with higher emission rates from two and three emitters [34]. The impact of additional emitters in the cavity can apparently be less detrimental than one may expect from $g^{(2)}(0)$ alone. In the future, it will be very interesting to further investigate emission properties beyond the well-established autocorrelation measurements on the basis of photon-number resolving detectors [36–40] or higher-order HBT setups [41].

1.2.2 *Lasing in the Presence of Strong Coupling in a Few-QD System*

The miniaturization limit of solid-state cavity-QED is given by a single QD coupled to a single mode of a microcavity. In order to achieve sufficient photon production from only a single emitter, the light-matter coupling must be as high as one can achieve by using emitters with large dipole moments and by placing them in the field maximum of the confined mode. Coupling-strength values for current micropillar cavity systems are close to 100 μeV, which is sufficiently high to be in the strong coupling regime even at increased excitation powers. While strong coupling is associated with weak excitation and the appearance of the vacuum Rabi doublet in the emission spectrum, lasing typically takes place in the weak coupling regime using strong excitation. It is therefore an interesting thought that the lasing threshold is crossed while the transition of a single QD is still in strong coupling with the mode. This was first discussed in [42] for a photonic-crystal cavity containing only few QD emitters, although in the past, neither was the influence of the background emitters quantified, nor the role of excitation-induced dephasing, which is the main reason for the transition to weak coupling at higher excitation, discussed.

The criterion for strong coupling is commonly accepted as the existence of two distinguishable peaks in the cavity emission spectrum, the so-called vacuum Rabi doublet. In general, the spectrum can be written as the modulus square of the difference between two poles [43]

$$S(\omega) \sim \left| \frac{1}{\omega - \omega_1} - \frac{1}{\omega - \omega_2} \right|^2 . \tag{1.9}$$

In the presence of dephasing, such as originating from cavity losses, spontaneous emission, or carrier relaxation processes following excitation, it is known [44, 45] that strong coupling persists as long as $4g > |\Gamma - \kappa|$, where g is the light-matter coupling strength, κ the cavity loss rate, and Γ gives the total exciton dephasing, before the Rabi doublet merges into a single line marking the transition to weak coupling [10]. This well-known strong-coupling criterion must be reviewed, however, when excitation is strong enough that states with higher total excitation in the Jaynes–Cummings ladder become populated, such as when approaching the laser threshold.

A nanolaser can be seen as a driven dissipative system that is defined by the usual Jaynes–Cummings Hamiltonian and Lindblad contributions for pump, relaxation, and losses. An analytic expression for the cavity emission spectrum can only be obtained by limiting the Hilbert space to a low-excitation subspace. Using the two-level formalism for simplicity, in [10] the three lowest states, namely the ground state in the zero-photon block $|g, 0\rangle$ and the states with one excitation $|e, 0\rangle$ and $|g, 1\rangle$ (Fig. 1.2 depicts these states and their relation to the dressed-state Jaynes–Cummings ladder), are used to derive the well-known expression for the cavity spectrum. While this "three-state approximation" (3SA) sufficiently describes the strong-coupling spectrum in the weak-excitation regime, higher states become realized under stronger excitation as one begins to climb the Jaynes–Cummings ladder. In this case, we

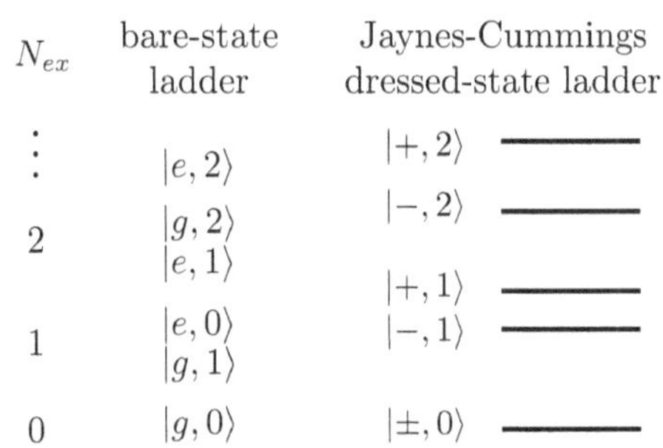

Fig. 1.2 Illustration of the implications of the 3- and 4-state approximation (SA). In the 4SA, the four lowest-energy states are explicitly considered including the state $|e, 1\rangle$ with a total excitation number of $N_{ex} = 2$. In a dressed-state picture, this allows to represent the lowest two rungs of the Jaynes–Cummings ladder and, thereby, to obtain the vacuum-Rabi doublet that results from the two possible transitions from the first rung to the ground state. The 3SA is limited to states with a maximum $N_{ex} = 1$. The Rabi-doublet is inaccurately described by the 3SA if the system pumped towards the lasing threshold, as the first JC rung cannot be represented by the $N_{ex} = 1$ states alone

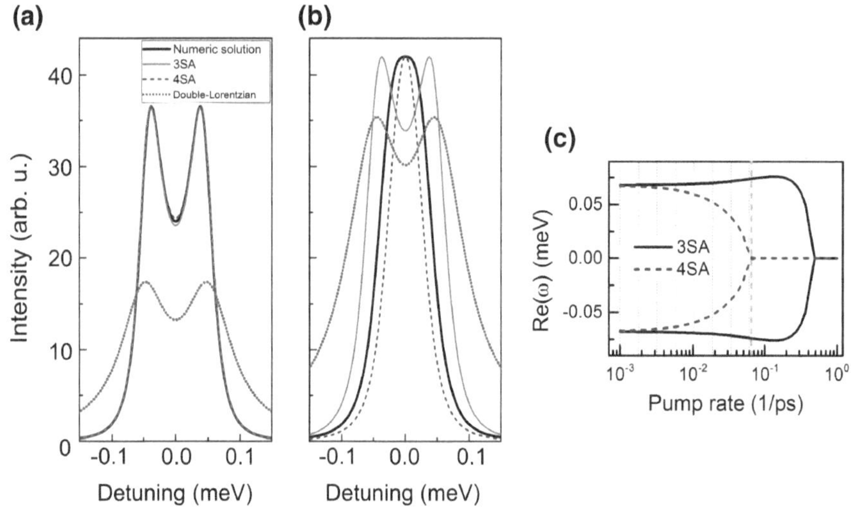

Fig. 1.3 Comparing different approximations to the cavity emission spectrum at low (**a**) and high (**b**) excitation power. **c** Real part of the poles that determine the peak positions in the 3SA and 4SA. Two separate values indicate the peak splitting of the strong regime

show that considering the next higher excited state $|c, 1\rangle$ with a total excitation of 2 extends the validity of the approximation into the transition regime to lasing. In this "four-state approximation" (4SA) the cavity emission spectrum can be written as a closed analytic expression of the form of (1.9), where the roots that determine the position of the peaks are changed by additional terms [46]. In Fig. 1.3 both levels of approximation are compared to the full numerical solution. While the additional higher excited state in the 4SA has little impact in the low-excitation regime (a), its absence can be noted at high excitation (b): The conventional 3SA textbook expression incorrectly predicts Rabi splitting, whereas the result obtained in the 4SA closely resembles the numerical result, which reveals that the transition to weak coupling has taken place. The remaining difference between the 4SA and the exact solution can be attributed to contributions from yet higher excited states. In systems that are driven with higher excitation-power densities, the 4SA-analytic formula for the cavity emission spectrum provides a new tool to evaluate experimental measurements and to extract parameters, such as the light-matter coupling strength, with greatly improved accuracy.

In Fig. 1.3c the eigenvalues that determine the peak-positions in the emission spectrum are compared for 3SA and 4SA. One can infer that the transition to weak coupling, indicated by the merging of the poles, takes places at much higher excitation-induced dephasing (i.e. higher pumping) if the approximation is restricted to only the three lowest states. This finding has severe implications for answering the question if lasing and strong coupling can coexist.

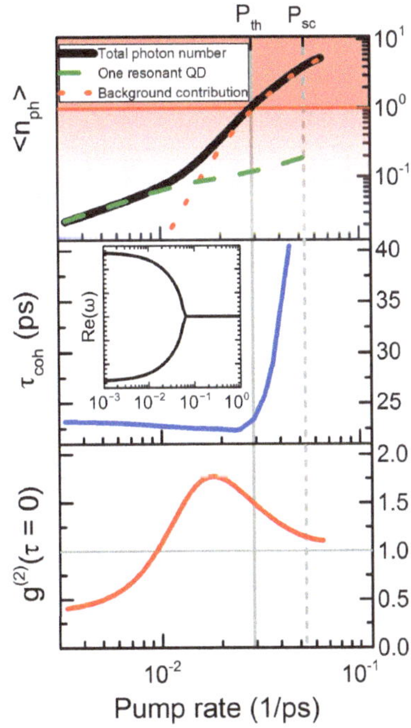

Fig. 1.4 Laser characteristics obtained from a numerical solution of a density-matrix approach that includes contributions from background emitters. From *top* to *bottom*, input-output curve, coherence time and second-order photon correlation function are shown, in their combination providing evidence that the system is a nanolaser that crosses the transition to lasing

As discussed in Sect. 1.1.2, the photon autocorrelation function $g^{(2)}(0)$ approaching unity is indicative for lasing, and so is a linewidth narrowing that begins to take place at the laser transition [47]. In Fig. 1.4 we show numerical results for both quantities together with the input-output curve. The calculation takes into account a single QD that is in strong coupling with the mode (see the inset) and several "background emitters" which only come into resonance at higher excitation. The underlying picture is that their exciton transition is detuned from the cavity mode, but transitions between higher multi-exciton states that become realized at stronger excitation are resonant. Non-resonant coupling via multi-exciton states is explained in Sect. 1.4.2 of this chapter. From the top to the bottom panel, we see a kink in the input-output curve that originates from background contributions (dotted curve) to the emission of the single QD (dashed curve). A mean intracavity photon number of 1 is reached at $P_{th} = 0.02$/ps. At this value, the coherence time shows a steep increase to moderately low values reaching 40 ps. The low coherence time reflects the significant fraction of spontaneous emission also in the lasing regime due to the small overall number of emitters. In contrast, in a many-emitter laser, spontaneous emission plays practically no role above threshold, and the coherence time can be 1 ns [21]. The autocorrelation function exhibits antibunching ($g^{(2)}(0) < 1$) of the strongly coupled single emitter in the low excitation regime. Antibunching disappears with the onset of contributions from the additional QDs, where spontaneous emission of their various emission channels leads to a thermal component ($g^{(2)}(0) > 1$). At high excitation, coherent lasing with $g^{(2)}(0) = 1$ is be reached.

These results are representative for the behavior of a few-emitter QD-microcavity system, which can operate in a variety of regimes depending on system parameters and excitation strength thereby crossing between single-emitter behavior, LED emission, and lasing in different regimes of light-matter coupling. An understanding of the intricate behavior requires to use a quantum-optical framework that takes into account a variety of effects that are fundamentally relevant for solid-state emitters and that are discussed in this chapter, such as non-resonant coupling, excitation-induced dephasing, and multi-exciton transitions.

1.2.3 Cavity-Enhanced Emission of Entangled Photon Pairs

Pairs of entangled photons are essential to many building blocks for quantum technologies. This includes quantum repeater stations that are needed to extend quantum communication networks beyond the limitations due to losses in optical fibres [48], as well as quantum teleportation and key distribution via the E91 protocol [49]. Today's applications mostly rely on parametric downconversion in non-linear crystals to generate entangled photons. Semiconductor QDs possess particular properties that make them promising candidates for integrated deterministic sources of entangled photons. The following discussion refers to the energy scheme in the left panel of Fig. 1.5. Prepared in the biexciton state, two decay channels are equally possible, in which an electron-hole pair of either spin direction recombines and leaves behind a remaining exciton with an electron and a hole of the opposite spin (X_H or X_V), emitting a photon that is either horizontally or vertically polarized. From there a second recombination is possible, resulting in the emission of another photon of the same polarization as the first one. If both intermediate exciton states are energetically indistinguishable, the "which-way" information is not revealed, and the result is the entangled photon state $(|HH\rangle + |VV\rangle)/2$, where $|12\rangle$ denotes the two-particle state of the two-photon system.

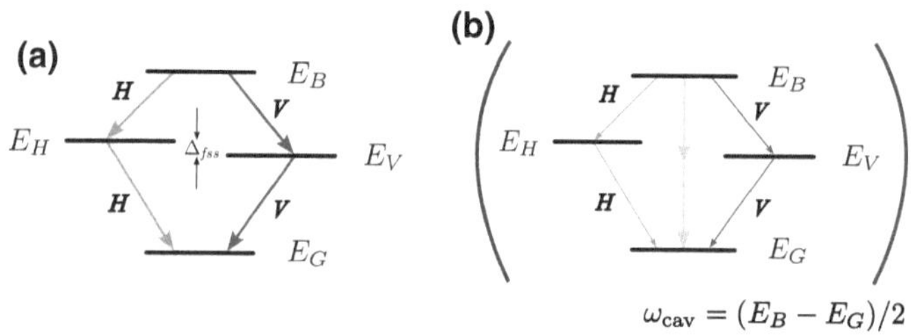

Fig. 1.5 **a** With energy displayed along the vertical direction, the so-called 'diamond' scheme of the biexciton-to-ground-state cascade of a semiconductor QD is schematically depicted. The intermediate excitonic states are energetically separated by the fine-structure (fss) splitting. **b** Emission of an entangled photon pair by cavity-enhancement of a direct two-photon emission process

Unfortunately, due to growth-related anisotropies, both excitons are not indistinguishable, but energetically separated by the fine-structure splitting (fss). The possibility to distinguish the two recombination channels by their spectral footprint provides the "which-way" information that compromises entanglement. Many attempts have been undertaken to minimize the fss, e.g. by applying external electrical fields or doping [50], and we refer the reader to Chap. 7 for a detailed discussion. We propose an elegant alternative approach that avoids the detrimental effect of the fss altogether.

In [3] the direct two-photon emission process from the biexciton to the ground state is employed, from which each photon carries half the energy of the biexciton. As higher-order process that is of second order in the Hamiltonian, the two-photon emission is as such highly unlikely and hardly relevant in previous discussions of biexciton emission of entangled photon pairs. If, however, the emitter is embedded in a cavity that is tuned exactly to the energy of the two-photon emission, the modified photonic density of states enhances the two-photon emission and suppresses emission through the cascade. This situation is depicted in the right panel of Fig. 1.5. Both processes compete with each other, and it depends sensitively on the parameters, in particular on width (Q-factor) and position of the cavity mode, whether entanglement can be preserved in the presence of fss.

For sufficiently high (but also typical) cavity-Q factors of about 20,000, we could demonstrate nearly complete independence of the degree of entanglement on the fss as shown in the left panel of Fig. 1.6. In the right panel, it is shown that in case of perfect resonance between cavity and two-photon emission process, the delay-time resolved autocorrelation function $g^{(2)}(\tau)$ exhibits bunching at $\tau = 0$, revealing that both photons are indeed most likely emitted simultaneously. If the cavity is detuned only slightly (0.25–0.5 meV) from that resonance, the bunching peak moves away

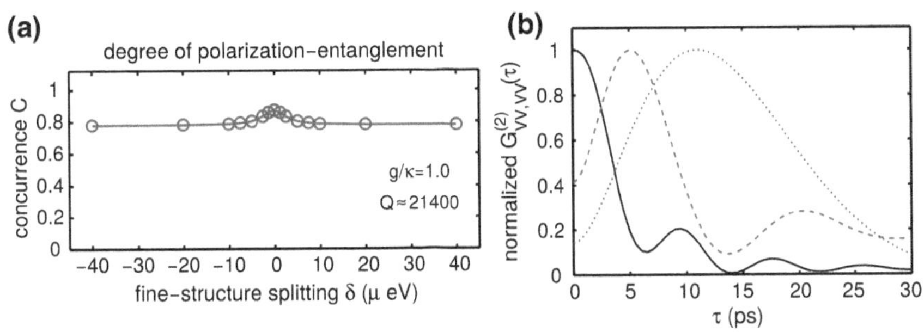

Fig. 1.6 **a** The degree of polarization entanglement (C = 1 representing a maximally entangled state) remains nearly constant as function of fine-structure splitting, if a Q = 21400 cavity mode is tuned exactly to the two-photon resonance of the biexciton transition. **b** The fingerprint of the two-photon emission is a bunching-peak in the two-photon correlation function at zero time delay, whereas emission through the biexction-exciton-ground state cascade exhibits bunching at a delayed time. In resonance (*solid curve*), two-photon emission is enhanced by the cavity and is the dominant process. For slight detuning (0.25 meV *dashed*, 0.5 meV *dotted curve*), emission through the cascade becomes increasingly more likely, indicated by the peak moving away from $\tau = 0$

from the origin $\tau = 0$, indicating that the two photons are emitted successively and the emission mechanism shifts from the direct higher-order process to the cascade.

An experimental verification of the feasibility of the proposed generation of a quantum-mechanical entangled state is yet to be provided. Both the direct creation of the biexciton state by a direct two-photon *absorption* process [51], and the possibility of the discussed two-photon emission process (although in the absence of a cavity) [52] have recently been demonstrated.

1.2.4 Single Photons from an Electrical Source with Long Pulses

As a final example for the generation of non-classical light we consider a single emitter without the influence of a cavity. In the experimental realization [53], a bottom DBR-mirror is used merely to enhance directionality of the emission.

The repetition rate of an on-demand single-photon source is limited by the recombination time of the emitter. When using short (picosecond) excitation pulses with sufficient temporal separation, each pulse triggers a single-photon emission event. The duration of the excitation pulse must remain shorter than the average recombination time in order to avoid re-excitation following the same trigger, which would result in the emission of successive photons. While short excitation pulses are easily realized in all-optical setups, electrical excitation by current injection is the device-relevant method of excitation and one key advantage of solid-state emitters over atomic systems. Picosecond-short excitation pulses are, however, more difficult to realize with electric pulse generators. Addressing this problem, we have proposed an excitation scheme that makes use of the multi-exciton landscape of QD excitations to realize strongly antibunched single-photon emission that is quite independent of the excitation-pulse duration [53].

Instead of using weak excitation pulses, the pump pulses are chosen strong enough to drive the system into saturation. Consequently, during the excitation the QD is rapidly filled with carriers and the emission is dominated by the recombination of higher multi-exciton configurations, which are spectrally detuned from the exciton-to-ground-state transition. After the excitation pulse has ended, carriers recombine until no carriers remain in the QD. The last decay in this cascade must result from the neutral or charged exciton with only one remaining electron-hole pair at a unique spectral position. It is the basic idea of the proposed scheme to trap the system in higher multi-exciton configurations *during* the excitation and to use spectral filtering to collect the single photon from the final exciton emission *after* the excitation pulse has ended. The detuned emission during the excitation is not collected, and the final exciton emission is independent of the duration of the excitation pulse. The dynamical behavior of this interplay of different recombination channels is depicted in Fig. 1.7 (theory in (a), experiment in (b)). As a central result, by using strong excitation a high degree of antibunching with $g^{(2)} < 0.1$ can be maintained quite

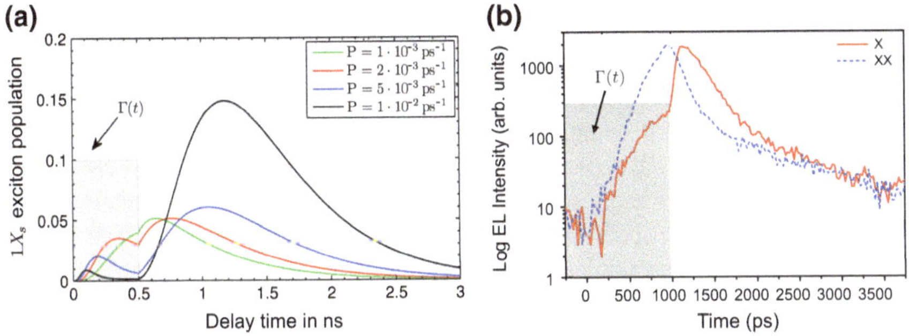

Fig. 1.7 **a** Time-dependence of the exciton realization probability following pulsed excitation of different excitation powers. The realization of the exciton configuration is suppressed with increasing pulse area of the excitation, as higher multi-exciton configurations form instead. After the pulse has ended, the exciton is realized and then decays on the typical nanosecond timescale. **b** Time-resolved measurements of the exciton (X) and biexciton (XX) emission intensity. In agreement with the microscopic model, during excitation exciton emission is suppressed and emission from the biexciton dominates. After the excitation pulse, the exciton transient reflects the theoretically predicted behavior. The duration of the excitation pulse $\Gamma(t)$ is indicated by the *shaded region*. Note the different scales (*top*: linear, *bottom*: logarithmic) when comparing the results. The *left figure* is taken from [22], the *right figure* is adapted from [53]

Fig. 1.8 Antibunching quantified in terms of $g^{(2)}(0)$ as function of the width of the electrical excitation pulse (EPW)

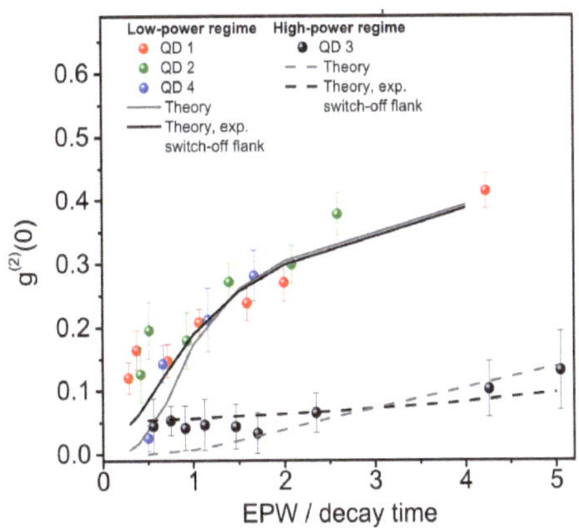

independently of the excitation-pulse duration, as shown in Fig. 1.8. Experimentally, this has been verified for nanosecond pulse durations exceeding the recombination time by more than 500% in excellent agreement with the theoretical results obtained from a solution of the von-Neumann equation [22].

1.3 Carrier Scattering and Dephasing

In solid-state physics, we usually deal with systems exhibiting a continuous density of states, where the full quantum dynamics of excited carriers is practically not accessible and theories have to be formulated in terms of averaged one- or two-particle quantities. On the other hand, semiconductor quantum dots (QDs) are quasi-atomic systems with a discrete spectrum of confined states, where the dynamics of the full density operator can be calculated via a von Neumann equation, as discussed in the introduction to this chapter. Unlike true atoms, QDs are not isolated but embedded into a semiconductor crystal and have to be treated as open quantum systems in interaction with their environment of lattice vibrations and charge carriers of the barrier material as well as the photon modes of free space. The dissipative processes, in particular carrier-carrier and carrier-phonon interaction, are typically described by many-particle methods that can be formulated such that the coherent von Neumann dynamics is supplemented by additional Lindblad terms, as we show below.

The density operator $\rho(t)$ of the QD system is expressed in a basis of many-particle states. This leads to a full description of the QD excitations, in which the Coulomb interaction of QD carriers can be directly included. As mentioned in the introduction, there are basically two ways to construct the many-particle basis and formulate the dynamics of the density operator. On the one hand, we can start from the single-particle states that are provided by the QD confinement potential and build many-particle configurations, or product states, by creating carriers in the single-particle states in all possible combinations. A given configuration $|\{n_i\}\rangle$ is then defined by specifying which single-particle states are occupied ($n_i = 1$) or empty ($n_i = 0$). The Coulomb configuration interaction can be included in the Hamiltonian of the von Neumann equation, thereby considering the formation of multi-exciton states, which are the fundamental QD excitations. Alternatively, we can take into account the configuration interaction in advance and select an appropriate set of multi-exciton states as a basis for the density operator, which is often done in quantum optics to formulate compact models of QD emitters. In this section, we will follow the first approach and stay with product states as the many-particle basis as discussed in [9].

The electronic many-particle basis can be augmented by photonic degrees of freedom to capture the full quantum dynamics of QD emitters inside a cavity. In this section, we focus on carrier scattering and dephasing by treating the electronic part of the density operator ρ_{el}. We can describe the dynamics of ρ_{el} by the von Neumann-Lindblad (vNL) equation [54–56]

$$\frac{\partial}{\partial t}\rho_{\text{el}} = -\frac{i}{\hbar}\left[H_S, \rho_{\text{el}}\right] + \sum_X \frac{\gamma_X}{2}\left[2s_X\rho_{\text{el}}s_X^\dagger - s_X^\dagger s_X\rho_{\text{el}} - \rho_{\text{el}}s_X^\dagger s_X\right]. \tag{1.10}$$

The commutator part represents the coherent time evolution of the density operator due to the system Hamiltonian H_S, which consists of a "free" part with the QD confinement energies and the Coulomb interaction between the QD carriers. It defines the eigenenergies of the QD many-electron system and takes into account all Coulomb renormalization effects caused by the interaction of QD carriers among each other. By accounting for the interaction between the QD system and its semiconductor environment, which is formally described by a number of reservoirs, the so-called Lindblad terms are obtained. They introduce dissipation into the time evolution, causing the QD to relax to a thermal equilibrium which is defined by the properties of the environment. This also includes interaction-induced dephasing of coherences that may be generated in the QD for example by light-matter interaction. The summation in (1.10) runs over all possible transitions between the eigenstates of the QD, the transitions being described by operators s_X. It is essential that both the interaction of QD carriers with phonons and with charge carriers of the barrier material can be formulated as Lindblad terms, where the transition rates γ_X can be calculated microscopically. But even if we choose phenomenological values of γ_X that are e.g. based on experimental results instead of a microscopic calculation, the structure of the vNL equation guarantees physically sensible results.

Due to their direct relevance for various QD applications, in the past two decades carrier-scattering processes in QD systems have been studied intensively both in experiment [57–63] and theory [64–74]. A microscopic description of carrier scattering requires the inclusion of carrier correlations that can be treated on different levels. For QD systems, due to the finite state space of the electronic excitations, approximate treatments of carrier correlations have been questioned and the importance of a configuration picture has been pointed out. [16] Consequences for QD laser threshold current densities, [75] or QD gain recovery dynamics [76] have been demonstrated. Moreover, a carrier-capture model consisting of several capture configurations in the QD has been used to explain signatures of a phonon bottleneck in time-resolved differential transmission measurements on InGaAs quantum dots [58]. In [77], the effect of full Coulomb configuration interaction on the carrier relaxation in weakly confined QDs due to electron-acoustic phonon interaction has been discussed. The description in terms of a many-particle basis, which is naturally chosen for quantum optical models of QD emitters, is complemented by the description of carrier scattering on a single-particle level, which is often used throughout the literature. Again, it is questionable if a single-particle description of carrier scattering may be used in QD systems with a small number of electronic states, as we illustrate in the following.

When the vNL equation (1.10) is used, relaxation processes occur as transitions between QD configurations, facilitating an exact treatment of the Pauli exclusion principle between the QD carriers. In a single-particle description, for example in terms of a Boltzmann equation, the central quantities are the occupation *probabilities* f_i of the single-particle states. These describe the population of a QD in the sense of an ensemble average, which corresponds to each carrier reacting not to the actual state of the collision partners in a scattering process, but to an independently averaged "mean-field" distribution of carriers, see Fig. 1.9. As discussed in [9], this corresponds to a mean-field-like treatment of Pauli blocking, which can lead to inaccurate results in the description of carrier relaxation: If the QD is prepared in a state where only one efficient relaxation channel is available, only one Lindblad term is present in (1.10), exhibiting a purely exponential time evolution with a constant rate due to the linear

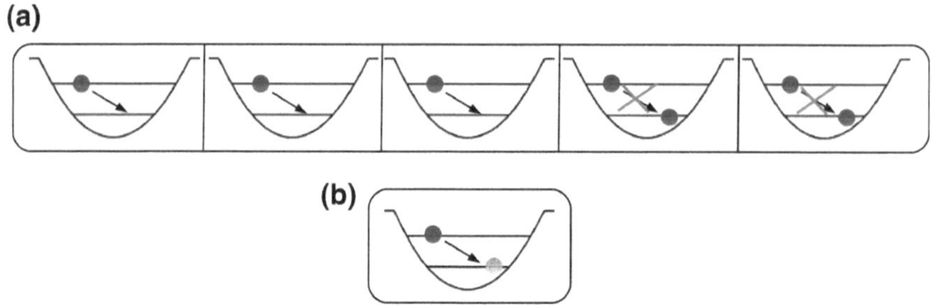

Fig. 1.9 Schematic of carrier relaxation in the conduction band of a QD ensemble. **a** In two out of five identical QDs, the lower confined state is populated, corresponding to a probability of $f = 0.4$, so that the relaxation of a second carrier from the upper confined state is blocked. At the same time, in the remaining three dots the relaxation is possible. In the Boltzmann description of the ensemble (**b**) the relaxation rate is weakened by a factor $(1 - f)$, which describes the average availability of the final state, indicated by a single QD with a "ghost" carrier present in the lower state

character of the vNL equation. In contrast, the nonlinear Boltzmann equation yields a power law behaviour, which is significantly slower than the exponential convergence and which can not be assigned a constant rate. In the regime of high excitation, where a large number of relaxation channels is available, carrier correlations are destroyed on a short time scale and a mean-field description of scattering becomes feasible again.

After these general considerations, we discuss the two relevant types of system-reservoir coupling, namely carrier-carrier Coulomb scattering and carrier scattering by phonons. First, the contact of the QD carriers with the carriers in delocalized states leads to various Auger-like scattering processes. These are beyond the interaction of QD carriers among each other, which is already taken care of in the system Hamiltonian H_S. Carriers are captured or ejected from the QD and they scatter between localized states due to the Coulomb interaction with delocalized carriers; the latter provide the necessary energy for the transition processes. A detailed analysis of these processes and the corresponding rates have been given in [9].

A second type of reservoir is provided by the phonons in the barrier material. In polar semiconductors one of the strongest contribution to carrier scattering processes is due to LO phonons, for which we assume a dispersionless spectrum $\omega_q = \omega_{LO}$. The interaction between QD system and phonon reservoir is described by the Hamiltonian

$$H_{SR} = \sum_{i,j,q} M_q^{i,j} a_i^\dagger a_j \left(D_q + D_{-q}^\dagger\right), \tag{1.11}$$

with the Fröhlich coupling matrix elements $M_q^{i,j}$ [78]. In (1.11) two scenarios can be distinguished. First, when both indices i, j refer to QD states, electronic transitions $|j\rangle \rightarrow |i\rangle$ inside the QD assisted by the emission or absorption of phonons are described, leading to carrier relaxation in the QD. The second case involves a QD and a delocalized state and corresponds to carrier capture from or reemission into the reservoir, again assisted by phonons. For the first case, treating all possible transitions X between configurations, which include the movement of a carrier from $|j\rangle$ to $|i\rangle$, one obtains in Born-Markov approximation the transition rate in the form

$$\gamma_X = \frac{2\pi}{\hbar^2} \sum_q |M_q^{i,j}|^2 \Big\{ (1 + N_{LO})\, \delta(\omega_X + \omega_{LO}) + N_{LO}\, \delta(\omega_X - \omega_{LO}) \Big\} . \quad (1.12)$$

N_{LO} is the phonon population at the lattice temperature. The first term corresponds to processes involving phonon emission, the second involving phonon absorption. In (1.12) one encounters a problem that is specific to the LO-phonon driven carrier kinetics in discrete electronic systems. The strict energy conservation expressed by the δ-functions is not generally met. In early theoretical considerations [79, 80], this observation has lead to the prediction of a "phonon bottleneck". However, only in lowest-order perturbation theory (Fermi's golden rule) does the scattering rate vanish. Equation (1.12) corresponds to this level, which results from applying the Born-Markov approximation. A better-suited non-perturbative description leads to the polaron picture. The quasiparticle renormalization effects of the non-perturbative treatment can be included in a generalized form of (1.12) via spectral functions combining the configuration picture with the formalism of many-particle Green's functions, as described in detail in [56]. In contrast to the phonon bottleneck predicted from perturbation theory, non-vanishing scattering rates are obtained from this approach. The Green's function formalism allows for a self-consistent treatment of carrier-carrier and carrier-phonon interaction, leading to a non-trivial co-action of both mechanisms and modifying all transition rates entering the vNL equation.

In [56], we use the theory introduced above to analyze experimental results for the carrier capture and relaxation dynamics in self-organized semiconductor QDs, which are obtained by time-resolved differential transmission (TRDT) measurements. Figure 1.10 contains examples of TRDT traces for the ground-state transition at different temperatures and fixed excitation power. From the TRDT signals, rise times can be extracted that reflect the efficiency of carrier capture and relaxation processes under different experimental conditions. The data can be understood from simulations of TRDT rise times under comparable conditions taking into account both carrier-carrier and carrier-phonon interaction. The results are collected in Fig. 1.11. In agreement with the experiment, we find a higher temperature sensitivity at lower carrier densities, which points to a dominant role of the phonon scattering in this regime. For low temperatures, the transition rate increases strongly with the carrier density due to more efficient carrier-carrier scattering (Auger-like processes assisted by delocalized carriers). The latter also causes larger broadening of the spectral functions, which in turn accelerates carrier-phonon scattering as well. For elevated temperatures, the density dependence is weak due to a dominating phonon contribution. As shown in [56], renormalization effects are of critical importance for this agreement, which can not be achieved by a perturbative treatment of system-reservoir interaction.

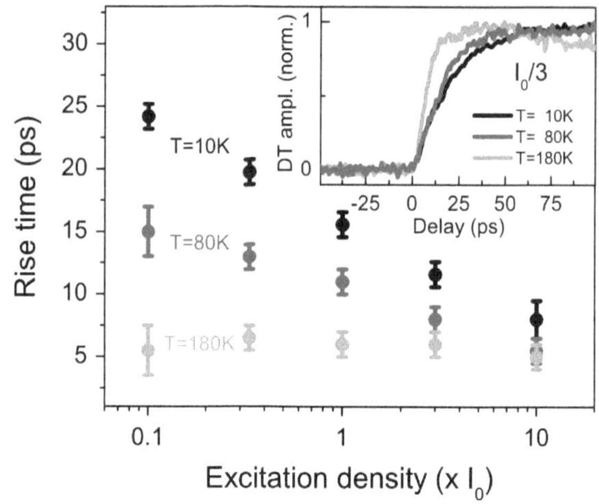

Fig. 1.10 Rise times extracted from monoexponential fits to time-resolved differential transmission (TRDT) transients of the QD ground state, obtained at various excitation intensities for temperatures T = 10, 80, and 180 K. The inset shows DT traces at an excitation density of $I_0/3$ with $I_0 = 40\,\text{W/cm}^2$

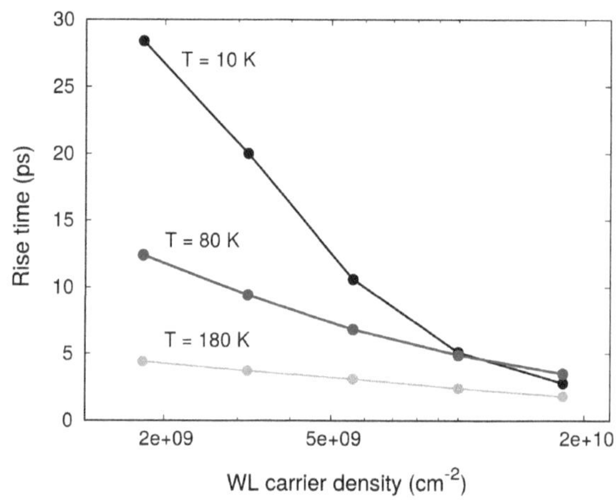

Fig. 1.11 Rise time of the TRDT signal for the QD ground state calculated as a function of the wetting-layer (WL) carrier density for different temperatures. Both carrier-carrier and carrier-phonon interaction are considered within a non-perturbative theory including joint quasiparticle renormalizations in the multi-exciton configuration-picture description

The many-particle interaction of the QD system with its environment induces not only transitions between QD configurations, but also dephasing of coherences in the QD that may be generated by resonant optical excitation. In the vNL equation (1.10), this is reflected by a decay of transition amplitudes between many-particle configurations. As widely discussed in the following chapters of this book, the amount of dephasing is crucial for quantum optical and quantum information applications of QD emitters, as it is directly connected to broadening of emission lines and to the lifetime of coherences that are exploited for the generation of entanglement. For example, dephasing affects the deterministic creation of photons by limiting the fidelity of resonant excitation of single QD transitions which is discussed in the chapters by *Rengstl*, *Jetter* and *Michler* and by *Portalupi* and *Michler*. This problem also appears in the context of spin preparation by optical pulses, which is the topic of the chapter by *Sun* and *Waks*. The limitation of indistinguishability of entangled photons due to dephasing is addressed in the chapters by *Weihs*, *Huber* and *Predojević* and by *Heindel*, *Rodt* and *Reitzenstein* while *Brash*, *Liu* and *Fox* very nicely express how strongly coherent control experiments in semiconductors rely on a reduction of dephasing for example due to carrier-phonon and carrier-carrier scattering.

In [5], the dephasing due to carrier-carrier and carrier-LO-phonon interaction in semiconductor QDs has been quantified focussing on the regime of elevated excitation density. Complementary to this is the low-density regime, where acoustic phonons and especially at low temperature Coulomb-mediated out-scattering of QD carriers lead to homogeneous broadening, see the contribution of P. Borri and W. Langbein in [23]. The pure dephasing associated with acoustic phonons can be comparable to scattering-induced dephasing when the carrier density is low. The homogeneous exciton linewidth has been studied in great detail experimentally [61, 81, 82] and theoretically using the independent Boson model, where the electronic system is described in a two-level approach [83–87]. Fewer experiments have investigated the dephasing of many-particle configurations [88–90].

1.4 Non-resonant QD-cavity Coupling

For weak excitation of self-assembled semiconductor QDs, the ground-state exciton is dominating, while at elevated excitation levels a rich structure of closely lying discrete optical transitions can be observed [63]. When QDs are placed in a high-Q microcavity, the narrow-linewidth mode singles out particular transitions. In contrast to atom-like isolated emitters, QDs exhibit an interesting peculiarity: Even for weak excitation and QD emission lines significantly detuned from the cavity resonance, photons can be emitted into the cavity mode. As a result of extensive experimental and theoretical investigations, different mechanisms are discussed in the literature. These involve phonon-assisted optical processes, non-resonant coupling mediated by multi-exciton transitions, and Coulomb-assisted non-resonant coupling.

The effect is mostly discussed in the context of individual QDs coupled to optical cavities. Beyond this, in cavity-QED lasers non-resonant QD-cavity coupling is also expected to be responsible for a frequently observed background emission contribution. In these cavity-QED lasers the number of resonant QD emitters is often too small to reach stimulated emission, which is nonetheless observed in corresponding experiments [8, 46, 91, 92].

1.4.1 Phonon-Assisted Non-resonant Coupling

The acoustic phonons have been connected [85, 86, 93] with a particular line shape of the excitonic transition. The dominant contribution to the carrier-phonon coupling is contained in the diagonal matrix elements with respect to the carrier-state index, which facilitates the application of the independent boson model [78]. The role of LA-phonons in the off-resonant cavity feeding was demonstrated [94–101]. It was shown [2, 102–104] that the appropriate frame work to describe the phonon-assisted energy transfer between the exciton and the cavity is provided by the polaron picture, in which the exciton generates an electric field to which the lattice ions react by displacements of their oscillation centers. A polaron transform can be used to connect the distorted lattice with the original one, and the QD-phonon interaction can be formally eliminated from the Hamiltonian [78]. As a result the Jaynes–Cummings Hamiltonian

$$\tilde{H}_{JC} = g\,(b^\dagger X\, B^\dagger + b\, X^\dagger\, B) \tag{1.13}$$

then includes phonon operators

$$B = \exp\left[\frac{M_q}{\omega_q}(D^\dagger_{-\mathbf{q}} - D_{\mathbf{q}})\right] \tag{1.14}$$

besides the usual carrier ($X = a_i^\dagger a_j$) and photon ones. The relevant coupling constant in the problem is the difference between those of the conduction and valance band states, $M_q = M_q^{i,i} - M_q^{j,j}$. With the phonons acting as a thermal bath, a system-reservoir treatment leads to Lindblad terms $\mathscr{L}_{b^\dagger X}$ and $\mathscr{L}_{bX^\dagger}$. In this case, the associated off-resonant cavity feeding rates as a function of detuning read

$$\gamma_{b^\dagger X}(\Delta) = 2g^2\,\langle B\rangle^2\,\mathrm{Re}\int_0^\infty \mathrm{d}t\, e^{-i\Delta t}(e^{\Phi(t)} - 1)\,, \tag{1.15}$$

with $\Phi(t)$ given by

$$\Phi(t) = \sum_q \left|\frac{M_q}{\omega_q}\right|^2 \left[(N_q + 1)e^{-i\omega_q t} + N_q e^{i\omega_q t}\right]. \tag{1.16}$$

The rate of the reverse transition is related to the above one by the thermal factor $e^{-\beta\Delta}$, as prescribed by the Kubo-Martin-Schwinger condition. When expanding the Φ exponential in (1.15), all multi-phonon processes of emission and absorption that generate a total energy of Δ can be seen to contribute. A more detailed discussion of this polaron master equation approach is given in the chapter of *Roy-Choudhury* and *Hughes*.

In Fig. 1.12 the phonon-assisted cavity-feeding rate is shown for typical InGaAs QD parameters [4]. There is a pronounced asymmetry between positive and negative detuning for low temperature. This is expected, since any thermal bath favors the process which lowers the system energy, all the more so at low temperatures. For a lattice temperature of 20 K efficient cavity feeding is obtained only up to detunings

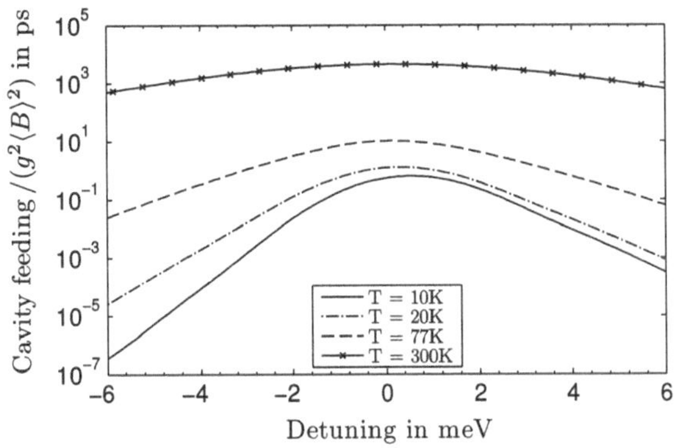

Fig. 1.12 Effective cavity feeding rates due to the QD-phonon coupling as a function of the detuning from the cavity resonance. Results are calculated from (1.15) and shown for 10, 20, 77 and 300 K

~2 meV between the transition and the mode. Nevertheless, we find that phonon-assisted recombination of QD excitations within this detuning range can lead to an emission enhancement in a cavity QED laser that can make the difference between thermal or coherent emission [4].

Many quantum optical application require state preparation with high fidelity. Phonon-assisted processes can be utilized to efficiently prepare e.g. the biexciton or exciton state by tuning the optical excitation on the corresponding phonon-sidebands as discussed in the chapters of *Portalupi* and *Michler* as well as *Roy-Choudhury* and *Hughes*.

1.4.2 Non-resonant Coupling Mediated by Multi-exciton Transitions

A quantification of cavity feeding effects is typically hindered in systems with many emitters. Here, various emission channels of different emitters overlap, masking individual emission properties. The few-emitter limit on the other hand offers the unique possibility to study non-resonant mode coupling in a highly controllable environment. Photonic crystal (PhC) [105] cavities in particular provide strong optical confinement with high Q factor and small mode volume [106], making them suitable to explore the miniaturization limit of lasing where the gain medium consists only of a few solid-state quantum emitters within a single mode cavity [91, 107].

Figure 1.13a shows the cavity-mode emission of a few (N ~ 4) QDs located in a PhC nanocavity. The QD ground-state exciton to cavity detuning is much larger than 2 meV. At first it may seem contradictory that lasing can be achieved in such a system. However, the discrete QD lines in Fig. 1.13a evolve into a broadband emission at elevated excitation intensities (highlighted by the gray shaded region). This is due to non-resonant coupling facilitated by a multitude of closely spaced excited multi-exciton states, some of which are overlapping with the cavity mode. More precisely, when a QD can accommodate many single-particle bound states, the number of

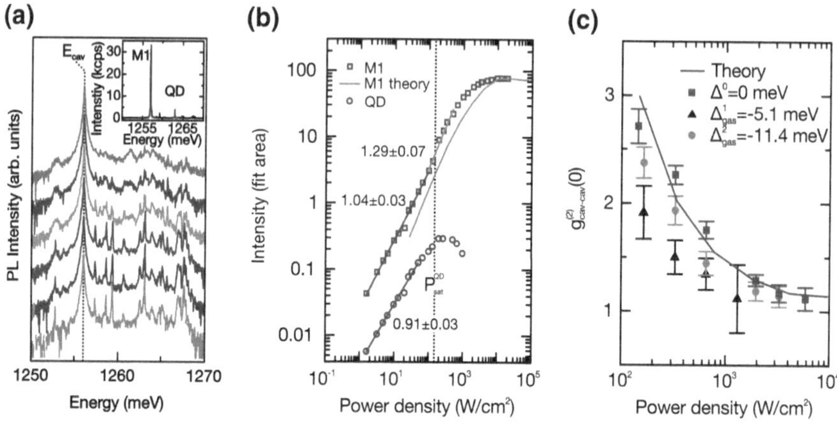

Fig. 1.13 **a** Experimental results for the emission spectrum from the cavity and increasing cw excitation power density. The spectra are plotted on a logarithmic scale with an offset to each other for clarity. The cavity mode is labeled E_{cav}. The inset shows a linear spectrum of the system for an excitation power density of $P_{\text{sat}}^{\text{QD}}$. **b** Integrated intensity of the cavity mode (*green*) and the QD (*blue*) as a function of excitation power density. *Black solid lines* represent power-law fits to the emission data. The *solid red line* depicts the intensity of the cavity mode emission calculated from theory. To connect the theoretical pump rate with the experimental power density, the *red curve* has been shifted along the power axis to ensure that the calculated exciton saturation coincides with $P_{\text{sat}}^{\text{QD}}$. **c** Second oder correlation $g^{(2)}(0)$ as function of the excitation power density. The colors represent three different cavity mode energies. The *solid green line* has been obtained from the microscopic model (color figure online)

carrier configurations becomes quite large, and their Coulomb interaction results in many closely spaced multiexcitonic transitions. When a QD is excited with multiple carriers, the subsequent relaxation dynamics of the carriers into a quasi-equilibrium state usually involves transitions between multiple configurations, and due to the Coulomb-configuration interaction, these configurations are associated with largely varying multi-exciton energies, which are then passed through the relaxation process. As soon as one of these configuration energies overlaps with the cavity mode, a Purcell-enhanced photon emission [63, 108, 109] can take place at larger detunings from the ground-state exciton. It is due to these transitions that lasing is possible even if the discrete ground-state excitons of the four QDs visible in the lower spectra in Fig. 1.13a are detuned from the mode by up to 17 meV.

The input-output characteristics at the cavity energy shows a slight superlinear increase for excitation-power densities and appears simultaneously with the saturation of the QD ground-state exciton, when multi-exciton states become increasingly populated with significant probability, see Fig. 1.13b. For all detunings between emitter transitions and mode, a clear transition from the spontaneous-emission regime with $g^{(2)}(0) > 1$ to coherent lasing with $g^{(2)}(0) = 1$ is observed with increasing excitation-power density, see Fig. 1.13c. Thus, in this example, the absolute energies of QD-transitions and cavity mode are of limited importance for the operation of the nanolaser.

To study the interplay of QD many-particle states and the resulting coupling to the cavity mode, we have solved the von Neumann-Lindblad equation for the system of four QDs and the quantized field of the cavity. Due to the large state space, calculations can only be performed for a subset of the QD many-particle configurations. When one of the excited configurations is in resonance with the cavity mode, we obtain results for the mean photon number similar to the experiment. Moreover, it turns out that in order to simultaneously describe the experimental results for intensity and $g^{(2)}(0)$, we need to assume that two of the excited configurations are resonant with the cavity mode. When performing independent configuration-interaction calculations for multi-exciton states, even for small QDs with a limited number of single-particle states, we find a very large number of excited multi-exciton states with largely varying configuration energies corresponding, e.g., to two or three electron-hole pairs (3X and 2X). Hence we assume that from each of the transitions within the manifold $3X^* \rightarrow 2X^*$ and $2X^* \rightarrow 1X^*$, one is resonant with the cavity mode. Here X^* represents an excited exciton state.

The theoretical results for the mean photon number and photon autocorrelation function are added as solid lines in Fig. 1.13b, c. Interestingly, in the low-excitation regime values up to $g^{(2)}(0) = 2.7$ are observed. From the theoretical model, we can attribute this enhanced probability of two- and multiple-photon emission events to two effects: One is the presence of competing resonant emission channels for each QD emitter, allowing for the simultaneous emission of photons into the mode. The second results from strong radiative coupling between different emitters associated with the effect of superradiance. This effect has been predicted in [110, 111] for QD nanolasers under continuous-wave excitation, and experimental proof in a system under pulsed excitation will be discussed in Sect. 1.5.

1.4.3 Coulomb-Assisted Non-resonant Coupling

In addition to the contributions from multi-exciton states, the role of the interaction with the delocalized states was recognized [109] and Coulomb hybridization of QD bound states with the delocalized states was demonstrated [112, 113]. In fact, carriers in the delocalized states can act as a thermal bath, which is able to compensate for the energy mismatch between a QD transition and the cavity resonance via Auger-like processes. This is an alternative mechanism to the Coulomb configuration interaction between carriers [109], as the Coulomb interaction involves other carriers outside of the QD. Importantly, its effect in opening a kinetic channel is present even for QDs hosting very few confined states, and holds even for a QD with single electron and hole levels.

To describe this mechanism, we start from the Hamiltonian containing the Jaynes–Cummings (JC) part for the interaction of the QD exciton with photons of the cavity mode as well as the Coulomb part for the interaction between QD and delocalized states. Two different techniques can be used to formally eliminate either (i) the exciton-photon or (ii) the Coulomb interaction part from the Hamiltonian. Specifically, in (i) the Schrieffer-Wolff transformation [78, 102] can be used that amounts to a perturbative diagonalization of the JC interaction part. The delocalized states remain unchanged, and the Coulomb interaction is felt by the photon subsystem due to the dressing of the QD states with the JC interaction.

A second approach considers a different scenario, in which the carrier-bath interaction is partly diagonalized and was used in Sect. 1.4.1 for the treatment of the phonon-assisted coupling. In [114] we extend the procedure to the case of the fermionic bath of carriers in delocalized states. The model is not exactly soluble but can be handled diagrammatically. The major difference to the first approach is that now the bath adapts itself to the presence or absence of the exciton.

A qualitative discussion can already be obtained within the Schrieffer-Wolff approach (SWA). Using this formalism, we obtain an effective JC interaction Hamiltonian

$$H'_{\text{int,SWA}} = -\frac{g}{\Delta} W\,(b^\dagger X + b\,X^\dagger) \tag{1.17}$$

that describes transitions between the QD exciton and the cavity photons, assisted by the Coulomb interaction with carriers in delocalized states. We consider the fermionic reservoir as being stationary and in thermal equilibrium and obtain Lindblad terms, $\mathcal{L}_{b^\dagger X}$ and $\mathcal{L}_{b\,X^\dagger}$, with rates given by

$$\gamma_{b^\dagger X} = 2\pi \frac{g^2}{\Delta^2} \sum_{\lambda,\mathbf{k},\mathbf{k}'} \left|W^{\lambda}_{\mathbf{k},\mathbf{k}'}\right|^2 f^{\lambda}_{\mathbf{k}}(1 - f^{\lambda}_{\mathbf{k}'})\,\delta(\Delta + \varepsilon^{\lambda}_{\mathbf{k}} - \varepsilon^{\lambda}_{\mathbf{k}'})\,. \tag{1.18}$$

Here, the occupancies $f^{\lambda}_{\mathbf{k}}$ are Fermi functions describing the carrier population of electrons and holes ($\lambda = e, h$) in the delocalized states. Similarly, $\gamma_{b\,X^\dagger}$ follows by changing Δ to $-\Delta$. In Fig. 1.14 we show results for the Coulomb-assisted feeding rate as a function of detuning. With increasing density n of carriers in delocalized states additional scattering channels can compensate for the energetic mismatch Δ between exciton and cavity, which leads to an increasing feeding rate. For comparison, we show the spontaneous emission rate caused by the JC coupling alone for typical QD-cavity parameter. At low carrier density, Coulomb assisted cavity feeding is negligible in comparison to the spontaneous emission rate, which is in agreement with previous experiments performed under low excitation condition [94, 97], in which only phonon signatures were found. However, for sufficiently high carrier densities ($n > 10^{10}/\text{cm}^2$) Coulomb assisted processes prevail even at large detuning and lead to a significant cavity feeding that is one order of magnitude stronger in comparison to the JC coupling alone.

Finally, we find a significant reduction of the off-resonant cavity feeding, if the wave functions for electrons and holes are similar (lines without dots). The reason is a large degree of compensation between the electrostatic (Hartree) Coulomb integrals contributing to (1.18). For identical electron and hole wave functions, the exciton

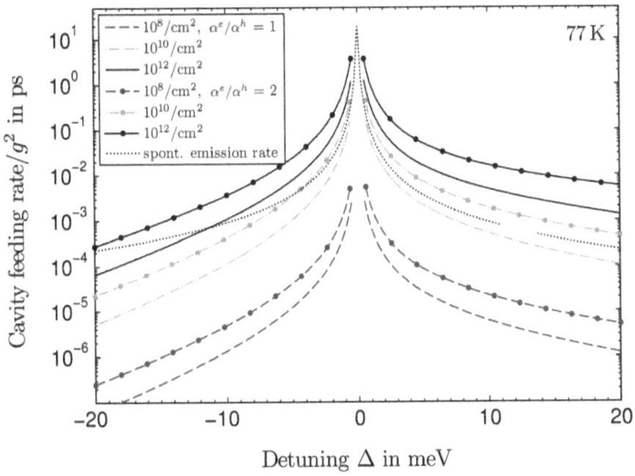

Fig. 1.14 Cavity feeding rate for a non-resonant QD coupled to the continuum of delocalized states at a temperature of 77 K obtained by using the SWA. We vary the carrier density in the delocalized states n from 10^8–10^{12}/cm^2 and consider electron and hole envelopes that are equal (*lines without dots*) or differ by a factor of two (*lines with dots*). In all calculations we use InGaAs parameters [9] and assume flat lens-shaped QDs. For comparison the spontaneous emission rate is shown (*dotted line*) for typical parameters ($\kappa = 0.1$/ps, $\Gamma = 0.01$/ps, $P = 0.1$/ps). Note that the rates are normalized to the square of the light-matter coupling strength

is not only globally but also *locally* neutral and there is no electrostatic interaction between the exciton and the carriers in the delocalized states. Classically, the system and the bath become uncoupled, only the exchange interaction is left. This points to an intrinsic difference between interband cavity assisted feeding and intraband scattering processes: In the latter case, electrons and holes can scatter independently, while in the former case the emission of a photon requires the presence of an exciton, i.e., a fully correlated electron-hole pair. As a consequence, any formalism describing the off-resonant cavity feeding, which relies on an interaction of Coulombian origin, like Auger interaction between the QD and carriers in delocalized states, or Fröhlich interaction of QD carriers with LO phonons, must obey a local neutrality condition: for locally neutral excitons and discarding the exchange terms the off-resonant process should vanish exactly.

1.5 Superradiant Emitter Coupling of Quantum Dots in Optical Resonators

In a conventional laser, the below-threshold spontaneous emission stems from independent emitters while the above-threshold stimulated emission – within a classical picture – results from a phase synchronization of the emitters with the radiation field. In terms of a quantum mechanical picture, the de-excitation of the active material, in semiconductors due to the recombination of electron-hole-pairs, is linked to photon generation in a way, that the expectation value $\langle b^\dagger \sigma_i^- \rangle$ has a finite value. Here $\sigma_i^- = v_i^\dagger c_i$ represents a microscopic polarization due to the de-excitation of the emitter i. $v_i^\dagger$ denotes a creation operator for a valence electron and c_i an annihilation operator for a conduction electron in the i-th QD. For independent emitters, the total emission rate is simply the sum of contributions from independent emitters, and quantum mechanical correlations are absent between the microscopic polarizations of different emitters. This is the typical situation for the above-threshold emission of conventional lasers. The phenomenon of radiative emitter coupling leads to the existence of correlations of the form $\langle \sigma_j^+ \sigma_i^- \rangle$ between the polarizations of different emitters i and j. The emitter i looses an excitation at the cost of adding an excitation in emitter j without changing the photon number in the system. Such a correlation can be driven by the exchange of a photon between the emitters. It leads to the phenomenon of superradiance, for which the emission rate of the ensemble can be enhanced or inhibited in comparison to independent emitters. For the latter effect, the term "subradiant emission regime" is also used.

Since the prediction of superradiance by Dicke in 1954, the effect has been extensively studied in a variety of systems [115, 116] including semiconductor QDs [117, 118]. In most cases the demonstration of superradiance relies on macroscopic emission properties, where the time-resolved intensity or emission linewidth changes in comparison to individual emitters. Most prominent is the transition from the exponential decay of independent emitters to a superradiant pulse for correlated emitters [115], even though most experiments resort to decay-time changes as function of the emitter number. The recent interest in superradiance of superconducting qubits [119], trapped atoms [120], and semiconductor magneto-plasmas [121] was motivated by the prospects to study directly the correlated state of the active material.

In this chapter, we demonstrate the influence of radiative emitter coupling, which leads to electronic correlations among different emitters, on the emission properties of QDs in optical resonators. We find a clear influence on the time-resolved emission (the occurrence of a superradiant pulse) as well as on the emission intensity. Especially at low excitation, dipole anti-correlations between pairs of emitters lead to "excitation trapping" in the so-called subradiant regime. As a consequence, we identify modified characteristic properties of nanolasers that exhibit radiative coupling effects, such as a non-constant β factor that reduces strongly in the low-excitation regime. Furthermore, electronic correlations have a direct influence on the statistical properties of the emitted photons. The effect of super-bunching has been predicted in [122, 123] and was recently demonstrated in direct comparison between theory and experiment in [6, 124].

In this section we consider cavity-QED lasers that operate with self-organized semiconductor QDs within a resonator acting as gain material. The QDs are excited by optical pumping and for the subsequent emission, the time-resolved and time-integrated intensity are analyzed. To characterize the statistical properties of the emission, the second-order photon correlation function $g^{(2)}(\tau = 0)$ is used (cf. Sect. 1.1.2). For the latter, a comparison of different emission regimes is presented in Fig. 1.15. In particular, a value of $g^{(2)} > 2$ is indicative for radiative coupling (case

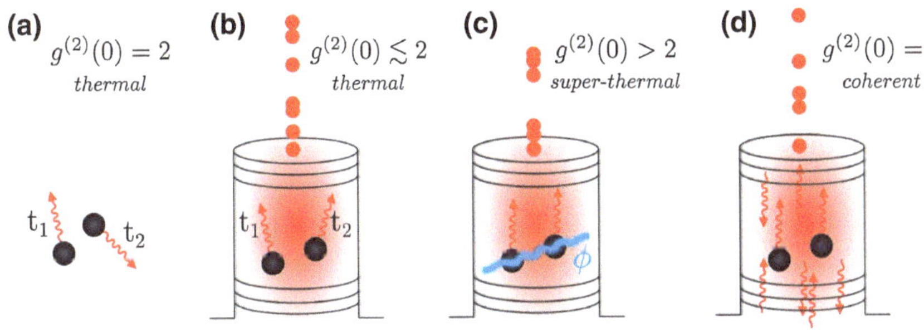

Fig. 1.15 Illustration of emission regimes. **a** Spontaneous recombination from independent emitters leads to thermal radiation. **b** Using three-dimensional photon confinement in a cavity-quantum electrodynamics laser, spontaneous emission is directed into a single resonator mode. For independent emitters, below threshold the photon emission is uncorrelated, producing thermal or close to thermal light. **c** The exchange of photons introduces correlations between the electronic states of different emitters. A relative phase information ϕ is spontaneously established, and the emission from this entangled many-particle state leads to a superradiant pulse with giant photon bunching. **d** Above threshold, stimulated emission dominates and leads to coherent radiation

c) between distant emitters [110, 122, 123]. In atomic systems, inhomogeneous distributions of the emitter energies and light-matter couplings have a strong detrimental effect on the radiative emitter coupling. For self-organized semiconductor QDs, the role of inhomogeneous broadening is substantially mitigated by the effect of non-resonant QD-cavity coupling discussed in the previous paragraph. It involves the ability to efficiently couple slightly detuned QDs to the cavity mode. Hence the cavity serves a two-fold purpose. It provides a channel for efficient photon exchange between emitters, which drives the inter-emitter correlations, and it enhances the coupling of slightly off-resonant emitters.

For the theoretical description of such a system, two options are available. When considering a small number of emitters, typically less than 10, a direct numerical solution of the von Neumann-Lindblad equation can be used [4–6, 122, 123]. This method has the advantage, that the full quantum dynamics of the coupled emitters and cavity-field system can be described. As a result, the density matrix of this system is available. It includes all existing quantum correlations and facilitates the calculation of various correlation functions. The drawback is, that only a small number of emitters can be included and each emitter is often only represented by a two-level system [122, 123, 125] or with few electronic configurations [4–6], since otherwise the Hilbert space (and the resulting dimension of the density matrix) becomes too large for a direct numerical solution. To describe larger systems, such as QD-ensemble lasers consisting of hundreds of emitters, the equation-of-motion technique can be used. By means of the cluster-expansion technique, a closed set of equations can be derived to include quantum correlations up to a given order, see Sect. 1.1.1. In [110, 124], a closed set of equations has been derived that contains inter-emitter correlations of the type $\langle \sigma_j^+ \sigma_i^- \rangle$, and the resulting changes in the cavity photon number $\langle b^\dagger b \rangle$ and in the second-order photon correlation function $g^{(2)}(\tau = 0)$.

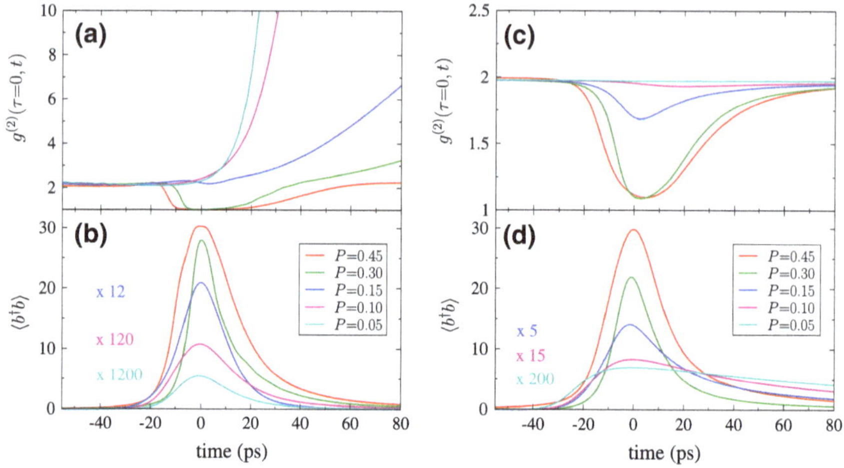

Fig. 1.16 Time evolution of the intensity autocorrelation function $g^{(2)}(\tau = 0, t)$ (*top*) and mean photon number (*bottom*) after pulsed excitation of a nanolaser with various pump-pulse areas P. *Different curves* correspond to below-threshold excitation ($P = 0.05$), the transition region ($P = 0.1$ and 0.15), and above-threshold excitation ($P = 0.3$ and 0.45). In the *left part*, superradiant coupling between different quantum dots is included, while in the *right part* it is omitted

Figures 1.16 and 1.17 show results of such a theory for a QD nanolaser working with 200 QD emitters within an optical resonator (cavity photon decay rate $\kappa = 0.4\,\text{ps}^{-1}$, coupling rate between QDs and cavity mode $g = 0.1\,\text{ps}^{-1}$, spontaneous emission rate into other modes $\gamma_{nl} = 0.005\,\text{ps}^{-1}$). In Fig. 1.16, the time evolution of the mean photon number and second-order photon correlation function are compared with (left part) and without (right part) superradiant coupling. In the presence of superradiant coupling, even for weak pump rates a short output pulse is obtained with a temporal width of about 20 ps, which is much shorter than the spontaneous lifetime of individual emitters in the cavity (about 200 ps). At the same time, $g^{(2)}(\tau = 0, t)$ is about 2 during the pulse maximum, indicating that the short pulse width is not linked to stimulated emission. Values for $g^{(2)}(\tau = 0, t)$ larger than 2 before and in particular after the pulse maximum are signatures of radiative emitter coupling. When omitting the radiative emitter coupling in the theory, $g^{(2)}(\tau = 0, t)$ remains 2 for weak pumping and the time-resolved emission shows a much slower exponential decay with the typical Purcell-enhanced lifetime of individual emitters (see also the left panel in Fig. 1.17). For stronger pump pulses, the system is driven above the laser threshold and a stimulated emission pulse can be identified with $g^{(2)}(\tau = 0, t)$ reaching a value of 1 during the pulse maximum. In this regime, the difference in the theoretical results with and without radiative emitter coupling becomes small, which indicates that stimulated emission dominates the photon generation and suppresses the role of inter-emitter correlations.

For the pulsed optical excitation discussed in the above examples, the time-integrated output intensity versus pump pulse area is shown in the right part of

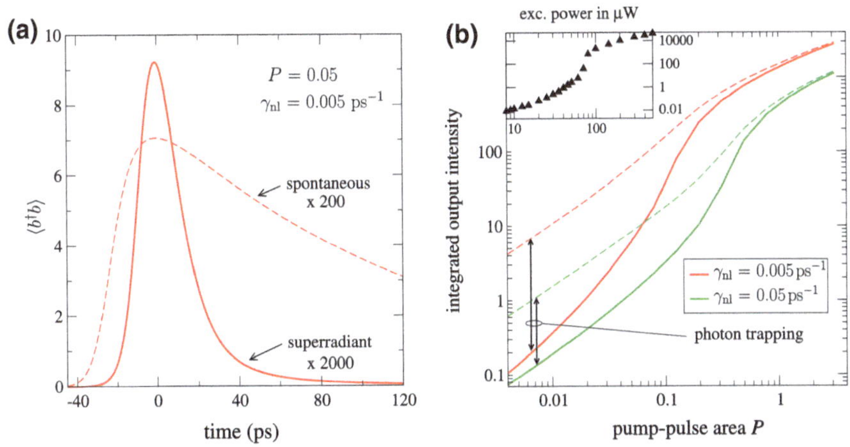

Fig. 1.17 Calculated time evolution of the cavity mean photon number after pulsed optical excitation for weak pumping (*left*) and time integrated output intensity versus pump-pulse area (*right*). Results including superradiant coupling of the emitters (*solid lines*) are compared to those for independent emitters (*dashed lines*). As indicated in the *right panel*, a stronger spontaneous emission rate into non-lasing modes γ_{nl} decreases the threshold modification due to superradiant coupling but corroborates the robustness of the effect. The inset confirms the presence of the threshold kink in the experiment despite a large β-factor as an indication of excitation trapping due to superradiant emitter coupling

Fig. 1.17. When comparing results with (solid lines) and without (dashed lines) superradiant coupling, a dramatic reduction of the output intensity for weak and intermediate pump rates is evident. In this regime, photons emitted into the cavity mode from the QDs are reabsorbed by other QDs in a way that inter-QD polarizations of the form $\langle \sigma_j^+ \sigma_i^- \rangle$ are driven. These inter-QD polarizations exist in addition to the photon-assisted polarization $\langle b^\dagger \sigma_i^- \rangle$. The build-up of the former clearly reduces the photon number in the cavity mode. On the other hand, above the laser threshold, when stimulated emission is present, the role of radiative emitter coupling is small. The combination of these two observations leads to a larger jump in the input-output curve at the laser threshold [110, 124]. In the standard rate-equation laser theory, this jump is solely determined by the β factor [126], which quantifies the fraction of the total spontaneous emission that is directed into the laser mode. From a rate-equation model it is found that β is solely determined by the relation of the rates associated with emission into the laser mode γ_l and into non-lasing modes or other loss channels γ_{nl}, i.e., $\beta = \gamma_l/(\gamma_l + \gamma_{nl})$ [127]. Correspondingly, for cavity-QED lasers the β factor is frequently estimated from the jump in the input-output curve. Our results demonstrate that in the presence of radiative coupling this strongly underestimates the value for β. Furthermore, without superradiant coupling a larger γ_{nl} leads to a larger kink as the β factor increases. Our calculation including radiative coupling leads to a smaller kink, as a larger γ_{nl} causes a reduction (damping) of inter-QD polarizations, thereby reducing the associated excitation trapping (right panel of Fig. 1.17).

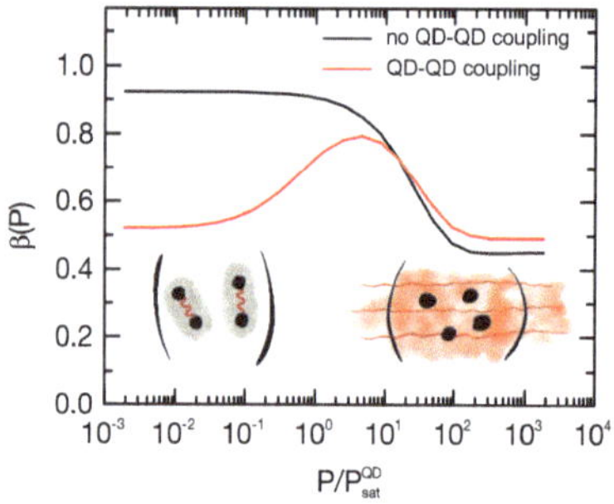

Fig. 1.18 (*Red curve*) pump-rate dependent β-factor obtained from the theoretical model with parameters applicable to experimentally studied system. (*Black curve*) calculation suppressing radiative coupling effects between emitters that are responsible for sub- and superradiant effects. Comparing *both curves* reveals that radiative coupling effects lead to a strong inhibition of spontaneous emission at low excitation (subradiance) and a slight enhancement of spontaneous emission above the laser threshold (superradiance)

This behavior actually raises the question whether a constant β factor is suited to characterize QD nanolasers that, due to the strong mode confinement, are likely to facilitate the formation of inter-emitter correlations by the light field. In the following, we suggest the idea of introducing a pump-rate dependent factor $\beta(P)$ that accounts for a reduction of the spontaneous emission rate due to excitation trapping, as well as for the interplay of different multi-exciton configurations and apply it to the few-QD nanolaser system discussed in the context of Fig. 1.13.

We first discuss the result for the pump-rate dependent $\beta(P)$ factor without the effects of radiative coupling, shown by the black curve in Fig. 1.18. The asymptotic values at low ($\beta(P) > 90\%$) and high excitation ($\beta(P) \approx 50\%$) reflect the conventional constant β factor associated with the multi-exciton transitions that dominate at low and high excitation. In between, a transition is seen as the system switches between multi-excitonic emission channels from different manifolds, cf. Sect. 1.4.2. To provide insight into the role of the radiative coupling in our system, the red curve shows the pump-rate dependent $\beta(P)$ factor calculated including correlations due to radiative coupling. In the low-excitation regime, where the super-thermal bunching is observed in $g^{(2)}(0)$ (Fig. 1.13c), inter-emitter coupling leads to a strong inhibition of the spontaneous emission rate resulting in $\beta(P) = 50\%$ instead of 90%. Thereby $\beta(P)$ reflects the photon-trapping effect discussed above.

In summary, radiative emitter coupling can influence classical emission properties like the intensity and its temporal evolution, but also the statistical properties of the light emission. The effects of inter-emitter coupling is most prominent in the spontaneous emission regime, but can also play a role in few-emitter lasers, where stimulated emission is weak. Accounting for radiative coupling effects in cavity-QED lasers can strongly modify the device characteristics and be of paramount importance for the correct identification of system parameters, opening up new research directions in cavity-QED nanolasers.

1.6 Summary and Outlook

Advances in the quantum information technologies have created a demand for fast, efficient, and integrable sources of non-classical light. Solid-state emitters are a promising platform, and quantum-dot systems have reached a technological maturity that permits a high level of control over their emission properties, beginning by tailoring the emission wavelength via material and growth design, but also altering spontaneous emission itself by using cavity-QED effects in microcavities. As a result, nearly many imaginable states of the light field can be realized with single or few-QD systems, from single-photon Fock-states to intense highly-bunched super-thermal light.

Theoretical models play an important role in the design of new devices, the definition of achievable specifications, and the exploration of new applications not yet thought of. Unlike atomic systems, QDs are not isolated but are embedded semiconductor systems. Theoretical modelling must combine quantum-optics with semiconductor physics. This chapter has given an overview over the constituents of such a theory and possible methods using examples from different applications.

The sophistication of technological possibilities will lead to even more advanced realizations of solid-state cavity-QED systems in the future. New materials such as two-dimensional semiconductors based on transition-metal dichalcogenides hold great promise for exploring excitonic effects up to room temperature. Individual cavities can already be combined to cavity arrays, so-called Jaynes–Cummings lattices, in which many-body systems such as the Bose–Hubbard model become realized. The potential of such systems for the quantum-information technologies is largely unexplored.

Acknowledgements The authors would like to thank Weng W. Chow (Albuquerque), Jan Wiersig (Magdeburg) and Paul Gartner (Bucharest) for many fruitful collaborations. This work has strongly profited from the ongoing collaboration regarding state-of-the-art experiments in the groups of Peter Michler (Stuttgart), Manfred Bayer (Dortmund), Stephan Reitzenstein (Berlin), and Johnathan Finley (Munich). We are grateful for financial support from the Deutsche Forschungsgemeinschaft (DFG) and the Bundesministerium für Bildung und Forschung (BMBF).

References

1. J.P. Dowling, G.J. Milburn, Philos. Trans. R. Soc. Lond. A Math. Phys. Eng. Sci. **361**(1809), 1655 (2003). doi:10.1098/rsta.2003.1227, http://rsta.royalsocietypublishing.org/content/361/1809/1655
2. C. Roy, S. Hughes, Phys. Rev. X **1**(2), 021009 (2011). doi:10.1103/PhysRevX.1.021009
3. S. Schumacher, J. Förstner, A. Zrenner, M. Florian, C. Gies, P. Gartner, F. Jahnke, Opt. Express **20**(5), 5335 (2012). doi:10.1364/OE.20.005335, http://www.opticsexpress.org/abstract.cfm?URI=oe-20-5-5335
4. M. Florian, P. Gartner, C. Gies, F. Jahnke, New J. Phys. **15**(3), 035019 (2013). doi:10.1088/1367-2630/15/3/035019

5. M. Florian, A. Steinhoff, C. Gies, F. Jahnke, Appl. Phys. B **122**(1), 1 (2016). doi:10.1007/s00340-015-6296-5
6. S. Lichtmannecker, M. Florian, T. Reichert, M. Blauth, M. Bichler, F. Jahnke, J.J. Finley, C. Gies, M. Kaniber, arXiv:1602.03998 [cond-mat] (2016)
7. S. Ritter, P. Gartner, C. Gies, F. Jahnke, Opt. Express **18**(10), 9909 (2010). doi:10.1364/OE.18.009909
8. C. Gies, M. Florian, P. Gartner, F. Jahnke, Opt. Express **19**(15), 14370–14388 (2011). doi:10.1364/OE.19.014370
9. A. Steinhoff, P. Gartner, M. Florian, F. Jahnke, Phys. Rev. B **85**(20), 205144 (2012). doi:10.1103/PhysRevB.85.205144
10. H.J. Carmichael, *Statistical Methods in Quantum Optics 1: Master Equations and Fokker-Planck Equations (Theoretical and Mathematical Physics)* (Springer, Berlin, 2003)
11. L. Mandel, E. Wolf, *Optical Coherence and Quantum Optics* (Cambridge University Press, Cambridge, 1995)
12. C. Gardiner, P. Zoller, *Quantum Noise A Handbook of Markovian and Non-Markovian Quantum Stochastic Methods with Applications to Quantum Optics* (Springer, Berlin, 2004)
13. C. Gies, M. Florian, P. Gartner, F. Jahnke, in *Quantum Optics With Semiconductor Nanostructures*, ed. by F. Jahnke (Woodhead Publishing Limited, Sawston, 2012)
14. M. Kira, S.W. Koch, Phys. Rev. A **78**(2), 022102 (2008). doi:10.1103/PhysRevA.78.022102
15. W. Hoyer, M. Kira, S.W. Koch, in *Nonequilibrium Physics at Short Time Scales*, ed. by K. Morawetz (Springer, Berlin, 2004)
16. M. Florian, C. Gies, F. Jahnke, H.A.M. Leymann, J. Wiersig, Physical Review B **87**(16), 165306 (2013). doi:10.1103/PhysRevB.87.165306
17. H.A.M. Leymann, A. Foerster, J. Wiersig, Physical Review B **89**(8), 085308 (2014). doi:10.1103/PhysRevB.89.085308
18. M. Richter, A. Carmele, A. Sitek, A. Knorr, Phys. Rev. Lett. **103**(8), 087407 (2009). doi:10.1103/PhysRevLett.103.087407
19. A. Carmele, A. Knorr, M. Richter, Phys. Rev. B **79**(3), 035316 (2009). doi:10.1103/PhysRevB.79.035316
20. J. Wiersig, Phys. Rev. B **82**(15), 155320 (2010). doi:10.1103/PhysRevB.82.155320
21. S. Ates, C. Gies, S.M. Ulrich, J. Wiersig, S. Reitzenstein, A. Löffler, A. Forchel, F. Jahnke, P. Michler, Physical Review B **78**(15), 155319 (2008). doi:10.1103/PhysRevB.78.155319
22. M. Florian, C. Gies, P. Gartner, F. Jahnke, Journal of the Optical Society of America B **29**(2), A31 (2012). doi:10.1364/JOSAB.29.000A31
23. P. Michler, *Single Quantum Dots - Fundamentals, Applications and New Concepts* (Springer, Berlin, 2003)
24. P. Michler, A. Kiraz, C. Becher, W.V. Schoenfeld, P.M. Petroff, L. Zhang, E. Hu, A. Imamoglu, Science **290**, 2282 (2000)
25. M. Bajcsy, A. Majumdar, A. Rundquist, J. Vučković, New J. Phys. **15**(2), 025014 (2013). doi:10.1088/1367-2630/15/2/025014
26. C. Lang, D. Bozyigit, C. Eichler, L. Steffen, J.M. Fink, A.A. Abdumalikov, M. Baur, S. Filipp, M.P. da Silva, A. Blais, A. Wallraff, Phys. Rev. Lett. **106**(24), 243601 (2011). doi:10.1103/PhysRevLett.106.243601
27. C.S. Muñoz, E. del Valle, A.G. Tudela, K. Müller, S. Lichtmannecker, M. Kaniber, C. Tejedor, J.J. Finley, F.P. Laussy, Nat. Photonics **8**(7), 550 (2014). doi:10.1038/nphoton.2014.114
28. N. Somaschi, V. Giesz, L.D. Santis, J.C. Loredo, M.P. Almeida, G. Hornecker, S.L. Portalupi, T. Grange, C. Antón, J. Demory, C. Gómez, I. Sagnes, N.D. Lanzillotti-Kimura, A. Lemaítre, A. Auffèves, A.G. White, L. Lanco, P. Senellart, Nat. Photonics **10**(5), 340 (2016). doi:10.1038/nphoton.2016.23
29. M. Pelton, C. Santori, J. Vučković, B. Zhang, G.S. Solomon, J. Plant, Y. Yamamoto, Physical Review Letters **89**(23), 233602 (2002). doi:10.1103/PhysRevLett.89.233602
30. S. Buckley, K. Rivoire, J. Vučković, Rep. Progr. Phys. **75**(12), 126503 (2012). doi:10.1088/0034-4885/75/12/126503

31. J.L. O'Brien, A. Furusawa, J. Vučković, Nat. Photonics **3**(12), 687 (2009). doi:10.1038/nphoton.2009.229
32. M. Nomura, N. Kumagai, S. Iwamoto, Y. Ota, Y. Arakawa, Opt. Express **17**(18), 15975 (2009). doi:10.1364/OE.17.015975
33. S. Reitzenstein, C. Böckler, A. Bazhenov, A. Gorbunov, A. Löffler, M. Kamp, V.D. Kulakovskii, A. Forchel, Opt. Express **16**(7), 4848–4857 (2008). doi:10.1364/OE.16.004848
34. C. Gies, F. Jahnke, W.W. Chow, Physical Review A **91**(6), 061804 (2015). doi:10.1103/PhysRevA.91.061804
35. A. Predojevic, M. Jezek, T. Huber, H. Jayakumar, T. Kauten, G.S. Solomon, R. Filip, G. Weihs, Opt. Express **22**(4), 4789 (2014). doi:10.1364/OE.22.004789, http://www.opticsexpress.org/abstract.cfm?URI=oe-22-4-4789
36. G. Reithmaier, S. Lichtmannecker, T. Reichert, P. Hasch, K. Müller, M. Bichler, R. Gross, J.J. Finley, Sci. Rep. **3**, 1901 (2013). doi:10.1038/srep01901. PMID: 23712624 PMCID: PMC3664895
37. B. Calkins, P.L. Mennea, A.E. Lita, B.J. Metcalf, W.S. Kolthammer, A. Lamas-Linares, J.B. Spring, P.C. Humphreys, R.P. Mirin, J.C. Gates, P.G.R. Smith, I.A. Walmsley, T. Gerrits, S.W. Nam, Opt. Express **21**(19), 22657 (2013). doi:10.1364/OE.21.022657
38. F.E. Becerra, J. Fan, A. Migdall, Nat. Photonics **9**(1), 48 (2015). doi:10.1038/nphoton.2014.280
39. M. Bell, A. Antipov, B. Karasik, A. Sergeev, V. Mitin, A. Verevkin, IEEE Trans. Appl. Supercond. **17**(2), 289 (2007). doi:10.1109/TASC.2007.898616
40. *Single-Photon Generation and Detection: Physics and Applications* (Academic Press, Dublin, 2013). Google-Books-ID: gERMqvh2OjAC
41. R. Filip, L. Lachman, Physical Review A **88**(4), 043827 (2013). doi:10.1103/PhysRevA.88.043827
42. M. Nomura, N. Kumagai, S. Iwamoto, Y. Ota, Y. Arakawa, Nat. Phys. **6**, 279 (2010). doi:10.1038/nphys1518
43. H.J. Carmichael, *Statistical Methods in Quantum Optics 2: Non-Classical Fields*, 2008th edn. (Springer, Berlin, 2007)
44. T. Yoshie, A. Scherer, J. Hendrickson, G. Khitrova, H.M. Gibbs, G. Rupper, C. Ell, O.B. Shchekin, D.G. Deppe, Nature **432**(7014), 200 (2004). doi:10.1038/nature03119
45. J.P. Reithmaier, G. Sk ek, A. Löffler, C. Hofmann, S. Kuhn, S. Reitzenstein, L.V. Keldysh, V.D. Kulakovskii, T.L. Reinecke, A. Forchel, Nature **432**(7014), 197 (2004). doi:10.1038/nature02969
46. F. Gericke, C. Gies, P. Gartner, S. Holzinger, C. Hopfmann, T. Heindel, J. Wolters, C. Schneider, M. Florian, F. Jahnke, S. Höfling, M. Kamp, S. Reitzenstein, arXiv:1606.05591 [cond-mat, physics:physics] (2016)
47. W.W. Chow, F. Jahnke, C. Gies, Light Sci. Appl. **3**(8), e201 (2014). doi:10.1038/lsa.2014.82
48. N. Sangouard, C. Simon, N. Gisin, J. Laurat, R. Tualle-Brouri, P. Grangier, J. Opt. Soc. Am. B **27**(6), A137 (2010). doi:10.1364/JOSAB.27.00A137
49. A.K. Ekert, Phys. Rev. Lett. **67**(6), 661 (1991). doi:10.1103/PhysRevLett.67.661
50. B.D. Gerardot, S. Seidl, P.A. Dalgarno, R.J. Warburton, D. Granados, J.M. Garcia, K. Kowalik, O. Krebs, K. Karrai, A. Badolato, P.M. Petroff, Appl. Phys. Lett. **90**(4), 041101 (2007). doi:10.1063/1.2431758
51. S. Bounouar, M. Müller, A.M. Barth, M. Glässl, V.M. Axt, P. Michler, Phys. Rev. B **91**(16), 161302 (2015). doi:10.1103/PhysRevB.91.161302
52. Y. Ota, S. Iwamoto, N. Kumagai, Y. Arakawa, Phys. Rev. Lett. **107**(23), 233602 (2011). doi:10.1103/PhysRevLett.107.233602
53. C.A. Kessler, M. Reischle, F. Hargart, W.M. Schulz, M. Eichfelder, R. Roßbach, M. Jetter, P. Michler, P. Gartner, M. Florian, C. Gies, F. Jahnke, Phys. Rev. B **86**(11), 115326 (2012). doi:10.1103/PhysRevB.86.115326
54. G. Lindblad, Commun. Math. Phys. **48**(2), 119 (1976). doi:10.1007/BF01608499
55. H.P. Breuer, *The Theory of Open Quantum Systems* (Oxford University Press, USA, 2007)

56. A. Steinhoff, H. Kurtze, P. Gartner, M. Florian, D. Reuter, A.D. Wieck, M. Bayer, F. Jahnke, Phys. Rev. B **88**(20), 205309 (2013). doi:10.1103/PhysRevB.88.205309
57. T.S. Sosnowski, T.B. Norris, H. Jiang, J. Singh, K. Kamath, P. Bhattacharya, Phys. Rev. B **57**, R9423 (1998)
58. J. Urayama, T.B. Norris, J. Singh, P. Bhattacharya, Phys. Rev. Lett. **86**, 4930 (2001). doi:10.1103/PhysRevLett.86.4930
59. F. Quochi, M. Dinu, N.H. Bonadeo, J. Shah, L.N. Pfeiffer, K.W. West, P.M. Platzman, Phys. B **314**, 263 (2002). doi:10.1016/S0921-4526(01)01383-7
60. F. Quochi, M. Dinu, L.N. Pfeiffer, K.W. West, C. Kerbage, R.S. Windeler, B.J. Eggleton, Phys. Rev. B **67**, 235323 (2003). doi:10.1103/PhysRevB.67.235323
61. P. Borri, W. Langbein, U. Woggon, V. Stavarache, D. Reuter, A.D. Wieck, Phys. Rev. B **71**(11), 115328 (2005). doi:10.1103/PhysRevB.71.115328
62. H. Kurtze, J. Seebeck, P. Gartner, D.R. Yakovlev, D. Reuter, A.D. Wieck, M. Bayer, F. Jahnke, Phys. Rev. B **80**(23), 235319 (2009). doi:10.1103/PhysRevB.80.235319
63. A. Laucht, M. Kaniber, A. Mohtashami, N. Hauke, M. Bichler, J.J. Finley, Phys. Rev. B **81**, 241302 (2010). doi:10.1103/PhysRevB.81.241302
64. U. Bockelmann, T. Egeler, Phys. Rev. B **46**, 15574 (1992). doi:10.1103/PhysRevB.46.15574
65. T. Inoshita, H. Sakaki, Phys. Rev. B **46**, 7260 (1992). doi:10.1103/PhysRevB.46.7260
66. I. Vurgaftman, Y. Lam, J. Singh, Phys. Rev. B **50**, 14309 (1994). doi:10.1103/PhysRevB.50.14309
67. A.L. Efros, V.A. Kharchenko, M. Rosen, Solid State Commun. **93**, 281 (1995)
68. T. Inoshita, H. Sakaki, Phys. Rev. B **56**, 4355 (1997). doi:10.1103/PhysRevB.56.R4355
69. A.V. Uskov, F. Adler, H. Schweizer, M.H. Pilkuhn, J. Appl. Phys. **81**, 7895 (1997). doi:10.1063/1.365363
70. M. Brasken, M. Lindberg, M. Sopanen, H. Lipsanen, J. Tulkki, Phys. Rev. B **58**, 15993 (1998). doi:10.1103/PhysRevB.58.R15993
71. H. Jiang, J. Singh, IEEE J. Quantum Electron. **34**, 1188 (1998)
72. R. Ferreira, G. Bastard, Appl. Phys. Lett. **74**, 2818 (1999). doi:10.1063/1.124024
73. T. Stauber, R. Zimmermann, H. Castella, Phys. Rev. B **62**, 7336 (2000). doi:10.1103/PhysRevB.62.7336
74. K. Lüdge, M.J.P. Bormann, E. Malić, P. Hövel, M. Kuntz, D. Bimberg, A. Knorr, E. Schöll, Phys. Rev. B **78**(3), 035316 (2008). doi:10.1103/PhysRevB.78.035316
75. M. Grundmann, D. Bimberg, physica status solidi (a) **164**(1), 297–300 (1997). doi:10.1002/1521-396X(199711)164:1<297::AID-PSSA297>3.0.CO;2-5
76. P. Borri, V. Cesari, W. Langbein, Phys. Rev. B **82**(11), 115326 (2010). doi:10.1103/PhysRevB.82.115326
77. J.I. Climente, A. Bertoni, M. Rontani, G. Goldoni, E. Molinari, Phys. Rev. B **74**(12), 125303 (2006). doi:10.1103/PhysRevB.74.125303
78. G.D. Mahan, *Many Particle Physics*, 3rd edn (Springer, Berlin, 2000)
79. U. Bockelmann, G. Bastard, Phys. Rev. B **42**, 8947 (1990). doi:10.1103/PhysRevB.42.8947
80. H. Benisty, C.M. Sotomayor-Torres, C. Weisbuch, Phys. Rev. B **44**, 10945 (1991). doi:10.1103/PhysRevB.44.10945
81. L. Besombes, K. Kheng, L. Marsal, H. Mariette, Phys. Rev. B **63**(15), 155307 (2001). doi:10.1103/PhysRevB.63.155307
82. P. Borri, W. Langbein, S. Schneider, U. Woggon, R.L. Sellin, D. Ouyang, D. Bimberg, Phys. Rev. Lett. **87**(15), 157401 (2001), http://link.aps.org/abstract/PRL/v87/e157401
83. S. Schmitt-Rink, D.A.B. Miller, D.S. Chemla, Phys. Rev. B **35**(15), 8113 (1987). doi:10.1103/PhysRevB.35.8113
84. T. Takagahara, Phys. Rev. B **60**(4), 2638 (1999). doi:10.1103/PhysRevB.60.2638
85. B. Krummheuer, V.M. Axt, T. Kuhn, Phys. Rev. B **65**(19), 195313 (2002). doi:10.1103/PhysRevB.65.195313
86. R. Zimmermann, E. Runge, Proceedings of the 26th ICPS (Edinburgh, 2002)
87. E. Pazy, Semicond. Sci. Technol. **17**(11), 1172 (2002). doi:10.1088/0268-1242/17/11/307

88. P. Borri, W. Langbein, S. Schneider, U. Woggon, R.L. Sellin, D. Ouyang, D. Bimberg, Phys. Rev. Lett. **89**(18), 187401 (2002). doi:10.1103/PhysRevLett.89.187401
89. H. Gotoh, H. Kamada, T. Saitoh, H. Ando, J. Temmyo, Phys. Rev. B **69**(15), 155328 (2004). doi:10.1103/PhysRevB.69.155328
90. G. Moody, R. Singh, H. Li, I.A. Akimov, M. Bayer, D. Reuter, A.D. Wieck, S.T. Cundiff, Phys. Rev. B **87**(4), 045313 (2013). doi:10.1103/PhysRevB.87.045313
91. S. Strauf, K. Hennessy, M.T. Rakher, Y.S. Choi, A. Badolato, L.C. Andreani, E.L. Hu, P.M. Petroff, D. Bouwmeester, Phys. Rev. Lett. **96**, 127404 (2006). doi:10.1103/PhysRevLett.96.127404
92. S.M. Ulrich, C. Gies, J. Wiersig, S. Reitzenstein, C. Hofmann, A. Löffler, A. Forchel, F. Jahnke, P. Michler, Phys. Rev. Lett. **98**, 043906 (2007)
93. I. Wilson-Rae, A. Imamoglu, Phys. Rev. B **65**(23), 235311 (2002). doi:10.1103/PhysRevB.65.235311
94. D. Press, S. Götzinger, S. Reitzenstein, C. Hofmann, A. Löffler, M. Kamp, A. Forchel, Y. Yamamoto, Phys. Rev. Lett. **98**(11), 117402 (2007). doi:10.1103/PhysRevLett.98.117402
95. G. Tarel, V. Savona, physica status solidi (c) **6**(4), 902 (2009). doi:10.1002/pssc.200880577
96. M. Kaniber, A. Laucht, A. Neumann, J.M. Villas-Bôas, M. Bichler, M.C. Amann, J.J. Finley, Phys. Rev. B **77**(16), 161303 (2008). doi:10.1103/PhysRevB.77.161303
97. S. Ates, S.M. Ulrich, A. Ulhaq, S. Reitzenstein, A. Löffler, S. Höfling, A. Forchel, P. Michler, Nat. Photonics **3**(12), 724 (2009). doi:10.1038/nphoton.2009.215
98. U. Hohenester, A. Laucht, M. Kaniber, N. Hauke, A. Neumann, A. Mohtashami, M. Seliger, M. Bichler, J.J. Finley, Phys. Rev. B **80**(20), 201311 (2009). doi:10.1103/PhysRevB.80.201311
99. D. Dalacu, K. Mnaymneh, V. Sazonova, P.J. Poole, G.C. Aers, J. Lapointe, R. Cheriton, A.J. SpringThorpe, R. Williams, Phys. Rev. B **82**(3), 033301 (2010). doi:10.1103/PhysRevB.82.033301
100. M. Calic, P. Gallo, M. Felici, K.A. Atlasov, B. Dwir, A. Rudra, G. Biasiol, L. Sorba, G. Tarel, V. Savona, E. Kapon, Phys. Rev. Lett. **106**(22), 227402 (2011). doi:10.1103/PhysRevLett.106.227402
101. A. Majumdar, E.D. Kim, Y. Gong, M. Bajcsy, J. Vučković, Phys. Rev. B **84**(8), 085309 (2011). doi:10.1103/PhysRevB.84.085309
102. U. Hohenester, Phys. Rev. B **81**(15), 155303 (2010). doi:10.1103/PhysRevB.81.155303
103. D.P.S. McCutcheon, A. Nazir, New J. Phys. **12**(11), 113042 (2010). doi:10.1088/1367-2630/12/11/113042, http://stacks.iop.org/1367-2630/12/i=11/a=113042
104. A. Würger, Phys. Rev. B **57**(1), 347 (1998). doi:10.1103/PhysRevB.57.347
105. J.D. Joannopoulos, P.R. Villeneuve, S. Fan, Nature **386**(6621), 143 (1997). doi:10.1038/386143a0
106. S. Noda, M. Fujita, T. Asano, Nat. Photonics **1**(8), 449 (2007). doi:10.1038/nphoton.2007.141
107. S. Strauf, F. Jahnke, Laser Photonics Rev. **5**(5), 607 (2011). doi:10.1002/lpor.201000039
108. E. Dekel, D. Gershoni, E. Ehrenfreund, D. Spektor, J.M. Garcia, P.M. Petroff, Phys. Rev. Lett. **80**(22), 4991 (1998). doi:10.1103/PhysRevLett.80.4991
109. M. Winger, T. Volz, G. Tarel, S. Portolan, A. Badolato, K.J. Hennessy, E.L. Hu, A. Beveratos, J. Finley, V. Savona, A. Imamoğlu, Phys. Rev. Lett. **103**(20), 207403 (2009). doi:10.1103/PhysRevLett.103.207403
110. H. Leymann, A. Foerster, F. Jahnke, J. Wiersig, C. Gies, Phys. Rev. Appl. **4**(4), 044018 (2015). doi:10.1103/PhysRevApplied.4.044018
111. E. Mascarenhas, D. Gerace, M.F. Santos, A. Auffèves, Phys. Rev. A **88**(6), 063825 (2013). doi:10.1103/PhysRevA.88.063825
112. K. Karrai, R.J. Warburton, C. Schulhauser, A. Högele, B. Urbaszek, E.J. McGhee, A.O. Govorov, J.M. Garcia, B.D. Gerardot, P.M. Petroff, Nature **427**(6970), 135 (2004). doi:10.1038/nature02109
113. N. Chauvin, C. Zinoni, M. Francardi, A. Gerardino, L. Balet, B. Alloing, L.H. Li, A. Fiore, Phys. Rev. B **80**(24), 241306 (2009). doi:10.1103/PhysRevB.80.241306

114. M. Florian, P. Gartner, A. Steinhoff, C. Gies, F. Jahnke, Phys. Rev. B **89**(16), 161302(R) (2014). doi:10.1103/PhysRevB.89.161302
115. M. Gross, S. Haroche, Phys. Rep. **93**(5), 301 (1982). doi:10.1016/0370-1573(82)90102-8
116. T. Brandes, Phys. Rep. **408**(5–6), 315 (2005). doi:10.1016/j.physrep.2004.12.002
117. M. Scheibner, T. Schmidt, L. Worschech, A. Forchel, G. Bacher, T. Passow, D. Hommel, Nat. Phys. **3**(2), 106 (2007). doi:10.1038/nphys494
118. M. Kozub, L. Pawicki, P. Machnikowski, Phys. Rev. B **86**(12), 121305 (2012). doi:10.1103/PhysRevB.86.121305
119. J.A. Mlynek, A.A. Abdumalikov, C. Eichler, A. Wallraff, Nature Communications **5** (2014). doi:10.1038/ncomms6186
120. J.G. Bohnet, Z. Chen, J.M. Weiner, D. Meiser, M.J. Holland, J.K. Thompson, Nature **484**(7392), 78 (2012). doi:10.1038/nature10920
121. G. Timothy, Noe Ii, J.H. Kim, J. Lee, Y. Wang, A.K. Wójcik, S.A. McGill, D.H. Reitze, A.A. Belyanin. J. Kono Nat. Phys. **8**(3), 219 (2012). doi:10.1038/nphys2207
122. V.V. Temnov, U. Woggon, Opt. Express **17**, 5774 (2009). doi:10.1364/OE.17.005774
123. A. Auffèves, D. Gerace, S. Portolan, A. Drezet, M.F. Santos, New J. Phys. **13**(9), 093020 (2011). doi:10.1088/1367-2630/13/9/093020
124. F. Jahnke, C. Gies, M. Aßmann, M. Bayer, H.a.M. Leymann, A. Foerster, J. Wiersig, C. Schneider, M. Kamp, S. Höfling, Nat. Commun. **7**, 11540 (2016). doi:10.1038/ncomms11540
125. D. Meiser, M.J. Holland, Physical Review A **81**(6), 063827 (2010). doi:10.1103/PhysRevA.81.063827
126. G. Björk, Y. Yamamoto, IEEE J. Quantum Electron. **27**, 2386 (1991). doi:10.1109/3.100877
127. H. Yokoyama, S.D. Brorson, J. Appl. Phys. **66**, 4801 (1989). doi:10.1063/1.343793

Chapter 2
Theory of Phonon Dressed Light-Matter Interactions and Resonance Fluorescence in Quantum Dot Cavity Systems

Kaushik Roy-Choudhury and Stephen Hughes

Abstract Electron-phonon coupling in semiconductor quantum dots plays a significant role in determining the optical properties of excited electron-hole pairs. This chapter describes the theory of phonon modified light-matter interactions, with a focus on the polaron master equation approach for open quantum systems. The theory is applied to study various light-matter interaction regimes and emerging experiments in the presence of electron–acoustic-phonon scattering, including phonon-modified vacuum Rabi splitting, spontaneous emission, off-resonance cavity feeding, photoluminescence intensity, and field-driven Mollow triplets.

2.1 Introduction

Quantum dots (QDs) as "artificial atoms" are excellent candidates for solid-state quantum bits and show promise for scalable quantum information processing at optical frequencies [1, 2]. However, QD excitons are intrinsically coupled to the underlying phonon "reservoir" [3], which can significantly reduce their coherence time on short time scales and open up interaction processes that are unique to the solid state system. In recent years, phonon dressing of semiconductor QD emission has manifested itself in a number of interesting experimental observations, including phonon-assisted inversion [4–10], damping and frequency shifting of driven Rabi oscillations [11–13] and excitation induced dephasing of Mollow triplet sidebands [14, 15], which distinguishes QDs from simple two-level atoms [16].

Over the past decade or so, various theories have been developed to incorporate electron-scattering processes in describing optically excited QDs [17, 18]. These include the independent Boson model (IBM) [3, 19, 20], cumulant expansion [21], correlation expansion [11], perturbative master equations (MEs) [22], polaron MEs [23–27], variational MEs [28] and path integral calculations [29]. In optical cavity structures, one can also couple the photonic environment to the QD,

K. Roy-Choudhury · S. Hughes (✉)
Department of Physics, Engineering Physics & Astronomy, Queen's University, Kingston, ON K7L 3N6, Canada
e-mail: shughes@queensu.ca

P. Michler (ed.), *Quantum Dots for Quantum Information Technologies*, Nano-Optics and Nanophotonics, DOI 10.1007/978-3-319-56378-7_2

exploiting such phenomena such as the Purcell effect [30], where the spontaneous emission (SE) rate of QD excitons is increased through an increased local density of photon states (LDOS), e.g., by coupling to a cavity mode [1, 31] or a slow-light waveguide mode [32–37]. In photonic crystal (PC) structures, one can also *suppress* the SE by coupling embedded QDs to a reduced LDOS within the photonic bandgap [38]; two example cavity structures or structured photonic reservoirs are shown schematically in Fig. 2.1a, b. Collective phonon coupling also affects how QDs couple to the photonic cavity structure, resulting in a complex interplay between the phonon coupling and the photon reservoir coupling. For example, in high-Q cavity structures, where Q is the quality factor, there are additional phonon-mediated scattering effects that can occur, including phonon-dressed vacuum Rabi splitting [39, 40] (which causes an asymmetry and a reduction of the Rabi field), and phonon-assisted cavity feeding [26, 40–46] (where a strong cavity mode appears even when far detuned from the exciton), which are now routinely measured experimentally.

One of the most useful approaches for modelling light-matter interactions in open quantum systems is through a quantum master equation (ME), where system-reservoir coupling can be treated in a controlled way. Coupled with a polaron transform (described below), where certain electron-phonon interactions are treated non-perturbatively, this facilitates a powerful starting point for modelling QD exciton processes in the presence of phonon and photon reservoir coupling. In this chapter we discuss the background theory underlying the polaron ME approaches and we apply the theory to several examples of phonon dressed light-matter interactions for QD cavity systems. We focus mainly on the underlying optical physics, with a brief connection to some applications which are presented in more detail in other chapters of this book.

2.2 General Theory for a Single QD Exciton with Exciton-Phonon and Exciton-Photon Interactions

2.2.1 Polaron Master Equation for a QD Exciton in a General Photonic Reservoir

Let us first consider a single neutral QD exciton (strong confinement limit, single electron-hole pair) that is modelled as a two-level system coupled to a photon reservoir and a phonon reservoir described by quantum field operators $\mathbf{f}(\mathbf{r}, \omega, t)$ and lowering operator b_q, respectively (see Fig. 2.1c). Such a simple two-level approximation is valid for small epitaxial QDs (e.g., self-assembled InGaAs/GaAs QDs), in a restricted frequency regime of interest. Although real QDs have many exciton levels over a broad band of frequencies, the one exciton model has successfully explained a number of experiments when probing single exciton dynamics, e.g., see [12, 16, 48–50]; in addition, one can extend such an approach to include other exciton levels [51, 52]. For example, Hargart et al. [53] has recently employed the polaronic

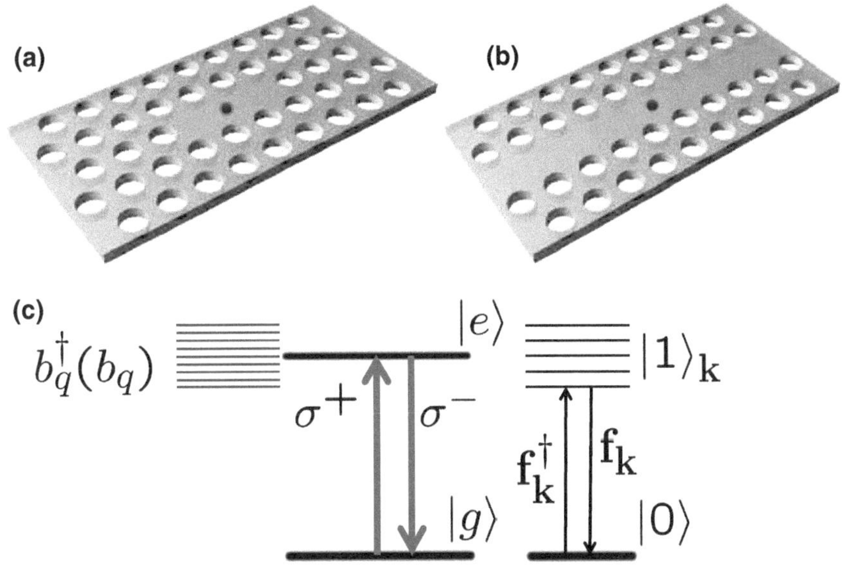

Fig. 2.1 Schematic of two optical cavity systems containing a quantum dot and an example energy level diagram used for some of the modelling. a, b Two selected QD-cavity schematics showing a QD in a structured photonic reservoir, including a photonic microcavity (**a**) and a PC waveguide (**b**). Although we show a semiconductor PC platform here, a wide variety of semiconductor cavity systems can be used here including the micropillar cavities. **c** Example energy level diagram for a single neutral QD exciton (electron-hole pair) interacting with a phonon bath and a photon bath. The operator $\mathbf{f}_{\mathbf{k}}^\dagger$ ($b_q^\dagger$) creates a photon (phonon). Figure adapted from in [47]

approach to derive a ME for describing the biexciton-exciton cascade in a QD coupled to a micropillar cavity, and used this model to successfully explain complex double-field-dressing experiments [53].

In a frame rotating at the frequency of the QD resonance frequency (ω_x), the coupled QD-phonon-photon system is described by the following Hamiltonian [47]:

$$\begin{aligned} H = \hbar \int d\mathbf{r} \int_0^\infty & d\omega\, \mathbf{f}^\dagger(\mathbf{r}, \omega, t)\mathbf{f}(\mathbf{r}, \omega, t) \\ & - \left[\sigma^+ e^{i\omega_x t} \int_0^\infty d\omega\, \mathbf{d} \cdot \mathbf{E}^+(\mathbf{r}_d, \omega, t) + \text{H.c.}\right] \\ & + \sum_q \hbar\omega_q b_q^\dagger b_q + \sigma^+\sigma^- \sum_q \hbar\lambda_q (b_q^\dagger + b_q), \end{aligned} \tag{2.1}$$

where the lowering operator σ^- describes a transition between the QD states $|e\rangle$ and $|g\rangle$, separated in frequency by ω_x, and the photon reservoir is described by the field operators $\mathbf{f}(\mathbf{r}, \omega, t)$, which are continuous in frequency ω and satisfy the Boson commutation rules. We have also applied a dipole and rotating wave approximation to describe coupling between the QD and the photon reservoir (second term in (2.1)). In the absence of photon interactions we recover the IBM [20], which models a reservoir of harmonic oscillator states coupled to a two-level system (last two terms in (2.1)). We consider a QD of dipole moment $\mathbf{d}$ to be located at some spatial position $\mathbf{r}_d$, and the exciton-phonon coupling strength λ_q is assumed to be real [24]. The positive frequency component of the electric field operator $\mathbf{E}^+(\mathbf{r}_d, \omega, t)$ can be expressed in terms of the electric-field Green function $\mathbf{G}(\mathbf{r}, \mathbf{r}'; \omega)$ as $\mathbf{E}^+(\mathbf{r}, \omega, t) = i \int d\mathbf{r}' \mathbf{G}(\mathbf{r}, \mathbf{r}'; \omega) \sqrt{\frac{\hbar}{\pi \epsilon_0} \epsilon_{\mathrm{I}}(\mathbf{r}', \omega)} \mathbf{f}(\mathbf{r}', \omega, t)$ [54]; in a medium described by the dielectric constant $\epsilon(\mathbf{r}, \omega) = \epsilon_{\mathrm{R}}(\mathbf{r}, \omega) + i\epsilon_{\mathrm{I}}(\mathbf{r}, \omega)$, $\mathbf{G}(\mathbf{r}, \mathbf{r}')$ connects to the photon reservoir and is the solution to Maxwell equations at $\mathbf{r}$ to a point dipole oscillating at $\mathbf{r}'$ (without any QD coupling). It should be noted that this expression for the electric field operator is quite general and fully satisfies the Kramers–Krönig relations in a general photonic medium including plasmonic structures [55].

We now present a derivation of the photon-reservoir polaron ME [47], and perform a polaron transform on the Hamiltonian in (2.1). This unitary transform includes phonons to all orders and also puts the Hamiltonian in an easier form to apply system-reservoir theory and perturbation theory, with polaron-shifted interaction terms. We are basically employing a convenient basis change using a unitary transform in which the bath-modified exciton becomes the polaronic quasiparticle. In this way, we fully recover the IBM without photon reservoir interactions and can include additional processes that give rise to the zero phonon line (ZPL), e.g., caused by effects such as spectral diffusion [56] and random charging effects [57] of the QD due to trapped carriers and electron-hole pairs.

Formally, the polaron transformation is given by $H' \to e^P H e^{-P}$, where $P = \sigma^+\sigma^- \sum_q \frac{\lambda_q}{\omega_q}(b_q^\dagger - b_q)$ [23], which yields a polaron-transformed Hamiltonian,

$$
\begin{aligned}
H' = \hbar \int d\mathbf{r} \int_0^\infty d\omega\, \mathbf{f}^\dagger(\mathbf{r}, \omega, t)\mathbf{f}(\mathbf{r}, \omega, t) + \sum_q \hbar\omega_q b_q^\dagger b_q, \\
- \left[B_+ \sigma^+ e^{i\omega_x' t} \int_0^\infty d\omega\, \mathbf{d} \cdot \mathbf{E}^+(\mathbf{r}_d, \omega, t) + \text{H.c.} \right],
\end{aligned}
\tag{2.2}
$$

where $B_\pm = \exp[\pm \sum_q \frac{\lambda_q}{\omega_q}(b_q - b_q^\dagger)]$ are the coherent phonon bath displacement operators [23]. The polaron shift, $\Delta_P = \int_0^\infty d\omega \frac{J_{\mathrm{pn}}(\omega)}{\omega}$, is caused by a bath-induced frequency shift and below we will assume this factor is absorbed in the polaron-shifted frequency of the QD, defined by ω_x' $(= \omega_x - \Delta_P)$. In the continuum limit, the phonon coupling is determined by the phonon spectral function $J_{\mathrm{pn}}(\omega)$ [58]. The deformation potential coupling with longitudinal acoustic (LA) phonons plays the strongest role QDs [11, 12, 14, 16, 19, 42] and coupling to longitudinal optical (LO) phonons is ignored here [59, 60]. The resultant phonon sidebands span a frequency range of $\approx\pm 5$ meV around the s-shell transition of interest (for example) which is assumed smaller than the energetic separation between s-shell and p-shell transitions ($\approx$25 meV) [16]. We assume bulk phonon interactions, and the expression for phonon coupling λ_q for deformation potential coupling to LA phonons is $\lambda_q = \sqrt{\frac{\hbar\omega_q}{2\rho_d c_s^2 V}}\frac{D}{\hbar}e^{-\frac{\omega^2}{4\omega_b^2}}$ [61], where ρ_d is the material density, V is the material volume, c_s is the sound velocity, $D = D_{val} - D_{con}$ is the difference in deformation potential between the valence and conduction bands and ω_b is the phonon cut-off frequency determined by the confinement length of the electron. The sum over the discrete phonon modes λ_q can be converted into an integral using the phonon density of states, $D(\omega) = \frac{V}{(2\pi)^3}\frac{4\pi\omega^2}{c_s^3}$, which allows us to introduce the continuous phonon spectral function, $J_{pn}(\omega) = \alpha_p\omega^3 e^{-\frac{\omega^2}{2\omega_b^2}}$ with the phonon coupling $\alpha_p = \frac{D^2}{4\pi^2\hbar c_s^5\rho_d}$ [58]. For polaron cavity-QED and polaron reservoir ME approaches, we use the continuum form of the phonon spectral function $J_{pn}(\omega)$ and with parameters for InAs QDs, with $\alpha_p/(2\pi)^2 = 0.06\,\text{ps}^2$ and $\omega_b = 1$ meV (unless stated otherwise), consistent with experiments [16]. Generally, we have found that the precise value of these phonon parameters will not change any of the qualitative findings below, and they can be used for fitting experiments on a particular QD (as we also do below).

If a weak interaction between the QD and the photon reservoir is assumed, a ME for the QD reduced density matrix ρ can be derived by retaining terms up to second order in the polaron-shifted interaction Hamiltonian, $H_I' = -[B_+\sigma^+ e^{i\omega_x' t}\int_0^\infty d\omega\,\mathbf{d}\cdot\mathbf{E}^+(\mathbf{r}_d,\omega,t) + \text{H.c.}]$. The time convolutionless [62]—or time-local—interaction picture ME is then

$$\frac{\partial\tilde{\rho}(t)}{\partial t} = -\frac{1}{\hbar^2}\int_0^t d\tau \text{Tr}_{R_{ph}}\text{Tr}_{R_{pn}}\{[\tilde{H}'_I(t), [\tilde{H}'_I(t-\tau), \tilde{\rho}(t)\rho_R]]\}, \tag{2.3}$$

where the Hamiltonian terms $\tilde{H}'_I(t) = \exp[iH_R' t/\hbar]H_I'\exp[-iH_R' t/\hbar]$ and $H_R' = \hbar\int d\mathbf{r}\int_0^\infty d\omega\,\mathbf{f}^\dagger(\mathbf{r},\omega,t)\mathbf{f}(\mathbf{r},\omega,t) + \sum_q \hbar\omega_q b_q^\dagger b_q$. The trace operators $\text{Tr}_{R_{ph}}$ and $\text{Tr}_{R_{pn}}$ denote a trace over the photon and phonon reservoirs, which are statistically independent with $\rho_R = \rho_{R_{ph}}\rho_{R_{pn}}$ [63]. The trace over the photon reservoir [64, 65] assumes thermal equilibrium and use the relations, $\text{Tr}_{R_{ph}}[\mathbf{f}(\mathbf{r},\omega), \mathbf{f}^\dagger(\mathbf{r}',\omega')] = [\tilde{n}(\omega)+1]\delta(\mathbf{r}-\mathbf{r}')\delta(\omega-\omega')$ and $\text{Tr}_{R_{ph}}[\mathbf{f}^\dagger(\mathbf{r},\omega), \mathbf{f}(\mathbf{r}',\omega')] = \tilde{n}(\omega)\delta(\mathbf{r}-\mathbf{r}')\delta(\omega-\omega')$ where $\tilde{n}(\omega)\approx 0$, which is valid for optical frequencies. For the problem of phonon-modified spontaneous emission, the final form of the polaron reservoir ME in the Schrödinger picture [47] and Markov limit ($t\to\infty$ in the integral of (2.3)) is

$$\frac{d\rho}{dt} = \frac{\tilde{\gamma}}{2}L(\sigma^-) - i\Delta_{\text{Lamb}}[\sigma^+\sigma^-, \rho], \tag{2.4}$$

where $L(O) = 2O\rho O^\dagger - O^\dagger O\rho - \rho O^\dagger O$ is the Lindblad operator [63], and the SE rate (the tilde indicates that it has been modified by phonon coupling) of the QD into the structured reservoir [47, 65] is derived to be

$$\tilde{\gamma} = 2\int_0^\infty \mathrm{Re}[C_{\mathrm{pn}}(\tau)J_{\mathrm{ph}}(\tau)]d\tau, \tag{2.5}$$

while the QD Lamb shift is $\Delta_{\mathrm{Lamb}} = \int_0^\infty \mathrm{Im}[C_{\mathrm{pn}}(\tau)J_{\mathrm{ph}}(\tau)]d\tau$. Thus the resulting SE rate and the Lamb shift display an interplay between photon and phonon bath dynamics, where $J_{\mathrm{ph}}(\tau)$ and $C_{\mathrm{pn}}(\tau)$ are the photon and the phonon bath correlation functions, respectively. This means that phonon coupling influences the SE decay, depending upon the dynamics of the photon and phonon reservoir functions.

The photon bath correlation function is defined as $J_{\mathrm{ph}}(\tau) = \int_0^\infty d\omega\, J_{\mathrm{ph}}(\omega)e^{i(\omega_x'-\omega)\tau}$, where the photon reservoir spectral function is

$$J_{\mathrm{ph}}(\omega) = \frac{\mathbf{d}\cdot\mathrm{Im}[\mathbf{G}(\mathbf{r}_d,\mathbf{r}_d;\omega)]\cdot\mathbf{d}}{\pi\hbar\epsilon_0}, \tag{2.6}$$

which is directly related to the photon Green function and projected LDOS. The phonon correlation function is defined from

$$C_{\mathrm{pn}}(\tau) = e^{[\phi(\tau)-\phi(0)]}, \tag{2.7}$$

where the IBM phase function is

$$\phi(t) = \int_0^\infty d\omega \frac{J_{\mathrm{pn}}(\omega)}{\omega^2}\left[\coth\left(\frac{\hbar\omega}{2k_BT}\right)\cos(\omega t) - i\sin(\omega t)\right], \tag{2.8}$$

which includes a sum over multiple phonon emission and absorption processes.

Note that the SE rate $\tilde{\gamma}$ in general includes contributions from photonic LDOS values at frequencies different from the ZPL frequency of the QD (where $\omega = \omega_x'$). Such non-local frequency-coupling effects for the SE is caused by a breakdown of Fermi's golden rule (even for weakly coupled photonic systems), which depends on the relative correlation times of both photon and phonon reservoirs [47]. Note that (2.5) is broadly applicable irrespective of the specific structure of the photon reservoir. However, this polaron reservoir ME approach is naturally restricted to weak-to-intermediate coupling between the QD and the photon reservoir, and it cannot treat effects such as the strong coupling regime (e.g., vacuum Rabi spitting). To treat the strong coupling regime of quantum optics, e.g., for a high-Q cavity, one can include a cavity photon operator at the level of a system operator which requires a different (but similar) approach based on the polaron cavity-QED ME, that we will also describe below.

2.2.2 *Spectrum Definitions*

The incoherent emission spectrum of a QD coupled to a structured photonic reservoir at a point detector at position $\mathbf{r}_D$ is given by [66]

$$S^{\mathrm{G}}(\mathbf{r}_{\mathrm{D}},\omega) = \langle (\mathbf{E}^{+}(\mathbf{r}_{\mathrm{D}},\omega))^{\dagger}\, \mathbf{E}^{+}(\mathbf{r}_{\mathrm{D}},\omega)\rangle, \tag{2.9}$$

where $\mathbf{E}^{+}(\mathbf{r}_{\mathrm{D}},\omega) = \int_0^{\infty} dt e^{i\omega t}\mathbf{E}^{+}(\mathbf{r}_{\mathrm{D}},t)$ is the Laplace transform of the positive frequency component of the electric field operator $\mathbf{E}(\mathbf{r}_{\mathrm{D}},t)$, defined as

$$\begin{aligned}\mathbf{E}(\mathbf{r}_{\mathrm{D}},t) &= \mathbf{E}^{+}(\mathbf{r}_{\mathrm{D}},t) + \mathbf{E}^{-}(\mathbf{r}_{\mathrm{D}},t) = \int_0^{\infty} d\omega'[\mathbf{E}^{+}(\mathbf{r}_{\mathrm{D}},\omega',t) + \mathbf{E}^{-}(\mathbf{r}_{\mathrm{D}},\omega',t)]\\ &= i\int_0^{\infty} d\omega' \int d\mathbf{r}'\mathbf{G}(\mathbf{r}_{\mathrm{D}},\mathbf{r}';\omega')\sqrt{\frac{\hbar}{\pi\epsilon_0}\epsilon_I(\mathbf{r}',\omega')}\mathbf{f}(\mathbf{r}',\omega',t) + \text{H.c.}\end{aligned} \tag{2.10}$$

In the frequency domain,

$$\mathbf{E}^{+}(\mathbf{r}_{\mathrm{D}},\omega) = i\int_0^{\infty} d\omega' \int d\mathbf{r}'\mathbf{G}(\mathbf{r}_{\mathrm{D}},\mathbf{r}';\omega')\sqrt{\frac{\hbar}{\pi\epsilon_0}\epsilon_I(\mathbf{r}',\omega')}\mathbf{f}(\mathbf{r}',\omega',\omega). \tag{2.11}$$

Starting from the original Hamiltonian H in (2.1), the electric field operator can be expressed in terms of the QD polarization using Laplace transform techniques, yielding [66]

$$\mathbf{E}(\mathbf{r}_{\mathrm{D}},\omega) = \mathbf{E}^{0}(\mathbf{r}_{\mathrm{D}},\omega) + \mathbf{G}(\mathbf{r}_{\mathrm{D}},\mathbf{r}_d;\omega)\cdot\mathbf{d}\frac{\sigma^{-}(\omega)}{\epsilon_0}, \tag{2.12}$$

where $\mathbf{E}^0$ denotes the free-field solution in the absence of a QD scatterer. The medium Green function $\mathbf{G}(\mathbf{r}_{\mathrm{D}},\mathbf{r}_d;\omega)$ includes all propagation effects [66], including radiative losses due to structured reservoirs [47]. In the following treatment, we assume that the Green function mainly accounts for radiative coupling to the structured photonic reservoir. Any additional radiative losses is thus accounted for by the phenomenological Lindblad term γ_0.

Assuming the initial vacuum state of the photonic reservoir, the incoherent spectrum from Green function theory can be derived using (2.9) and (2.12) so that,

$$S(\mathbf{r}_{\mathrm{D}},\omega) = \left|\frac{\mathbf{G}(\mathbf{r}_{\mathrm{D}},\mathbf{r}_d;\omega)\cdot\mathbf{d}}{\epsilon_0}\right|^2 \langle\sigma^{+}(\omega)\sigma^{-}(\omega)\rangle = \alpha_{\mathrm{prop}}(\mathbf{r}_{\mathrm{D}},\mathbf{r}_d;\omega)\, S_0(\omega), \tag{2.13}$$

where $S_0 = \langle\sigma^{+}(\omega)\sigma^{-}(\omega)\rangle$ is the polarization spectrum and $\alpha_{\mathrm{prop}} = \frac{1}{\epsilon_0}|\mathbf{d}\cdot\mathbf{G}(\mathbf{R}_{\mathrm{D}},\mathbf{r}_d;\omega)|^2$ accounts for light propagation and filtering from the QD (at $\mathbf{r}_d$) to the detector (at $\mathbf{r}_{\mathrm{D}}$). The polarization spectrum is obtained from a first-order quantum correlation function for the QD operators,

$$S_0(\omega) = \langle\sigma^{+}(\omega)\sigma^{-}(\omega)\rangle = \int_0^{\infty} dt_1 \int_0^{\infty} dt_2\, e^{i\omega(t_2-t_1)}\langle\sigma^{+}(t_1)\sigma^{-}(t_2)\rangle. \tag{2.14}$$

Using a rotating frame at the exciton frequency (ω'_x), denoting $\tau = t_2 - t_1$ and taking the limit of $t_1 = t \to \infty$, we derive the *steady-state* polarization spectrum,

$$S_0(\omega) = \lim_{t\to\infty} \mathrm{Re}\left[\int_0^\infty d\tau \langle\sigma^+(t+\tau)\sigma^-(t)\rangle e^{i(\omega'_x-\omega)\tau}\right], \quad (2.15)$$

which can be calculated, in the usual way, using the ME and the quantum regression theorem [63].

When calculating the spectrum using a polaronic approach, a transformation is required to obtain S_0 in the lab frame. For example, if σ_{P}^- denotes the QD lowering operator in the polaron frame, then a transformation of the correlation function ($\langle\sigma_{\mathrm{P}}^+(t+\tau)\sigma_{\mathrm{P}}^-(t)\rangle$) from the polaron to the lab frame,

$$\langle\sigma_{\mathrm{P}}^+(t+\tau)\sigma_{\mathrm{P}}^-(t)\rangle \to \langle\sigma^+(t+\tau)B_+(t+\tau)B_-(t)\sigma^-(t)\rangle, \quad (2.16)$$

which produces a phase relaxation term $e^{\phi(\tau)}$ [58] that accounts for the phonon-induced pure dephasing of the QD polarization. The overall decay is thus clearly non-Markovian and this is a major advantage of the polaronic approaches, i.e., although they use a Born–Markov approximation for the equations of motion, non-Markovian coupling effects due to the phonon reservoir are captured through the polaron transform (indeed, the approach fully includes the IBM solution), even though the equations of motion are time local. The final polarization spectrum calculated using the polaron approach is then

$$S_0(\omega) = \lim_{t\to\infty} \mathrm{Re}[\int_0^\infty d\tau \langle\sigma_{\mathrm{P}}^+(t+\tau)\sigma_{\mathrm{P}}^-(t)\rangle e^{\phi(\tau)} e^{i(\omega'_x-\omega)\tau}]. \quad (2.17)$$

It is important to note that the main expression for the emission spectrum (2.13) is exact and is only limited by the approximations made in calculating S_0 from the above theories. Thus no phenomenological input-output formalism is required as the Green function is already a solution for the scattering problem and the propagation of light in the medium is fully accounted for.

In the case of a high-Q resonator, the dynamics of the photon reservoir can be described by a single mode (lowering) operator a in the system Hamiltonian H (shown later). The damping of the cavity mode to the environment is then described using the phenomenological decay κ (cavity decay rate). In this case, a calculation of the reservoir/cavity emitted spectrum can be obtained using

$$S_{\mathrm{cav}}^{\mathrm{CM}}(\omega) = F(\mathbf{r}_{\mathrm{d}}, \mathbf{r}_{\mathrm{D}})\frac{\kappa}{\pi} \lim_{t\to\infty} \mathrm{Re}\left[\int_0^\infty d\tau \langle a^\dagger(t+\tau)a(t)\rangle e^{i(\omega'_x-\omega)\tau}\right], \quad (2.18)$$

where $F(\mathbf{r}_{\mathrm{d}}, \mathbf{r}_{\mathrm{D}})$ is a frequency-independent geometrical factor to account for the propagation from the QD to the detector position. The superscript CM denotes a "coupled mode" formalism which treats the QD and cavity as coupled modes and the photonic reservoir in this case is assumed to be described a single cavity mode a. The resonance width of the cavity mode determines its damping rate and any such reservoir effects arising due to finite cavity lifetime is incorporated using phenomenological decay terms κ for the cavity operator. Note that $\langle a^\dagger(t+\tau)a(t)\rangle = \langle a_{\rm P}^\dagger(t+\tau)a_{\rm P}(t)\rangle$, so the IBM phase factor does not affect the dynamics of the cavity-mode two time correlation function, which is valid for a high-Q resonator. Naturally, this approach is restricted to cavities with a Lorentzian lineshape, whereas the photonic reservoir theory above can include any arbitrarily shaped LDOS (though is restricted to the weak coupling regime). Having two different polaronic MEs allows to model various QD photonic systems.

2.3 Phonon-Modified Spontaneous Emission and the Breakdown of Fermi's Golden Rule

From the general expression for the SE rate (2.5), it is clear that a dynamical interplay between the photon and phonon reservoir dynamics can affect the QD SE rate, since this expression now includes contributions from the photonic LDOS at non-local frequencies (i.e., $\omega \neq \omega_x$). In the Markov limit, to obtain a long-time SE decay rate ($t \to \infty$), the spectral width of the phonon reservoir determines the frequency bandwidth of the photon reservoir which contributes to the SE rate $\tilde{\gamma}$ ($=\tilde{\gamma}(\infty)$), resulting in a breakdown of Fermi's golden rule for SE. The SE rate without phonon interaction is given by $\gamma(t) = 2\int_0^t {\rm Re}[J_{\rm ph}(\tau)]d\tau$, [65] and in the Markov limit, then $\gamma(t\to\infty) \propto {\rm LDOS}(\omega_x)$, which is the usual result for a two-level atom. The non-local frequency contribution $\tilde{\gamma}_{\rm nl}$ of the SE rate due to phonon coupling can be clarified by rewriting (2.5) as follows:

$$\begin{aligned}\tilde{\gamma} &= 2\int_0^t {\rm Re}[e^{\phi(\tau)-\phi(0)}J_{\rm ph}(\tau)]d\tau,\\ &= 2e^{-\phi(0)}\int_0^t {\rm Re}[(e^{\phi(\tau)}-1)J_{\rm ph}(\tau)+J_{\rm ph}(\tau)]d\tau,\\ &= \tilde{\gamma}_{\rm nl} + \langle B\rangle^2\gamma, \end{aligned}\tag{2.19}$$

where $\langle B\rangle^2 = e^{-\phi(0)}$ and $\tilde{\gamma}_{\rm nl} = 2\langle B\rangle^2\int_0^t {\rm Re}[C'_{\rm pn}(\tau)J_{\rm ph}(\tau)]d\tau$, where $C'_{\rm pn}(\tau) = (e^{\phi(\tau)}-1)$. Hence in regions where the photonic LDOS is large (e.g., at the peak of a Lorentzian cavity mode), the renormalized local frequency component $\langle B\rangle^2\gamma$ dominates, and in regions where the photonic LDOS is small, $\tilde{\gamma}_{\rm nl}$ leads to an enhancement of the SE rate. The corresponding Lamb shifts, $\Delta_{\rm Lamb}$ (see (2.5)) are found to be negligible, and can be safely neglected [58].

To help explain the different between SE decay with and without phonon coupling, we define the phonon-mediated SE enhancement factor:

$$\chi = \tilde{\gamma}/\gamma, \tag{2.20}$$

so the phonon-modified Purcell factor is

$$\mathrm{PF} = \tilde{\gamma}/\gamma_b, \tag{2.21}$$

where $\gamma_b = d^2 n_b \omega^3/(6\pi\hbar\epsilon_0 c^3)$ is the SE rate of a QD in the background semiconductor material with refractive index n_b.

2.3.1 *Lorentzian Cavity*

Consider first the stereotypical example of a Lorentzian cavity, where an analytic expression for the photon bath relaxation function can be obtained by Fourier transforming the photon reservoir spectral function: $J_{\mathrm{ph}}(\omega) = g^2 \frac{1}{\pi} \frac{\frac{\kappa}{2}}{(\omega-\omega_c)^2+(\frac{\kappa}{2})^2}$, where g is the QD-cavity coupling rate, κ is the cavity decay rate and ω_c is the cavity peak frequency. The corresponding cavity relaxation function is a damped oscillatory function. Figure 2.2a, c plot the relaxation functions of the phonon and photon baths (for $\omega_x = \omega_c$), respectively, and Fig. 2.2b, d show the corresponding reservoir spectral functions. For the cavity parameters chosen, the photon damping time is similar to the phonon reservoir time and SE rate is expected to have a strong non-local compo-

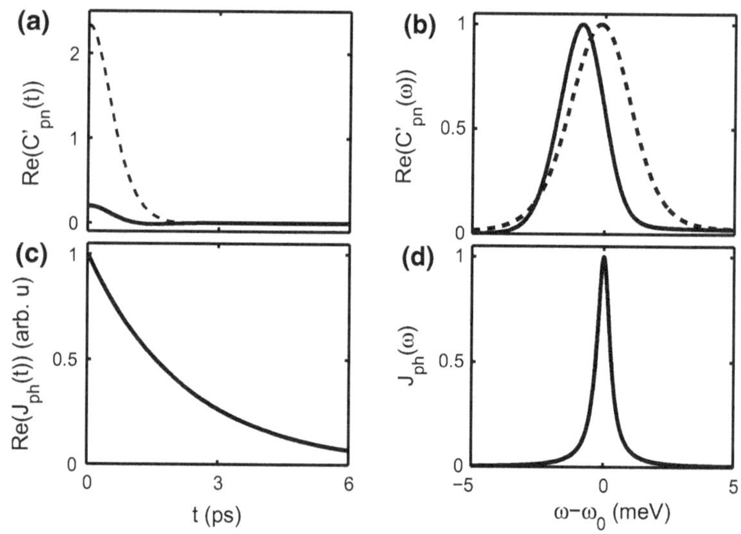

Fig. 2.2 Example phonon and photon reservoir functions in time and frequency domains. Real part of phonon bath relaxation function $C'_{\mathrm{pn}}(t)$ (**a**) and phonon reservoir spectral function $C'_{\mathrm{pn}}(\omega)$ (**b**) at T = 4 K (*solid line*) and 40 K (*dashed line*) for phonon coupling $\alpha_{\mathrm{p}}/(2\pi)^2 = 0.06\mathrm{ps}^2$ and phonon cutoff frequency, $\omega_{\mathrm{b}} = 1\,\mathrm{meV}$. Photon reservoir relaxation function $J_{\mathrm{ph}}(t)$ (real part) (**c**) and photon reservoir spectral function $J_{\mathrm{ph}}(\omega)$ (**d**) for a cavity with decay rate $\kappa = 0.6\,\mathrm{meV}$ and $\omega_0 = \omega_c$ in (**d**); this rate corresponds to $Q \sim 2300$ at $\omega_c/2\pi = 1440\,\mathrm{meV}$

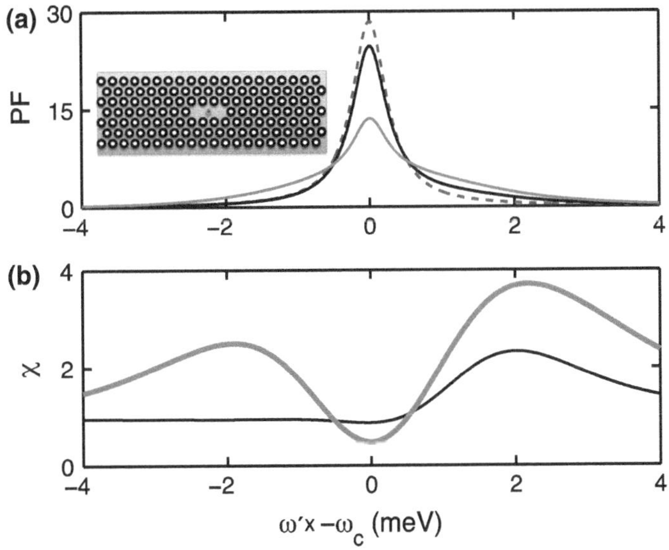

Fig. 2.3 Purcell factor (a) and SE enhancement factor χ (b) for a simple Lorentzian cavity. The cavity decay rate is $\kappa = 0.6$ meV and the QD-cavity coupling $g = 0.08$ meV (weak coupling regime). The *solid dark* and *light lines* show calculations with phonons at T = 4 and 40 K, respectively, and the dashed line in (**a**) is PF without phonon modification of the SE rate (γ). *Inset* of **a** shows a schematic of a QD coupled to a single PC cavity. Figure adapted from [47]

nent $\tilde{\gamma}_{\rm nl}$. In the frequency domain [67], the spectral widths of the interacting phonon (Fig. 2.2b) and photon reservoir (Fig. 2.2d) should be comparable to obtain a large non-local SE rate $\tilde{\gamma}_{\rm nl}$. It is also clear from Fig. 2.2a, c, that when the relaxation time $\tau_{\rm pn}$ of the phonon reservoir is much larger compared to the photon reservoir lifetime $\tau_{\rm ph}$ (i.e., an unstructured reservoir), phonons do not influence the SE rate [65]; and in the opposite limit ($\tau_{\rm ph} \gg \tau_{\rm pn}$), phonons simply renormalize the original SE rate by a mean field value that depends on the temperature ($\langle B\rangle^2\gamma$) [68]. The renormalization factor $\langle B\rangle^2$ decreases with temperature, and for our parameters, $\langle B\rangle$ is equal to 0.91, 0.85 and 0.55 for $T = 4$, 10 and 40 K, respectively.

In Fig. 2.3, we plot the Purcell factor, PF, (a) and the phonon mediated SE enhancement factor $\chi = \tilde{\gamma}/\gamma$ (b) for the cavity considered in Fig. 2.2d. The dashed line in Fig. 2.3a is the PF without phonons. These calculations assume excitation by a weak resonant drive ($\omega_L = \omega'_x$) and show that the SE rate is reduced by phonons near the LDOS peak and enhanced by phonons away from the LDOS peak, as mentioned in the discussion following (2.19). These phonon-mediated effects also increase with temperature as shown the calculations at T = 4 K (dark solid) and 40 K (light solid). The slight asymmetry of the SE modification with respect to the LDOS peak is caused due to unequal phonon emission and absorption at low temperatures [16, 47].

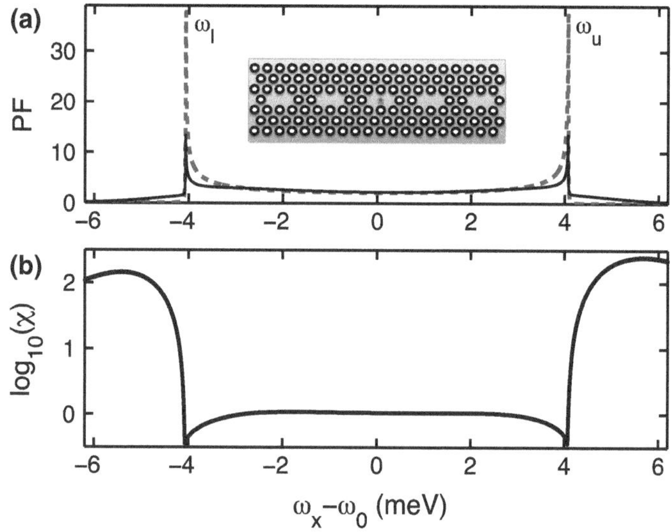

Fig. 2.4 Purcell factor (a) and SE enhancement factor χ (b) for a coupled-cavity waveguide. The *solid lines* show calculations with phonons at T = 40 K and the *dashed line* in (**a**) is PF without phonon modification of SE rate (γ) and ω_0 is the waveguide band-center frequency. *Inset* of **a** shows a schematic of QD coupled to a PC coupled cavity waveguide. Figure adapted from [47]

2.3.2 Slow-Light Coupled-Cavity Waveguide

We next consider a more complex photonic LDOS structure. In Fig. 2.4, we show calculations of the PF (a) and SE enhancement χ (b) for a slow-light coupled cavity waveguide [47], known as a CROW structure. The spectral bath function for the coupled-cavity waveguide can be calculated using an analytic tight-binding model [69, 70] and is given by, $J_{\rm ph}(\omega) = \frac{-d^2\omega}{2\hbar\epsilon_0 n_b^2 V_{\rm eff}} \frac{1}{\pi} {\rm Im}\left[\frac{1}{\sqrt{(\omega-\tilde{\omega}_u)(\omega-\tilde{\omega}_l^*)}}\right]$, where $\tilde{\omega}_{u,l} = \omega_{u,l} \pm i\kappa_{u,l}$[70] and $\omega_{u,l}$ is the waveguide mode-edge frequencies (Fig. 2.4a) and $\kappa_{u,l}$ are effective damping rates; and $V_{\rm eff}$ is the effective mode volume of a single cavity of the coupled-cavity waveguide. The parameters are taken from [71]. As in the case of the single cavity mode, the phonon-modified PF near the waveguide mode-edges is strongly dominated by the local component $\langle B\rangle^2\gamma$ (solid line, Fig. 2.4a) and is reduced compared to the phonon unmodified PF (dashed line). However away from the mode-edge and outside the waveguide bandwidth, phonons cause a strong enhancement of the SE rate. In the case of the waveguide, the spectral width of the photonic reservoir ($\propto$ PF, Fig. 2.4a) is comparable to that of the phonon reservoir (Fig. 2.2b) and presents a richer spectral structure compared to a simple cavity. The SE enhancement is hence much stronger than the case of the cavity, increasing by orders of magnitude in some frequency regimes. The SE rate is also enhanced inside the waveguide band, but the enhancement factor is not as large as outside and this difference is caused by the asymmetric structure of the mode-edge LDOS compared to a symmetric Lorentzian cavity [47]. These results demonstrate that phonon coupling can have a significant influence on the SE rate and that phonon interactions cause Fermi's golden rule to breakdown in general.

2.4 Independent Boson Model Lineshape with a Broadened Zero Phonon Line

It is instructive to connect the polaron ME model to the well known IBM model for linear field excitation, since this is an *exactly solvable model* for the polarization phase, where phonon effects are included to all orders. However, the model must be complimented by processes to broaden the ZPL since the IBM phase factor does not decay to zero in the long time limit. Nevertheless, the QD background radiative decay, pure dephasing and incoherent pumping can easily included in the ME. For example, to obtain the simple linear optical properties of the system, we consider a QD that is weakly excited by an incoherent pump P [63, 72], which maps on to a range of typical experiments that are performed to measure the emitted spectrum for a QD photonic structure [73]. In a ME approach, the incoherent pumping term is included as a Lindblad operator, $\frac{P}{2}L(\sigma^+)$, where P is the pump rate; the ME can also include other Lindblad operators to account for additional incoherent processes such as background spontaneous decay, $\frac{\gamma_0}{2}L(\sigma^-)$, and QD pure dephasing, $\frac{\gamma_d}{2}L(\sigma^+\sigma^-)$. Neglecting coupling to a structured photonic reservoir and phonons, these *incoherent* processes determine the lineshape of the QD emission spectrum, manifesting in a simple Lorentzian broadening of the ZPL. The ME with incoherent pumping and

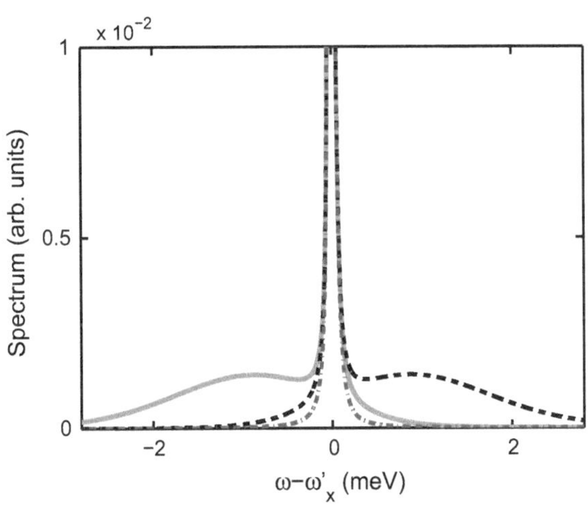

Fig. 2.5 Characteristic IBM spectral lineshape plus the zero phonon line. Normalized QD emission (polarization) spectrum (*thick orange solid*) and linear absorption spectrum (*thick black dashed*) at T = 4 K for a single uncoupled QD. The *red dash-dotted line* denotes the polarization spectrum S_0^P in the polaron frame. The ZPL parameters are $\gamma_0 = 5\,\mu$eV and $\gamma_d = 5\,\mu$eV (pure dephasing rate). Figure from [74] (color figure online)

additional ZPL decay processes then becomes:

$$\frac{d\rho}{dt} = \frac{\tilde{\gamma}}{2}L(\sigma^-) - i\Delta_{\rm Lamb}[\sigma^+\sigma^-, \rho] + \frac{\gamma_0}{2}L(\sigma^-) + \frac{\gamma_{\rm d}}{2}L(\sigma^+\sigma^-) + \frac{P}{2}L(\sigma^+). \tag{2.22}$$

Numerically it is straightforward to solve the above ME, and we solve this ME (and all the MEs below) using the Quantum Optics Toolbox [75] for Matlab, and in some cases it is also possible to obtain analytical solutions. Using a weak incoherent pump field, the resulting emission spectrum (thick orange solid line) is plotted in Fig. 2.5 and shows the appearance of the phonon sidebands arising due to the IBM phase function ϕ. The clear asymmetry of the sidebands is due to the fact that phonon emission is more probable than absorption at low temperatures, so the phonon emission is more probable on the lower energy side of the ZPL. The Lorentzian ZPL (red dash-dotted line in Fig. 2.5) corresponds to the polarization spectrum $S_0^{\rm P}$ in the polaron frame.

Similar to Sect. 2.2.2, the QD susceptibility function [58] can be defined as

$$\chi(\omega) \propto i \lim_{t\to\infty} \int_0^\infty d\tau \langle \sigma_{\rm P}^-(t+\tau)\sigma_{\rm P}^+(t)\rangle e^{\phi(\tau)} e^{-i(\omega_x'-\omega)\tau}, \tag{2.23}$$

and the linear absorption spectrum (Im(χ)) is also plotted in Fig. 2.5 (thick black dashed line). The linear absorption spectrum is simply a reflection of the emitted spectrum about the QD ZPL [76], where the phonon sidebands are now more enhanced to the right, since phonon absorption is more probably on the higher energy side of the ZPL. A photon at higher frequencies ($\omega_L > \omega_x'$), can excite the QD more easily by phonon emission process at low temperature; since the reverse process ($\omega_L < \omega_x'$) requires phonon absorption, which is less probable at low temperatures, so absorption (black dashed line) is stronger on the higher energy side of the ZPL.

2.5 Cavity-QED Polaron Master Equation: Vacuum Rabi Splitting and Cavity-Assisted Feeding

In this section, we derive the time-local polaron cavity-QED ME for a QD coupled to a high-Q cavity [58]. We first replace the photon reservoir term in the Hamiltonian given by (2.1) with a single cavity mode, so that

$$\begin{aligned} H = {} & \hbar\Delta_{cx}a^\dagger a + \sum_q \hbar\omega_q b_q^\dagger b_q + g(\sigma^+ a + a^\dagger\sigma^-) \\ & + (\sum_m \Omega_m a^\dagger a_m + \text{H.c.}) + \sigma^+\sigma^- \sum_q \hbar\lambda_q(b_q^\dagger + b_q), \end{aligned} \tag{2.24}$$

where $\Delta_{cx} = \omega_c - \omega_x$ is the cavity-QD detuning, g is the QD-cavity coupling for a single cavity mode described by lowering operator a, and Ω_m represents the coupling to the photon environment that causes decay of the cavity mode. The time-dependence of a is kept implicit in (2.24). The QD-cavity coupling g can be expressed in terms of the dipole moment $\mathbf{d} = d\hat{\mathbf{n}}$ where $g = \eta(\mathbf{n}, \mathbf{r}_d)\left[\frac{d^2\omega_c}{2\hbar\epsilon_0\epsilon V_{\rm eff}}\right]^{\frac{1}{2}}$, where $V_{\rm eff}$ is effective mode volume of a dielectric cavity with dielectric constant ϵ, and η accounts for any deviation from the field antinode position and misalignment in polarization coupling (i.e., for optimal coupling it is simply 1). A dipole and a rotating wave approximation is used in describing the QD-cavity interaction. As before, we perform a polaron transform on H and now obtain [58]

$$H' = \hbar\Delta_{cx'}a^\dagger a + \sum_q \hbar\omega_q b_q^\dagger b_q + g'(\sigma^+ a + a^\dagger\sigma^-) + X_g\zeta_g + X_u\zeta_u, \tag{2.25}$$

where $g' = \langle B\rangle g$, $X_g = g[\sigma^+ a + a^\dagger\sigma^-]$, $X_u = ig[\sigma^+ a - a^\dagger\sigma^-]$, and the phonon fluctuation operators are defined through $\zeta_g = \frac{1}{2}(B_+ + B_- - 2\langle B\rangle)$ and $\zeta_u = \frac{1}{2i}(B_+ - B_-)$ [58] where $\langle B\rangle = \langle B_+\rangle = \langle B_-\rangle = e^{-\phi(0)/2}$. The polaron shift Δ_P is again absorbed into the QD frequency ω_x' and $\Delta_{cx'} = \omega_c - \omega_x'$. We can then derive a time-local ME for the reduced density matrix ρ of the QD-cavity system; following [58], we use a second-order Born approximation with the polaron transformed interaction $H_{\rm I}' = X_g\zeta_g + X_u\zeta_u$. The cavity-QED ME in the interaction picture, is again derived from

$$\frac{\partial\tilde{\rho}}{\partial t} = -\frac{1}{\hbar^2}\int_0^t dt' {\rm Tr}_{\rm B}\{[\tilde{H}_{\rm I}'(t), [\tilde{H}_{\rm I}'(t'), \tilde{\rho}(t)\rho_{\rm B}]]\}, \tag{2.26}$$

where $\tilde{H}_{\rm I}'(t) = e^{i(H_{\rm S}'+H_{\rm B}')t/\hbar}H_{\rm I}'e^{-i(H_{\rm S}'+H_{\rm B}')t/\hbar}$, with $H_{\rm S}' = \hbar\Delta_{cx'}a^\dagger a + g'[\sigma^+ a + a^\dagger\sigma^-]$ and $H_{\rm B}' = \sum_q \hbar\omega_q b_q^\dagger b_q$. Performing the trace and transforming back into the Schrödinger picture, we obtain the time-local polaron cavity-QED ME [58],

$$\begin{aligned}\frac{d\rho}{dt} &= \frac{1}{i\hbar}[H_{\rm S}', \rho] + \frac{\gamma_0}{2}L(\sigma^-) + \frac{\gamma_{\rm d}}{2}L(\sigma^+\sigma^-) + \frac{P}{2}L(\sigma^+) + \frac{\kappa}{2}L(a)\\ &- \frac{1}{\hbar^2}\int_0^\infty d\tau \sum_{m=g,u}(G_m(\tau)[X_m, e^{-iH_{\rm S}'\tau/\hbar}X_m e^{iH_{\rm S}'\tau/\hbar}\rho(t)] + {\rm H.c.}),\end{aligned} \tag{2.27}$$

where κ is the cavity decay rate (defined earlier), and $G_g = \langle B\rangle^2(\cosh(\phi(t)) - 1)$ and $G_u = \langle B\rangle^2\sinh(\phi(t))$ are the polaron Green functions [58]; as before, we have extended the upper limit of the integral in (2.26) to $t \to \infty$ to obtain a Markov form for the ME (though this is not a requirement; however, we have found that there is no loss in accuracy in doing this). Note that additional incoherent processes are included in the ME using the respective Lindblad terms (see Sect. 2.4) and we also now include phenomenological damping of the cavity (through $\frac{\kappa}{2}L(a)$). For the Born–Markov approximation to be valid, in the polaron frame, the system dynamics should be much slower compared to the rate of relaxation of the phonon bath. Although the bath relaxation time is only a few ps for typical QDs, this approximation may be restrictive in certain regimes, e.g., when dealing with vacuum Rabi oscillations at large g [74].

Before concluding this section it should be noted that (2.27) can be further simplified to a simpler effective ME [58], if a weak excitation approximation (WEA) is made (i.e., a single quantum excitation), which is exact for the linear spectrum; however, one should note that (2.27) can easily include strong pump fields as well, if modeling multi-photon effects [24]. Using a WEA, the polaron cavity-QED ME now takes the more transparent analytical form,

$$\begin{aligned}\frac{d\rho}{dt} &= \frac{1}{i\hbar}[H_{\rm S}^{\rm eff}, \rho] + \frac{\Gamma^{\sigma^+ a}}{2}L(\sigma^+ a) + \frac{\Gamma^{a^\dagger \sigma^-}}{2}L(a^\dagger \sigma^-) + \frac{\kappa}{2}L(a) \\ &+ \frac{\gamma_0}{2}L(\sigma^-) + \frac{\gamma_{\rm d}}{2}L(\sigma^+\sigma^-) + \frac{P}{2}L(\sigma^+) + \gamma_{\rm cd} a^\dagger \sigma^- \rho a^\dagger \sigma^- + \gamma_{\rm cd}^* \sigma^+ a \rho \sigma^+ a \\ &+ \left\{ M_1[(a^\dagger\sigma^- + \sigma^+ a), (2\sigma^+\sigma^- a^\dagger a + \sigma^+\sigma^- - a^\dagger a)\rho] + {\rm H.c.}\right\} \\ &+ \left\{ M_2[(a^\dagger\sigma^- - \sigma^+ a), (2\sigma^+\sigma^- a^\dagger a + \sigma^+\sigma^- - a^\dagger a)\rho] + {\rm H.c.}\right\}, \end{aligned} \tag{2.28}$$

where $H_{\rm S}^{\rm eff} = H_{\rm S}' + \hbar\Delta^{a^\dagger\sigma^-}\sigma^+ a a^\dagger \sigma^- + \hbar\Delta^{\sigma^+ a} a^\dagger \sigma^- \sigma^+ a$. If we denote the QD-cavity system Rabi frequency as $\Omega = \sqrt{\Delta_{cx'}^2 + 4g'^2}$, then the phonon-mediated cavity/exciton scattering rates are given, analytically, by

$$\begin{aligned}\Gamma^{a^\dagger\sigma^-/\sigma^+ a} &= 2{\rm Re}\Bigg[\int_0^\infty d\tau \frac{2g'^2}{\Omega^2}(1-\cos(\Omega\tau))(e^{-\phi(\tau)}-1) \\ &+ (\frac{2g'^2}{\Omega^2}(1-\cos(\Omega\tau)) + \cos(\Omega\tau))\,(e^{\phi(\tau)}-1)\Bigg] \\ &\pm 2g'^2{\rm Im}\left[\int_0^\infty d\tau \frac{\Delta_{cx'}}{\Omega}\sin(\Omega\tau)(e^{\phi(\tau)}-1)\right], \end{aligned} \tag{2.29}$$

and similar expressions can be obtained for the phonon-mediated Lamb shifts, which are given in [74]; we do not give these terms here as they turn out to be negligible. We also have a cross-dephasing term [77]

$$\begin{aligned}\gamma_{\rm cd} &= 2g'^2{\rm Re}\Bigg[\int_0^\infty d\tau \Big(\frac{2g'^2}{\Omega^2}(1-\cos(\Omega\tau)) \\ &+ \cos(\Omega\tau)\Big)(e^{-\phi(\tau)}-1) + \frac{2g'^2}{\Omega^2}(1-\cos(\Omega\tau))(e^{\phi(\tau)}-1)\Bigg] \\ &- 2ig'^2{\rm Re}\left[\int_0^\infty d\tau \frac{\Delta_{cx'}}{\Omega}\sin(\Omega\tau)(e^{-\phi(\tau)}-1)\right], \end{aligned} \tag{2.30}$$

and the M terms are defined from

$$M_1 = -2g'^2 \int_0^\infty d\tau \frac{g'\Delta_{cx'}}{\Omega^2}\big(\cos(\Omega\tau) - 1\big)(\cosh\phi - 1), \tag{2.31}$$

$$M_2 = -2ig'^2 \int_0^\infty d\tau \frac{g'}{\Omega}\sin(\Omega\tau)\sinh\phi. \tag{2.32}$$

More insight into the $M_{1,2}$ scattering terms can be obtained by deriving the optical Bloch equations for the system operators a and σ^-, using the WEA. For example,

$$\begin{aligned}\frac{d\langle a\rangle}{dt} &= -g_c\sigma^- - i(\Delta_{cx'} + \Delta^{\sigma^+a})a - \frac{\Gamma_c^{\rm eff}}{2}a,\\ \frac{d\langle\sigma^-\rangle}{dt} &= -g_x a - i\Delta^{a^\dagger\sigma^-}\sigma^- - \frac{\Gamma_x^{\rm eff}}{2}\sigma^-,\end{aligned} \tag{2.33}$$

where $\Gamma_c^{\rm eff} = \kappa + \Gamma^{\sigma^+a}$ and $\Gamma_x^{\rm eff} = \gamma_0 + \gamma_{\rm d} + P + \Gamma^{a^\dagger\sigma^-}$ are the effective dephasing rates and $g_c = ig' - M_1 - M_2$ and $g_x = ig' + M_1 - M_2$ are the complex couplings of the cavity and QD, respectively, in the presence of phonon coupling. Thus the processes denoted by the scattering terms $M_{1,2}$, result in a complex coupling between the QD and cavity. At resonance ($\Delta_{cx'} = 0$), since $M_1 = 0$, the complex QD-cavity coupling is given by $g_{c/x} = ig' - M_2$.

Note that the above ME has solved the incoherent scattering terms exactly (within the WEA) and one could use such an approach, e.g., to investigate the strong coupling regime as a function of temperature. The ensuing vacuum Rabi oscillations appearing in the integrals ensure that the phonon bath is correctly coupled to the dressed resonances of the system. In the weak coupling limit, specifically when $\Delta_{cx'} \gg g'$, the Rabi frequency $\Omega \to \Delta_{cx'}$, and the incoherent cavity and exciton scattering rates are simply given by

$$\Gamma_0^{a^\dagger\sigma^-/\sigma^+a} = 2g'^2{\rm Re}\left[\int_0^\infty d\tau e^{\mp i\Delta_{cx'}\tau}(e^{\phi(\tau)} - 1)\right], \tag{2.34}$$

and the corresponding phonon-induced Lamb shifts are given by $\Delta_0^{a^\dagger\sigma^-/\sigma^+a} = g'^2{\rm Im}[\int_0^\infty d\tau e^{\mp i\Delta_{cx'}\tau}(e^{\phi(\tau)} - 1)]$. These phonon-mediated scattering rates and Lamb shifts are the same as those derived in earlier work [58] (see also [26]), where an effective Lindblad form for the cavity-QED polaron ME was introduced. This simple phonon-mediated scattering rates between the QD exciton and cavity mode have already been widely adopted to connect to various experiments with Q-cavity system, including a recent study of ultrafast polariton-phonon dynamics [78].

A connection between the polaron reservoir ME approach (Sect. 2.2.1) and the polaron cavity-QED theory can be made by expressing the SE rate (2.5) in terms of these Lindblad decay rates [47],

$$\tilde{\gamma} = \tilde{\gamma}_{\mathrm{P}} = \Gamma_0^{a^\dagger\sigma^-} + 2g'^2 \frac{\left(\frac{\kappa+\Gamma_0^{\sigma^+a}-\Gamma_0^{a^\dagger\sigma^-}}{2}\right)}{\Delta_{cx'}^2 + \left(\frac{\kappa+\Gamma_0^{\sigma^+a}-\Gamma_0^{a^\dagger\sigma^-}}{2}\right)^2}, \tag{2.35}$$

fix formatting which, however, is only valid when the spectral width of the cavity is much smaller than the width of the phonon bath function ($\approx$5 meV). This condition is satisfied by high Q cavities with $\kappa \leq 0.1$ meV. Moreover, when $\kappa \gg \Gamma_0^{\sigma^+a} - \Gamma_0^{a^\dagger\sigma^-}$, then

$$\tilde{\gamma}_{\mathrm{P}} = \Gamma_0^{a^\dagger\sigma^-} + 2g'^2 \frac{(\frac{\kappa}{2})}{\Delta_{cx'}^2 + (\frac{\kappa}{2})^2}, \tag{2.36}$$

which can be interpreted as a cavity-feeding term (studied in more detail below) plus a phonon-modified (via $g \to g'$) cavity-induced SE rate. Note this phonon-assisted coupling will also reduce the QD-cavity coupling rate for increasing temperatures (recall $g' = \langle B \rangle g$). Since $\langle B \rangle$ reduces from a value of unity with increasing temperature, it is important to note that the g' value that is assessed and used in experiments is indeed temperature dependent. It is also worth noting that phonon-coupling effects can even lead to stronger photon correlation effects, e.g., causing the WEA to breakdown at even lower field values [79].

2.5.1 *Phonon Dressed Vacuum Rabi Splitting*

Now we employ the above cavity-QED polaron ME to study strong coupling between a QD and a high-Q cavity, where the cavity is described by the cavity mode operator a. This is shown in Fig. 2.6 for a bath temperature of $T = 4$ K. In the strong coupling regime, the coherently coupled QD-cavity system undergoes vacuum Rabi oscillations, when a single quanta of energy is coherently exchanged between the

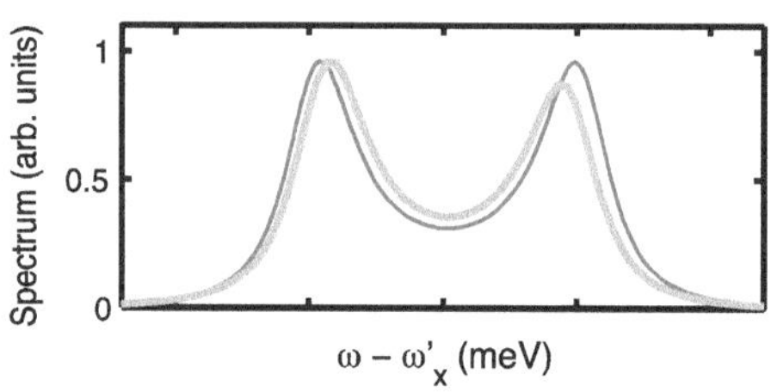

Fig. 2.6 Example spectrum in the on-resonance strong coupling regime. Cavity emission spectra at T = 4 K for a strongly coupled QD-cavity system at resonance ($\omega_x' = \omega_c$). The *magenta* (*thin solid*) and the *orange* (*thick*) *line* plots the coupled mode spectrum, $S_{\mathrm{cav}}^{\mathrm{CM}}$ (2.18) calculated using polaron cavity-QED ME approach, in the absence and presence of phonon coupling, respectively. The main parameters used are $g = 100\,\mu$eV, $\kappa = 65\,\mu$eV, $\gamma_0 = 5\,\mu$eV, and $\gamma_{\mathrm{d}} = 55\,\mu$eV (color figure online)

QD and the cavity [1, 2]. For a simple two-level atom model, coupled to a symmetric cavity without phonon effects, the emission spectrum shows the two hybridized polariton states of equal intensity (magenta, thin solid) line, Fig. 2.6, separated in frequency by twice the QD-cavity coupling constant g. Such a coherent transfer of energy between light and matter is potentially important for building quantum light-matter interfaces, which can be used for long distance quantum communication with photons [2]. Phonon interactions however affects the coherence of the system. In the presence of phonon coupling, the polaritons appear with different intensities and reduced vacuum Rabi-splitting ($\approx 2\langle B\rangle g$, orange (thick solid) line, Fig. 2.6) [39, 40]. The QD (ω_x') and cavity (ω_c) are assumed to be at a detuning of 0 meV (ZPL) and the frequency of the QD, ω_x' includes the polaron shift. The parameters used for the simulations are close to typical experimental numbers, which show vacuum Rabi splitting of $2g = 200$ μeV [40] ($2g' = 183$ μeV) in Fig. 2.6. In general, since the polaron cavity-QED ME approach is derived using a Born–Markov approximation for the polaron-transformed interaction terms, it is valid as long as the system dynamics (i.e., the vacuum Rabi period $\pi/g \approx 20$ ps (Fig. 2.6)) is longer than phonon relaxation time (≈3–5 ps) (Fig. 2.5).

2.5.2 *Off-Resonant Cavity Feeding*

In this subsection we consider the case of phonon-mediated cavity "feeding" from an off-resonant QD [41]. A simple schematic in Fig. 2.7, shows the simplest case of cavity feeding by a QD through emission of a single phonon. Because of the off-resonant condition ($\Delta_{cx'} > 1$ meV), an approximation of weak-coupling between the QD and cavity is not very restrictive. This allows us to use the polaron reservoir ME theory (Sect. 2.2.1) or calculate the spectrum with the polaron cavity-QED ME approach if the Q is sufficiently large. A comparison between these two approaches and a correlation expansion approach is given in [74].

Using the polaron cavity-QED ME, we compute the cavity emission spectrum for a range of different cavity detunings in Fig. 2.8, where normalized spectra (orange solid line) are plotted for different detunings in the presence of phonon coupling. The

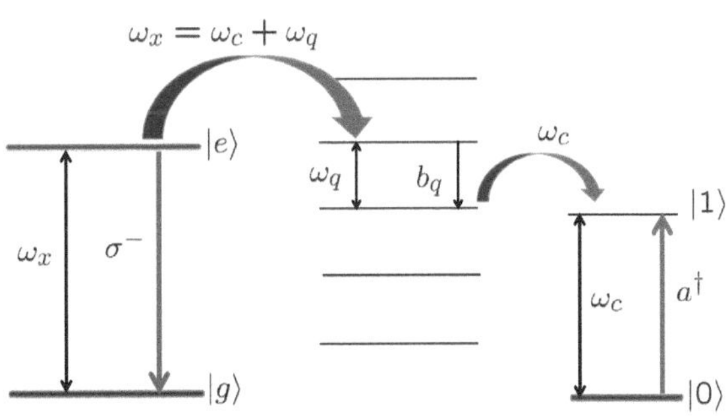

Fig. 2.7 Simple schematic for the phonon-mediated off-resonant feeding process. The a cavity (at resonance ω_c) is excited by a QD (ω_x) through emission of phonons (ω_q), when here $\omega_x = \omega_c + \omega_q$. Figure from [74]

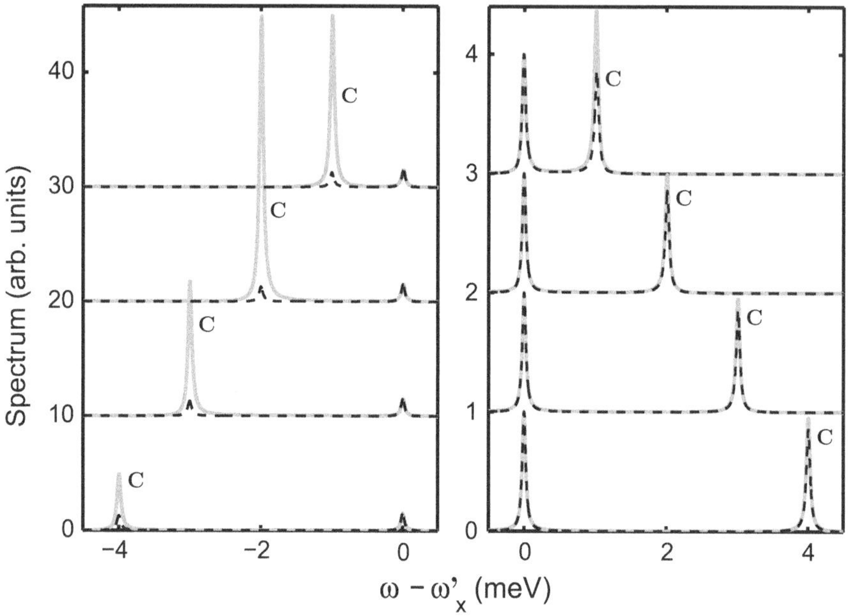

Fig. 2.8 **Off-resonance cavity feeding regime**. Normalized linear emission spectra for a QD-cavity system without phonons (*black dashed*) and with phonons (T = 4 K) (*solid orange*). The QD-cavity detunings are 1, 2, 3, 4 meV and −1, −2, −3, −4 meV (*top to bottom*) in *left* and *right* panels, respectively. Individual spectra are normalized by peak ZPL intensity and data in *bottom-left* panel is multiplied by 1.5 for better visibility. The main parameters are $g = 100\,\mu$eV, $\kappa = 65\,\mu$eV, $\gamma_0 = 5\,\mu$eV, $\gamma_{\rm d} = 55\,\mu$eV. Figure adapted from [74] (color figure online)

QD-cavity coupling rate to cavity decay rate ratio is $g/\kappa > 1$, so it is appropriate to treat the cavity mode as the level of a system operator. The dark dashed lines represent the corresponding normalized spectra in the absence of phonon coupling, and the individual spectra are normalized with respect to the peak ZPL intensity. Left (right) panels plots spectra for positive (negative) QD-cavity detunings ($\Delta_{x'c}$). Cavity feeding can be estimated from the ratio of peak heights as both Lorentzians have comparable linewidths ($\Gamma_x = 60\,\mu$eV and $\kappa = 65\,\mu$eV) and it increases in the presence of phonons. The feeding is asymmetric and stronger when the cavity mode is to the right (higher spectral energy). At the low temperature considered (4 K), phonon emission is more probable than phonon absorption; thus if a cavity has lower energy than an excited QD, the cavity can be excited more easily in a two-step quantum process where a cavity photon is created along with a phonon emission. When $\omega_c > \omega_x'$, exciting the cavity will require phonon absorption which is less feasible at low temperatures.

Theoretical attempts have been made to partly describe cavity feeding using a linear susceptibility model where the IBM lineshape is added to the QD spectral function and coupled to a Green function approach for the medium [43, 44]. However, it turns out that this model greatly underestimates the amount of feeding over large detunings and can result in drastically different spectra [74]; basically the linear susceptibility theory misses the two step quantum process which is correctly incorporated in the SE rate (2.36) and through the Lindblad QD/cavity scattering rates $\Gamma^{\sigma^+ a/a^\dagger \sigma^-}$. The linear susceptibility technique is however successful in explaining the on-resonance ($\Delta_{cx'} = 0$) phonon-induced asymmetric vacuum Rabi doublet [43]. This is because at resonance, these two-step quantum feeding processes have negligible contributions to the emission spectra [74].

2.6 Coherent Driving and Nonlinear Excitation

In this final section of our chapter, we describe how to include a coherent pump field in the polaron ME theories, which can connect to a wide range of experimental conditions with coherent driving fields. Consider the case of a general photon reservoir with a coherent pump field exciting the QD [80], and working in a rotating frame at the laser frequency, the Hamiltonian becomes

$$H = \hbar \int d\mathbf{r} \int_0^\infty d\omega\, \mathbf{f}^\dagger(\mathbf{r}, \omega)\mathbf{f}(\mathbf{r}, \omega) + \hbar\Delta_{xL}\sigma^+\sigma^- + \Sigma_q \hbar\omega_q b_q^\dagger b_q + \eta_x(\sigma^- + \sigma^+)$$
$$- [\sigma^+ e^{i\omega_L t} \int_0^\infty d\omega\, \mathbf{d} \cdot \mathbf{E}(\mathbf{r}_d, \omega) + \text{H.c}] + \sigma^+\sigma^- \Sigma_q \hbar\lambda_q (b_q^\dagger + b_q), \quad (2.37)$$

where the QD with exciton frequency ω_x is excited by a cw laser of amplitude η_x (half of the Rabi frequency, $\eta_x = \frac{\Omega_x}{2}$) and central frequency ω_L, and $\Delta_{xL} = \omega_x - \omega_L$ is the QD-laser detuning. As before, the polaron transformed Hamiltonian H' can be separated into system (H'_S), reservoir (H'_R) and interaction (H'_I) components,

$$H' = H'_\text{S} + H'_\text{R} + H'_\text{I},$$
$$H'_\text{S} = \hbar\Delta_{x'L}\sigma^+\sigma^- + \hbar\langle B\rangle\eta_x[\sigma^+ + \sigma^-],$$
$$H'_\text{R} = \hbar \int d\mathbf{r} \int_0^\infty d\omega\, \mathbf{f}^\dagger(\mathbf{r}, \omega)\mathbf{f}(\mathbf{r}, \omega) + \Sigma_q \hbar\omega_q b_q^\dagger b_q,$$
$$H'_\text{I} = -[B_+\sigma^+ e^{i\omega_L t} \int_0^\infty d\omega\, \mathbf{d} \cdot \mathbf{E}(\mathbf{r}_d, \omega) + \text{H.c}] + X_g\zeta_g + X_u\zeta_u, \quad (2.38)$$

where terms X_g and X_u are defined as $X_g = \hbar\eta_x(\sigma^- + \sigma^+)$ and $X_u = i\hbar\eta_x(\sigma^+ - \sigma^-)$, and $\zeta_g = \frac{1}{2}(B_+ + B_- - 2\langle B\rangle)$ and $\zeta_u = \frac{1}{2i}(B_+ - B_-)$ denote phonon induced fluctuation operators [23]. Using exactly the same techniques as before, we derived the polaron-transformed ME with coherent pumping,

$$\frac{d\rho}{dt} = \frac{1}{i\hbar}[H'_S, \rho] + \mathcal{L}_\text{ph}(\rho) + \mathcal{L}^\text{D}_\text{pn}(\rho) + \gamma_0 L(\sigma^-) + \gamma_\text{d} L(\sigma^+\sigma^-), \quad (2.39)$$

where γ_0 and $\gamma_{\rm d}$ ZPL terms have been added and $\mathcal{L}_{\rm ph}(\rho)$ is similar to before and can be written in terms of a general SE rate than is influenced from phonons. The additional (new) phonon term from the coherent drive is [81]

$$\mathcal{L}^{\rm D}_{\rm pn}(\rho) = \Gamma^{\sigma^+} L[\sigma^+] + \Gamma^{\sigma^-} L[\sigma^-] - \Gamma^{\rm cd}(\sigma^+\rho\sigma^+ + {\rm H.c.}) \\ - (\Gamma_u(\sigma^+\sigma^-\rho(\sigma^+ - \sigma^-) + \sigma^-\rho) + {\rm H.c.}), \tag{2.40}$$

where the relevant phonon-mediated scattering rates are given by the following analytical expressions,

$$\Gamma^{\sigma^+/\sigma^-} = \frac{\Omega_{\rm R}^2}{2}\int_0^\infty \Big({\rm Re}[(\cosh(\phi(\tau)) - 1)f(\tau) \\ + \sinh(\phi(\tau))\cos(\eta\tau)] \mp {\rm Im}[(e^{\phi(\tau)} - 1)\frac{\Delta_{Lx}\sin(\eta\tau)}{\eta}]\Big)d\tau,$$

$$\Gamma^{\rm cd} = \frac{\Omega_{\rm R}^2}{2}\int_0^\infty {\rm Re}[\sinh(\phi(\tau))\cos(\eta\tau) - (\cosh(\phi(\tau)) - 1)f(\tau)]d\tau,$$

$$\Gamma_u = i\frac{\Omega_{\rm R}^3}{2\eta}\int_0^\infty \sinh(\phi(\tau))\sin(\eta\tau)d\tau, \tag{2.41}$$

with $f(\tau) = \frac{\Delta_{Lx}^2\cos(\eta\tau) + \Omega_{\rm R}^2}{\eta^2}$, $\eta = \sqrt{\Omega_{\rm R}^2 + \Delta_{Lx}^2}$, and $\Omega_{\rm R} = 2\langle B\rangle\eta_x$. The above rates incorporate the spectral shape of the phonon bath by accounting for phonon damping during Rabi oscillations of the driven QD and are valid for weak and strong drives [81]. If the drive field is not very strong ($\eta_x \ll \omega_{\rm b}$), simpler expressions [58] for the scattering terms $\Gamma^{\sigma^+/\sigma^-}$ and $\Gamma^{\rm cd}$ for use in an effective phonon ME are recovered starting from (2.41), so that

$$\mathcal{L}^{D}_{\rm pn}(\rho) = \Gamma^{\sigma^-} L[\sigma^-] + \Gamma^{\sigma^+} L[\sigma^+] - \gamma_{\rm cd}(\sigma^+\rho\sigma^+ + \sigma^-\rho\sigma^-), \tag{2.42}$$

where [58]

$$\Gamma^{\sigma^{+/-}} = 2\langle B\rangle^2\eta_x^2{\rm Re}\Big[\int_0^\infty d\tau e^{\pm i(\omega_L - \omega'_x)\tau}(e^{\phi(\tau)} - 1)\Big], \tag{2.43}$$

represents additional SE from the pump (Γ^{σ^-}) and pump-induced incoherent excitation (Γ^{σ^+}), and the term

$$\gamma_{\rm cd} = 2\langle B\rangle^2\eta_x^2{\rm Re}\Big[\int_0^\infty d\tau\cos(\Delta_{x'L}\tau)(1 - e^{-\phi(\tau)})\Big], \tag{2.44}$$

represents a cross-dephasing process (which is in the form of a squeezing operator) that can influence the spectral lineshape of the QD.

Subsequently, we obtain a simple effective phonon ME,

$$\frac{d\rho}{dt} = \frac{1}{i\hbar}[H'_S, \rho] + \Gamma^{\sigma^-} L[\sigma^-] + \Gamma^{\sigma^+} L[\sigma^+] - \gamma_{\rm cd}(\sigma^+\rho\sigma^+ + \sigma^-\rho\sigma^-)$$
$$+ \tilde{\gamma} L(\sigma^-) + \gamma_{\rm d} L(\sigma^+\sigma^-), \qquad (2.45)$$

where we only include one photon-reservoir SE term and focus more on the drive dependent scattering terms with a coherent pump field.

2.6.1 *Photoluminescence Lineshapes and Phonon-Mediated Population Inversion*

The photoluminescence intensity (PLI) of the continuous wave (cw) laser excited QD is proportional to the population of the QD, $n_x = \langle\sigma^+\sigma^-\rangle$. Using the ME (2.45), an analytical expression of this quantity can be obtained from the optical Bloch equations, given by [16]

$$n_x = \langle\sigma^+\sigma^-\rangle = \frac{1}{2}\left[1 + \frac{\Gamma^{\sigma^+} - \Gamma^{\sigma^-} - \tilde{\gamma}}{\Gamma^{\sigma^+} + \Gamma^{\sigma^-} + \tilde{\gamma} + \frac{4\langle B\rangle^2 \eta_x^2 (\Gamma_{\rm pol} + \gamma_{\rm cd})}{\Gamma_{\rm pol}^2 + \Delta_{x'L}^2 - \gamma_{\rm cd}^2}}\right], \qquad (2.46)$$

where $\Gamma_{\rm pol} = \frac{1}{2}(\Gamma^{\sigma^+} + \Gamma^{\sigma^-} + \tilde{\gamma} + \gamma_{\rm d})$. Using this expression, it is easy to show that population inversions [4, 10] can be realized for sufficiently large drives, and this has been shown for both cw [82] and pulsed QD systems [81] using the polaron ME approaches. Experimentally, phonon-assisted QD inversion has recently been reported by various groups [7–9] and successfully described using path integral techniques [6, 7] and polaron ME techniques [81]. An advantage of the ME approach is that is can easily compute emission spectra and quantum optical properties, e.g., to assess figures-of-merit for single photon sources [81, 83], where recent work suggests that the phonon-assisted scheme is likely not as good as single photons created through direct on-resonance Rabi oscillation [81, 84] (in part as the interaction with the phonon bath is less pronounced at the lower pump values required with on resonance excitation).

We will not explore the polaron ME calculations of population inversion in this chapter (for details, see [81, 82]), but rather connect to the PLI lineshape that is predicted from the same model and show how this relates to recent experiments. We show two example PLI from Weiler et al., who studied self-assembled In(Ga,As)/GaAs QDs grown by metal-organic vapor-phase epitaxy, embedded in a planar cavity (distributed Bragg reflectors (DBR) on the top and bottom). Figure 2.9 shows example PLI calculations and measurement for different temperatures. The depicted profiles consists of two parts, including a sharp ZPL (Lorentzian profile) at the QD resonance and a broader phonon-assisted excitation feature around the ZPL. In the latter case, a distinct asymmetry for $\Delta > 0$ is clearly visible, which is a direct signature of the unequal probabilities for LA phonon absorption and emission at low temperatures. To

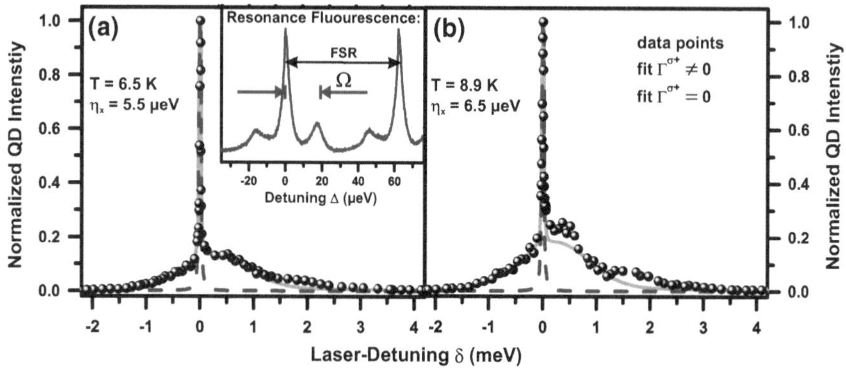

Fig. 2.9 Theoretical and experimental PLI for a driven In QD in a planar cavity system. a, b Intensity profiles: Integrated QD intensity derived from a frequency scan plotted as a function of laser-QD detuning. The scans have been performed at two different temperatures. Experimental data indicated by *black circles* and theory (*red solid lines*). The corresponding values used to fit the data with our theory (with phonon-induced processes Γ^{σ^+} and Γ^{σ^-}: *solid* (*red*) *line*, with only the process Γ^{σ^-}: *dashed* (*blue*) *line*) are the cut-off frequency $\omega_b = 0.6$ meV, coupling parameter $\alpha_p/(2\pi)^2 = 1.5 \times 0.06$ ps^2, radiative decay rate $\tilde{\gamma} = 0.82$ μeV (800 ps) and pure dephasing rate $\gamma_d = 0.6$ μeV. The thermally-averaged bath displacement operator is calculated to be $\langle Bs \rangle = 0.91$ for the conditions in (**a**) and $\langle B \rangle = 0.87$ for (**b**). *Inset* PL spectrum with the characteristic RF spectrum (Mollow triplet) in the frequency domain, revealing a Rabi energy (center to sideband) of $\Omega = (16.7 \pm 0.7)$ μeV. Figure adapted from Weiler et al. [16] (color figure online)

help identify this process further, we also plot the calculation when the Γ^{σ^+} process is turned off, which confirms that the laser-driven incoherent excitation process is the dominant phonon scattering process in this experiment.

Figure 2.10 illustrates the effect of phonon-assisted incoherent excitation on the QD PLI of a QD, as a function of laser detuning. In each case, we recognize substantially more PLI when the laser is blue-detuned from the exciton line, which allows phonon-assisted excitation of the exciton state. To gain more insight into the effect of the phonon-induced incoherent coupling, we have systematically studied theoretically the effects of ω_b, T, α_p, η_x on the resulting intensity profiles in Fig. 2.9a–d. An increase in the cut-off frequency ω_b (i.e., a decrease in QD size) leads to a blue shift of the phonon reservoir replica of the QD intensity profile. In contrast, increasing temperature T, excitation intensity η_x or coupling factor α_p overall increases the QD intensity for off-resonant excitation conditions. This can also be seen in Fig. 2.9a, b, where increased temperature and excitation strength leads to a higher emission efficiency in Fig. 2.9b as compared to Fig. 2.9a. For increasing temperature, these features become more symmetric due to increasing phonon state occupations. Parameters α_p and η_x have similar effects on the shape of the intensity profile but keep the asymmetry unchanged. Variation of $\tilde{\gamma}$, γ_d (not shown) mainly affects the width of the ZPL and has almost negligible effect on the broader intensity profile.

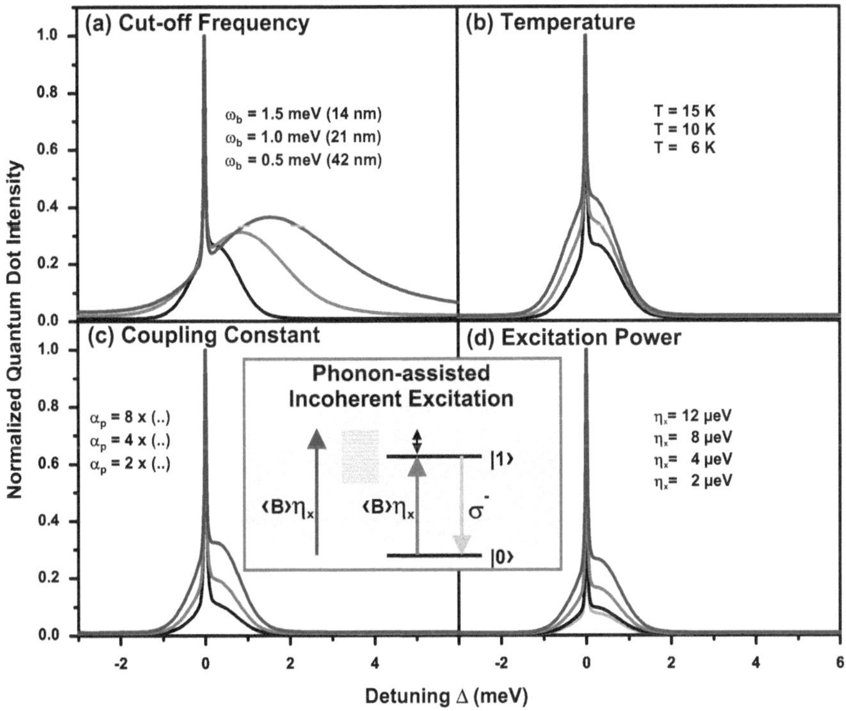

Fig. 2.10 Example PLI lineshapes for a coherently driven QD. Theoretical study of the influence of the relevant parameters on the intensity profile of the QD emission. While keeping the other parameters fixed ($\omega_b = 0.5$ meV, $T = 6$ K, $\gamma = 0.8$ μeV, $\gamma_d = 0.8$ μeV, $\eta_x = 12$ μeV), the **a** phonon bath cut-off frequency ω_b, **b** temperature T, **c** coupling factor α_p, and **d** pump power $\eta_x = 2\Omega$ have been systematically increased from *bottom to top* respectively. *Inset* Illustration of the incoherent excitation process (*red arrow*), Γ^{σ^+} scattering, mediated by the acoustic phonon bath (*green lines*). Figure from Weiler et al. [16] (color figure online)

The above results are computed for a low Q cavity, and in the presence of a structured reservoir one can obtain much richer PLI profiles, as shown in [45, 85]. The example of a high-Q cavity profiles (using the cavity-QED polaron ME with multiple photon states) are shown in [86].

2.6.2 *Phonon-Dressed Mollow Triplets*

In this final subsection, we describe how to compute the well known Mollow triplet for an excited QD system, which is a well known quantum optics effects that originates from the emission of a field-driven two level system, where three distinct spectral lines result appear from photon decay between pairs of dressed states. Resonant excitation of single QDs has recently gained significant interest in the semiconductor optics community [87–89], partly because coherent excitation is promising for the generation of single photons with excellent coherence properties [83] and it is also of fundamental interest in solid state quantum optics. Resonance fluorescence (RF) emission below saturation of the quantum emitter has also revealed close-to-Fourier limit single photons with record-high emission coherence and two-photon interference visibility [90]. Recent experiments have even been able to beat the Fourier limit for single-photon emission coherence in the Heitler regime, where the excitation strengths are well below the saturation of the quantum emitter [91, 92]. Another achievement with respect to RF is the demonstration of single- and cascaded photon emission between the Mollow sidebands above saturation of the QD [93].

For simplicity, we will focus on a low Q cavity system, suitable for describing driven QDs in a planar cavity, and we will connect directly to the data of Ulhaq et al. [77]. The experimental Mollow triplets in a high Q cavity have also been studied experimentally (e.g., [14, 94]) and theoretically (e.g., [24, 41, 95]), and many of the same scattering rates above can directly explain this data as well. Since we are not using the full cavity-QED ME (with a coherent pump field), it is possible to derive the Mollow triplet spectrum analytically, which is more useful for fitting experimental data. Using the ME (2.45) and the property $\langle \dot{O} \rangle = \mathrm{tr}[\dot{\rho} O]$ [63], we obtain the following optical Bloch equations:

$$\frac{d\langle \sigma^- \rangle}{dt} = -(\gamma_{\rm pol} - i\Delta)\langle \sigma^- \rangle - \gamma_{\rm cd}\langle \sigma^+ \rangle + i\frac{\Omega_{\rm R}}{2}\langle \sigma^z \rangle, \tag{2.47a}$$

$$\frac{d\langle \sigma^+ \rangle}{dt} = -(\gamma_{\rm pol} + i\Delta)\langle \sigma^+ \rangle - \gamma_{\rm cd}\langle \sigma^- \rangle - i\frac{\Omega_{\rm R}}{2}\langle \sigma^z \rangle, \tag{2.47b}$$

$$\frac{d\langle \sigma^z \rangle}{dt} = i\Omega_{\rm R}\langle \sigma^- \rangle - i\Omega_{\rm R}\langle \sigma^+ \rangle - \gamma_{\rm pop}\langle \sigma^z \rangle - \gamma'_{\rm pop}, \tag{2.47c}$$

where again $\gamma_{\rm pol} = \frac{1}{2}(\Gamma^{\sigma^+} + \Gamma^{\sigma^-} + \tilde{\gamma} + \gamma_{\rm d})$, the population decay $\gamma_{\rm pop} = (\Gamma^{\sigma^+} + \Gamma^{\sigma^-} + \tilde{\gamma})$, and we have also introduced $\gamma'_{\rm pop} = \gamma_{\rm pop} - 2\Gamma^{\sigma^+}$. The laser-exciton detuning $\Delta = \omega_L - \omega'_x$ The phonon-modified Rabi frequency ($\Omega_{\rm R}$) is defines after (2.41).

The incoherent spectrum can be computed from a time integration of the appropriate two-time correlation function [63]:

$$S(\mathbf{r}, \omega) \equiv F(\mathbf{r})S(\omega) \propto \lim_{t\to\infty}\mathrm{Re}\left\{\int_0^\infty d\tau \langle \delta\sigma^+(t)\delta\sigma^-(t+\tau)\rangle e^{i(\omega-\omega_{\rm L})\tau}\right\}, \tag{2.48}$$

where $\langle \delta O \rangle = \langle O \rangle - O$ and $F(\mathbf{r})$ is a geometrical factor. In the spectrum calculation, we use $\langle \delta\sigma^+(t)\delta\sigma^-(t+\tau)\rangle$ rather than $\langle \sigma^+(t)\sigma^-(t+\tau)\rangle$ to subtract off the coherent contribution from the cw pump field (which simply yields a Dirac delta function response at the laser frequency). Note also that we do not need to add in the phonon correlation phase ($e^{-i\phi(\tau)}$) when computing the two-time correlation function, as the emitted spectrum is detected via a *weakly-coupled* planar cavity mode, in which case $\langle \delta a^\dagger(t) \delta a(t+\tau)\rangle \propto \langle \delta\sigma^+(t)\delta\sigma^-(t+\tau)\rangle$; so we are actually obtaining the cavity emission which requires no change in the aforementioned correlation functions when coming out of the polaron frame. We then define the steady-state expectation values $f(0) \equiv \langle \delta\sigma^+\delta\sigma^-\rangle_{\rm ss}$, $g(0) \equiv \langle \delta\sigma^+\delta\sigma^+\rangle_{\rm ss}$, and $h(0) \equiv \langle \delta\sigma^+\delta\sigma^z\rangle_{\rm ss}$, and keep the explicit laser-exciton detuning dependence in the solution. Using the frequency detuning $\delta\omega = \omega - \omega_L$, we finally obtain the spectrum lineshape analytically,

$$S(\omega) \equiv \\ \mathrm{Re}\left\{\frac{-f(0)D(\omega) + ih(0)C(\omega)D(\omega) - [\gamma_{\rm cd} + \Omega_{\rm R}C(\omega)]\,[g(0) + ih(0)C(\omega)]}{(D(\omega)+i2\Delta)D(\omega) - [\gamma_{\rm cd} + \Omega_{\rm R}C(\omega)]^2}\right\}, \tag{2.49}$$

where $C(\omega) - \frac{\Omega_{\rm R}}{2(i\delta\omega - \gamma_{\rm pop})}$ and $D(\omega) = i\delta\omega - \gamma_{\rm pol} - i\Delta + \frac{\Omega_{\rm R}^2}{2(i\delta\omega - \gamma_{\rm pop})}$.

The corresponding steady-state inversion and polarization components are calculated to be

$$\langle\sigma^z\rangle_{ss} = -\frac{\gamma'_{\rm pop}}{\gamma_{\rm pop} + \frac{\Omega_{\rm R}^2(\gamma_{\rm pol}+\gamma_{\rm cd})}{(\gamma_{\rm pol}^2 + \Delta^2 - \gamma_{\rm cd}^2)}}, \quad \langle\sigma^-\rangle_{ss} = \frac{i\Omega_{\rm R}\left(\gamma_{\rm pol} + i\Delta + \gamma_{\rm cd}\right)}{2(\gamma_{\rm pol}^2 + \Delta^2 - \gamma_{\rm cd}^2)}\langle\sigma^z\rangle_{\rm ss}, \tag{2.50}$$

from which we can obtain the following steady-state values for f, g, and h:

$$f(0) = \frac{1}{2}\left(1 + \langle\sigma^z\rangle_{\rm ss} - 2\langle\sigma^+\rangle_{\rm ss}\langle\sigma^-\rangle_{\rm ss}\right), \tag{2.51a}$$

$$g(0) = -\langle\sigma^+\rangle_{\rm ss}^2, \tag{2.51b}$$

$$h(0) = -\langle\sigma^+\rangle_{\rm ss}(1 + \langle\sigma^z\rangle_{\rm ss}). \tag{2.51c}$$

These equations are used to obtain $S(\omega)$, which is an exact solution to the given effective phonon ME. The full-width at half-maximum (FWHM) of spectral resonances can be obtained from (2.49), though these are rather complicated to write down analytically. However, one can simply fit the analytical spectrum to a sum of Lorentzian line shapes and easily extract the broadening parameters. In the high-field limit, the on-resonance ($\Delta = 0$) FWHM values are $\gamma_{\rm side} \approx \frac{3}{2}(\gamma_0 + \gamma_{\rm ph}) + \frac{1}{2}\gamma_{\rm d} - \gamma_{\rm cd}$ and $\gamma_{\rm center} \approx \gamma_0 + \gamma_{\rm ph} + \gamma_{\rm d} + \gamma_{\rm cd}$ for the sideband and center resonances, respectively; thus the cross dephasing term acts to *squeeze* the sidebands while broadening the center line or vice versa.

From the analytical spectra above, we can investigate when the Mollow sidebands will become asymmetric and whether the detuning dependence will exhibit broadening or narrowing of the sideband resonances. To make these regimes clearer, we define the following ratio:

$$r = \frac{\gamma_{\rm pol} + \gamma_{\rm cd}}{\gamma_{\rm pop}} = \frac{1}{2}\left[1 + \frac{\gamma_{\rm d} + 2\gamma_{\rm cd}}{\gamma + \gamma_{\rm ph}}\right], \tag{2.52}$$

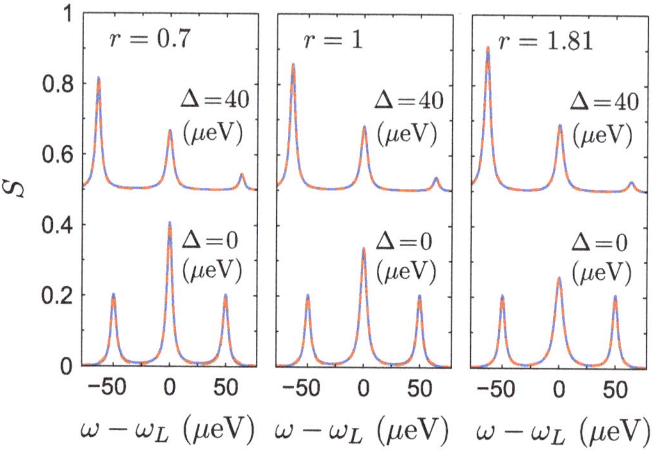

(a) Fluorescence spectra for different r values (see text).

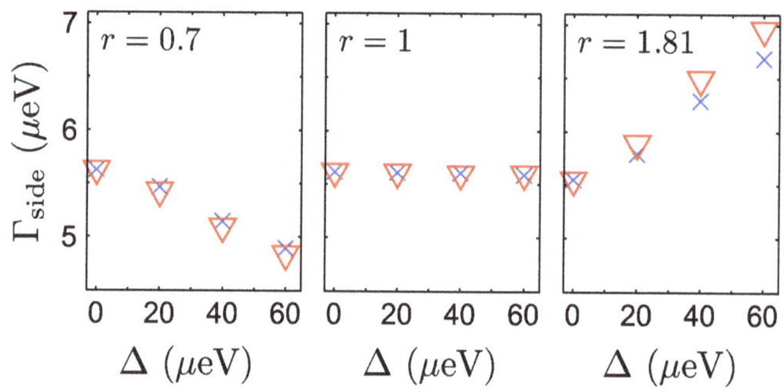

(b) Corresponding values of the sideband FWHM.

Fig. 2.11 Example the coherently driven QD Mollow triplets in the presence of electron-phonon coupling and different laser detuning. a Analytically computed spectrum as a function of detuning for three different values of r. The *solid red curve* shows the analytical solution (2.49) and the *blue dashed curve* shows the three-Lorentzian fit. Positive and negative detunings $|\Delta|$ reveal simply a mirror image of each other. The phonon parameters are given in [77] (and discussed below) at a bath temperature of $T = 6\,\text{K}$, yielding $\gamma_{\text{cd}} \approx 0.6$ μeV and $\gamma_{\text{ph}} \approx 1.6$ μeV for the chosen Rabi field. Here we adjust γ and γ_d to maintain the same on-resonance FWHM value of $\gamma_{\text{side}}(\Delta = 0) \approx \frac{3}{2}(\gamma + \gamma_{\text{ph}}) + \frac{1}{2}\gamma_{\text{d}} - \gamma_{\text{cd}} = 5.6$ μeV: for $r = 0.7, 1.0, 1.81$, we use $\gamma_{\text{d}}(\gamma)$ as 0.4(2.4), 2.2(1.8), and 5.1(0.8) μeV, respectively. **b** Extracted FWHM of the lower (*blue crosses*) and higher energy sideband (*red inverted triangles*) as a function of detuning Δ. One clearly sees a trend of either increasing or decreasing sideband line width as a function of laser detuning, depending on the value of r, where $r \approx 1$ denotes the crossover. The center Mollow line exhibits the opposite trend. Figure from Ulhaq et al. [77]

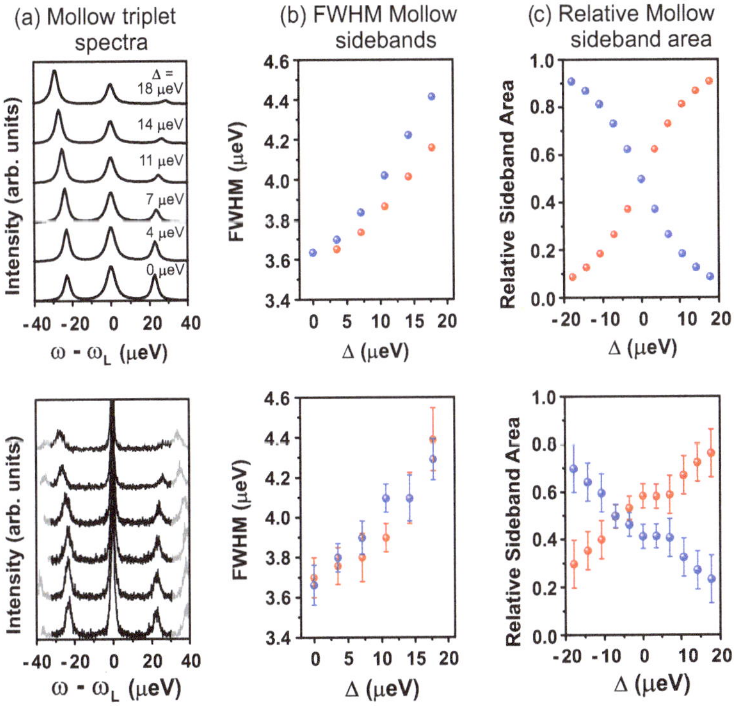

Fig. 2.12 Experimental and theoretical QD driven Mollow triplets. Detuning-dependent Mollow triplets for a pump power of $P = 500\,\mu$eV, showing theoretical predictions versus experimental results for a system with $r = 2.01$. **a** Mollow triplet spectra for increasing negative detunings, Δ, the spectra are plotted with respect to the energetic laser position set to zero. **b** FWHM of the *blue and red Mollow sideband* reveal distinct sideband broadening with increasing laser-detuning. **c** Change of the relative Mollow sideband area with Δ. The theory is seen to match the complex experimental trends without changing any fitting parameters. Figure from Ulhaq et al. [77]

where $\gamma_{\rm ph} = \Gamma^{\sigma^+} + \Gamma^{\sigma^-}$. It is also worth noting that for off-resonant driving and $\gamma_{\rm d} = 0$, a completely symmetric Mollow triplet is expected *only* if all phonon terms are neglected. Under systematic increase of the excitation-detuning Δ, sideband spectral broadening or narrowing can be achieved depending upon the value of r. In Fig. 2.11a, b we plot the Mollow triplet as a function of Δ, and extract the FWHM of the sidebands for three values of r. As can be seen, $r < 1$ (for a sufficiently small $\gamma_{\rm d}$) leads to spectral sideband narrowing, whereas for $r > 1$ the effect of spectral sideband broadening occurs. Interestingly, the reverse trend occurs for the center resonance (not shown), namely when the sidebands broaden (narrows) then the center line narrows (broadens); so depending on the r value, one can observe squeezing or anti-squeezing of the spectral resonances with increasing Δ (in addition to the squeezing that already occurs from a finite $\gamma_{\rm cd}$). Similar results have been obtained by McCutcheon and Nazir [65]. For these calculations, we use a cut-off frequency and electron-phonon coupling strength, we use $\omega_b = 1$ meV and $\alpha_p/(2\pi)^2 = 0.15 \pm$

0.01 ps^2; the deformation potential constant here is somewhat higher compared to the value used above, but these can vary for different QD samples, and we use this value to fit the experiments below.

Finally, we show example spectra calculations and experiments for an excited QD system, where the experiments were first shown in [77]. Figure 2.12a shows a direct comparison of the Mollow triplet spectra for increasing negative detuning $\Delta < 0$, from which the FWHM and relative intensities are extracted. The discrepancy between the expected and measured central Mollow line intensity results from contributions of scattered laser stray-light to the true QD emission that can not experimentally be differentiated due to the equal emission frequency. For the detuning $\Delta \neq 0$, the spectral resolution of the high-resolution spectroscopy is not sufficient to distinguish between laser-excitation and QD Rayleigh line emission. The grey shaded peaks in Fig. 2.12a (lower panel) belong to a higher order interference of the Fabry-Pérot interferometer. The extracted FWHM values are depicted in Fig. 2.12b. For the system under investigation, r is calculated to be around 2.01, and therefore an increase in the sidebands' width is expected according to the theoretical model. Indeed, we observe a systematic increase with increasing negative detuning $\Delta < 0$. Moreover, we observe spectral narrowing (squeezing) of the center line though we do not attempt to fit this resonance as it has a large contribution from coherent scattering. Additionally, the relative sideband areas $A_{\text{red/blue}}$ in dependence on Δ are plotted in Fig. 2.12c. As becomes already visible from the Mollow spectra, for positive detunings the blue sideband gains intensity whereas the red sideband area decreases, and vice versa. The crossing between relative intensities is expected to occur at $\Delta = 0$. Interestingly, we observe crossings at moderate negative laser-detuning values for all different QDs under study. Note the high value of pure dephasing which causes $r > 1$ in the sample is likely due to the fact that the samples were manufactured using metal organic chemical vapor depositions, which are supposed to incorporate more impurities compared to sample grown by molecular beam epitaxy. The higher magnitude of pure dephasing results in broadening of Mollow sidebands.

2.7 Conclusions and Outlook

We have described a general quantum optics theory of phonon-dressed light matter interactions in QD cavity systems with a focus on using a polaron master equations, which captures certain electron-phonon coupling effects nonperturbatively, and presents a convenient starting point for deriving system-reservoir master equations where residual phonon-exciton-photon interaction terms (which are now much smaller in the polaron frame) are included at the level of a 2nd-order Born–Markov approximation. This powerful open-systems approach also allows one to identify certain scattering rates analytically, such as phonon-mediated cavity feeding and incoherent excitation, which can be used to connect to and explain a number of emerging experiments in the field, and easily lends itself to connecting to a wide range of experimental observables. The chapter has focused on a single exciton (or electron-holer pair) coupled to a reservoir of phonons and a photonic environment, since such a regime already covers a wide range of recent QD-cavity experiments, including regimes of phonon-dressed strong coupling, phonon-mediated cavity feeding, and the breakdown of Fermi's golden rule. However, the theory can also be extended to account for more exciton levels. Indeed, with resonantly excited QDs, it is possible to observe five spectral peaks resulting in a "Mollow quintuplet," which is caused by the simultaneous excitation of both x- and y- polarized excitons [51]; in addition, even more spectral peaks are possible in the driven biexciton-cascade system [53], and both of these effects have been successfully explained using an extended polaron master equation. In this chapter, we have also treated the excitation laser as cw, though the theory can include optical pulses as well, which for example has been used to describe the phonon-induced population inversion experiments of Quilter et al. [7, 81]. For the Mollow triplet section, we have focused on low Q cavities, but is it also possible to model high Q cavities [24] and even completely general photonic bath functions such as those that represent slow-light photonic crystal waveguides [80]. The general methodologies presented are thus widely applicable to many QD cavity experiments. As more sophisticated experiments and fabricated samples continue to be developed, one can expect to gain a deeper understanding of the intrinsic electron-phonon and electron-photon interactions in these QD cavity systems, which on the one hand leads to fascinating quantum optical physics in a solid state platform, and on the other hand can help with the design of emerging QD optical devices, such as coherently driven single and entangled photon pair sources. In this regard, we anticipate a bright future for realizing all-integrated QD quantum optical devices on a semiconductor chip.

Acknowledgements This work was supported by the National Sciences and Engineering Research Council of Canada and Queen's University. It is a pleasure to acknowledge Peter Michler, Stefanie Weiler, Ata Ulhaq, Sven Ulrich, Fabian Hargart, and Michael Jetter for ongoing collaborations and for supplying some of the experimental data shown in this chapter. We acknowledge Chiranjeeb Roy and Ronchun Ge for discussions and for some of their contributions on the polaron ME work. We also thank Andreas Knorr and Mark Fox for useful discussions.

References

1. T. Yoshie, A. Scherer, J. Hendrickson, G. Khitrova, H.M. Gibbs, G. Rupper, C. Ell, O.B. Shchekin, D.G. Deppe, Nature **432**, 200 (2004)
2. R. Bose, T. Cai, K. Roy Choudhury, G.S. Solomon, E. Waks, Nat. Photonics **8**, 858 (2014)
3. B. Krummheuer, V.M. Axt, T. Kuhn, Phys. Rev. B **65**, 195313 (2002)
4. T.M. Stace, A.C. Doherty, S.D. Barrett, Phys. Rev. Lett. **95**, 106801 (2005)
5. S. Hughes, H.J. Carmichael, Phys. Rev. Lett. **107**, 193601 (2011)
6. M. Glässl, A. Vagov, S. Lüker, D.E. Reiter, M.D. Croitoru, P. Machnikowski, V.M. Axt, T. Kuhn, Phys. Rev. B **84**, 195311 (2011)
7. J.H. Quilter, A.J. Brash, F. Liu, M. Glässl, A.M. Barth, V.M. Axt, A.J. Ramsay, M.S. Skolnick, A.M. Fox, Phys. Rev. Lett. **114**, 137401 (2015)
8. S. Bounouar, M. Müller, A.M. Barth, M. Glässl, V.M. Axt, P. Michler, Phys. Rev. B **91**, 161302(R) (2015)

9. P.-L. Ardelt, L. Hanschke, K.A. Fischer, K. Müller, A. Kleinkauf, M. Koller, A. Bechtold, T. Simmet, J. Wierzbowski, H. Riedl, G. Abstreiter, J.J. Finley, Phys. Rev. B **90**, 241404(R) (2014)
10. S. Hughes, H.J. Carmichael, Viewpoint: crystal vibrations invert quantum dot exciton. Physics **8**, 29 (2015)
11. J. Förstner, C. Weber, J. Danckwerts, A. Knorr, Phys. Rev. Lett. **91**, 127401 (2003)
12. A.J. Ramsay, T.M. Godden, S.J. Boyle, E.M. Gauger, A. Nazir, B.W. Lovett, A.M. Fox, M.S. Skolnick, Phys. Rev. Lett. **105**, 177402 (2010)
13. L. Monniello, C. Tonin, R. Hostein, A. Lemaitre, A. Martinez, V. Voliotis, R. Grousson, Phys. Rev. Lett. **111**, 026403 (2013)
14. S.M. Ulrich, S. Ates, S. Reitzenstein, A. Löffler, A. Forchel, P. Michler, Phys. Rev. Lett. **106**, 247402 (2011)
15. H. Kim, T.C. Shen, K. Roy-Choudhury, G.S. Solomon, E. Waks, Phys. Rev. Lett. **113**, 027403 (2014)
16. S. Weiler, A. Ulhaq, S.M. Ulrich, D. Richter, M. Jetter, P. Michler, C. Roy, S. Hughes, Phys. Rev. B **86**, 241304(R) (2012)
17. T. Takagahara, Phys. Rev. B **60**, 2638 (1999)
18. F. Grosse, E.A. Muljarov, R. Zimmermann, *Phonons in Quantum Dots and Their Role in Exciton Dephasing* (Springer, Berlin, 2008), pp. 165–187
19. L. Besombes, K. Kheng, L. Marsal, H. Mariette, Phys. Rev. B **63**, 155307 (2001)
20. G.D. Mahan, *Many-Particle Physics* (Plenum, New York, 1990)
21. E.A. Muljarov, R. Zimmerman, Phys. Rev. Lett. **93**, 237401 (2004)
22. A. Nazir, Phys. Rev. B **78**, 153309 (2008)
23. I. Wilson-Rae, A. Imamoğlu, Phys. Rev. B **65**, 235311 (2002)
24. C. Roy, S. Hughes, Phys. Rev. Lett. **106**, 247403 (2011)
25. U. Hohenester, A. Laucht, M. Kaniber, N. Hauke, A. Neumann, A. Mohtashami, M. Seliger, M. Bichler, J.J. Finley, Phys. Rev. B **80**, 201311 (2009)
26. U. Hohenester, Phys. Rev. B **81**, 155303 (2010)
27. P. Kaer, T.R. Nielsen, P. Lodahl, A.-P. Jauho, J. Mørk, Phys. Rev. B **86**, 085302 (2012)
28. D.P.S. McCutcheon, N.S. Dattani, E.M. Gauger, B.W. Lovett, A. Nazir, Phys. Rev. B **84**, 081305(R) (2011)
29. A. Vagov, M.D. Croitoru, M. Glässl, V.M. Axt, T. Kuhn, Phys. Rev. B **83**, 094303 (2011)
30. E.M. Purcell, Phys. Rev. **69**, 681 (1946)
31. J.P. Reithmaier, G. Sk1, A. Löffler, C. Hofmann, S. Kuhn, S. Reitzenstein, L.V. Keldysh, V.D. Kulakovskii, T.L. Reinecke, A. Forchel, Nature **432**, 197 (2004)
32. S. Hughes, Opt. Lett. **29**, 2659 (2004)
33. V.S.C. Manga Rao, S. Hughes, Phys. Rev. B **75**, 205437 (2007)
34. A. Laucht, S. Pütz, T. Günthner, N. Hauke, R. Saive, S. Frédérick, M. Bichler, M.-C. Amann, A.W. Holleitner, M. Kaniber, J.J. Finley, Phys. Rev. X **2**, 011014 (2012)
35. A. Schwagmann, S. Kalliakos, I. Farrer, J.P. Griffiths, G.A.C. Jones, D.A. Ritchie, A.J. Shields, Appl. Phys. Lett. **99**, 261108 (2011)
36. T. Lund-Hansen, S. Stobbe, B. Julsgaard, H. Thyrrestrup, T. Sünner, M. Kamp, A. Forchel, P. Lodahl, Phys. Rev. Lett. **101**, 113903 (2008)
37. G. Lecamp, P. Lalanne, J.P. Hugonin, Phys. Rev. Lett. **99**, 023902 (2007)
38. D. Englund, E. Waks, G. Solomon, B. Zhang, T. Nakaoka, Y. Arakawa, Y. Yamamoto, J. Vučković, Phys. Rev. Lett. **95**, 013904 (2005)
39. F. Milde, A. Knorr, S. Hughes, Phys. Rev. B **78**, 035330 (2008)
40. Y. Ota, S. Iwamoto, N. Kumagai, Y. Arakawa, (2009). arXiv:0908.0788
41. A. Majumdar, E.D. Kim, Y. Gong, M. Bajcsy, J. Vučković, Phys. Rev. B **84**, 085309 (2011)
42. M. Calic, P. Gallo, M. Felici, K.A. Atlasov, B. Dwir, A. Rudra, G. Biasiol, L. Sorba, G. Tarel, V. Savona, E. Kapon, Phys. Rev. Lett. **106**, 227402 (2011)
43. S. Hughes, P. Yao, F. Milde, A. Knorr, D. Dalacu, K. Mnaymneh, V. Sazonova, P.J. Poole, G.C. Aers, J. Lapointe, R. Cheriton, R.L. Williams, Phys. Rev. B **83**, 165313 (2011)
44. G. Tarel, V. Savona, Phys. Rev. B **81**, 075305 (2010)

45. K.H. Madsen, P. Kaer, A. Kreiner-Møller, S. Stobbe, Phys. Rev. B **88**, 045316 (2013)
46. M. Florian, P. Gartner, C. Gies, F. Jahnke, New J. Phys. **15**, 035019 (2013)
47. K. Roy-Choudhury, S. Hughes, Optica **2**, 434 (2015)
48. A. Zrenner, E. Beham, S. Stufler, F. Findeis, M. Bichler, G. Abstreiter, Nature **418**, 612 (2002)
49. P. Michler, A. Kiraz, C. Becher, W.V. Schoenfeld, P.M. Petroff, L. Zhang, E. Hu, A. Imamoğlu, Science **290**, 2282 (2000)
50. Y.-J. Wei, Y.-M. He, M.-C. Chen, H. Yi-Nan, Y. He, W. Dian, C. Schneider, M. Kamp, S. Höfling, L. Chao-Yang, J.-W. Pan, Nano Lett. **14**, 6519 (2014)
51. Rong-Chun Ge, S. Weiler, A. Ulhaq, S.M. Ulrich, M. Jetter, P. Michler, S. Hughes, Opt. Lett. **10**, 1691 (2013)
52. U. Hohenester, G. Pfanner, M. Seliger, Phys. Rev. Lett. **99**, 047402 (2007)
53. F. Hargart, M. Müller, K. Roy-Choudhury, S.L. Portalupi, C. Schneider, S. Höfling, M. Kamp, S. Hughes, P. Michler, Phys. Rev. B **93**, 115308 (2016)
54. S. Scheel, L. Knöll, D.-G. Welsch, Phys. Rev. A **60**, 4094 (1999)
55. Rong-Chun Ge, C. Van Vlack, P. Yao, J.F. Young, S. Hughes, Phys. Rev. B **87**, 205425 (2013)
56. M. Bayer, A. Forchel, Phys. Rev. B **65**, 041308(R) (2002)
57. Ł. Dusanowski, A. Musiał, A. Maryński, P. Mrowiński, J. Andrzejewski, P. Machnikowski, J. Misiewicz, A. Somers, S. Höfling, J.P. Reithmaier, G. Sęk, Phys. Rev. B **90**, 125424 (2014)
58. C. Roy, S. Hughes, Phys. Rev. B **85**, 115309 (2012)
59. J. Seebeck, T.R. Nielsen, P. Gartner, F. Jahnke, Phys. Rev. B **71**, 125327 (2005)
60. E.A. Muljarov, R. Zimmermann, Phys. Rev. Lett. **98**, 187401 (2007)
61. J. Förstner, C. Weber, J. Danckwerts, A. Knorr, Physica status solidi (b) **238**(3), 419 (2003)
62. H.-P. Breuer, F. Petruccione, *The Theory of Open Quantum Systems* (Oxford University Press, Oxford, 2002)
63. H.J. Carmichael, *Statistical Methods in Quantum Optics 1: Master Equations and Fokker-Planck Equations* (Springer, Berlin, 2003)
64. A. Kowalewska-Kudłask, R. Tanaś, J. Mod. Opt. **48**, 347 (2001)
65. D.P.S. McCutcheon, A. Nazir, Phys. Rev. Lett. **110**, 217401 (2013)
66. R.C. Ge, C. Van Vlack, P. Yao, J.F. Young, S. Hughes, Phys. Rev. B **87**, 205425 (2013)
67. D. Valente, J. Suffczyński, T. Jakubczyk, A. Dousse, A. Lemaître, I. Sagnes, L. Lanco, P. Voisin, A. Aufféves, P. Senellart, Phys. Rev. B **89**, 041302(R) (2014)
68. C. Roy, S. John, Phys. Rev. A **81**, 023817 (2010)
69. A. Yariv, Y. Xu, R.K. Lee, A. Scherer, Opt. Lett. **24**, 711 (1999)
70. D.P. Fussell, M.M. Dignam, Phys. Rev. A **76**, 053801 (2007)
71. E. Kuramochi, M. Notomi, S. Mitsugi, A. Shinya, T. Tanabe, T. Watanabe, Appl. Phys. Lett. **88**, 041112 (2006)
72. P. Yao, P.K. Pathak, E. Illes, S. Hughes, S. Munch, S. Reitzenstein, P. Franeck, A. Löffer, T. Heindel, S. Höfling, L. Worschech, A. Forchel, Phys. Rev. B **81**, 033309 (2010)
73. S. Reitzenstein, C. Böckler, A. Loöffler, S. Hoöfling, L. Worschech, A. Forchel, P. Yao, S. Hughes, Phys. Rev. B **82**, 235313 (2010)
74. K. Roy-Choudhury, S. Hughes, Phys. Rev. B **92**, 205406 (2015)
75. S.M. Tan, J. Opt. B: Quantum Semiclass. Opt. **1**, 424 (1999)
76. K.J. Ahn, J. Förstner, A. Knorr, Phys. Rev. B **71**, 153309 (2005)
77. A. Ulhaq, S. Weiler, C. Roy, S.M. Ulrich, M. Jetter, S. Hughes, P. Michler, Opt. Express **21**, 4382 (2013)
78. K. Müller, K.A. Fischer, A. Rundquist, C. Dory, K.G. Lagoudakis, T. Sarmiento, Y.A. Kelaita, V. Borish, J. Vučković, Phys. Rev. X **5**, 031006 (2015)
79. J. Iles-Smith, A. Nazir, Optica **3**, 207 (2016)
80. K. Roy-Choudhury, N. Mann, R. Manson, S. Hughes, Phys. Rev. B **93**, 245421 (2015)
81. R. Manson, K. Roy-Choudhury, S. Hughes, Phys. Rev. B. **93**, 155423 (2016)
82. S. Hughes, H.J. Carmichael, New J. Phys. **15**, 053039 (2013)
83. A. Kiraz, M. Atatüre, A. Imamoğlu, Phys. Rev. A **69**, 032305 (2004)
84. N. Somaschi, V. Giesz, L. De Santis, J.C. Loredo, M.P. Almeida, G. Hornecker, S.L. Portalupi, T. Grange, C. Anton, J. Demory, C. Gomez, I. Sagnes, N.D. Lanzillotti-Kimura, A. Lemaıtre, A. Auffeves, A.G. White, L. Lanco, P. Senellart, Nat. Photon. **10**, 340 (2016)

85. K. Roy-Choudhury, S. Hughes, Opt. Lett. **40**, 1838 (2015)
86. C. Roy, S. Hughes, Phys. Rev. X **1**, 021009 (2011)
87. A. Muller, E.B. Flagg, P. Bianucci, X.Y. Wang, D.G. Deppe, W. Ma, J. Zhang, G.J. Salamo, M. Xiao, C.K. Shih, Phys. Rev. Lett. **99**, 187402 (2007)
88. E.B. Flagg, A. Muller, J.W. Robertson, S. Founta, D.G. Deppe, M. Xiao, W. Ma, G.J. Salamo, C.K. Shih, Nat. Phys. **5**, 203 (2009)
89. A. Nick Vamivakas, Y. Zhao, C.-Y. Lu, M. Atatüre, Nat. Phys. **5**, 198 (2009)
90. S. Ates, S.M. Ulrich, S. Reitzenstein, A. Löffler, A. Forchel, P. Michler, Phys. Rev. Lett. **103**, 167402 (2009)
91. C. Matthiesen, A.N. Vamivakas, M. Atatüre, Phys. Rev. Lett. **108**, 093602 (2012)
92. H.S. Nguyen, C. Voisin, P. Roussignol, C. Diedrichs, G. Cassabois, App. Phys. Lett. **99**, 261904 (2011)
93. A. Ulhaq, S. Weiler, S.M. Ulrich, R. Roßbach, M. Jetter, P. Michler, Nat. Photonics **6**, 238 (2012)
94. H. Kim, T.C. Shen, K. Roy-Choudhury, G.S. Solomon, E. Waks, Phys. Rev. Lett. **113**, 027403 (2014)
95. C. Roy, H. Kim, E. Waks, S. Hughes, Photon. Nanostruct. Fundam. Appl. **10**, 359 (2012)

Part II
Excitons and Single Photon Emission

Chapter 3
Resonantly Excited Quantum Dots: Superior Non-classical Light Sources for Quantum Information

Simone Luca Portalupi and Peter Michler

Abstract In this contribution, we briefly recall the fundamental optical and quantum optical properties of single photons and photon pairs, like coherence, purity, indistinguishability and entanglement, which are necessary to understand their huge potential for quantum information applications. We put special emphasis on resonant excitation schemes of excitons and biexcitons in semiconductor quantum dots since these provide photon wave packets with superior properties. This includes continuous-wave and pulsed excitation, rapid adiabatic passage, spin-flip Raman transitions, two-photon excitation and phonon-assisted excitation methods for excitons and biexcitons. We then review the recent progress on the generation of single and entangled photon states under these different resonant excitation schemes and discuss the pro and cons of the different methods.

3.1 Introduction

Since the first demonstration of single-photon emission for semiconductor quantum dots (QDs) in 2000 [1] a steadily increasing number of research activities has been initiated in the field of QD-based non-classical light generation. It is already foreseen that semiconductor quantum emitters will be key elements of modern optics based quantum technologies. For example, light sources which are able to deliver triggered single photons, entangled photon pairs or even photon cluster states are building blocks for several applications e.g. in quantum cryptography, quantum computing, quantum sensing and quantum metrology. Soon after the demonstration of a few other milestone achievements like electrical triggering of single-photon emission [2] and the generation of entangled photon pairs [3–5] with QDs, it became clear that most of the new quantum technologies require non-classical sources with superior properties. This includes high single-photon emission probabilities (brightness), low multi-photon probabilities (purity), identical photon wave packets

S.L. Portalupi (✉) · P. Michler
Institut für Halbleiteroptik und Funktionelle Grenzflächen,
University of Stuttgart, Stuttgart, Germany
e-mail: s.portalupi@ihfg.uni-stuttgart.de

P. Michler (ed.), *Quantum Dots for Quantum Information Technologies*,
Nano-Optics and Nanophotonics, DOI 10.1007/978-3-319-56378-7_3

(photon indistinguishability) i.e. ultimately Fourier transform-limited photons. For example, it has been recently shown that linear optical quantum computing could work with indistinguishable photons if the product of the detector efficiency and the single-photon emission probability is larger than 2/3 [6]. A detailed comparison of the source requirements for various applications can be found in [7].

The photon wave packet properties critically depend on the type of optical or electrical excitation of the charge carriers inside the QD. One has to distinguish between *non-resonant* and *resonant* excitation schemes. Early work in this research field has exclusively focused on non-resonant pumping schemes, i.e. charge carriers have been generated in the QD barriers, wetting layer or higher electronic shells (p-, d-shells) of the QD due to the relative simplicity of this method. This excitation type leads to the generation of additional charge carriers in the vicinity of the QD which act via carrier-carrier scattering processes as sources of decoherence for the subsequently emitted photons. Moreover, non-resonantly excited carriers have to relax to the lowest electronic shell (s-shell) before they can recombine radiatively under emission of the desired photon. This non-radiative relaxation process introduces some uncertainty into the photon emission process which eventually introduces some time jitter and limits the photon indistinguishability in a two-photon interference process.

However, it has been soon realized that truly *resonant* excitation techniques would produce photon wave packets with superior properties, e.g. long coherence times and high indistinguishabilities. Resonant excitation techniques allow the generation of the electron-hole pairs, i.e. excitons (one bound electron-hole pair) and biexcitons (two electron-hole pairs) directly in the s-shell of the QD, thereby mostly avoiding the generation of unwanted extra charge carriers in the vicinity of the QD. As a consequence, nearly ideal photons are emitted which are perfectly suited for quantum information applications.

First efforts of resonant QD excitation were not successful due to large laser background since the emitted desired photon and the excitation laser possess the same wavelength. In 2007, Müller and co-workers [8] solved the problem by introducing a newly developed side-excitation technique where the emitted photons are collected in orthogonal geometry from top of the sample. A few years later, Vamivakas and coworkers have shown that even top excitation and collection is possible by utilizing a cross-polarized setting in a confocal microscope [9].

This chapter reviews the recent achievements of the resonantly excited QDs and discusses the pros and cons of the different resonant excitation methods. Section 3.2 gives a short introduction into important photon properties and the different resonant excitation methods. Section 3.3 reviews in details the various methods for the resonant pumping of the exciton and biexciton. The last section reports phonon-assisted approaches for the excitation of excitons and biexcitons.

3.2 Fundamental Optical Properties of Photons: Coherence, Purity, Indistinguishability and Entanglement

In order to be useful for quantum information, computation or cryptography applications, the light emitted by the source must fulfill various specific requirements. To clarify this statement, let us consider the sketch in Fig. 3.1.

The first row (a) depicts a stream of single, identical photons from an ideal light source; in comparison, the other rows show what happen when one of the properties, which make the source ideal, is lacking. In the case of an ideal source, the emission efficiency is equal to 1. This means that one excitation pulse will trigger the emission: this property is often referred as brightness and in the ideal case it is equal to 100%. An efficiency smaller than 1 will result in a photon stream with some "missing" photons with respect to the ideal example as sketched in the row (b) of Fig. 3.1. It becomes clear that a reduced brightness will affect the computational complexity. For quantum cryptography applications, it is fundamental that the photon stream carrying the information is composed by single photons in order to improve the security of the network. This signifies that in the wave packet only one photon must be present. This property is usually quantified via a Hanbury-Brown and Twiss measurement that defines the values of the second-order correlation function: for a pure single-photon source it gets $g^{(2)}(0) = 0$ [10]. When the second-order correlation function differs

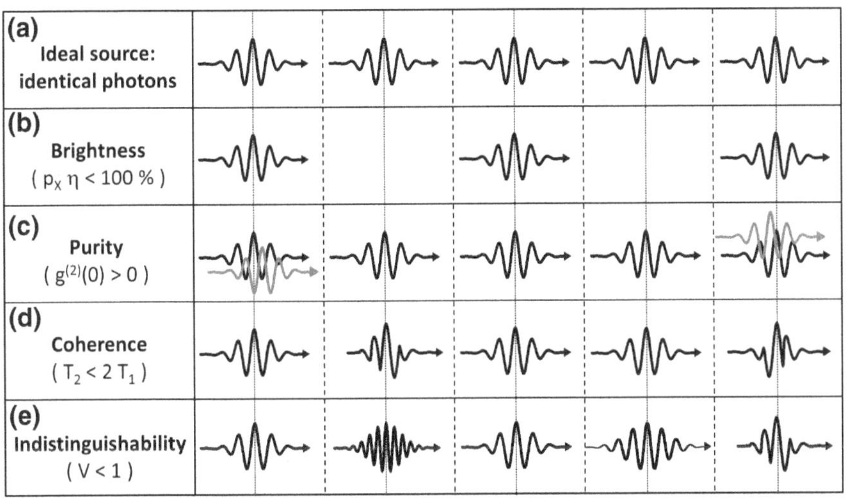

Fig. 3.1 **a** Ideal source: a stream of single, identical photons are generated per excitation laser pulse. **b** Example of source with brightness smaller than 1. Only at some laser pulses correspond a photon emission. **c** Case of non-zero single-photon purity. In some cases, two photons are present. **d** Loss of photon coherence schematized as *broken lines*. **e** Non-ideal indistinguishability: the photon wave packet changes due to phase variations or spectral wandering

from zero ($g^{(2)}(0) > 0$) more photons can be present in the time slot resulting in a modified stream of photons as in row (c). Similarly to classical light, also for single photons the coherence can be ascribed as a measurement of the phase stability of the light. The photon linewidth can be written as:

$$\gamma = \frac{1}{T_2} = \frac{1}{2T_1} + \frac{1}{T_2^*}, \tag{3.1}$$

where T_1 indicates the transition radiative lifetime and T_2^* corresponds to the pure dephasing time (i.e. a loss of coherence without recombination). The lifetime T_1 and the coherence time T_2 characterize the photon wave packet. Since the electron-hole pair is subject to interaction with the surrounding, this may lead to a loss of coherence during the radiative emission process. This case is depicted in row (d) where the broken lines indicate a photon phase change. In the case that the environmental fluctuations are faster than the radiative transition, the pure dephasing term T_2^* comes into play. Only when decoherence mechanisms are absent (or anyhow negligible) the (3.1) becomes:

$$\frac{T_2}{2T_1} = 1. \tag{3.2}$$

The aforementioned condition is usually referred as Fourier transform-limit. This property is one fundamental requirement in order to have highly indistinguishable photons. To perform low error quantum computation, the utilized photons need to be identical, in other words have an indistinguishability $V = 100\%$. When two photons are indistinguishable, that means they are identical in terms of spectral bandwidth, pulse width, polarization, carrier frequency, mode profile and they arrive at the same time at the two input ports of a beamsplitter (BS). Quantum mechanics then predicts that only the output configurations with the two photons leaving the BS at the same side are possible. This photon coalescence arising from the two-photon interference on a beamsplitter was first measured by Hong-Ou-Mandel [11] and constitutes the basis for low error-rate quantum computation. It is then clear that in presence of pure dephasing, the non-perfect Fourier transform wave packet will strongly affect the indistinguishability. Together with that, environmental fluctuations that are slower than the transition radiative lifetime will produce a variation of wavelength of the emitted photons. This spectral wandering mechanism may also limit the measurable indistinguishability, since the photons arriving on a BS can have different wavelengths and then being distinguishable. Pure dephasing and spectral wandering are depicted in Fig. 3.1e as limiting factors of the photon indistinguishability. The Hong-Ou-Mandel (HOM) effect is often measured by feeding an unbalanced Mach-Zehnder interferometer with two timely separated photons that will then interfere on a beam-splitter. The measured correlation times, recorded on two detectors placed at the BS outputs, are then used to estimate the HOM visibility [10]. The large majority of the experiments described in this chapter made use of this experimental configuration to determine the photon indistinguishability. It is important at this stage to underline that spectral wandering (or spectral diffusion) can set strong limitation for

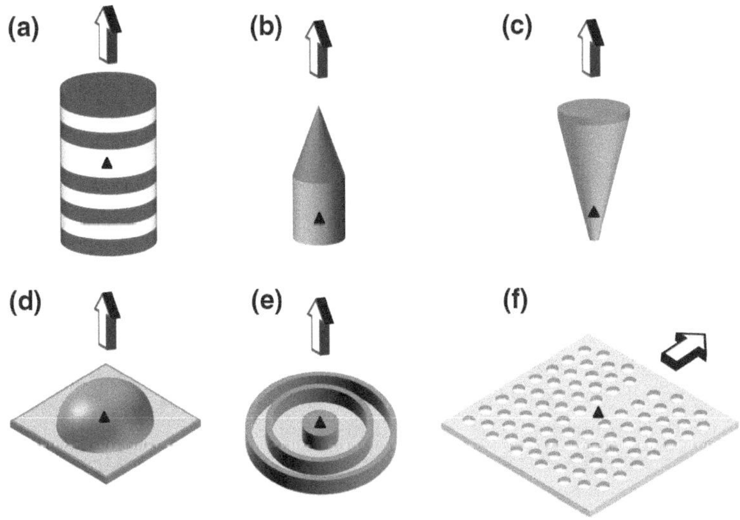

Fig. 3.2 Different types of cavities or photonic structures. **a** Micropillar cavity. **b** Photonic nanowire. **c** Photonic trumpet. **d** Microlens. **e** Bull's eye cavity. **f** Photonic crystal waveguide. The *arrows* indicate the main emission direction

two-photon interference in case the photons are coming from two different, separate sources, since in this case, the environmental fluctuations are independent for the two sources. Several techniques have been employed in order to stabilize the QD environment: from applying an electric field, to the use of two-color pumping scheme [12] or by controlling the nuclear spins [13]. In a recent paper [14], theoretical limits of the second-order correlation function $g^{(2)}(0)$ as well as the photon indistinguishability have been discussed.

It is worth mentioning that the source brightness for semiconductor QDs is generally limited by the strong refractive index mismatch between the host material (~3.4 for GaAs) and the air surrounding it: the light is mostly confined in the semiconductor and only a brightness of few percent can be achieved. This value can anyway be easily improved up to 10–20% by simply sandwiching the QD layer between two distributed Bragg reflectors (DBRs) to form a $1D$ photonic crystal cavity (generally called planar cavity) [15]. This value can be further increased by use of photonic cavities or engineered structures to enhance the light extraction. In Fig. 3.2 some examples between the most developed geometries are shown. In Fig. 3.2a, e two cavities based on photonic crystals are sketched, namely micropillars (a) and circular Bragg gratings (e), sometimes referred to as bull's eye cavity: these structures make use of the photonic crystal bandgap together with total internal reflection (TIR) to strongly confine the light in the cavity region. An engineering of the material surrounding gives rise to photonic nanowires (b), trumpets (c) or lenses ((d), see Chap. 6) with various geometries. In more detail, the brightness can be defined as the product

$$Brightness = p \times \eta, \tag{3.3}$$

where p is state occupation factor while the photon extraction efficiency η is given by:

$$\eta = \eta_{out} \times \beta = \eta_{out} \times \frac{\Gamma_M}{\Gamma_M + \Gamma_{Rad}} \tag{3.4}$$

with η_{out} the out-coupling coefficient (photons through defined loss channels over total photon emission) and β the spontaneous emission of the quantum emitter in the optical mode under investigation (with Γ_m and Γ_{Rad} the spontaneous emission rate in the mode and in the radiation modes respectively) [16]. The probability p is in some cases limited by the presence of charged exciton states (trions) or dot blinking. Some techniques have been anyhow implemented to maximize p, like applying an electric field or using combination of lasers with different wavelength [12, 17, 18]. From (3.4) it can be seen that two strategies are possible to enhance the β factor: one is by suppressing the spontaneous emission contribution to radiation modes, as employed by photonic nanowires and photonic crystal waveguides. The other is by increasing the spontaneous emission rate Γ_M by means of cavity quantum electrodynamics (CQED): in the weak coupling regime, the transition radiative lifetime can indeed be shortened via the Purcell effect [10]. State-of-the-art values for different systems are here reported: for micropillars (Fig. 3.2a) a brightness around 0.8 was reached [19, 20]. For nanowires (Fig. 3.2b) and photonic trumpets (Fig. 3.2b) values of around 0.72 [21] and 0.75 [22] were reported. For microlenses (Fig. 3.2d) an extraction efficiency as high as 0.23 was measured [23], while for circular Bragg gratings a collection efficiency of 0.48 was obtained (brightness was not included in the last cited papers). Photonic crystal waveguides (Fig. 3.2f) achieved very high β factors [24, 25], showing that photons were emitted into a specific mode up to $\beta = 0.98$ [26].

The last property summarized in here is the photon entanglement, theorized first in 1935 [27]. Two particles are defined as entangled if their wave function cannot be factorized in terms of product of the wave functions of the two individual particles. This implies that one measurement on one particle determines the quantum state of the second particle. Several properties are related to the degree of entanglement such as the negativity [28], concurrence [29, 30], tangle [31] and fidelity. The latter, gives a measurement on how the received signal compares with the emitted one, after transmitting into a quantum channel [32]. In formulas it writes as $F = \langle\psi|W|\psi\rangle$ where ψ is a pure state and W is the density matrix that represents the pure state emerging as mixed after transmission in the quantum channel. Despite the strong progresses achieved in the last years [33–36], the properties of the emitted photons are still limited by the mostly-used non-resonant pumping schemes: this motivates the search for excitation schemes to improve the photon properties.

3.3 Resonant QD Excitation Methods for Exciton and Biexciton

3.3.1 Sources of Decoherence

From a spectroscopic point of view, in order to generate electrons and holes that can be trapped into the quantum dot for successively recombine radiatively, several excitation wavelengths can be used. The different pumping regimes can be understood considering the band structure of a semiconductor quantum dot: as a matter of example, we consider in the following In(Ga)As QDs grown in Stranski-Krastanov mode [37]. The band structure will schematically look like as depicted in Fig. 3.3. The small size of the QD and the lower band gap allow the formation of quantized levels, here showing p- and s-shells only. In red it is displayed the host material, i.e. the GaAs, while in blue the so-called wetting layer (WL): this comes from the growth method and constitutes a very thin layer on which the QD is consecutively formed. It has a band gap that differs from the GaAs as well as from the discrete energy levels of the QD. The electron and holes that are created because of the pumping can diffuse and get trapped into the QD, relaxing to the lowest energetic state called s-shell. When confined into the QD, they will Coulomb interact and they can form a new quasi-particle called *exciton*. The exciton can radiatively recombine with a sub-nanosecond lifetime (i.e. the probability of the spontaneous emission). It is important to mention that different states can be identified accordingly to the particle spins. Describing the state in terms of angular momentum $|J, J_z\rangle$ results in

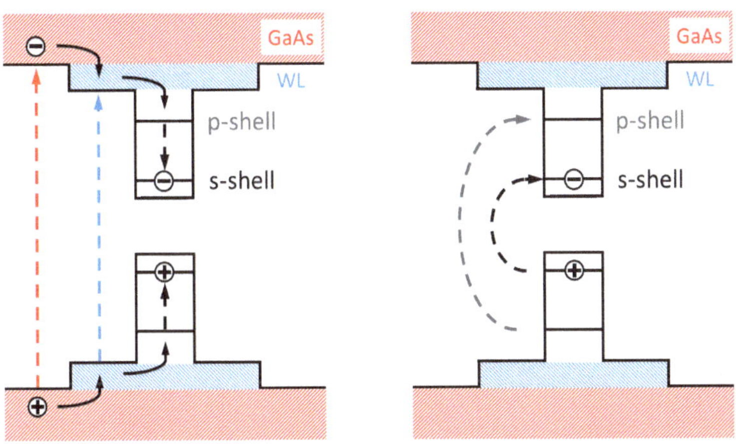

Fig. 3.3 Sketch of the band structure of a QD (i.e. In(Ga)As) hosted in GaAs. (*Left*) Non-resonant excitation schemes are depicted, i.e. above band gap or wetting layer excitation. In the present case the charges created can be trapped into the QD. (*Right*) Quasi-resonant and resonant excitation scheme. The electron and hole pair that will form the exciton are created by resonantly address the p- or s-shell respectively

the following spin states for the electron and heavy holes which are typically the energetically lowest levels in strained QDs:

$$|e,\uparrow\rangle = \left|\frac{1}{2},+\frac{1}{2}\right\rangle;\ |e,\downarrow\rangle = \left|\frac{1}{2},-\frac{1}{2}\right\rangle$$
$$|h,\Uparrow\rangle = \left|\frac{3}{2},+\frac{3}{2}\right\rangle;\ |h,\Downarrow\rangle = \left|\frac{3}{2},-\frac{3}{2}\right\rangle \tag{3.5}$$

In the total angular momentum basis, the exciton state can be then described as:

$$|e\uparrow, h\Downarrow\rangle \text{ and } |e\downarrow, h\Uparrow\rangle$$
$$|e\uparrow, h\Uparrow\rangle \text{ and } |e\downarrow, h\Downarrow\rangle \tag{3.6}$$

where only the states in the first row with total angular momentum of ± 1 are coupled to light, then referred as "bright excitons". The other states, with total angular momentum of ± 2 are on the contrary named "dark" states (see Chap. 4). When two electron-hole pairs are present into the QD in the lowest energetic state, a *biexciton* is formed. It is interesting to notice that the total spin of this particle is zero. In case of simultaneous presence of one exciton plus an additional charge, the so-called trion is formed: it is named as negatively charged trion if the additional charge is an electron or vice versa positively charged trion in presence of an excess hole. The exchange interaction is responsible for the fine-structure splitting of the QD, observed as a polarization splitting of the spin degenerate bright excitons. Such polarization splitting is indeed not observed in presence of a trion, since the remaining particle (having the two electrons or holes a local spin density of zero) cannot experience any exchange interaction [38]. Due to selection rules, only recombination of particles between the same shells are allowed (i.e. $s \rightarrow s$ or $p \rightarrow p$). Several mechanisms, dependent on the excitation energy, can be used in order to feed the QD with charges. If the charges are created into the GaAs layer (Fig. 3.3 (left)), they can relax into the QD and the excess energy is exchanged with the solid state environment. Similarly, it is also possible to directly address the wetting layer, resulting in a similar effect. These non-resonant excitation schemes have the advantage to be easy to implement; from the spectroscopic point of view, the laser is spectrally well separated from the emitted photons, being then relatively easy to filter. The disadvantage of these pumping schemes relates to the fact that the non-radiative recombination from excited states to the s-shell introduces time jitter in the emission, one of the factor limiting the photon indistinguishability. In addition to that, multi-exciton state formation also further limits the coherence of the emitted photons [39]. In order to diminish the influence of the environment, so decreasing the number of relaxations that the charges need to reach the lowest energy level, resonant schemes can be used. An intermediate step between non-resonant and resonant schemes is the so called p-shell pumping (or quasi-resonant pumping). In this case, the laser is tuned in resonance with a higher energy level of the QD and the charges have only one relaxation step for reaching the s-shell (see Fig. 3.3 (right)). This method permits to improve the emitted

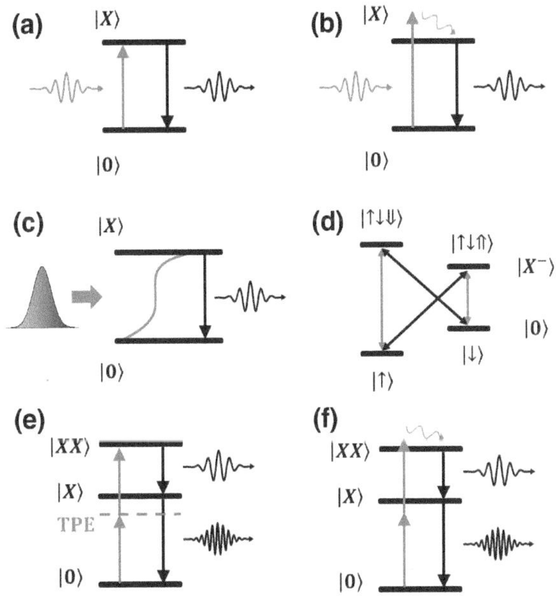

Fig. 3.4 Schemes for resonant X or XX excitation. **a** Resonant exciton pumping (in CW or at the π pulse). **b** Phonon-mediated X preparation. **c** Adiabatic rapid passage scheme. **d** Spin-flip Raman transition. **e** Resonant two-photon biexciton excitation. **f** Phonon-assisted two-photon XX excitation

photon coherence without renouncing to the excitation-emission spectral separation. This method allowed reaching reasonably high photon coherence [19, 40] but it still cannot compete with a purely resonant excitation scheme of the s-shell.

Pumping the QD resonantly, i.e. tuning the laser in resonance with the s-shell of the dot, would then allow populating the quantum dot without interacting with the host material: this would lead to a decrease of the detrimental effect of pumping, while populating directly the two-level system with the pump laser. There are several different methods to resonantly excite the QD s-shell that will be described along this chapter and they are pictorially depicted in Fig. 3.4. The exciton level can be populated by addressing the exciton-to-ground state transition resonantly with a CW laser or a single laser pulse (Fig. 3.4a). The radiative recombination will result in emission of one single photon with superior properties (Sects. 3.3.2 and 3.3.3). Phonon-mediated processes are also permitted as schematized in Fig. 3.4b, where one laser photon (blue detuned from the exciton transition) together with the emission of a phonon brings the system in the exciton state (see Sect. 3.4.1). Chirped pulses constitute the basis for the adiabatic rapid passage scheme (Fig. 3.4c) further described in Sect. 3.3.4. An alternative way to generate tunable single-photons is based on spin-flip Raman transitions that will be described in Sect. 3.3.5 (Fig. 3.4d). Due to optical selection rules, the biexciton transition cannot be directly resonantly addressed but a two-photon process, via a virtual state, can allow the direct excitation of the biexciton state, as detailed in Sect. 3.3.6. Similarly to one phonon-mediated exciton population, also the XX can be populated using a two-photon process mediated by phonon emission (Fig. 3.4f and Sect. 3.4.2).

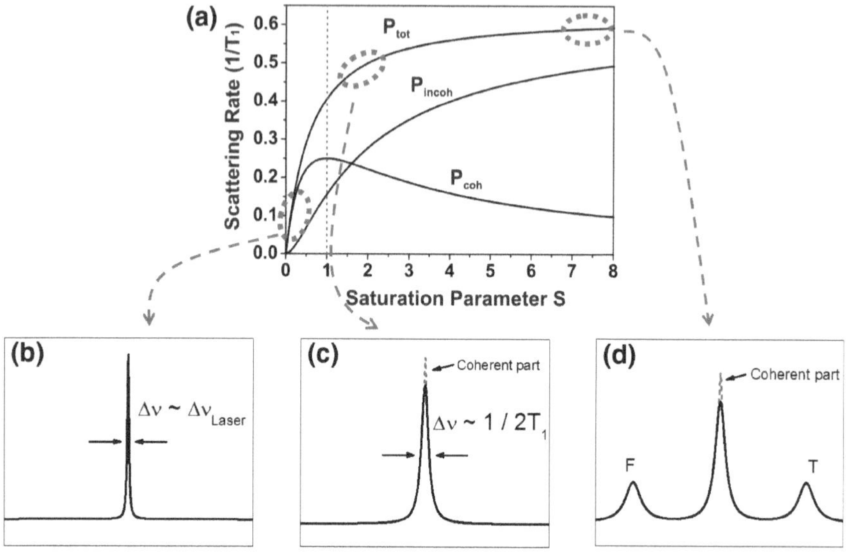

Fig. 3.5 **a** Emitted power versus the saturation parameter: total power as sum of the coherently and incoherently scattered light is displayed. **b** Low power regime: the emitted photon linewidth is mainly given by the laser linewidth. **c** Medium power regime: the emitted photon linewidth is close to Fourier transform limit plus the coherently scattered part, sketched in *gray*. **d** Under high pumping, the dressing of the states gives rise to the Mollow-triplet spectrum, where also the coherently scattered components is shown in *gray*

As it will be explained in details in the next section, the resonant pumping under CW excitation results in very different regimes according to the pump power. The total emitted power P_S can then be represented as sum of coherently and incoherently scattered parts, as schematized in Fig. 3.5. In order to set the most general definition, the scattering rate in Fig. 3.5a is defined as a function of the saturation parameter S. This is given by the formula $S = (\Omega^2/(\gamma\Gamma))/(1 + \Delta^2/\gamma^2)$, where Ω is the Rabi frequency, Δ the laser-to-QD detuning, $\gamma = 1/T_2$ as in (3.1) and Γ the radiative linewidth (i.e. $\Gamma = 1/T_1$). In the case of low pumping power (i.e. for the case where the saturation parameter $S \ll 1$), the emission linewidth displays a FWHM mostly set by the excitation laser linewidth (see Fig. 3.5b). These emitted photons consequently showed high coherence values. Fourier transform-limited photons can be obtained while exciting at medium powers ($S \sim 1$): the FWHM can reach values around $1/2T_1$. In Fig. 3.5c the coherently scattered light component is also shown in gray, superimposed to the resonantly generated photons. At high pumping powers ($S \gg 1$), the two-level system becomes "dressed" and the radiative recombination between adjacent multiplets is responsible for the spectrum displayed in Fig. 3.5d. A detailed explanation will be given in the following sections. Similarly, by using a resonant pulsed laser excitation, Fourier transform-limited photons are also achievable. In the case of pulsed operation, the coherent control of the two-level system reflects in the observation of oscillations in the detected intensity: they are called Rabi oscillations

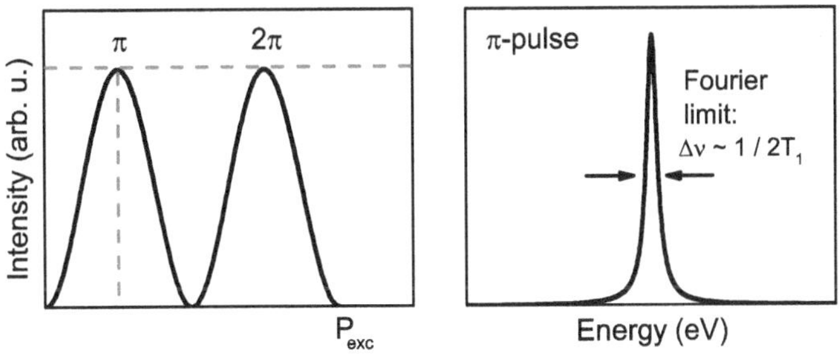

Fig. 3.6 (*Left*) Emitted intensity behavior versus pulses pump power: Rabi oscillations are sketched with particular attention for the π-pulse conditions. (*Right*) Spectral profile of an ideal Fourier transform-limited photon at the π-pulse

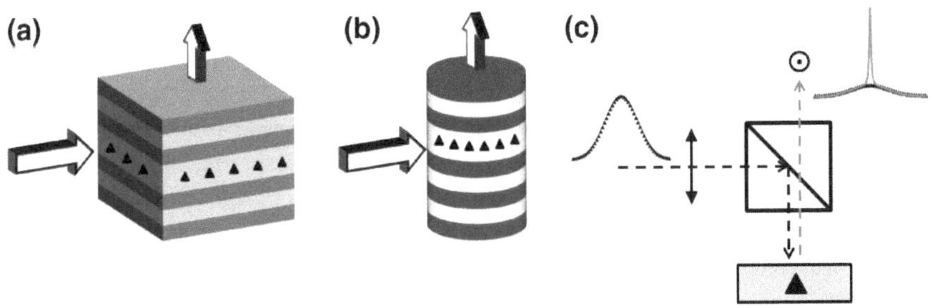

Fig. 3.7 **a** Sketch of the side excitation pumping scheme: when using a planar microcavity, the 1D photonic crystal helps in guiding the laser light. The emitted photons are then collected from the *top*. **b** Same scheme as in **a** but implemented with a micropillar cavity type. **c** Collinear *top* excitation based on cross-polarized detection. One polarizer is placed in the excitation arm (laser pulse shown by black symbols) while a cross-polarized analyzer is used in the detection arm (RF photons plotted in *gray*). The selected QD into solid GaAs matrix is symbolized by the *black triangle* in the *gray square*

and, in the ideal case, represent the different population of the two-level system. In particular, at the π-pulse, a full population inversion is achieved (Fig. 3.6), which is the basis for the on-demand generation of single photons.

Despite the advantage of decreasing the pump-induced decoherence, the resonant s-shell pumping carries an intrinsic experimental challenge, i.e. the discrimination between the photon emitted from the QD (resonance fluorescence, RF) and the laser photons which have the same wavelength. One effective scheme to perform that is based on the geometrically orthogonal configuration of excitation and detection (see Fig. 3.7a, b) and it has been first successfully utilized in [8]. It demonstrated to be very effective for planar photonic crystals, where the two distributed Bragg reflectors help in guiding the laser beam (see sketch in Fig. 3.7a). The same scheme can also be implemented for micropillars (Fig. 3.7b) with the only limitation that the investigated cavity needs to be accessible to side excitation (i.e. located near the sample edge). The orthogonal orientation between excitation and collection "channels" strongly helps in reducing the collected laser light, making the discrimination between RF photons

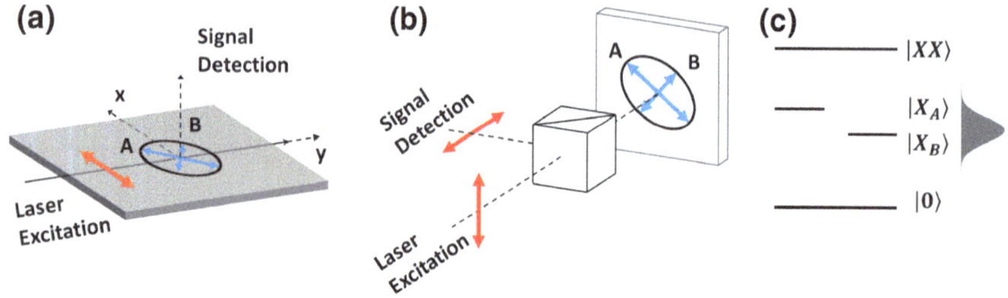

Fig. 3.8 **a** Scheme of the side excitation pumping scheme. **b** Sketch of a collinear crossed-polarized setup: excitation and detection linear polarizers are indeed placed respectively at 90°. An elliptical QD is also drawn, 45° oriented with respect to excitation and detection. **c** Pictorial example of the XX, X (with non-zero FSS) and ground states: the pulse stands for indicating that for a correct operation the laser must contemporary excite both exciton FSS components

and pump quite effective. An alternative pumping scheme that can be implemented also for samples not accessible from the side, is the collinear top pumping, based on cross-polarized detection. In this configuration, the discrimination between laser photons and QD emission is made considering the different light polarization. It is schematized in Fig. 3.7c. One polarizer is placed in the excitation arm and results in a linearly polarized laser beam. Along the detection arm, a second polarizer (sometimes referred as analyzer) is placed and oriented in cross-polarization with respect to the first one. In this way, the laser light, that upon reflection on the sample mainly maintains the original polarization, is fully suppressed by the analyzer (Fig. 3.7c). For this type of experiments, a high polarization suppression needs to be achieved, generally in the range of 10^5 or 10^6. A linearly polarized laser beam is sent through a beamsplitter on the sample. Currently this method has been intensively used for QD resonant pumping but it intrinsically limits the number of collected photons to 50%. Despite the collinear method is very effective, as it will be explained in Sect. 3.3.3, the side excitation shows a benefit for the overall source brightness, since no polarization selection is theoretically required in the light collection. In both excitation schemes, i.e. side or top pumping, it is necessary to consider the reciprocal orientation of the QD and the incident laser field. In Fig. 3.8 the two different experimental configurations are depicted. The QD is considered as elliptical, in the most general case, meaning that it can be considered as sum of two different dipoles (named A and B in the figure). The incoming laser can then excite one or the superposition of both dipoles accordingly to its orientation. In the case of side pumping (Fig. 3.8a) the laser orientation must be in the same plane of the QD in order to efficiently excite it and the dipole superposition is set by the specific orientation of the dot with respect to the sample side. On the contrary, in top excitation (Fig. 3.8b) the laser orientation with respect to the QD can be adjusted, then releasing the limit set by the emitter orientation. When working with a neutral exciton and in presence of a non-zero fine structure splitting (FSS), the mutual orientation of the laser beam and the QD axes has to be taken into account. Indeed, if the laser polarization is oriented along one specific QD axis, only one dipole component will be excited and the QD will emit linearly polarized photons along the excited dipole direction. Being parallel to the excitation laser beam, the QD photons will be fully suppressed by the analyzer. For this reason the laser beam must excite both dipole moments, and also the laser pulse must be energetically in resonance with both FSS components. As a consequence, a superposition state is generated and light with timely alternating polarizations is emitted. The oscillation period is determined by the FSS. (Fig. 3.8c). If using a charged exciton, less care is required, but in all circumstances this excitation mechanism intrinsically limits the source brightness to 50%, while the side excitation previously described does not suffer from this limitation.

3.3.2 CW-pumping Methods: Low, Medium and High Excitation Powers

The first section regarding the resonant pumping deals with continuous wave (CW) resonant excitation. The low power regime (in CW pumping) has been used to generate ultra-coherent single photons [41]. When the condition given by the (3.2) is fulfilled, the photon linewidth of the incoherent part will correspond to the lower limit given by: $2\hbar/T_2 = \hbar/T_1$. The resonant excitation signal is composed by two terms, namely the resonant Rayleigh scattering (RRS) and the incoherent resonant photoluminescence (RPL). While exciting the QD with low pumping power, the emission is mainly dominated by the coherent Rayleigh scattering. It is interesting to notice that in this regime, while the linewidth is mainly given by the classical (Poissonian) resonant laser light, the emission statistics result to be non-classical, with values of the correlation function $g^{(2)}(0) \approx 0$. In [41] the authors used Fourier transform spectroscopy to demonstrate that the radiated electric field has the classical character of the incident resonant laser beam. The three spectra in Fig. 3.9a–c were taken with three different powers, respectively $\Omega^2 = 155\,\text{ns}^{-2}$, $\Omega^2 = 64.6\,\text{ns}^{-2}$ and $\Omega^2 = 11.6\,\text{ns}^{-2}$ and allowed to distinguish between different regimes in which the coherent or incoherent laser scattering are dominating the overall emission spectrum. The inverse Fourier transform of the theoretically estimated first-order coherence function (solid curves in Fig. 3.9a–c) allowed a direct comparison of the resonant excitation signals in the different regimes. Figure 3.9d–f show that the coherent component increases lowering the excitation power and its linewidth is well below the radiative limit (displayed as the shaded region). The ratio between the two signals is given by:

$$\frac{I_{RRS}}{I_{RRS} + I_{RPL}} = \frac{T_2}{2T_1} \times \frac{1}{1 + \Omega^2 T_1 T_2} \tag{3.7}$$

Based on this equation, when pure dephasing comes into play, the coherence time becomes $T_2 < 2T_1$ meaning that the highest fraction of coherent resonant Rayleigh scattering component is given by $T_2/2T_1$. In [41], considering that $T_2 = 1.3T_1$, the ratio between RRS over the total emission could not exceed 65%. The interesting aspect is that for an excitation power of $\Omega^2 = 2\,\text{ns}^{-2}$ such a ratio was equal to 50%

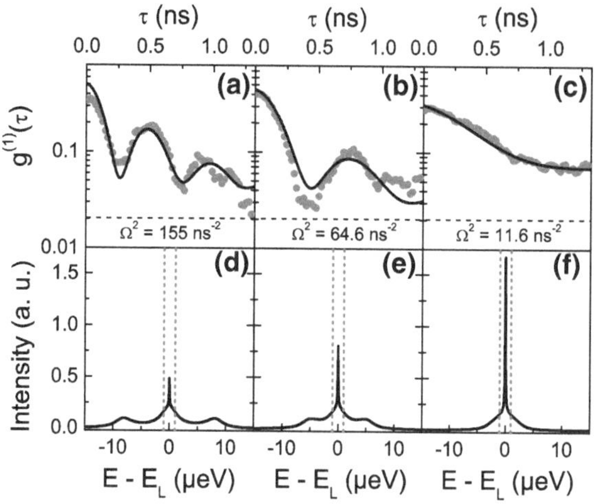

Fig. 3.9 **a**, **b** and **c** High resolution measurements made by Fourier transform spectroscopy of one exciton line under resonant pumping (labeled as B in [41]) for decreasing powers ($\Omega^2 = 155\,\text{ns}^{-2}$, $\Omega^2 = 64.6\,\text{ns}^{-2}$ and $\Omega^2 = 11.6\,\text{ns}^{-2}$ respectively). The *solid lines* correspond to fit to the first order correlation functions with $T_1 = 320\,\text{ps}$, $T_2 = 410\,\text{ps}$ and the coherence time of the laser $T_L = 10\,\text{ns}$. The *dashed lines* mark the background levels. **d**, **e** and **f** Calculated resonant emission spectra corresponding to **a**, **b** and **c**. These were extracted from the inverse Fourier transform of the first-order correlation function. The *dotted* regions indicate the radiative limit (width $= 2\,\mu\text{eV}$) [41]

and the measured second-order correlation function resulted to be $g^{(2)}(0) = 0.3$, only limited by the setup time response. This proved that, even being the resonantly emitted photon linewidth entirely determined by the excitation laser, the light presented a single-photon nature. This motivated the authors to refer to such single-photons with a sub-radiative limit linewidth as "ultra-coherent". Recent results demonstrated that using micropillar cavities, it was possible to obtain ratios $T_2/2T_1 \cong 1.0$ with a fraction of RSS photons over total close to unity [42].

In this low excitation level regime, also referred as Heitler regime, the two processes of photon absorption and emission become one coherent single event [43]. This leads to another fundamental property of the coherently generated single-photons: the mutual coherence between excitation laser and QD photons can exceed timescales of 3 s. This allowed the use of engineered modulated laser beams to produce single-photons with arbitrary synthesized waveforms, which displayed a degree of indistinguishability around 100% [44]. Furthermore, these single photons were generated taking advantage from the coherent nature of the elastically scattered photons in the Heitler regime. The great advantage of this method relates on the possibility to manipulate the single-photon waveform at the generation process without the loss of photons connected to the use of spectral filtering. In [44] an electro-optic modulator (EOM) driven by a variable amplitude (200 MHz) radio-frequency source was used to encode the excitation laser field. Two examples are reported in Fig. 3.10 with the corresponding laser spectrum in (a) and (c). The more complex

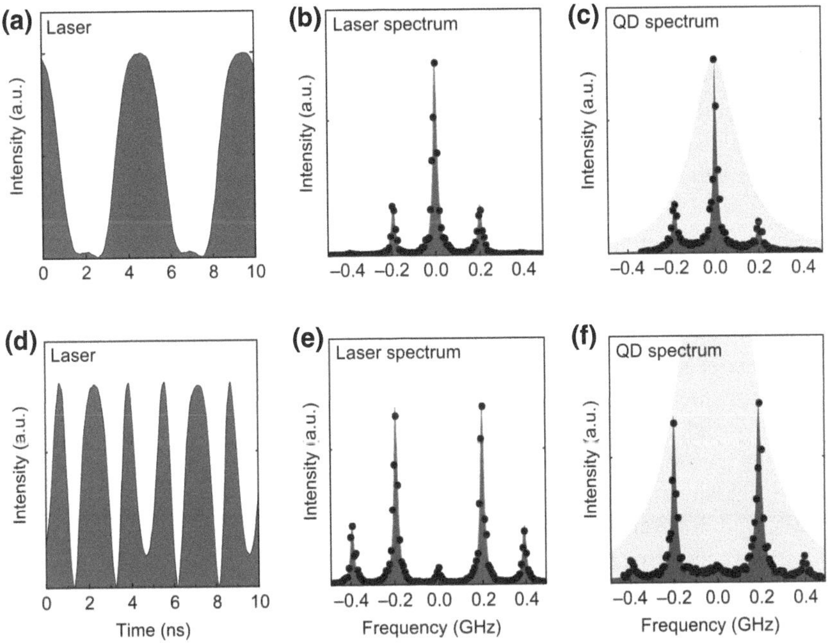

Fig. 3.10 **a** Excitation laser intensity over time, modulated via an electro-optic modulator (200 MHz sine wave driving). **b** High resolution spectrum (measured with a Fabry-Pérot cavity of 20 MHz resolution) of the modulated laser. **c**, **d** Data as in **a** and **b** but with different driving of the electro-optic modulator. The waveform was designed in order to lead to the suppression of the spectral component corresponding to the original carrier frequency. **e**, **f** QD emission spectra corresponding to the synthesized waveforms shown in **a** and **c**. The *shaded areas* represent the transition linewidth inside which the photons that are elastically scattered replicated the driving laser spectra [44]

encoding in Fig. 3.10c resulted in the suppression of the laser spectral component at zero detuning (see Fig. 3.10d), i.e. at the carrier frequency. These two laser spectra produced the emitted QD spectra in Fig. 3.10e, f: the distance and relative strength of the spectral components were imposed by the encoded laser beam, being in addition weighted by the QD transition linewidth (in [44] 650 ps and depicted by the shaded region in Fig. 3.10e, f). The presented technique allows to deterministically synthesize single-photons waveforms, taking advantage of the mutual coherence between QD photons and laser beam: the single-photon nature of the emission is totally preserved and phase encoding is also possible, without any photon loss. In addition to that the authors demonstrated that these single-photons can exhibit a degree of indistinguishability close to 100% (a conservative value obtained performing only the corrections arising from an imperfect polarization control gave a visibility equal to $V_{HOM} = 0.96 \pm 0.04$).

Using optical heterodyning measurements the mutual coherence of the QD photons and the excitation laser was measured. This measurement technique is based on the interference between the resonant fluorescence photons coming from the QD and a strong local oscillator originating from the excitation laser but shifted in frequency: in the case of [44] the QD emitted at 315.315 GHz (951 nm) and the laser was shifted by $\delta\nu \approx 210$ kHz. Using a fast Fourier transform algorithm the spectrum analyzer could calculate the power spectrum. The measured linewidth and lineshape directly reflected the phase stability, meaning the mutual coherence between the QD photons and the excitation laser beam. The observed Gaussian profile had a FWHM of 299 ± 9 MHz, only limited by the system response resolution (200 MHz). Despite the setup resolution, the observed linewidth demonstrated that the emitted QD photons were phased-locked with the excitation laser beam, with mutual coherence to be more that 3 s. It is worth noting that in the referenced paper, the photon emission was not limited by spectral wandering, since a measured broadening of the linewidth became noticeable only on a second timescale.

These results opened an interesting perspective in the field of quantum information, since the emitted photons with arbitrary synthesized waveforms are ideally suited for quantum interference applications. On the other hand, the price to pay for this fundamentally indistinguishable and coherent single photons is the process efficiency. A coherent fraction of around 0.9 in pulsed excitation is obtained probabilistically generating photons with an efficiency around 5–10%. Accordingly to the quantum information application intended to realize, one could privilege the use of a deterministic excitation schemes as reported in next paragraphs.

In 2009, S. Ates and coauthors demonstrated the possibility to generate highly indistinguishable single photons by resonantly pumping QDs into micropillars cavities [45], studying the transition of the photon properties while passing from medium to high pump power regimes. The fundamental condition achieved was the generation of photons that fulfill the Fourier transform-limit, written in (3.2). Since a fundamental requirement for resonant QD spectroscopy is discerning between emitted photons and laser, a side excitation scheme was utilized (see Fig. 3.7 and [45]). The use of microcavities (micropillars in the described case) presents two main advantages for the generation of highly indistinguishable photons. On one hand, the modification of the density of states leads to an increased probability of light emission into the cavity mode. On the other hand, funneling the QD emission into the cavity mode, which can be optimized to be highly directional, further enhances the light extraction. The lifetime reduction induced by the Purcell effect [46] is also beneficial for the photon coherence: in presence of pure dephasing, one efficient way to restore the Fourier transform-limit $T_2/2T_1 = 1$ from (3.1) is indeed to decrease the radiative lifetime T_1. In [45], the Purcell factor of the investigated pillar was around 13 giving a lifetime reduction from 850 ± 10 ps to 65 ± 10 ps. The FWHM of the fluorescence peak should depend on T_1, T_2 and the Rabi energy Ω via the formula $2/T_2\sqrt{1 + T_1T_2\Omega^2}$ [8]. In the medium excitation regime, $\Omega^2 \ll 1/(T_1T_2)$, the FWHM tends to $2/T_2$. From the bottom trace of Fig. 3.11a the linewidth was measured to be $1.15 \pm 0.05\,\mu$eV that resulted in a coherence time of $T_2 = 1150 \pm 50$ ps. Once more, this 7-fold reduction of the linewidth in comparison to p-shell excitation (for a comparably low temperature of 10 K) indicated a reduction of the dephasing processes. Measuring the transition lifetime ($T_1 = 630 \pm 20$ ps in the presently

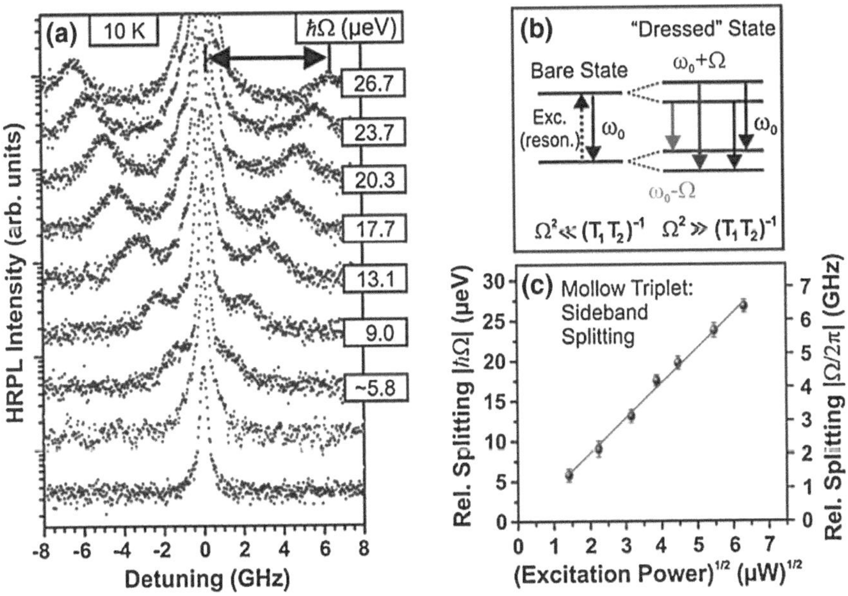

Fig. 3.11 **a** High resolution photoluminescence spectra from a QD versus power of the CW resonant pumping **b** Sketch of the evolution from bare states to "dressed" states; the three transitions forming the Mollow triplet are also schematized. **c** Evolution of the Rabi splitting over the excitation power: data were extracted from spectra in **a** [45]

discussed conditions) the ratio $T_2/2T_1$ was estimated to be 0.91 ± 0.05, showing a value very close to the Fourier transform-limit for CW pumping. It is worth noting that the measured value of T_1 came from time resolved measurements under p-shell excitation: this means that the *real* lifetime was supposed to be slightly shorter, since the p-shell to s-shell relaxation may increase measured value (by around 10–50 ps). The measured indistinguishability gave values up to $V_{HOM}(0) = 0.90 \pm 0.05$: Such high visibility value allowed to infer that the generated single photons were highly (post-selectively) indistinguishable within their coherence time and that the wave functions overlapped nearly perfectly. The presented work pioneered the generation of highly indistinguishable photons using a QD-cavity system as it will be shown in the next paragraph.

When addressing a two-level system with a strong resonant excitation, a two-level picture is no longer valid and it has to be replaced by a dressed state approach [47]. Under this condition, the two-level system evolves into a quadruplet made of two excited and two ground levels, as depicted in Fig. 3.11b. The two transitions from the same level of different multiplets are spectrally degenerate at the laser frequency $\omega_0 = \omega_{laser}$ and they are accompanied by two sidebands at the frequencies $\omega_0 \pm \Omega$, where Ω is the Raby frequency. The spectrally central feature, called Rayleigh line, is formed by two components: the elastic (coherent) and the inelastic (incoherent) part. Between these two components, the first should dominate at low power and its linewidth should match the excitation laser one. Going from the low power regime to higher pumping, the typical sidebands of the Mollow triplet starts to appear (as shown in Fig. 3.11a). The splitting between these spectral features is indeed proportional to the square root of the excitation power ($\Omega \propto \sqrt{P_{exc}}$, see Fig. 3.11c). Increasing the power, the laser background contribution becomes more and more influential, up to when it dominates the central Rayleigh peak. This behavior can be understood considering the non-linear absorption of a resonantly-excited two-level system: raising the power, while the QD emission reaches the saturation, the laser contribution increases linearly.

As already briefly mentioned, a strong resonant excitation of the QD s-shell allows entering in the so-called dressed state regime. In such regime, where the Rabi frequency is larger than the natural linewidth of the exciton transition ($\Omega \gg \Gamma_{rad}$), the

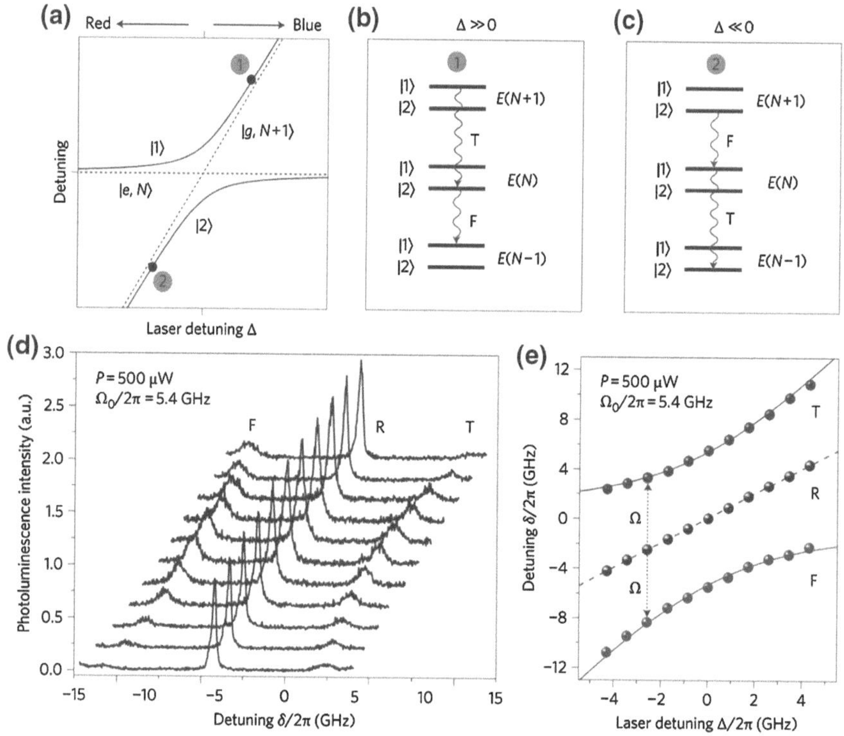

Fig. 3.12 **a** Dressed states eigenenergies (*solid lines*) over laser-to-QD detuning. *Dashed lines* stand for the uncoupled QD plus laser eigenstates. **b, c** Photon emission sequence of the two sidebands for the case of strong blue-detuned laser ($\Delta \gg 0$) and strong red-detuned ($\Delta \ll 0$) laser respectively. (d) High resolution photoluminescence measurements showing the laser detuning dependence of the resonance fluorescence spectra. **e** Spectral position deduced from **d** of the three Mollow triplet components as a function of the laser detuning. Fits ($v = \Delta \pm \sqrt{\Omega_0^2 + \Delta^2}$ and $v = \Delta$) are shown with *solid lines* [48]

two-level system description is no longer valid and a more complex dressed state description has to be used. The previous discussion was limited to the case of a laser on resonance with the two-level system while a complete description of the dressed states includes the effects of a non-zero laser-to-QD detuning, indicated as Δ. The single line spectrum under strong CW pumping is replaced by a complex spectrum (the so-called Mollow triplet) formed by three distinct peaks: a central line, called *Rayleigh* line R, and two symmetric side peaks referred as *three photon* line (at higher energies with respect to R) and *fluorescence* line F (as low energy component). These complex spectral features come from the spontaneous emission descending the ladder of "dressed states" doublets: these doublets arise from the coupling between the laser beam and the two-level system. The two rungs of the ladder are defined as $|1\rangle = c\,|g, N+1\rangle + s\,|e, N\rangle$ and $|2\rangle = c\,|g, N+1\rangle - s\,|e, N\rangle$, with N the number of quanta and $|g, N+1\rangle$, $|e, N\rangle$ representing the eigenstates of the uncoupled QD laser system. The separation of these components is set by the generalized Rabi frequency $\Omega = \sqrt{\Omega_0^2 + \Delta^2}$ where Ω_0 is the bare Rabi frequency and Δ the laser-QD detuning. The aforementioned amplitudes c and s of the dressed eigenstates are defined as $c = \sqrt{(\Omega + \Delta)/2\Omega}$ and $s = \sqrt{(\Omega - \Delta)/2\Omega}$. Looking at the detuning dependence of the eigenenergies (Fig. 3.12a), it can be noticed that for large detunings $\Delta \gg 0$ the steady state solution of the dressed states shows that the state $|1\rangle$ has a strong $|g, N+1\rangle$ character, meaning that the system results to be mainly prepared in the state $|1\rangle$ (as $c^2 \gg s^2$). Under this condition, the first emitted photon will be T and the system will be, as a consequence, in the state $|2\rangle$ of the lower successive manifold (Fig. 3.12b): the following emission can then be of a F photon. This cascaded emission follows the described time-order and results in a "T photon heralding the F transition". A clear signature of this unique behavior was observed firstly in [48] through photon-correlation experiments between the different components of the Mollow triplet. A strong photon bunching was observed if cross-correlating the emission from T and F lines, while photon antibunching was recorded for the autocorrelation of the individual sidebands. The central Rayleigh line, on the other hand, is composed by photons which are totally time-uncorrelated and should follow a Poissonian distribution. Emitting a photon in the R line, i.e. following the transition between the same states $|1\rangle$ or $|2\rangle$, does not modulate the population, since the emitter remains in the same dressed state. One important feature of the Mollow triplet is shown in Fig. 3.12d where spectra for the same power but at different laser-to-dot detunings are plotted: the central R line just follows the laser spectral behavior and the sidebands F and T remain symmetric around it. The measured linewidth of the Mollow side peaks $\Delta\nu = 730 \pm 20$ MHz, differed from the Fourier-limited linewidth given by $\Delta\omega/2\pi = (3/2T_1)(1/2\pi) = 300$ MHz due to power induced dephasing [49]. Using a Michelson interferometer and successively a spectrometer, Ulhaq et al. enabled measurements with a high spectral purity of the individual components of the Mollow triplet. An average photon count rate of 70,000 counts per second (cps) was obtained after the spectral filtering, meaning that 5.9×10^6 photons were extracted from the first lens. The count rate obtained in [48] could in principle be further improved with photonic structures or microcavities that will be touched upon in the following paragraphs. The autocorrelation of the central Rayleigh peak showed a long-timescale bunching, while expecting a pure Poissonian statistics, but this was explained as due to *blinking*: under pure resonant excitation, the blinking of the excitonic competing states arise because of two different spin configurations, namely the dark exciton, that is non-radiative. This blinking behavior observed on several tens of nanoseconds was observed in all photon correlation measurements and taken into account in [48]. A clear photon antibunching was otherwise observed if autocorrelation of each Mollow sideband was measured. The cascaded-photon emission was verified performing photon cross-correlations between the two Mollow satellites, always using the T photon as "start" and the F photon as "stop" triggers. Clear bunching was observed at zero time delays for both positive and negative detunings Δ, superimposed to the previously mentioned long-timescale blinking. Another remarkable feature of the Mollow triplet sidebands is their high indistinguishability as proven in [50].

3.3.3 Pulsed Resonant Pumping of the Exciton

Different physical regimes as well as important fundamental proof-of-principles have been investigated and achieved using continuous wave resonant excitation. On the other hand, the use of pulsed resonant excitation would enable the generation of *on-demand* single photons, in combination with the aforementioned advantages carried out with a resonant pumping scheme.

It comes from the year 2013 the demonstration of pulsed resonant excitation used to achieve deterministic generation of time-tagged single-photons with a record indistinguishability of $0.97 \pm 0.02\%$ [51]. Together with a higher brightness achievable in comparison with CW pumping in the Heitler regime, the pulsed and triggered operation is an essential requirement for future applications like quantum interference between distant QDs and entanglement of independent and remote QD spins [52, 53]. In the experiment of [51], the QDs were embedded into a planar microcavity, in order to slightly improve the light extraction with respect to a single slab of GaAs. In the cited work, the excitation and collection was performed with a confocal geometry: the laser beam was sent perpendicularly to the planar cavity (i.e. along the periodicity direction of the 1D photonic crystal) and the QD emission was collected as well along the same axis.

The pioneering work of [51] carried out a very important information, that is the capability to generate on-demand single photons with an extremely high degree of indistinguishability by using resonantly-excited semiconductor QDs. On the other hand, in order to reach such high photon "performances", high epitaxy standards were necessary. A way to push even further the QD emission properties, enhancing the photon extraction, shortening the radiative lifetime and improving the photon indistinguishability to $\approx$100% is based on the use of photonic cavities. On one hand, the light extraction is usually strongly limited by the high refractive index of the host material and it needs to be enhanced. Tailoring the geometry of the environment, like using nanowires [21], demonstrated a very high extraction efficiency but a currently limited indistinguishability probably due to a significant charge noise at the wire surface. On the other hand, the use of micro- and nano- resonators allows for an improved extraction efficiency together with a shortening of the radiative lifetime and a consequent improvement of the photon indistinguishability (see as a comparison (3.1)). The shortening of the lifetime is due to the so called Purcell effect, happening when the emitter-cavity system is working in the "weak coupling" regime. The modification of the lifetime when the emitter is in resonance with the resonator is given by the Purcell factor:

$$F_p = \frac{\tau_{off}}{\tau_{on}} - 1 \tag{3.8}$$

where τ_{off} and τ_{on} are the lifetime off and on resonance respectively (in some cases F_p is defined only as ratio τ_{off}/τ_{on}). The aforementioned factor is proportional to Q/V, i.e. the ratio of the cavity quality factor and mode volume, meaning that the cavity geometry must be chosen accordingly to the intended application (ranging from planar 2D photonic crystals, to bull's eye cavities and micropillars). The Purcell enhancement is maximum when the emitter is in spectral resonance with one of the cavity modes, it is placed in the maximum of the corresponding electric field and the dipole is aligned parallel to the E-field.

At the beginning of the year 2016, Ding and coauthors used QDs embedded in micropillar cavities to deterministically generate single photons with an indistinguishability as high as 98.5% and with an extraction efficiency of 66% [54].

Later the same year, Somaschi and coauthors further improved the source performances, with a more complex and optimized cavity design [55]. First of all, the cavity was fabricated using the so-called "in-situ" lithography technique: it is based on a low-temperature laser lithography, deterministically aligned on a single emitter. The spatial matching is achieved by determining the emitter position with nanometric accuracy. The QD photoluminescence (PL) is simultaneously recorded, allowing for the adjustment of the cavity design to spectrally match the emitter after the etching. Together with a high fabrication yield, i.e. a large number of working cavities with a QD spectrally and spatially matched, the mentioned technique was used to lithographically write an advanced pillar geometry: the pillar was connected with four one-dimensional wires to the surrounding, allowing for the application of an external electric field to the structure (Fig. 3.13a). The sample was doped during the MBE growth in order to obtain an optimized p-i-n diode with a defined Fermi level around the QD and minimizing the free carrier absorption in the mirrors. Thanks to the lithographic technique, several working structures were produced around one macroscopic metal contact (Fig. 3.13b). The spatial matching can be observed in the PL map in Fig. 3.13c, where the strong emission from the Purcell enhanced dot is coming from the central part of the structure. The diode structure embedding the micropillar carried two advantages: on one hand, the QD emission could be Stark-tuned in and out the cavity mode (Fig. 3.13d). On the other hand, it stabilized the charge environment, leading to an indistinguishability of $\approx$100%. The average quality factor of the utilized pillars was around 12000, leading to a Purcell fac-

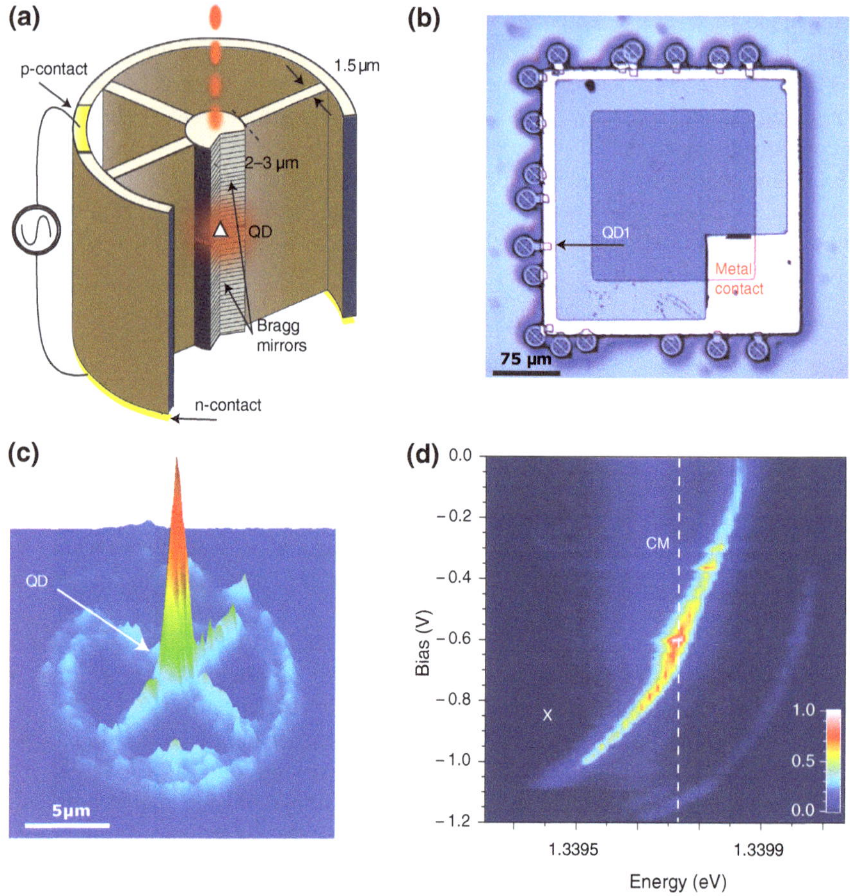

Fig. 3.13 **a** Sketch of the deterministic QD-cavity system realized with an electrically-controlled connected micropillar. The n-contact was deposited on the back side of the sample chip. **b** Optical microscope picture of the metal mesa forming the *top* p-contact. Up to 18 devices were connected to it. **c** Micro-PL map of the working device showing the successful deterministic placement of the QD in the middle of the connected pillar. **d** Emission intensity map versus the applied bias and the photon energy: the X was Stark tuned into the cavity mode (CM, indicated with a *dashed line*) [55]

tor of $F_p = 7.6$. The conducted study reported on the sources characteristics under quasi-resonant and purely resonant excitation, showing interesting properties in such pumping scheme transition. Under p-shell excitation, the brightness was measured to be equal to 0.65 ± 0.07, defined as the product $\beta \eta_{out} p_x$, where $\beta = F_p/(F_p + 1)$ represents the fraction of light emitted in the mode, $\eta_{out} = \kappa_{top}/\kappa$ the outcoupling efficiency (ratio between the photon escape rate from the top of the cavity over the total one) and p_x the occupation factor of the dot state. The measured brightness was consistent with $p_x = 1$, $\beta = 0.88$ and $\eta_{out} = 0.70$ (as was measured via reflectivity experiments). It is important to clarify that the presented values were found for

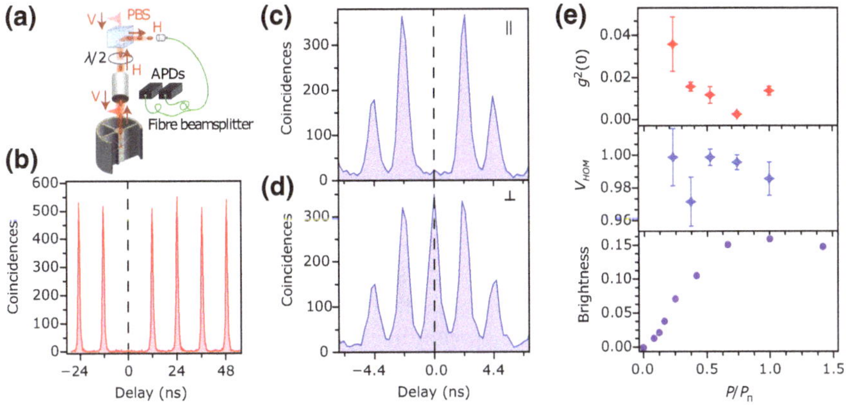

Fig. 3.14 **a** Illustration of the cross-polarized, confocal, setup used for resonant fluorescence. The temporally shaped laser pulses were sent from the *top*, focused on the micropillar by means of a microscope objective. An effective polarization suppression was achieved by means of a half-wave plate and a polarizing beam splitter. **b** Second-order correlation measurement showing a purity as high as $g^{(2)}(0) = 0.0028 \pm 0.0012$ at $0.75P_{\pi}$. **c**, **d** Correlation histograms showing the interference of two successively emitted photons with parallel and orthogonal polarization respectively. **e** Second-order correlation function at different pumping powers (*top*). Indistinguishability V_{HOM} (*middle*) and source brightness (*bottom*) versus laser power. Error bars were deducted starting from the assumption of a Poissonian statistics in the events that were detected [55]

an unpolarized single-photon source, while the polarized brightness is reduced by a factor of 2 (as for many applications, polarized single photons are required). The measured HOM visibility (photon delay = 3 ns) reached $V_{HOM} = 0.78 \pm 0.07$, corrected by the non-zero $g^{(2)}(0) = 0.024 \pm 0.07$ (without corrections $V_{HOM} = 0.74 \pm 0.07$). Comparable results were also observed for other structures, indicating the reduced noise from charge fluctuations for these electrically controlled pillars. The effect of the pump-induced time jitter for these Purcell enhanced structures had an effect on the maximum observable indistinguishability [56]. In the case of [55] the time jitter limited the indistinguishability to around 70–80% for p-shell pumping, further indicating that the charge noise was not the limiting factor for a unity indistinguishability. To prove this statement, measurements under pure resonant excitation were conducted. In the present case, a neutral exciton with a FSS of 10–15 μeV was under investigation. The pillar ellipticity introduced a polarization splitting in the cavity mode of around 90 μeV. The linearly polarized laser was set in parallel along one cavity axis (set as V), roughly 45° oriented with respect to the QD axes (see Fig. 3.14a). The V polarized laser created then a superposition of both excitonic states that evolved toward the orthogonal state, coupled with the H-polarized cavity mode. The described evolution took place on a timescale that is inversely proportional to the dot FSS. A confocal crossed-polarized microscope was used for these experiments, so that the detection was along the H-polarization direction (extinction ratio 10^5). An etalon with 15 μeV bandwidth was used to further suppress the unwanted scattered laser light. Together with a $g^{(2)} = 0.0028 \pm 0.0012$ (Fig. 3.14b) the measured HOM visibility reached $V_{HOM} = 0.9956 \pm 0.0045$ ($M = 0.989 \pm 0.004$ without corrections) revealing a nearly perfect two-photon interference (Fig. 3.14c). As a counter check, for fully distinguishable photons the visibility dropped to $V_{HOM} = 0.057 \pm 0.084$ (Fig. 3.14d). In Fig. 3.14e the full dataset of second-order correlation, indistinguishability and brightness are reported: the M values were never below 0.973 in the full power range. Independent reflectivity measurements (under CW excitation) showed a radiatively limited exciton linewidth, further proving the charge noise cancellation on a millisecond timescale. It is important to say that the used etalon was not necessary to obtain high indistinguishability values, meaning that the phonon-assisted emission was suppressed thanks to the Purcell acceleration of the spontaneous emission of the zero-phonon line and the cavity filtering effect. Estimating once more the polarized brightness, the authors found it to be 0.16 ± 0.02 at the π-pulse. For the device under study, the photon extraction $\beta\eta_{out} = 0.64$. The lower brightness for the resonant case was due to the cross-polarized detection: the described polarization rotation was limited by the value of the exciton FSS and the Purcell enhancement in the V mode. With the described FSS and F_p, the occupation factor of the H-emitting exciton was $p_x = 0.23$. The authors also suggested that using a different excitation mechanisms, like the side excitation, the brightness would increase up to 0.65 without affecting purity and indistinguishability.

Following the line of [55] the mentioned device is setting a new standard in the non-classical light sources for quantum information. Making a comparison between the existing devices resulted in the graph in Fig. 3.15, where only the results for

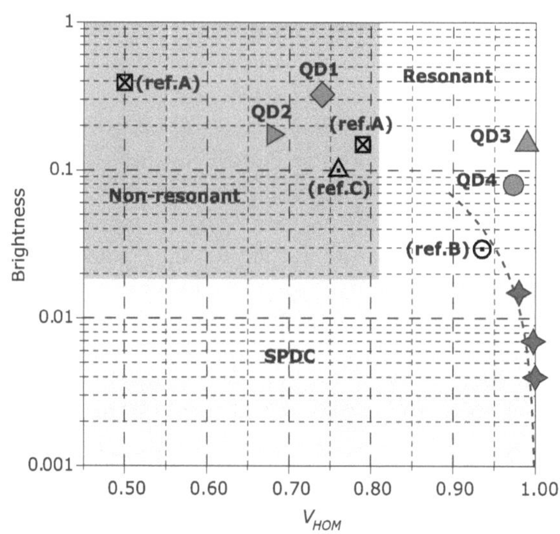

Fig. 3.15 The graph shows brightness and indistinguishability for different SPDC and QD single-photon sources. *Open* symbols refer to state-of-the-art QD-based sources (*ref. A* indicates [20], *ref. B* means [51] and *ref. C* indicates [23]). High quality heralded SPCD sources are represented by *gray stars* while data from [55] are summarized by full symbols (labelled as QD). QD1, 2, 3, 4 refer to the four devices under study: QD1 and 2 in p-shell excitation while 3 and 4 in resonance excitation

which the $g^{(2)}(0) < 0.05$ are shown (the M values are plotted without corrections for non-ideal $g^{(2)}$). Heralded spontaneous parametric down conversion sources (SPDC) are up to now the most used sources in quantum information: despite the fact that the brightness is limited by multiphoton generation, the near ideal indistinguishability made them suitable for quantum information experiments. For the SPDC the brightness is defined as the average photon number for every single spatial mode (making the assumption that both collection and detection efficiencies are equal to one). This comparison was motivated by the will of making a fair comparison between different kind of sources, separately from the photon bandwidth as well as from the driving repetition rate. It is also necessary to specify that the mentioned two-photon indistinguishability refers to photons coming from the same downconversion event, while for photons coming from different independent heralding are limited to much lower values of indistinguishability and brightness [57, 58]. It can be seen that the devices shown in [55] are 20 times brighter than state-of-the-art SPDC, while having ultimate values of $g^{(2)}(0) = 0.0028 \pm 0.0012$ and $M = 0.9956 \pm 0.0045$. More details on this comparison can be found in [55] and supplementary materials.

Another demonstration that the QD-based technology is currently ready to be used on the "marketplace" of quantum information is that high extraction efficiency and indistinguishability have been demonstrated independently by other groups. Unsleber et al. proved that, by using a similar deterministic lithography as the previously described to fabricate standard micropillars (i.e. without electrical control) [59]. The extraction efficiency was as high as $\eta = 74 \pm 4\%$ meaning an extraction efficiency of linearly polarized photons of $\eta_{lin} = 37 \pm 2\%$, in good agreement with theoretical simulations. With a Purcell factor of around 6, the measured raw HOM visibility was as high as $V_{raw} = 84 \pm 3\%$ that corrected with the actual $g^{(2)}(0) = 0.0092 \pm 0.0004$ and the setup imperfections yielded a value of $V_{corr} = 88 \pm 3\%$. This value, obtained for a pumping power equal to $\pi/4$-pulse, dropped to $V_{corr} = 73 \pm 1\%$ at the π-pulse, probably due to power induced dephasing effects (i.e. coupling to longitudinal acoustic phonons).

The final objective would be the generation of single photons that can perfectly coalesce independently from the time separation. Spectral diffusion can indeed be responsible for the degradation from near-unity to lower values on a long interference timescale. Once more, QD embedded in micropillar cavities can fulfill this requirement, having demonstrated the possibility to generate long streams of single photons with very high mutual indistinguishability [60]. Even if a non-deterministic fabrication technique was utilized, pure and highly indistinguishable single-photons were produced. Time-dependent HOM experiments were conducted, showing that an indistinguishability of $95.9 \pm 0.2\%$ for 13 ns time separation between photons (Fig. 3.16a) only drops to $92.1 \pm 0.5\%$ for all delays between 2 and 14.7 µs (Fig. 3.16b). These measurements demonstrated that a stream of more than 1000 photons with very high mutual indistinguishability can be generated. In addition to that, Fig. 3.16c shows that the time-dependent HOM revealed a slow dephasing process happening on a time scale of around 0.7 µs that is 4000 times longer than the radiative lifetime of the emitted single photons. Photons emitted on time separation shorter than 0.7 µs should than perceive the same electrical environment,

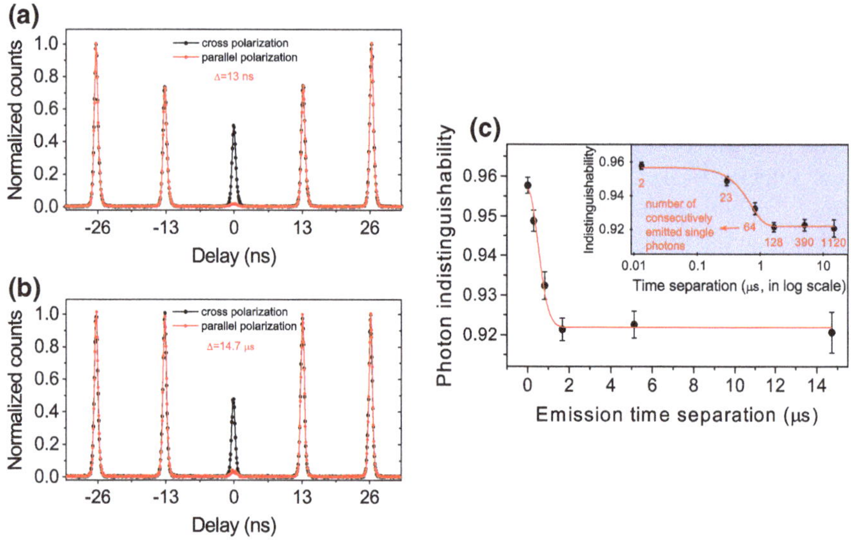

Fig. 3.16 **a, b** Cross (*black*) and parallel (*red*) input photons for the HOM experiments. In **a** the time separation was 13 ns while in **b** it was increased to 14.7 ns. The resonant excitation was performed at the π-pulse in all cases. No background was subtracted and the integration time was around 5 min. **c** Photon indistinguishability extracted from measurements at various photon separation time. The high count rate allowed to eliminate the shot noise, allowing a precise determination of the error bars. The *inset* shows the same data in a log x-scale, also clearly reporting the number of consecutively emitted photons [60]

while on a longer time separation the perceived electric field should be different due to spectral diffusion. This time averaged wave packet was then determined by the spectral wandering in comparison to the intrinsic linewidth. The observed plateau in the indistinguishability was attributed to the nearly Fourier transform-limited emitted photons: $T_2/2T_1 = 0.91 \pm 0.05$. The authors also proved that, likewise for the sample in use, the photon properties are strongly influenced by the pumping scheme: while the $g^{(2)}(0)$ dropped from 0.007 ± 0.001 in s-shell pumping to 0.027 ± 0.002 in p-shell, the value of indistinguishability dramatically decreased to 0.21 ± 0.02. Power induced dephasing as well as the time jitter induced by the p- to s-shell relaxation were attributed as factors limiting the photon coalescence. Under pure s-shell excitation, Wang and coauthors also proved that an increase of temperature is also detrimental for the two-photon interference, meaning that a device operating at the lowest possible temperature is desirable for best performances.

3.3.4 Adiabatic Rapid Passage

An alternative way to generate deterministic and indistinguishable photons is using the Adiabatic Rapid Passage (ARP) [61]. The main advantage relates on the robustness of the generation, with respect to standard pulsed photon generation. In fact,

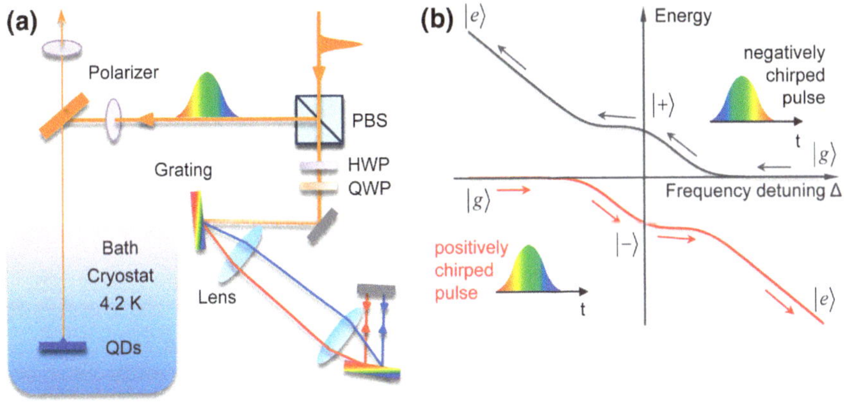

Fig. 3.17 **a** Scheme of the experimental setup. The stretching of a 3 ps transform-limited laser pulse up to 30 ps was made by two symmetrical gratings combined with a telescope. The initials in the picture stand for polarizing beam splitter (PBS), half-wave plate (HWP) and quarter-wave plate (QWP) **b** Schematic in the rotation frame of the eigenstates of the two-level system while interacting with the chirped pulse. The coherent superposition of the atomic states $|g\rangle$ and $|e\rangle$ gives the dresses states $|\pm\rangle$. The scheme also shows that for a *blue* detuned laser, the dressed states become $|e\rangle$ and $|g\rangle$ (while for a *blue* detuned the states become $|g\rangle$ and $|e\rangle$). By means of a positively chirped pulse, the QD evolves adiabatically to the excited passing along the $|-\rangle$ state. Conversely, the evolution happens along the state $|+\rangle$ for a negatively chirped pulse (laser sweep from high to low frequencies). This evolution rate is slower than the peak Rabi frequency (equal to 100 GHz) in the described experiment [62]

the use of a laser π-pulse is sensitive to the variation of the pulse area as well as the dipole moments of the emitter. In [62], the authors show how to use frequency-chirped pulses to implement this ARP method. A negatively charged QD X^- embedded in a planar cavity (with 5 top and 24 bottom DBR mirrors) was used and resonantly excited with three different methods: transform-limit hyperbolic secant pulses of 3 ps from a (mode-locked) Ti:sapphire laser, negatively chirped pulses with 30 ps of duration and positively chirped pulses with the same 30 ps duration. The negatively chirped pulses were generated with a stretcher made with two parallel-placed gratings (1200 lines/mm) while the introduction of a telescope between the two symmetrical gratings was used to generate positively chirped pulses. The excitation/collection was performed with a confocal microscope in cross-polarization configuration (laser extinction 10^6). A sketch of the setup is shown in Fig. 3.17a. The intensity behavior as a function of the power displayed Rabi oscillations for transform-limited resonant laser pulses. The maximum signal was found at the π-pulse, as expected, and the observed damped oscillations were assumed to be related to excitation induced dephasing (Fig. 3.18a). The same intensity in the emitted signal was also obtained with chirped pulses, for a laser power equivalent to roughly 1.5π. However, negatively chirped pulses displayed a signal decrease for increasing power (in agreement with previous results and current simulations). This could be due to a relaxation from the $|+\rangle$ to the $|-\rangle$ eigenstate (dressed states) assisted by the emission of an acoustic

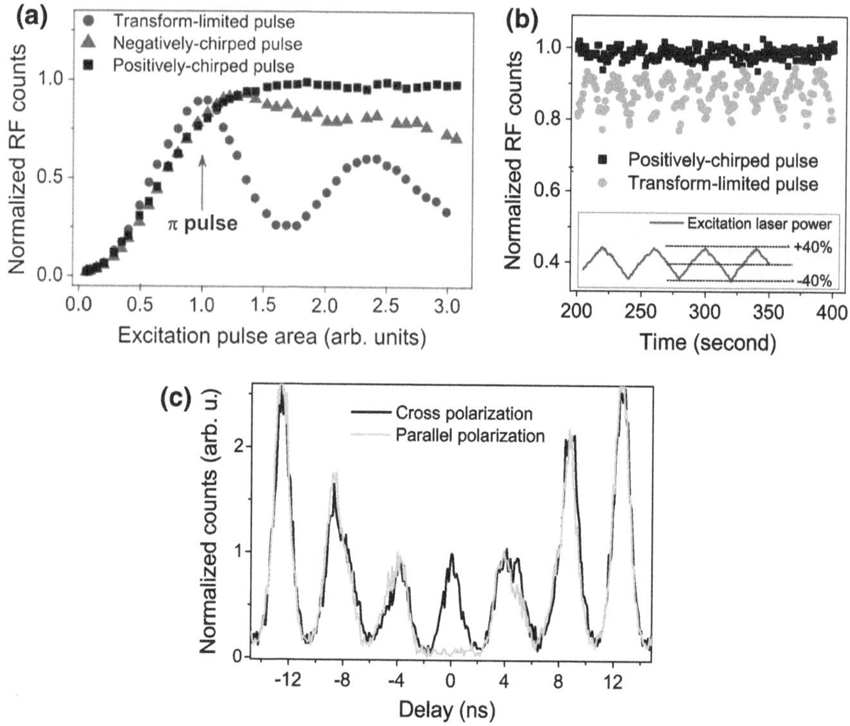

Fig. 3.18 **a** Comparison of the count rate behavior versus pump power for the three considered methods. **b** Temporal dependence of the resonant fluorescence counts under externally modulated laser power (time trace in the *inset*) **c** Non-postselective two-photon interference. Two chirped pulses with a pulse area of 1.6π were used to deterministically generate the photons. For the HOM measurements, two positively chirped pulses, delayed by 4 ns where used. Raw (corrected) visibility of 0.979 (0.995) was found [62]

phonon (3.17b): this effect can break down the ARP, showing that negatively chirped pulsed can be subject to a photon emission efficiency reduction induced by fluctuation in the excitation laser power. Conversely, using positively chirped pulses, the intensity remained constant increasing the laser power (see Fig. 3.18a): this can be explained thinking that the relaxation from $|-\rangle$ to $|+\rangle$ associated with a phonon absorption is less probable than phonon emission at low temperature. The improved stability with respect to the laser power was also demonstrated with the deliberate introduction of a pump power fluctuation, with a 50 MHz triangle modulation. This modulation, with 80% peak amplitude, was centered around the π-pulse for the transform-limited pulses and at 1.9π-pulse for the positively chirped pulses (Fig. 3.18b). The recorded time traces of the resonant fluorescence photons showed an intensity fluctuation of 15% for the transform-limited pulses while only 2.2% was recorded for the positively chirped ARP. The measured resonant fluorescence photons generated via ARP or with transform-limited pulses displayed a similar linewidth and a multiphoton probability generation of $g^{(2)}(0) = 0.003 \pm 0.002$ (the authors used the ratio between the intensity integrated in a 3.2 ns time window around zero delay and the average counts of the seven peaks besides). Interestingly, the recorded two-photon interference visibility reached values as high as 0.979 ± 0.006 (raw) and 0.995 ± 0.007 ($g^{(2)}(0)$ and setup imperfections corrected), calculated from the integrated counts in a 3.2 ns time window around the zero delay peak (ratio between parallel and crossed polarization). This measurement is reported in Fig. 3.18c. Unfortunately, the brightness (intended as photons collected at the first lens per excitation laser pulse) is not reported but the authors wrote that the product of the overall collection efficiency and detection efficiency gave 0.2%, far away for the needed 67% efficiency threshold for loss-tolerant quantum computing. Therefore, they suggested the use of cavities to increase the overall efficiency and underlined the importance of the described method for the resonant pumping of the biexciton state (further described in Sect. 3.3.6).

3.3.5 Spin-Flip Raman Transition

Another method to generate tunable single-photons with high indistinguishability relies on a driven single quantum dot spin. Fernandez et al. demonstrated the generation of tunable Raman photons using an optically driven Λ system, formed placing a single electron-charged QD into a magnetic field (in Voigt geometry) [63]. One of the advantages of such a method is related to the emitted photon linewidth that for Raman photons can in principle be very narrow and limited only by the laser linewidth and by the low-energy spin coherence (the interaction of the electron spin with the nuclear spins is one example of decoherence mechanism). The presence of the magnetic field lifts the degeneracy of the ground and excited trion states, and for large enough Zeeman splitting each of the two excited states can form an independent Λ system with the two ground states (see Fig. 3.19a). The Zeeman splitting results to be $g_e \mu_B B$ for the ground state and $g_h \mu_B B$ for the trion state, where g_e and g_h are the electron and hole in-plane g factors. The polarization of each transition is determined by the optical selection rules and helped the authors to discriminate between the emitted photons and the pump laser (further suppressed with crossed polarizers and with a scanning Fabry-Pérot interferometer). The recombination from excited to ground state happens on a ns timescale and for the Voigt geometry takes place with equal probability (differently from the Faraday configuration). Driving the QD resonantly on one of the optical transitions turns in the spin pumping in the other ground state: this prevents further photon emission and absorption. For this reason, the spin state must be restored on a short time scale to allow efficient photon generation. In case the sample is provided with a back contact (divided by a tunneling barrier from the QD layer), operating the source at the edge of the voltage for which the dot is singly charged results in spontaneous spin-flip. Otherwise, a second laser can be used to optically pump back the electron spin. In the particular case for which $g_h = -g_e$ the two diagonal transitions are degenerate and the pump laser could act as pump and repump simultaneously, so no spin restoration would be required. Under

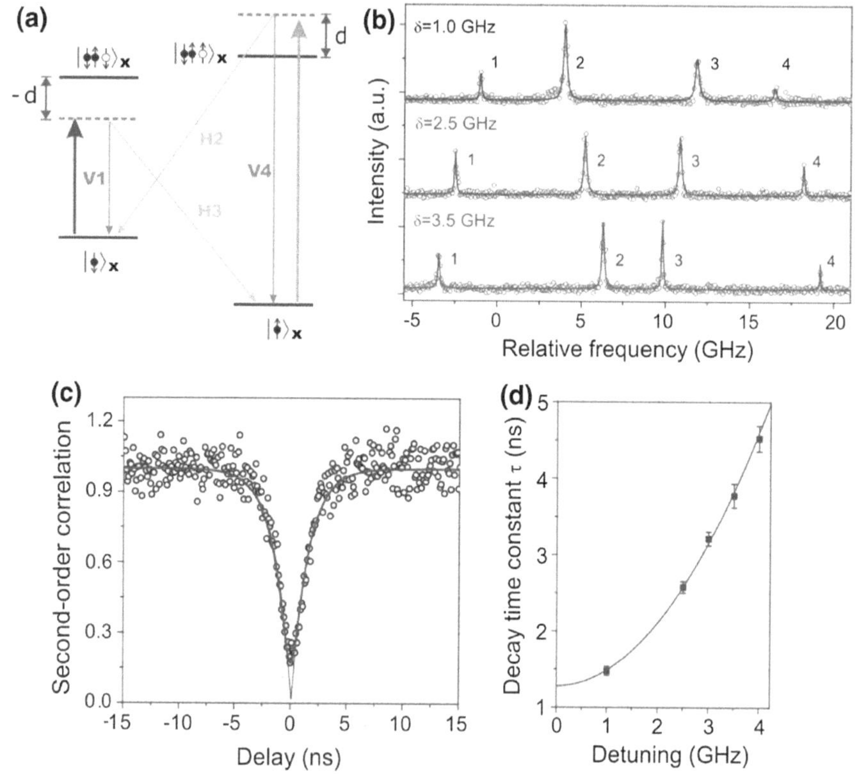

Fig. 3.19 **a** Diagram of the energy levels: the polarization of vertical and diagonal transitions is set by optical selection rules. **b** Raman spectra for different laser detunings. **c** Raman photons' second-order correlation: laser delay of $\delta = 1$ GHz. The *solid black* corresponds to theoretical fit, convolved with the system response time (detector resolution around 400 ps), while the *solid gray* is the deconvolved curve. **d** Increasing decay time constant for increasing laser detuning with a theoretical curve shown by *solid curve* [64]

the aforementioned configuration, Fernandez and coauthors showed that setting the laser resonantly with one of the diagonal transitions gave rise to two distinct peaks detuned from the laser frequency (photon emission from the two vertical transitions) and that this resonantly scattered photons could be tuned varying the magnetic field (see [63] for further experimental details). In addition to that, the authors demonstrated that tuning the laser close to the (energetically higher) outer transition it is possible to generate off-resonance Raman scattered photons. The achieved tuning range was around 2.5 GHz, with an additional variation of the emitted intensity as well as photon linewidth.

It was also demonstrated that these optically tunable Raman photons can also be highly indistinguishable [64]. With an experimental configuration similar to the previously mentioned one, the authors demonstrated subnatural linewidth of the emitted photons under CW operation and high indistinguishability with a near to zero multiphoton emission for pulsed excitation. In CW experiments, two lasers where used: one red detuned by $-\delta$ from the $|\downarrow\rangle \leftrightarrow |\downarrow\uparrow\Downarrow\rangle$ transition and the second blue detuned by δ from the transition $|\uparrow\rangle \leftrightarrow |\uparrow\downarrow\Uparrow\rangle$ (Fig. 3.19a). These lasers were used for the optical pumping and repumping of the electron spin: the fast spin restoration is indeed necessary for high repetition rate operation. It was demonstrated that the emission from the diagonal transitions was tunable with the excita-

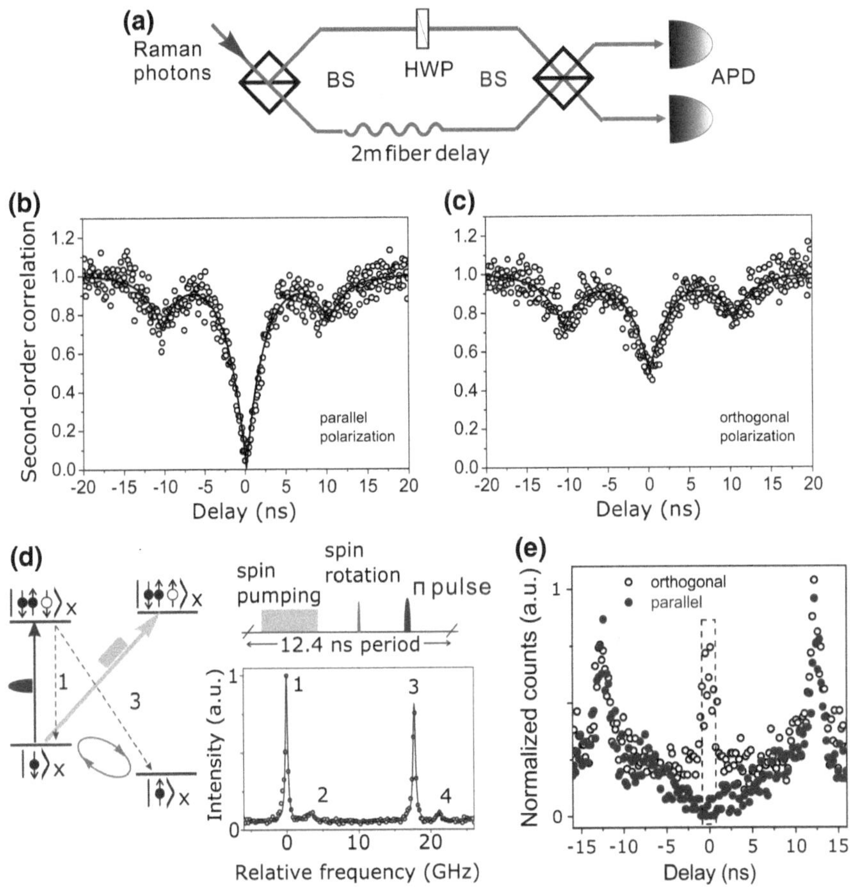

Fig. 3.20 **a** Unbalanced Mach-Zehnder interferometer used for HOM experiments: the used delays were 10 ns for CW operation and 12.4 ns for pulsed operation. **b**, **c** Two-photon interference measurements in parallel and orthogonal polarization configuration respectively. **d** Sequence of laser sequence for the pulse Raman photon generation. Electro-optics modulators were used to generate the two 4 ns and 105 ps pulses from CW lasers. A filtered Ti:sapphire laser was used to generate the 20 ps spin-rotation laser pulse. The synchronization signal came from the Ti:sapphire laser. A Raman fluorescence spectrum, for a magnetic field of 2.8 T is shown in the *bottom-right* panel of **d**. **e** Two-photon interference from Raman-single photons (π-pulse excited): the *dashed* region is the 2.2 ns window used in the visibility estimation [64]

tion laser detuning δ (Fig. 3.19b), giving a knob to control the Zeeman splitting (or equivalently the hole g factor). After deconvolution, the experimental second-order correlation was found to be $g^{(2)}(0) = 0.01 \pm 0.02$ (Fig. 3.19c). Complementary to the spectral narrowing, the decay time constant was found to follow the expected expression $\tau = (2 + \Gamma^2/\Omega^2 + 4\delta^2/\Omega^2)/\Gamma$, with Γ the spontaneous emission rate and Ω the Rabi frequency (Fig. 3.19d). These findings shown the possibility to generate single Raman photons with tunable frequency, bandwidth and lifetime, important properties for hybrid quantum networks. In addition to that, the HOM interference for two off-resonant Raman photons was measured (setup sketched in Fig. 3.20a). From the measurement of the coincidence probability for identical ($g_{||}^{(2)}(0) = 0.01 \pm 0.02$) and distinguishable ($g_{\perp}^{(2)}(0) = 0.50 \pm 0.02$) photons the visibility ($V(t) = 1 - g_{||}^{(2)}(t)/g_{\perp}^{(2)}(t)$) was estimated to be at zero delay 0.98 ± 0.03 (Fig. 3.20b, c).

In pulsed operation, a different laser configuration was used (Fig. 3.20d). A laser pulse of 4 ns resonant with the diagonal transition $| \downarrow\rangle \leftrightarrow | \uparrow\downarrow\Uparrow\rangle$ gave the optical pumping to the spin state $| \uparrow\rangle$. The first laser pulse was followed by a 20 ps laser pulse (red detuned from the transition $| \downarrow\rangle \leftrightarrow | \downarrow\uparrow\Downarrow\rangle$) to coherently rotate the electron spin from $| \uparrow\rangle$ to $| \downarrow\rangle$: this two pulses sequence was used for a deterministic spin recycling. The resonant π-pulse (105 ps) was used to bring the system from $| \downarrow\rangle$ to $| \downarrow\uparrow\Downarrow\rangle$, then followed by the spontaneous emission in the two channels marked as 1 and 3 in Fig. 3.20d. The overall source efficiency was estimated from the observed count rate to be 1.2%, mainly limited by the actual photon extraction. Another limitation relates to the not possible on-demand emission due to the simultaneous emission of two spectrally distinct lines. The single-photon emission was demonstrated by a $g^{(2)}(0)$ as low as 0.01 ± 0.01. To estimate the HOM visibility, the two-photon events in a coincidence window of 2.2 ns were analyzed, resulting in a visibility of 0.95 ± 0.03 (Fig. 3.20e). These highly indistinguishable and tunable photons were also used to perform two-photon HOM with photons coming from remote and distinct sources.

3.3.6 Two-Photon Excitation of the Biexciton

So far, SPDC sources are currently widely used for entangled photon pair generation. As for the case of indistinguishable photons, QDs represent a new powerful alternative for the deterministic (and not probabilistic as for SPDC) generation of entangled photon pairs. For this purpose, the radiative cascade $XX \rightarrow X \rightarrow O$ can be utilized for the generation of polarization-entangled photons. Trapping two electron-hole pairs in the QD leads to the formation of the XX (see Sect. 3.3) and after the radiative recombination of one exciton, the system has one remaining exciton that can emit a second photon then driving the system in the ground state. For the efficient generation of polarization-entangled photons, the exciton FSS must be energetically smaller than the radiative linewidth: in this case, the two decay paths become indistinguishable, resulting in a two-photon Bell state as $|\Psi^+\rangle = 1/\sqrt{2}(|H_{XX}\rangle|H_X\rangle + |V_{XX}\rangle|V_X\rangle)$. In order to make sure that only one entangled photon pair is emitted per excitation laser pulse, the biexciton must be driven at saturation. Similarly to the excitonic case, the use on non-resonant excitation methods limits the coherence and the indistinguishability of the emitted photons and, in addition, due to different charge configurations the emission is not on-demand. For this reason, the entangled photon pair should be emitted by resonantly pumping the XX state. On the other hand, the optical selection rules prevent the possibility to directly excite the biexciton with one resonant photon. To overcome this issue, a two-photon excitation (TPE) scheme can be used [65], which at the π-pulse can coherently lead to the biexciton deterministic preparation with near unity fidelity. As schematized in Fig. 3.21 the virtual TPE state energetically lies between the exciton and biexciton emission lines and it is used to "reach" the biexciton state with the use of two excitation photons. The main advantages rely on one hand on the reduced decoherence: no phonon relaxation processes are needed for the biexciton preparation and no charge carriers are created in the semiconductor, then avoiding the carrier-carrier scattering processes. On the other hand, the laser is spectrally well separated with respect to exciton and biexciton radiative emission (see Fig. 3.21b, c), meaning that from an experimental point of view, a polarization suppression scheme is not necessary for the laser suppression, but only a spectral selection (using a spectrometer or narrow notch filters) is required. This excitation scheme, successfully used for bulk crystals and single quantum dots, was adopted to generate on-demand indistinguishable polarization-entangled photon pairs [66]. In the cited case, the exciton (1.4212 eV) and biexciton (1.4189 eV) lines were separated by 2.3 meV (the biexciton binding energy). In Fig. 3.21b a trion line is observed in the non-resonant spectrum but it disappears under TPE excitation (Fig. 3.21c). In some cases, the excitation laser may need to be spectrally shaped in order to minimize the unwanted phonon-mediated pumping as well as the impact of scattered laser light in the collected QD emission. In [66] a shaped laser ($\tau_{pulse} = 21.4$ ps, FWHM = 95 μeV) linearly polarized was used to address the virtual biexciton TPE state. A footprint of the successful TPE pumping was given by the observed equal emitted intensity from XX and X lines (Fig. 3.21c), since the XX is deterministically prepared after each pump laser pulse and the X population derives only from the radiative cascade $XX \rightarrow X$. Given that, as explained, there is no need of any polarization suppression to filter the laser light, the authors successfully performed polarization-dependent cross-correlation measurements on the photons coming from the $XX \rightarrow X$ cascade. In addition to that, the purity of the emitted photons was found to be almost perfect with a recorded $g^{(2)}(0) < 0.004$ for both X and XX transitions. Signature of the coherent nature of the excitation process was found observing Rabi oscillations in the emitted intensities over the laser power for both transitions. The biexciton occupation probability was estimated to be $\epsilon_{XX} = 0.98 \pm 0.07$. A similar behavior was also observed for the X state (with $\epsilon_X = 0.86 \pm 0.08$), fed via the $XX \rightarrow X$ cascade.

Quantum-state tomography measurements were performed (at the π-pulse) in order to evaluate how the degree of entanglement of the emitted photon pair was affected by this described QD coherent control method. A good approximation of the fidelity of the Bell state was obtained by measuring the degrees of polarization

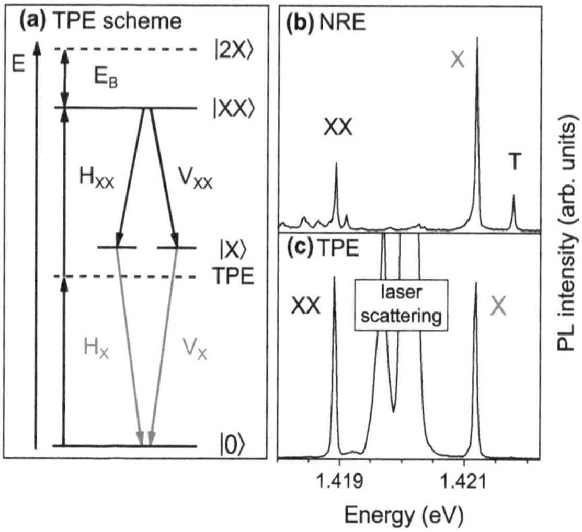

Fig. 3.21 **a** Energy scheme: the shaped laser used for the excitation lies resonant with the TPE virtual state. The two possible recombination path (through the two fine-structure split exciton states) are displayed. The intermediate exciton states fixes the photon polarization. **b** Non-resonant excitation (NRE) spectrum, evidencing the X, XX and T line (the last due to carrier capture in the QD). **c** Shaped laser resonant with the TPE virtual state (in between X and XX). The trion disappeared and the X and XX intensities are now equal. The exciton population cannot be transferred to a trion (by charge capture) before the radiative cascade is over: X and XX are then emitted as a pair. The biexciton generation efficiency increases by a factor of 4 passing from NRE to TPE [66]

correlation between exciton and biexciton in different bases: linear, diagonal and circular.

While in linear and diagonal bases (Fig. 3.22a, b) a clear bunching was observed for parallel polarization of the X and XX photons (and antibunching in the opposite, orthogonal cases), in the circular basis (Fig. 3.22c) the behavior was opposite (antibunching for parallel and bunching for orthogonal polarization): this is a clear signature of the polarization entanglement. The polarization correlation contrast can be defined as:

$$C_{\mu} = \frac{g^{(2)}_{xx,x}(0) - g^{(2)}_{xx,\bar{x}}(0)}{g^{(2)}_{xx,x}(0) + g^{(2)}_{xx,\bar{x}}(0)} \qquad (3.9)$$

with $g^{(2)}_{xx,x}(0)$ the second-order correlation function, at zero delay, between the first and the second emitted photons in the selected polarization basis, while $g^{(2)}_{xx,\bar{x}}(0)$ is the respective cross-polarized second-order correlation function. Extracting the contrasts from the previously shown measurements, the authors found the following contrasts: $C_{linear} = 0.87 \pm 0.02$, $C_{diagonal} = 0.67 \pm 0.04$ and $C_{circular} = -0.69 \pm 0.02$. From these values, the fidelity was calculated as $f = (1 + C_{linear} + C_{diagonal} - C_{circular})/4 = 0.81 \pm 0.02$. Performing the same tomography measure-

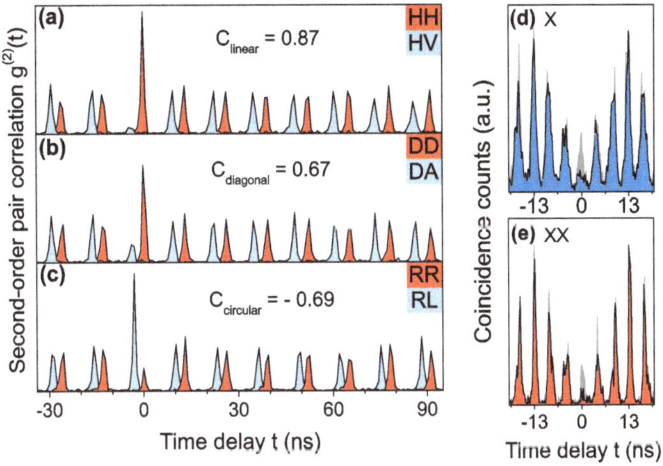

Fig. 3.22 **a** Cross-correlation histograms for linear basis, **b** diagonal basis and **c** circular basis. It is clear from the data that antibunching was observed for orthogonal polarizations in the basis **a** and **b**, while bunching was found for parallel polarization. Opposite behavior was found for circular basis. The crossed-polarization contrast was derived considering the relative areas at zero delay for **a**, **b** and **c** histograms. A time-shift in the orthogonal polarization data was added for clarity. **d**, **e** HOM interference for X and XX for two consecutive photons. Parallel polarization configuration is shown in *dark color*, while in *gray* the orthogonal polarization is depicted: 0.71 ± 0.04 for **d** and 0.86 ± 0.03 was estimated (see text for further explanations) [66]

ments for above-bandgap excitation led to a fidelity of 0.72 ± 0.01: the improved fidelity for the TPE scheme was attributed to the improved coherence time (T_2) that passed from 229 ps (non-resonant) to 357 ps (resonant) for the X photon and from 114 ps (non-resonant) to 192 ps (resonant) for the XX photon. The non-perfect fidelity under TPE pumping was attributed to eventual cross-dephasing events involving polarization dependent phonon interaction.

To measure the indistinguishability, a standard unbalanced Mach-Zehnder interferometer (MZI) was used, exciting the QD every 13 ns with two π-pulses delayed by 4 ns. For every excitation cycle (i.e. two laser pulses), the QD emitted a pair of photons. The X and XX photons were then spectrally separated, their polarization was projected in horizontal orientation and then used to feed the MZI (with an unbalance of 4 ns). If X photons were used, a raw visibility of 0.44 ± 0.03 was found while for the XX photons a higher visibility was obtained (0.58 ± 0.01). Correcting from experimental imperfections (i.e. detector dark counts, reduced mode overlap on the beamsplitter and non-zero $g^{(2)}(0)$) gave visibilities of 0.69 ± 0.04 for the X and 0.84 ± 0.05 for the XX. These data were taken comparing the zero-delay peak for indistinguishable (parallel polarization) or distinguishable (orthogonal polarization) photons but comparable results were obtained also checking the intensities of the side peaks and their known peak area ratios (Fig. 3.22). The difference in the observed visibilities, i.e. a lower indistinguishability for the X with respect to the XX, was attributed to time jitter introduced in the exciton population by the biexciton lifetime. Once more, using a cavity to Purcell enhance the XX radiative decay, i.e. shortening the lifetime, could improve the X indistinguishability by reducing the XX lifetime-induced time jitter. Recent results showed that the use of the described TPE method allowed for the generation of highly indistinguishable (0.93 ± 0.07) and highly entangled (0.94 ± 0.01) single photons, showing once more the strength of this pumping mechanism [67].

In [68] a "full-coherent control of the ground-biexciton qubit" was achieved using a TPE scheme and the coherence time was indeed improved. Such mechanism makes the source suitable for the generation of time-bin entanglement (see Chap. 8).

3.4 Phonon-Assisted Excitation Methods for Exciton and Biexciton

In the following section, phonon-assisted processes will be described. Up to now, phonons have been considered as "detrimental": they are responsible of the intensity damping of the Rabi oscillations and of the ARP signal; the phonon-assisted emission, around the zero phonon line, limits the photon indistinguishability making the emission spectrum not-perfectly Fourier transformed (reason why they are often spectrally filtered out). Despite that, here we will show how to use phonon-mediated processes in order to excite X and XX states.

3.4.1 Phonon-Assisted Exciton Excitation

In early 2015, it was demonstrated that by using a strong laser pulse tuned on the phonon sideband (i.e. the spectral region where longitudinal-acoustic phonon-assisted processes can take place) of a neutral exciton it is possible to invert the population of the excitonic two-level system [69]. This population inversion was achieved working in an incoherent regime with the dephasing time which was shorter than the pulse duration. The mechanism behind this phonon-assisted population inversion was triggered with a circularly polarized laser pulse, used to excite an InGaAs/GaAs QD. The laser was positively detuned inside the phonon sideband which has, typically, a maximum at 1 meV above the exciton zero phonon line. In the mentioned experiment, the laser bandwidth (0.2 meV) was much larger than the exciton FSS (13 μeV), meaning that the exciton spin dynamics were neglected. Also from a theoretical point of view, the QD could be treated as a two-level system, being the laser far away from the biexciton transition. Applying a strong laser pulse on the QD, the system is not any longer composed by the two bare states $|0\rangle$ and $|X\rangle$ (with relative populations C_0 and C_X) but by two dressed states (see Fig. 3.23a). In the rotating frame, the system could be described by $|0_R\rangle$, composed by the ground state and the N excitation laser photons, and by $|X_R\rangle$ which is made of the exciton state and the laser field with $N-1$ photons. The reported Hamiltonian had the form:

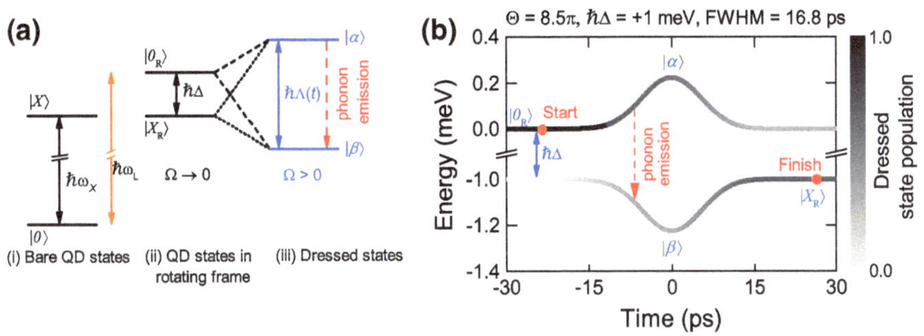

Fig. 3.23 **a** Sketch of the energy diagram for QD bare states in lab frame (*i*) and rotating frame (*ii*). $|0\rangle$ ($|0_R\rangle$) and $|X\rangle$ ($|X_R\rangle$) refer to ground and exciton state in lab (rotating) frame. The exciton transition energy is indicated as $\hbar\omega_X$ while the laser angular frequency is marked as ω_L. The positive detuning from the exciton transition is indicated as $\hbar\Delta$. Optically dressed states are indicated as $|\alpha\rangle$ and $|\beta\rangle$ (*iii*). **b** Time evolution of the dressed QD states during the absorption of a laser pulse (8.5π and $\hbar\Delta = +1$ meV. The instantaneous population of the states is indicated by the *gray* scale [69]

$$H_{QD} = -\hbar\Delta|X_R\rangle\langle X_R| + \frac{\hbar\Omega(t)}{2}|0_R\rangle\langle X_R| + h.c. \tag{3.10}$$

In the previous equation, the laser detuning with respect to the exciton is $\Delta = \omega_L - \omega_X$, while $\Omega(t)$ is the Rabi frequency which follows in time the envelope of the laser pulse. The splitting of the two eigenstates $|\alpha\rangle$ and $|\beta\rangle$ is given by the effective Rabi energy $\hbar\Lambda(t) = \hbar\sqrt{\Delta^2 + \Omega(t)^2}$ and described by the admixing angle defined as $2\theta(t) = arctan(\Omega(t)/\Delta)$. The authors further defined the pulse area as $\Theta = \int_{-\infty}^{+\infty}\Omega(t)dt$. Being the QD into a solid state matrix, it interacts with the environment, and through the deformation potential with the longitudinal acoustic (LA) phonons. While the exciton-phonon interaction introduces pure dephasing of the excitonic dipole (see reference 12 in [69]), when driving with a strong laser field, both dressed states have an excitonic component. This would make possible a relaxation between the states assisted by the emission of a phonon with energy $\hbar\Delta$ (see Fig. 3.23a (iii)), in case of positive detunings. This described process is responsible for the different population inversion results that can be obtained using a positively or negatively chirped ARP [62]. A scheme of the phonon-assisted population inversion, in the rotating frame, is shown in Fig. 3.23b. For the case under investigation of a positively detuned laser ($\Delta > 0$), the initial state coincides with the dressed state $|\alpha\rangle$: when the laser is turned on, relaxation to the dressed state $|\beta\rangle$, which has a lower energy, is activated. Looking at the same process in the lab frame, the absorption of a laser photon brings a consequent formation of an exciton plus the emission of a phonon. In [69] the authors used a path-integral method to model the experiment and the gray scale in Fig. 3.23b shows the calculated evolution over time of the population between the dressed states $|\alpha\rangle$ and $|\beta\rangle$. The laser activated admixture of the states drops with the decrease of the laser pulse intensity. In the case that the phonon relaxation, integrated over time, is intense enough, population inversion can occur, with the lower dressed state being populated. The advantage of the described method relies on the robustness against fluctuations of the pulse area or of the laser detuning. The maximum value of inverted population that one can reach is limited by the thermal occupation of the two states, given by $C_X - C_0 = tanh(\hbar\Delta/2k_B T)$: in [69] a maximum of 0.91 was estimated. Measurements for different pulse areas (not shown) demonstrated that an exciton population of 0.67 ± 0.06 was achieved for the maximally available experimental power.

3.4.2 Phonon-Assisted Biexciton Excitation

In a similar manner, also the biexciton state can be prepared using a phonon-assisted two-photon excitation scheme. In [70], a controlled two-photon excitation scheme was used for this purpose. A meticulous choice of the laser energy as well as of the pulse duration can optimize the state preparation working at the point where the relaxation processes due to the carrier-phonon coupling are at the maximum efficiency. A standard Ti:sapphire laser was used to generate the pulses in combination with a pulse shaper to control their temporal profile. Linearly polarized photons were sent from the side of the sample with a crossed-polarization collection from the top (see Fig. 3.7). Figure 3.24a shows a PL spectrum for above GaAs bandgap excitation. Using a resonant two-photon excitation scheme with a shaped laser beam (similar to the one described in [66]) the PL spectrum changed as in Fig. 3.24b, where XX and X appear with the same intensity. Power dependent measurements for two-photon resonant pumping (at a laser pulse width of 13 ps) led to the well-known Rabi oscillations (Fig. 3.25a), where the solid lines came from a numerically exact real-time path-integral simulation. From the measured data, the biexciton occupation factor at the π-pulse was found to be $C_{XX} = 0.94 \pm 0.01$. The observed decrease of the Rabi frequency with increasing pulse area is known to be due to the two-photon

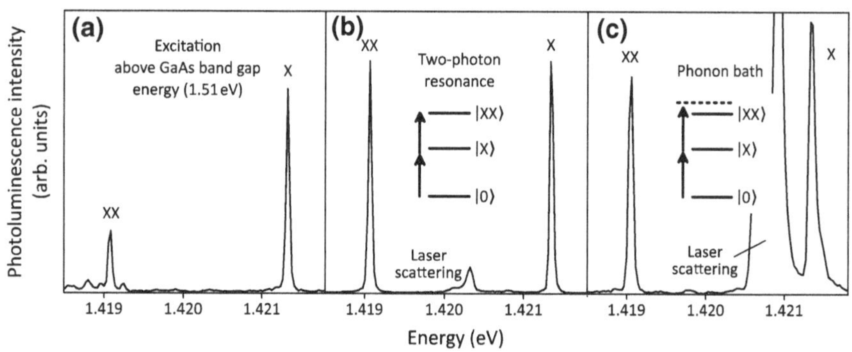

Fig. 3.24 **a**, **b** and **c** Emission spectra of a single QD under different excitation schemes: non-resonant (above GaAs band gap) pumping **a**, resonant two-photon pumping **b** and phonon-assisted pumping (13 ps pulses, 0.65 meV positively detuned) **c**. The *insets* in **b** and **c** show the system energy levels [70]

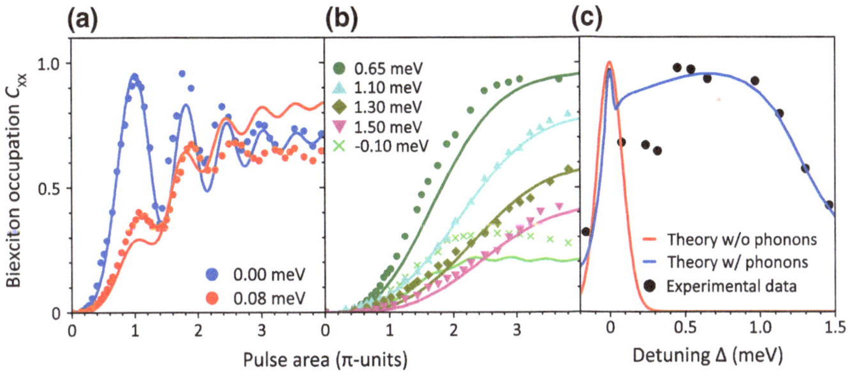

Fig. 3.25 **a** Behavior as a function of the pump area (13 ps pulses) of the biexciton occupation. *Solid lines* come from path-integral calculations. Data for zero detuning (*blue*) and 0.08 meV detuning (*red*) are shown. **b** Additional data for various detunings (see legend). **c** Maximal biexciton occupation obtained at different laser detunings. Experimental data (*dots*) are also supported by theoretical calculations (*solid lines*), for the case without phonons (*red*) and with phonons (*blue*) [70]

excitation process, while the decrease in intensity was due to coupling with acoustic phonons. It is worth noting that the mean oscillations above 0.5 were attributed to a slight chirp in the laser pulse but accordingly to simulations this chirp had a very small effect for off-resonant laser pulses. The spectrum for off-resonant pumping is shown in Fig. 3.24c, for a laser detuning of $\Delta = 0.65$ meV (this quantity Δ is related to the two-photon biexciton resonance). In such a configuration, i.e. a system formed by few isolated and discrete energy levels excited with a detuned laser, it is commonly known that the Rabi oscillations decreased with increasing laser frequency. This behavior was also confirmed and observed as shown in Fig. 3.25a, b where the Rabi oscillations decrease up to a complete disappearance. The highest biexciton population was found for values of detuning between 0.45 and 1.00 meV reaching up to $C_{XX} = 0.95 \pm 0.02$, value very close to the one measured at a resonant π-pulse. The advantage relies on the large, flat plateau which shows the robustness of the off-resonant phonon-mediated pumping with respect to power fluctuations. On the other hand, for too large detunings, the efficiency of the process started to degrade, leading to a decrease of the XX occupation. At this point it is also necessary to underline that the phonon-induced relaxation is only possible in case of a dressed two-level system and that such relaxation is responsible for the thermal occupation of the dressed state. From the physical point of view, the phonon-assisted biexciton excitation is equivalent to the excitonic case, while the representation is not as simple as the one depicted for the X case. In the case of positive detunings, this reflects to the population of the XX while for negative detunings the biexciton is no longer the final state for the thermalization at low temperatures, since the energetic order of the dressed states changes. Intuitively, the excess energy provided by the pump laser will result in emission of a phonon (see sketch in Fig. 3.24c), while for negative detunings the energy gap should be filled by absorption of a phonon that is less probable than phonon emission at low temperature. This explanation is reflected in the observation of Fig. 3.25c, where the XX occupation dropped for $\Delta < 0$. The peak at zero detuning was given by the direct, resonant, two-photon pumping, while the drop at small positive detunings relates to the region in which the Rabi oscillations smear out giving way to the off-resonant pumping behavior. The maximum in the biexciton occupation is obtained when the most pronounced phonons are in resonance with the relevant dressed-state transition. A further increase of the detuning only lead to leaving this resonance condition, resulting in a less efficient biexciton population. Measurements performed with different pulse duration demonstrated that the pulse has to be long enough to allow the relaxation of the system in the energetically lowest dressed state. A too short pulse does not allow such relaxation, resulting in a less efficient XX preparation. Finally, the observed coherence time of photons generated with this phonon-assisted scheme was comparable with the coherence of the resonantly generated photons: 277 ± 8 ps and 271 ± 7 ps respectively. Together with the aforementioned advantage related to the robustness against power fluctuations, another advantage of off-resonance, phonon-mediated pumping scheme is given by a relatively easy possibility to filter out the pump laser, particularly important feature for schemes based on two-photon emission.

3.5 Summary and Outlook

In the last decade, research and technological advances made semiconductor QDs highly attractive as non-classical light sources for quantum information and computation applications. The on-demand nature of the emission together with highly pure single-photon emission fulfill the needs of quantum information protocols. In addition to that, in this chapter we discussed the fundamental properties of QD-based quantum light sources, while showing the latest achievements in terms of photon indistinguishability, brightness and coherence. Different excitation methods have been described giving as a take-home message that the resonant excitation, either CW or pulsed, produces sensitive improvements in the source performances. Depending on the experimental needs, i.e. generation of entangled photon pairs, long coherence time, highly bright sources, one can choose the most appropriate pumping scheme. A short summary of the state-of-the-art results discussed in this chapter is reported in Table 3.1. Currently, pulsed resonant excitation at the π-pulse is one of the most used methods since it is relatively straightforward (except for the need of high laser suppression) and allows for the generation of highly indistinguishable and on-demand single photons. For the same reason, also the two-photon excitation of the biexciton carries the same advantages together with an easier suppression of the pump laser (being spectrally separated) and even if high indistinguishability still needs to be achieved, this method is highly attractive for the generation of polarization-entangled photons with superior properties.

The near-unity indistinguishability reached by several independent groups, together with a source brightness one order of magnitude higher than previous state-

Table 3.1 Summary of state-of-the-art properties achieved under different pumping conditions. Advantages and disadvantages are schematized. For pulsed resonant pumping, only results with reported values are cited. Within the remarks section, some of the advantages related to the excitation schemes are also mentioned

Pumping scheme	Coherent preparation	Photon indist.	On-demand emission	Brightness (polarized)	Technical challenges	Remarks
CW low power	Yes	0.926[a] [44]	No	Low		Waveform engineering. Linewidth of exc. laser
CW medium power	Yes	0.90[a] [45]	No	N/A		Fourier transform-limited
CW high power	Yes	0.98[a] [50]	No	N/A		Cascaded photons
Pulsed	Yes	0.996 [55]	Yes	0.16	Require good exc. laser/RF photon discrimination	Brightness limited to 50% for crossed-polarized configuration
	Yes	0.97 [51]	Yes	0.03		
	Yes	0.96 [60]	Yes	> 0.03		
Phonon assisted (X)	No	N/A	Yes	N/A		Easy laser discrimination. Robust against laser fluctuations (power and energy)
Adiabatic rapid passage	Yes	0.995 [62]	Yes	N/A	Generation of chirped excitation pulses	Easy laser discrimination. Robust against laser fluctuations (power and energy)
Spin-flip Raman transitions	Yes	0.95 (pulsed) [64]	Yes[b]	N/A	Several exc. lasers used.	Narrow linewidth (Raman photons). Spin manipulation possible. Generation of photons with different wavelengths
TPE	Yes	0.69 (X) 0.85 (XX) [66]	Yes	N/A	High power required	Easy laser discrimination
Phonon assisted (XX)	No	N/A	Yes	N/A	High power required	Easy laser discrimination. Robust against laser fluctuations (power and energy)

[a]post-selected indistinguishability, defined by detector resolution
[b]deterministic photon emission is achievable using and external magnetic field or a two-lasers excitation scheme

of-the-art demonstrate the reachable quality of QD-based sources paving the way to unprecedented achievable quantum computation schemes. A fundamental task for the future will be to transfer the excellent properties of the InGaAs QDs, which have been here summarized, to other QD materials emitting in the visible and telecom-wavelength regime.

As a final remark, we have also reported about the effectiveness of QD-cavity systems. It is worth noting that huge steps forward have been made in the fabrication of such systems: at present, several different techniques showed their potentials in the deterministic fabrication of single QDs coupled with cavities. Currently, high source brightness has been reached under non-resonant pumping making use of photonic cavities or structures to enhance light extraction. Under pure resonant pumping, despite the highly attractive achieved results, there is still room for improving the source brightness.

Several groups are nowadays focusing their efforts in realizing integrated photonic circuits based on GaAs platform, so taking advantages from the superior QD properties (see Chap. 13). Following a similar argument as before, it is of great interest to transfer such a technology to other wavelengths, in particular with the perspective of utilizing other semiconductor platforms as the highly developed Si-based technology.

References

1. P. Michler, A. Kiraz, C. Becher, W.V. Schoenfeld, P.M. Petroff, L. Zhang, E. Hu, A. Imamoğlu, A quantum dot single-photon turnstile device. Science **290**, 2282 (2000)
2. Z. Yuan, B.E. Kardynal, R.M. Stevenson, A.J. Shields, C.J. Lobo, K. Cooper, N.S. Beattie, D.A. Ritchie, M. Pepper, Electrically driven single-photon source. Science **295**, 102 (2002)
3. N. Akopian, N.H. Lindner, E. Poem, Y. Berlatzky, J. Avron, D. Gershoni, B.D. Gerardot, P.M. Petroff, Entangled photon pairs from semiconductor quantum dots. Phys. Rev. Lett. **96**, 130501 (2006)
4. R.M. Stevenson, R.J. Young, P. Atkinson, K. Cooper, D.A. Ritchie, A.J. Shields, A semiconductor source of triggered entangled photon pairs. Nature **439**, 179 (2006)
5. R. Hafenbrak, S.M. Ulrich, P. Michler, L. Wang, A. Rastelli, O.G. Schmidt, Triggered polarization-entangled photon pairs from a single quantum dot up to 30 K. New J. Phys. **9**, 315 (2007)
6. M. Varnava, D.E. Browne, T. Rudolph, How good must single photon sources and detectors be for efficient linear optical quantum computation? Phys. Rev. Lett. **100**, 060502 (2008)
7. I. Aharonovich, D. Englund, M. Toth, Solid-state single-photon emitters. Nat. Photonics **10**, 631 (2016)
8. A. Muller, E.B. Flagg, P. Bianucci, X.Y. Wang, D.G. Deppe, W. Ma, J. Zhang, G.J. Salamo, M. Xiao, C.K. Shih, Resonance fluorescence from a coherently driven semiconductor quantum dot in a cavity. Phys. Rev. Lett. **99**, 187402 (2007)
9. A.N. Vamivakas, Y. Zhao, C.-Y. Lu, M. Atatüre, Spin-resolved quantum-dot resonance fluorescence. Nat. Phys. **5**, 198 (2009)
10. P. Michler, *Single Semiconductor Quantum Dots* (Springer, Berlin, 2009)
11. C.K. Hong, Z.Y. Ou, L. Mandel, Measurement of subpicosecond time intervals between two photons by interference. Phys. Rev. Lett. **59**, 2044 (1987)
12. H.S. Nguyen, G. Sallen, C. Voisin, P. Roussignol, C. Diederichs, G. Cassabois, Optically gated resonant emission of single quantum dots. Phys. Rev. Lett. **108**, 057401 (2012)

13. A.V. Kuhlmann, J.H. Prechtel, J. Houel, A. Ludwig, D. Reuter, A.D. Wieck, R.J. Warburton, Transform-limited single photons from a single quantum dot. Nat. Commun. **6**, 8204 (2015)
14. K.A. Fischer, K. Müller, K.G. Lagoudakis, J. Vučković, Dynamical modeling of pulsed two-photon interference. New J. Phys. **18**, 113053 (2016)
15. H. Benisty, H. De Neve, C. Weisbuch, Impact of planar microcavity effects on light extraction-Part I: basic concepts and analytical trends. IEEE J. Quantum Electron. **34**, 1612 (1998)
16. N. Gregersen, P. Kaer, J. Mørk, Modeling and design of high-efficiency single-photon sources. IEEE J. Sel. Top. Quantum Electron. **19**, 9000516 (2013)
17. J. Finley, P. Fry, A. Ashmore, A. Lemaître, A. Tartakovskii, R. Oulton, D. Mowbray, M. Skolnick, M. Hopkinson, P.D. Buckle, P. Maksym, Observation of multicharged excitons and biexcitons in a single InGaAs quantum dot. Phys. Rev. B **63**, 161305 (2001)
18. M. Metcalfe, S.M. Carr, A. Muller, G.S. Solomon, J. Lawall, Resolved sideband emission of InAs/GaAs quantum dots strained by surface acoustic waves. Phys. Rev. Lett. **105**, 037401 (2010)
19. S.L. Portalupi, G. Hornecker, V. Giesz, T. Grange, A. Lemaître, J. Demory, I. Sagnes, N.D. Lanzillotti-Kimura, L. Lanco, A. Auffèves, P. Senellart, Bright phonon-tuned single-photon source. Nano Lett. **15**, 6290 (2015)
20. O. Gazzano, S. Michaelis de Vasconcellos, C. Arnold, A. Nowak, E. Galopin, I. Sagnes, L. Lanco, A. Lemaître, P. Senellart, Bright solid-state sources of indistinguishable single photons. Nat. Commun. **4**, 1425 (2013)
21. J. Claudon, J. Bleuse, N.S. Malik, M. Bazin, P. Jaffrennou, N. Gregersen, C. Sauvan, P. Lalanne, J.-M. Gérard, A highly efficient single-photon source based on a quantum dot in a photonic nanowire. Nat. Photonics **4**, 174 (2010)
22. M. Munsch, N.S. Malik, E. Dupuy, A. Delga, J. Bleuse, J.-M. Gérard, J. Claudon, N. Gregersen, J. Mørk, Dielectric GaAs antenna ensuring an efficient broadband coupling between an InAs quantum dot and a gaussian optical beam. Phys. Rev. Lett. **110**, 177402 (2013)
23. M. Gschrey, A. Thoma, P. Schnauber, M. Seifried, R. Schmidt, B. Wohlfeil, L. Krüger, J.-H. Schulze, T. Heindel, S. Burger, F. Schmidt, A. Strittmatter, S. Rodt, S. Reitzenstein, Highly indistinguishable photons from deterministic quantum-dot microlenses utilizing three-dimensional in situ electron-beam lithography. Nature Commun. **6**, 7662 (2015)
24. A. Schwagmann, S. Kalliakos, I. Farrer, J.P. Griffiths, G.A.C. Jones, D.A. Ritchie, A.J. Shields, On-chip single photon emission from an integrated semiconductor quantum dot into a photonic crystal waveguide. Appl. Phys. Lett. **99**, 261108 (2011)
25. A. Laucht, S. Pütz, T. Günthner, N. Hauke, R. Saive, S. Frédérick, M. Bichler, M.-C. Amann, A.W. Holleitner, M. Kaniber, J.J. Finley, Awaveguide-coupled on-chip single-photon source. Phys. Rev. X **2**, 011014 (2012)
26. M. Arcari, I. Söllner, A. Javadi, S. Lindskov Hansen, S. Mahmoodian, J. Liu, H. Thyrrestrup, E.H. Lee, J.D. Song, S. Stobbe, P. Lodahl, Near-unity coupling efficiency of a quantum emitter to a photonic crystal waveguide. Phys. Rev. Lett. **113**, 093603 (2014)
27. A. Einstein, B. Podolsky, N. Rosen, Can quantum-mechanical description of physical reality be considered complete? Phys. Rev. **47**, 777 (1935)
28. G. Vidal, R.F. Werner, Computable measure of entanglement. Phys. Rev. A **65**, 032314 (2002)
29. S. Hill, W.K. Wootters, Entanglement of a pair of quantum bits. Phys. Rev. Lett. **78**, 5022 (1997)
30. V. Coffman, J. Kundu, W.K. Wootters, Distributed entanglement. Phys. Rev. A **61**, 052306 (2000)
31. D.F.V. James, P.G. Kwiat, W.J. Munro, A.G. White, Measurement of qubits. Phys. Rev. A **64**, 052312 (2001)
32. C.H. Bennett, H.J. Bernstein, S. Popescu, B. Schumacher, Concentrating partial entanglement by local operations. Phys. Rev. A **53**, 2046 (1996)
33. A. Dousse, J. Suffczynski, A. Beveratos, O. Krebs, A. Lemaître, I. Sagnes, J. Bloch, P. Voisin, P. Senellart, Ultrabright source of entangled photon pairs. Nature **466**, 217 (2010)
34. K.D. Jöns, R. Hafenbrak, P. Atkinson, A. Rastelli, O.G. Schmidt, P. Michler, Quantum state tomography measurements on strain-tuned InxGa1-xAs/GaAs quantum dots. Phys. Status Solidi B **249**, 697 (2012)

35. Y. Chen, J. Zhang, M. Zopf, K. Jung, Y. Zhang, R. Keil, F. Ding, O.G. Schmidt, Wavelength-tunable entangled photons from silicon-integrated III-V quantum dots. Nat. Commun. **7**, 10387 (2016)
36. R. Trotta, J. Martn-Snchez, J.S. Wildmann, G. Piredda, M. Reindl, C. Schimpf, E. Zallo, S. Stroj, J. Edlinger, A. Rastelli, Wavelength-tunable sources of entangled photons interfaced with atomic vapours. Nat. Commun. **7**, 10375 (2016)
37. I.N. Stranski, L. Krastanow, Sitz. Ber. Akad. Wiss., Math.-naturwiss. Kl. Abt. IIb **146**, 797 (1938)
38. B. Patton, W. Langbein, U. Woggon, Trion, biexciton, and exciton dynamics in single self-assembled CdSe quantum dots. Phys. Rev. B **68**, 125316 (2003)
39. A.J. Bennett, D.C. Unitt, A.J. Shields, P. Atkinson, D.A. Ritchie, Influence of exciton dynamics on the interference of two photons from a microcavity single-photon source. Opt. Express **13**, 7772 (2005)
40. V. Giesz, S.L. Portalupi, T. Grange, C. Antn, L. De Santis, J. Demory, N. Somaschi, I. Sagnes, A. Lematre, L. Lanco, A. Auffèves, P. Senellart, Cavity-enhanced two-photon interference using remote quantum dot sources. Phys. Rev. B **92**, 161302 (2015)
41. H.S. Nguyen, G. Sallen, C. Voisin, P. Roussignol, C. Diederichs, G. Cassabois, Ultra-coherent single photon source. Appl. Phys. Lett. **99**, 261904 (2011)
42. A.J. Bennett, J.P. Lee, D.J.P. Ellis, T. Meany, E. Murray, F.F. Floether, J.P. Griffths, I. Farrer, D.A. Ritchie, A.J. Shields, Cavity-enhanced coherent light scattering from a quantum dot. Sci. Adv. **2**, e1501256 (2016)
43. C. Matthiesen, A.N. Vamivakas, M. Atatüre, Subnatural Linewidth Single Photons from a Quantum Dot. Phys. Rev. Lett. **108**, 093602 (2012)
44. C. Matthiesen, M. Geller, C.H.H. Schulte, C. Le Gall, J. Hansom, Z. Li, M. Hugues, E. Clarke, M. Atatüre, Phase-locked indistinguishable photons with synthesized waveforms from a solid-state source. Nat. Commun. **4**, 1600 (2013)
45. S. Ates, S. Ulrich, S. Reitzenstein, A. Löffler, A. Forchel, P. Michler, Post-selected indistinguishable photons from the resonance fluorescence of a single quantum dot in a microcavity. Phys. Rev. Lett. **103**, 167402 (2009)
46. E.M. Purcell, Spontaneous emission probabilities at radio frequencies. Phys. Rev. **69**, 681 (1946)
47. B.R. Mollow, Power spectrum of light scattered by two-level systems. Phys. Rev. **188**, 1969–1975 (1969)
48. A. Ulhaq, S. Weiler, S.M. Ulrich, R. Roßbach, M. Jetter, P. Michler, Cascaded single-photon emission from the Mollow triplet sidebands of a quantum dot. Nat. Photonics **6**, 238 (2012)
49. S.M. Ulrich, S. Ates, S. Reitzenstein, A. Löffler, A. Forchel, P. Michler, Dephasing of Mollow triplet sideband emission of a resonantly driven quantum dot in a microcavity. Phys. Rev. Lett. **106**, 247403 (2011)
50. S. Weiler, D. Stojanovic, S.M. Ulrich, M. Jetter, P. Michler, Postselected indistinguishable single-photon emission from the Mollow triplet sidebands of a resonantly excited quantum dot. Phys. Rev. B **87**, 241302 (2013)
51. Y.-M. He, Y. He, Y.-J. Wei, D. Wu, M. Atatüre, C. Schneider, S. Höfling, M. Kamp, C.-Y. Lu, J.-W. Pan, On-demand semiconductor single-photon source with near-unity indistinguishability. Nat. Nanotechnol. **8**, 213 (2013)
52. H.J. Kimble, The quantum internet. Nature **453**, 1023 (2008)
53. S.D. Barrett, P. Kok, Efficient high-fidelity quantum computation using matter qubits and linear optics. Phys. Rev. A **71**, 060301 (2005)
54. X. Ding, Y. He, Z.-C. Duan, N. Gregersen, M.-C. Chen, S. Unsleber, S. Maier, C. Schneider, M. Kamp, S. Höfling, C.-Y. Lu, J.-W. Pan, On-demand single photons with high extraction efficiency and near-unity indistinguishability from a resonantly driven quantum dot in a micropillar. Phys. Rev. Lett. **116**, 020401 (2016)
55. N. Somaschi, V. Giesz, L. De Santis, J.C. Loredo, M.P. Almeida, G. Hornecker, S.L. Portalupi, T. Grange, C. Antón, J. Demory, C. Gómez, I. Sagnes, N.D. Lanzillotti-Kimura, A. Lemaître, A. Auffèves, A.G. White, L. Lanco, P. Senellart, Near-optimal single-photon sources in the solid state. Nat. Photonics **10**, 340 (2016)

56. A. Kiraz, M. Atatüre, A. Imamoglu, Quantum-dot single-photon sources: prospects for applications in linear optics quantum-information processing. Phys. Rev. A **69**, 032305 (2004)
57. J.-W. Pan, Z.-B. Chen, C.-Y. Lu, H. Weinfurter, A. Zeilinger, M. Żukowski, Multiphoton entanglement and interferometry. Rev. Mod. Phys. **84**, 777838 (2012)
58. M. Barbieri, Effects of frequency correlation in linear optical entangling gates operated with independent photons. Phys. Rev. A **76**, 043825 (2007)
59. S. Unsleber, Y.-M. He, S. Gerhardt, S. Maier, C.-Y. Lu, J.-W. Pan, N. Gregersen, M. Kamp, C. Schneider, S. Höfling, Highly indistinguishable on-demand resonance fluorescence photons from a deterministic quantum dot micropillar device with 74% extraction efficiency. Opt. Express **24**, 8539 (2016)
60. H. Wang, Z.-C. Duan, Y.-H. Li, S. Chen, J.-P. Li, Y.-M. He, M.-C. Chen, Y. He, X. Ding, C.-Z. Peng, C. Schneider, M. Kamp, S. Höfling, C.-Y. Lu, J.-W. Pan, Near-transform-limited single photons from an efficient solid-state quantum emitter. Phys. Rev. Lett. **116**, 213601 (2016)
61. K. Bergmann, N.V. Vitanov, B.W. Shore, Perspective: stimulated Raman adiabatic passage: the status after 25 years. J. Chem. Phys. **142**, 170901 (2015)
62. Y. Wei, Y.-M. He, M. Chen, Y. Hu, Y. He, D. Wu, C. Schneider, M. Kamp, S. Höfling, C.-Y. Lu, J.-W. Pan, Deterministic and robust generation of single photons from a single quantum dot with 99.5% indistinguishability using adiabatic rapid passage. Nano Lett. **14**, 6515 (2014)
63. G. Fernandez, T. Volz, R. Desbuquois, A. Badolato, A. Imamoglu, Optically tunable spontaneous Raman fluorescence from a single self-assembled InGaAs quantum dot. Phys. Rev. Lett. **103**, 087406 (2009)
64. Y. He, Y.M. He, Y.J. Wei, X. Jiang, M.C. Chen, F.L. Xiong, Y. Zhao, C. Schneider, M. Kamp, S. Höfling, C.Y. Lu, J.W. Pan, Indistinguishable tunable single photons emitted by spin-flip Raman transitions in InGaAs quantum dots. Phys. Rev. Lett. **111**, 237403 (2013)
65. K. Brunner, G. Abstreiter, G. Bhm, G. Trnkle, G. Weimann, Sharp-line photoluminescence and two-photon absorption of zero-dimensional biexcitons in a GaAs/AlGaAs structure. Phys. Rev. Lett. **73**, 1138 (1994)
66. M. Müller, S. Bounouar, K.D. Jöns, M. Glässl, P. Michler, On-demand generation of indistinguishable polarization-entangled photon pairs. Nat. Photonics **8**, 224 (2014)
67. D. Huber, M. Reindl, Y. Huo, H. Huang, J.S. Wildmann, O.G. Schmidt, A. Rastelli, R. Trotta, Highly indistinguishable and strongly entangled photons from symmetric GaAs quantum dots (2016). arXiv:1610.06889v1
68. H. Jayakumar, A. Predojević, T. Huber, T. Kauten, G.S. Solomon, G. Weihs, Deterministic photon pairs and coherent optical control of a single quantum dot. Phys. Rev. Lett. **110**, 135505 (2013)
69. J.H. Quilter, A.J. Brash, F. Liu, M. Glässl, A.M. Barth, V.M. Axt, A.J. Ramsay, M.S. Skolnick, A.M. Fox, Phonon-assisted population inversion of a single InGaAs/GaAs quantum dot by pulsed laser excitation. Phys. Rev. Lett. **114**, 137401 (2015)
70. S. Bounouar, M. Müller, A.M. Barth, M. Glässl, V.M. Axt, P. Michler, Phonon-assisted robust and deterministic two-photon biexciton preparation in a quantum dot. Phys. Rev. B **91**, 161302 (2015)

Chapter 4
Coherent Control of Dark Excitons in Semiconductor Quantum Dots

E.R. Schmidgall, I. Schwartz, D. Cogan, L. Gantz, Y. Don and D. Gershoni

Abstract We review studies of the quantum dot confined dark exciton and demonstrate its use as a matter qubit. The dark exciton is an optically forbidden semiconductor electronic excitation, in which an electron-hole pair is generated with parallel spin projections. This optcally inactive excitation lives orders of magnitude longer than the corresponding optically active excitation, the bright exciton, in which the pair has anti-parallel spins. We show that despite its optical inactivity, the dark exciton can be deterministically generated in any desired coherent superposition of its two eigenstates using a single picosecond optical pulse. We provide lower bounds for the dark exciton life and coherence times and show that its coherent state can be fully controlled using short optical pulses. We also study its behavior in an externally applied magnetic field and present a method for its optical depletion from the quantum dot. Our results demonstrate that the dark exciton is an excellent matter spin qubit.

4.1 Bright and Dark Excitons in Quantum Dots

In semiconductors, the absorption of a photon promotes an electron from the full valence band to the empty conduction band. This fundamental excitation is particularly efficient because the valence band is formed from molecular p-like orbitals while the conduction band is formed from s-like orbitals. The dipole moment between these orbitals strongly interacts with the electric field of the light. Consequently, this photoexcitation does not alter the spin of the promoted electron. The excitation leaves an excited electron in the conduction band and a missing electron in the valence band.

E.R. Schmidgall (✉) · I. Schwartz · D. Cogan · L. Gantz · Y. Don · D. Gershoni (✉)
Department of Physics and the Solid State Institute, Technion Israel Institute of Technology, 32000 Haifa, Israel
e-mail: eschmid@uw.edu

D. Gershoni
e-mail: dg@physics.technion.ac.il

E.R. Schmidgall
Department of Physics, University of Washington, Seattle, WA 98195, USA

P. Michler (ed.), *Quantum Dots for Quantum Information Technologies*, Nano-Optics and Nanophotonics, DOI 10.1007/978-3-319-56378-7_4

When the missing electron is treated as a heavy hole with opposite quantum numbers (positive rather than negative charge, spin up instead of spin down), this electron-heavy hole combination can be treated as a two-body system. Absorption of a photon thus results, in this picture, in the creation of an electron-heavy hole (e-h) pair with antiparallel spins, or a bright exciton (BE).

Heavy holes are holes where their spin is aligned with the orbital molecular momentum for a total angular momentum projection of $\pm 3/2$ on the quantum dot (QD) growth axis. There are also light holes, where the hole spin is antiparallel to the orbital momentum for a total angular momentum projection of $\pm 1/2$ on the QD growth axis. In InAs/GaAs self-assembled QDs, the quantum size and lattice mismatch strain result in a considerable energy difference between the heavy holes and the light holes. Consequently, the lowest energy BE states are composed of an electron-heavy hole pair, and this state has total integer spin projections of ± 1 on the QD growth axis, reflecting the difference in orbital momentum between the ground valence and excited conduction states. The spin of these BEs can be straightforwardly coherently "written" [2], "read," and manipulated [3, 4] using the polarization of laser light, due to the fact that this orbital momentum is aligned with the electronic spin (Chap. 10). Thus, for example, optical recombination of an electron hole pair in which the electron spin projection is $+1/2$ and that of the hole is $-3/2$ ($\uparrow\Downarrow$ BE with total spin -1) results in emission of a left-hand circularly polarized (σ^-) photon along the symmetry axis of the QD, carrying the energy and orbital momentum released by the pair recombination. Similarly, rectilinear horizontal (H) and vertical (V) polarizations are linear combinations of σ^+ and σ^-, circular polarizations.

A dark exciton (DE) is an electronic excitation in which the heavy-hole spin is parallel to that of the electron. Thus, in QDs, the DE has a total integer spin of 2 [5], with projections of ± 2 on the QD growth axis. This reflects the difference in both angular momentum and spin between the valence band and conduction band electron states. These DEs are almost optically inactive since photons barely interact with electronic spin. Thus the lifetime of the DE is orders of magnitude longer than that of the BE [6, 7].

In the absence of any external magnetic field or exchange interaction, all spin configurations are degenerate in energy. However, the exchange interaction between the electron and hole removes the degeneracy between the four possible electron-hole spin pairs [1, 8, 9]. For epitaxially grown semiconductor QDs on [001] oriented substrates one generally assumes combined lattice and structural symmetry of C_{2v} (i. e., symmetry under rotations of π radians around the structural symmetry axis [001], and under two reflections about perpendicular planes which contain the symmetry axis: the [110]–[001] and the [$1\bar{1}0$]–[001] planes [10–12]).

From general symmetry considerations, it can be shown that for C_{2v} symmetrical QDs the exchange interaction Hamiltonian written in the basis $|+1\rangle$, $|-1\rangle$, $|+2\rangle$, $|-2\rangle$ has the following form [10–12]:

$$\mathcal{H}_{C_{2v}} = \frac{1}{2}\begin{pmatrix} \Delta_0 & \Delta_1^* & 0 & 0 \\ \Delta_1 & \Delta_0 & 0 & 0 \\ 0 & 0 & -\Delta_0 & \Delta_2^* \\ 0 & 0 & \Delta_2 & -\Delta_0 \end{pmatrix} \tag{4.1}$$

Here Δ_i, $i = 0, 1, 2$ are parameters that one either measures [5, 7, 8, 13, 14] or calculates using simplified models [10, 12, 15, 16].

For these QDs, there is no mixing between the DE and BE eigenstates. The two subspaces are energetically separated by $\Delta_0 \sim 300\,\mu\text{eV}$ [13].

$\Delta_{1,2}$ are in general complex numbers [10] and can be expressed as $\Delta_{1,2} = \delta_{1,2}e^{2i\theta_{1,2}}$, where $\delta_{1,2}$ are positive numbers. Thus the eigenvalues of the Hamiltonian are expressed as

$$E_{\text{BE}_\pm} = \tfrac{1}{2}(\Delta_0 \pm \delta_1) \tag{4.2a}$$

$$E_{\text{DE}_\pm} = \tfrac{1}{2}(-\Delta_0 \pm \delta_2) \tag{4.2b}$$

and the eigenvectors as

$$\upsilon_{\text{BE}_\pm} = \frac{1}{\sqrt{2}}\begin{pmatrix} e^{-i\theta_1} \\ \pm e^{i\theta_1} \end{pmatrix} \tag{4.3a}$$

$$\upsilon_{\text{DE}_\pm} = \frac{1}{\sqrt{2}}\begin{pmatrix} e^{-i\theta_2} \\ \pm e^{i\theta_2} \end{pmatrix} \tag{4.3b}$$

The parameter $\delta_1 \sim 30\,\mu\text{eV}$, which removes the degeneracy between the two BE states, is closely related to the oscillator strength for optical transitions to these fundamental excitations [11, 16]. Thus, using these relations, one finds that the positive (υ_{BE_+}) and negative (υ_{BE_-}) eigenstates of the BE have dipole matrix elements linearly polarized along the $(\cos\theta_1, -\sin\theta_1, 0)$ and $(\sin\theta_1, \cos\theta_1, 0)$ directions respectively, where θ_1 is measured from the [100] crystallographic direction.

Atomistic calculations [17, 18] and accumulated experimental data [7, 19–21] imply that most often the lowest (highest) energy BE emission spectral line is polarized along the $[1\bar{1}0]$ ([110]) direction, even for a circularly symmetric QD. If one defines the lowest energy line polarization as horizontal polarization (i.e. symmetrical superposition of right and left hand circular polarizations) this situation is described by $\theta_1 = 135°$.

The value of Δ_2 is mostly determined by the short range e-h exchange interaction [15] which has the symmetry of the unit cell. This implies that Δ_2 must be a real number [8], thus compelling θ_2 to be either 0° or 90°. Atomistic model simulations [18, 22], as well as recent experimental data [7] indicate that $\theta_2 = 90°$, and that $\delta_2 \sim 1.5\,\mu\text{eV}$, is rather small.

The excitonic energy fine structure, the BE and DE spin eigenstates, and the optical transitions resulting from the C_{2v} Hamiltonian are schematically depicted in Fig. 4.1.

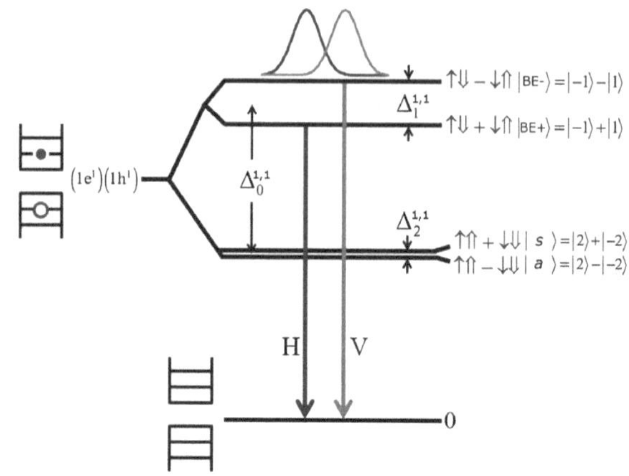

Fig. 4.1 Spin states and fine structure of a QD-confined exciton, showing both BE and DE eigenstates and energy splittings. Optical recombination is represented by an arrow labeled with the corresponding photon polarization. Image from [1]

4.1.1 The Optical Activity and Oscillator Strength of the Dark Exciton

From the dipole approximation, it follows that the DEs are completely dark. However, despite its name–even for C_{2v} symmetric QDs–the DE is not completely dark. Residual heavy-hole light-hole mixing gives one of the DE eigenstates a small $\hat{z}$-polarized dipole moment, while the other one is totally dark [10, 15, 23]. Realistic atomistic model calculations of InAs/GaAs self assembled QDs yield $\hat{z}$-polarized optical activity of one DE eigenstate that is 3–6 orders of magnitude weaker than the BE, and the optical activity of the second DE eigenstate is much weaker [17, 18, 24, 25]. This DE emission was experimentally observed in PL collected orthogonal to the QD growth direction [25].

In reality, ideally symmetrized systems of macroscopic scale are extremely rare. Recent theoretical studies of epitaxial growth of strained heterostructures [26] show that indeed self-assembled QDs can actually grow highly asymmetrical, largely deviating from C_{2v} symmetry. In an asymmetrical QD, the subspaces of the BEs and DEs are no longer separated and their eigenstates are mixed [8].

We model the symmetry reduction by introducing a small angle φ by which the symmetry axis of the QD is tilted relative to the [001] crystallographic direction. As a result, the quantization axis of the QD potential is no longer aligned with the underlying semiconductor lattice. If for simplicity, one also assumes that the electron and heavy-hole envelope wavefunction symmetry axes are inclined by the same angle φ relative to the [001] crystallographic direction, it follows that the Hamiltonian of a C_{2v} QD as expressed by (4.2) and (4.3) is modified to the following reduced symmetry (C_s) Hamiltonian:

$$\mathcal{H}'_{C_s} = \frac{1}{2}\begin{pmatrix} \Delta_0+\delta_1 & 0 & 0 & 0 \\ 0 & \Delta_0-\delta_1 & 0 & 2\delta_1\varphi \\ 0 & 0 & -\Delta_0+\delta_2 & 0 \\ 0 & 2\delta_1\varphi & 0 & -\Delta_0-\delta_2 \end{pmatrix} \tag{4.4}$$

This Hamiltonian can be expressed in terms of the original basis $|+1\rangle, |-1\rangle, |+2\rangle, |-2\rangle$:

$$\mathcal{H}_{C_s} = \frac{1}{2}\begin{pmatrix} \Delta_0 & \delta_1 e^{-2i\theta_1} & i\delta_1\varphi e^{-i(\theta_1-\theta_2)} & -i\delta_1\varphi e^{-i(\theta_1+\theta_2)} \\ \delta_1 e^{2i\theta_1} & \Delta_0 & -i\delta_1\varphi e^{i(\theta_1+\theta_2)} & i\delta_1\varphi e^{i(\theta_1-\theta_2)} \\ -i\delta_1\varphi e^{i(\theta_1-\theta_2)} & i\delta_1\varphi e^{-i(\theta_1+\theta_2)} & -\Delta_0 & \delta_2 e^{-2i\theta_2} \\ i\delta_1\varphi e^{i(\theta_1+\theta_2)} & -i\delta_1\varphi e^{-i(\theta_1-\theta_2)} & \delta_2 e^{2i\theta_2} & -\Delta_0 \end{pmatrix} \tag{4.5}$$

Noting that $\Delta_0 \gg \delta_1 \gg \delta_2$ and $|\varphi| \ll 1$, one can approximately express the eigenenergies and eigenstates of the reduced symmetry C_s Hamiltonian in terms of the eigenstates of the symmetric one ((4.2) and (4.3)):

$$E_1 = \frac{+\Delta_0+\delta_1}{2}, \qquad v_1 = \Big(1, 0, 0, 0\Big) \tag{4.6a}$$

$$E_2 = \frac{+\Delta_0-\delta_1}{2} + \frac{(\delta_1\varphi)^2}{\Delta_0}, \qquad v_2 = N\Big(0, 1, 0, \frac{\delta_1\varphi}{\Delta_0}\Big) \tag{4.6b}$$

$$E_3 = \frac{-\Delta_0+\delta_2}{2}, \qquad v_3 = \Big(0, 0, 1, 0\Big) \tag{4.6c}$$

$$E_4 = \frac{-\Delta_0-\delta_2}{2} - \frac{(\delta_1\varphi)^2}{\Delta_0}, \qquad v_4 = N\Big(0, -\frac{\delta_1\varphi}{\Delta_0}, 0, 1\Big) \tag{4.6d}$$

where $N = [1 + (\delta_1\varphi/\Delta_0)^2]^{-0.5}$ is a normalization factor.

The symmetry reduction therefore results in a small mixing between the BE and DE states. For the lowest energy BE state and the lowest energy DE state, this mixing is mainly constructive, while for the other pair it is destructive. This small mixing hardly affects the BE state [27], but greatly enhances the optical activity of one of the DE states such that it is linearly H polarized like the lowest BE state. A realistic estimation of $|\varphi| \sim 0.2$ yields a DE oscillator strength which is 2500 times smaller than that of the BE, in quantitative agreement with recent experimental reports [7], which we review below.

4.2 The Dark Exciton as a Spin Qubit

The fields of quantum information processing (QIP) and quantum communication have generated substantial interest in the past several decades. This is because quantum computers can potentially solve certain problems much faster than what is currently possible with classical computing, the best known examples being Shor's quantum algorithm for factorizing large numbers [28] and Grover's algorithm for unstructured search [29].

In classical computing, information is stored as a binary bit, taking either of two discrete values: 0 or 1. QIP relies on quantum bits, or "qubits" [30, 31], which are simply quantum two-level systems. These qubits have the same two-level structure as classical bits, but they differ from classical bits in two main ways. The first is *superposition*. Qubits can be generated in a coherent superposition of the basis $|0\rangle$ and $|1\rangle$ logical states,

$$|\psi\rangle = \alpha|0\rangle + \beta|1\rangle. \tag{4.7}$$

The second is *entanglement* between multiple qubits, where a multi-qubit state cannot be written as a simple product of single qubit states (Chaps. 7, 8 and 12). It is these two features, superposition and entanglement, that provide the parallelism underlying many quantum algorithms, which use qubits to perform calculations by applying a series of gate operations and measuring the resulting qubit state [28–31].

A candidate qubit system needs to meet several criteria, summarized by David DiVincenzo [31] as follows:

1. The qubit must be well-characterized and scalable.
2. It must be possible to initialize the qubit in a known pure state ("coherent writing").
3. The qubit must have a long coherence time, relative to the time required to perform gate operations.
4. It must be possible to develop a universal system of quantum gates for the qubit.
5. It must be possible to measure a selected qubit ("reading").

For applications in quantum communication, two more features are necessary [31]:

1. The ability to transfer a qubit from a stationary qubit (i.e. a matter qubit) to a flying qubit (i.e. a polarized photon).
2. The ability to faithfully transmit the flying qubits to distributed locations.

There are several candidates for a physical implementation of a qubit, reviewed by Ladd et al. in [32] for quantum computing applications and by Kimble in [33] for quantum communications applications. The most promising candidates are ions in electrical traps [34, 35], neutral atoms in optical lattices [36, 37], nuclear magnetic resonances of molecules in liquid solutions [38, 39], superconducting circuits [40–42], nitrogen-vacancy centers in diamond [43, 44], and individual spins in both gate-defined [45] and self-assembled QDs [46, 47]. Self-assembled semiconductor QDs are considered one of the best candidates for interfacing matter qubits (such as spins) with flying qubits (photons).

QD-confined carrier spins such as the electron [48–50], heavy hole [51–53], and BE [2–4] have been considered as candidate qubits for QIP. For all of these carriers, techniques for initialization of a spin state and its coherent control have been demonstrated (Chaps. 9 and 10).

For the single charge (electron or heavy hole) spin qubits, initialization is by optical pumping, a process which requires a pulse of a few nsec duration. Additionally, complete coherent control of these spins requires a set of two sequential optical pulses. The coherent evolution of the precessing spins between the pulses is an essential part of the control scheme. This is due to the fact that both the single spin and the trion, composed of the single carrier and an additional exciton, are composed of two Kramers degenerate states. Each carrier spin couples to one of the trion states through either a σ^+ or σ^- circularly polarized photon. The result is that optical control via a laser pulse can be performed only in the circular polarization basis, and timed precession is necessary to enable control about a second axis on the Bloch sphere. Separating the Kramers degenerate spin states with a magnetic field is necessary for this spin state precession, and thus these single carrier qubits require a magnetic field. Complete control by a single ultrashort optical pulse is also impossible.

The BE, in contrast, has a whole integer total spin. Its initialization does not require optical pumping and gating. It can be photogenerated and initialized in any desired coherent superposition of its two eigenstates using a single short polarized optical pulse [2]. In contrast to the single spins, it is optically coupled to vacuum and biexcitonic levels with zero angular momentum, which have no degeneracy, forming a Λ system [54] even in the absence of a magnetic field. In this Λ system, the optical control can be in any arbitrary polarization, corresponding to rotation about an arbitrary axis on the Bloch sphere. Thus, the BE spin state can be controlled with a single optical pulse [4, 55]. The duration of the control operation is thus limited to the duration of the laser pulse and is not related to the precession period of the qubit two-level system.

BEs have other advantages over single half-integer carrier spins. One of these advantages is that the BE is electrically neutral, making it less susceptible to decoherence due to electrostatic fluctuations in the vicinity [52, 56–59]. Additionally, the BE is partially protected from decoherence due to nuclear magnetic field fluctuations due to its zero field fine structure splitting and its heavy hole content [51–53, 59, 60]. However, the typically short BE radiative lifetimes (<1 ns in the QD studied here) limit the usefulness of BEs as qubits.

The DE has the same advantages as the BE in terms of its electric neutrality and its protection from decoherence due to nuclear magnetic field fluctuations. However, its optical activity is substantially weaker. We demonstrate here that, despite this weakness, the DE meets several of the requirements to be a good qubit, including the requirements for coherent writing, reading, and spin state control.

4.3 Experimental Techniques

4.3.1 Sample Structure

The samples used in this work were grown by molecular beam epitaxy on a (001) oriented GaAs substrate. One layer of strain-induced $In_xGa_{1-x}As$ QDs was deposited in the center of a one-wavelength microcavity formed by two unequal stacks of alternating quarter-wavelength layers of AlAs and GaAs to form a distributed Bragg

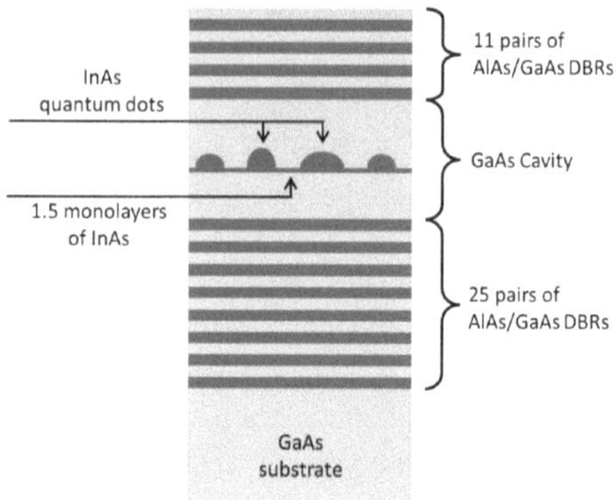

Fig. 4.2 The sample used in this research. Image from [61], reprinted with permission

reflector (DBR) both above and below the microcavity. The height and composition of the QDs were controlled by partially covering the InAs QDs with a 3 nm layer of GaAs and subsequent growth interruption [62]. To improve photon collection efficiency, the microcavity was designed to have a cavity mode which matches the QD emission due to ground-state *e*-*h* pair recombinations. During the growth of the QD layer the sample was not rotated, resulting in a gradient in the density of the formed QDs [62]. The estimated QD density in the sample areas that were measured is $10^8\,\mathrm{cm}^{-2}$; however the density of the QDs that emit in resonance with the microcavity mode is more than two orders of magnitude lower [63]. Thus, single QDs separated by a few tens of micrometers were easily located by scanning the sample surface during micro-PL measurements. Strong antibunching in intensity autocorrelation measurements was then used to verify that the isolated QDs are single ones and that they form single photon sources (Chaps. 3 and 6). A schematic of the sample is shown in Fig. 4.2.

4.3.2 Experimental System

Figure 4.3 shows the low temperature polarization sensitive micro-PL (μ-PL) setup used in these experiments. The sample is located in a cryostat at liquid He temperature (4 K). The PL is collected by a $\times 60$ microscope objective with a numerical aperture of 0.85. This objective is also used to focus excitation lasers on the sample. Laser excitation in this system can be either via synchronously-pumped, cavity-dumped dye lasers with a repetition rate of up to 76 MHz. The temporal width of the dye laser pulses is $\sim$12 ps, and their spectral width is $\sim$100 μeV. The pulses from these dye lasers can be temporally ordered by a series of translating retro-reflection (TRR) stages, such that a timed sequence of several pulses of differing energies is possible. Wavelength tunable cw lasers are also incorporated into the system, and these lasers can be modulated by either an acousto-optic modulator (AOM) or an electro-optic modulator (EOM) to produce longer (few nsec) windows of cw excitation.

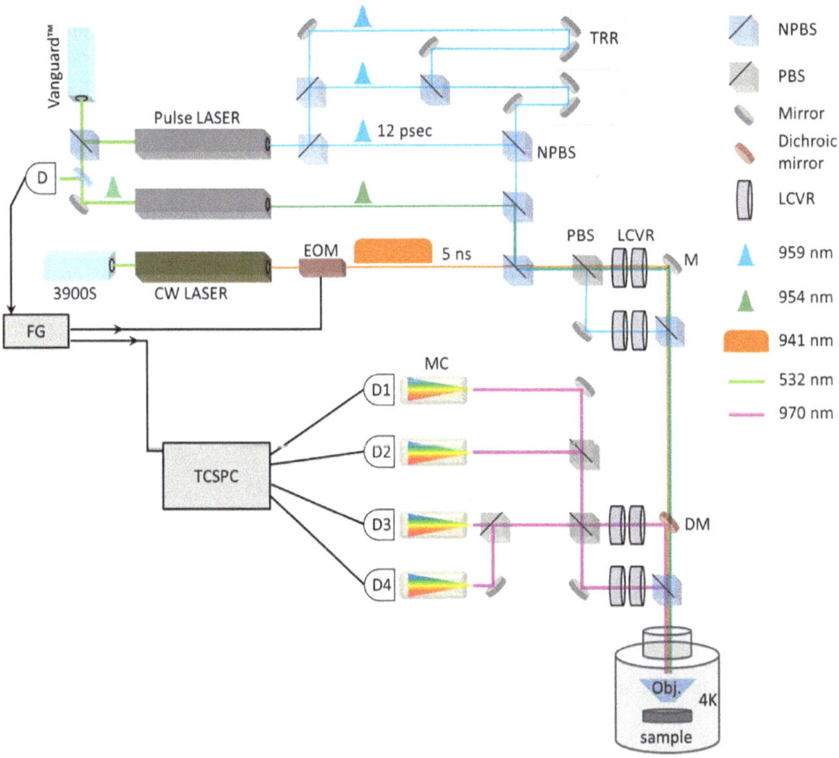

Fig. 4.3 The experimental setup used in this research. NPBS, non-polarizing beamsplitter; PBS, polarizing beamsplitter; TRR, translating retro-reflector; M, mirror; DM, dichroic mirror; TCSPC, time-correlated single-photon counter; MC, monochromator; OBJ, microscope objective; LCVR, liquid crystal variable retarder; Dn, detector n; cw, continuous wave; FG, function generator. Image from [61], reprinted with permission

There are two dedicated channels for laser excitation (green line in Fig. 4.3), where a dichroic mirror prevents PL emission via these channels while allowing excitation lasers to pass through to the sample. In each channel, complete polarization control of the exciting laser is achieved by a pair of liquid crystal variable retarders (LCVRs) [9].

There are two channels for PL emission, as shown in Fig. 4.3 by a magenta line. Here, emitted PL is split into the two channels by a non-polarizing beamsplitter (NPBS). Two pairs of LCVRs are used to analyze the polarization of the PL, rotating the polarization of the emitted light to the axes of a polarizing beamsplitter, which is used to split the PL signal into four detection channels. These channels correspond to bottom/transmit, bottom/reflect, side/transmit, side/reflect in Fig. 4.3. In this manner, two independent polarization projections and their complementary polarizations can be analyzed by the four detection channels. The PL in each detection channel is sent to either a 1 or 0.5 m monochromator and detected by a silicon avalanche photodetector

(APD) or a charged coupled device (CCD) camera. The APDs were connected to a time-correlated single photon counter (TCSPC), which could be synchronized to the pulsed lasers. This way the arrival times of up to four photons of differing energies and polarizations could be recorded relative to each other or to the laser pulse sequence [64].

The overall spatial resolution of this experimental setup is $\sim 1\,\mu m$, the overall spectral resolution is about 10 μeV, and the overall temporal resolution due to the timing jitter of the APDs and TCSPC is about 400 ps.

4.3.3 Measurement Techniques

Various types of measurements are used to characterize the QD spin state.

Polarization-Sensitive Photoluminescence Spectroscopy

To perform polarization-sensitive PL spectroscopy, the QD is optically excited with one or several lasers. The excitation gives rise to light emission from various long-lived states that do not relax to lower energy states within their radiative lifetime. The emission is analyzed using either the CCD camera (when several emission lines are studied) or the APDs (when a specific emission line is studied). The PL signal from the QD is then studied as a function of polarization for some combination of the excitation lasers and/or collected PL. For the DE experiments, PL spectroscopy was used to identify DE optical transitions and to measure their power dependence (Sect. 4.5).

Photoluminescence Excitation Spectroscopy

Photoluminescence excitation (PLE) spectroscopy is used to identify higher-energy states whose lifetime is shorter than the corresponding radiative recombination time. In order to probe optical transitions to excited levels in the QD, the excitation energy of a pulsed or cw laser is varied while emission at a certain recombination energy, usually corresponding to a ground state emission line, is monitored. Enhancement in the recorded signal with respect to the laser energy reveals the transition energies of excited levels. These absorption resonances can be used to optically excite higher energy levels, some of which relax quickly to the corresponding ground state. The polarization of both excitation and emission reveal the selection rules of the optical transition [1, 65]. Here, PLE measurements were used primarily to identify absorption resonances to the DE (Sect. 4.5) and for optical reset of the DE from the QD (Sect. 4.8).

Time-Resolved Spectroscopy

In some experiments, the temporal evolution of the optically excited QD is investigated. This can be done in two ways. In the first case, the PL signal is measured as a function of time relative to the laser pulse sequence, determined by the laser clock pulse, using the TCSPC. This technique was used in the optical reset experiments (Sect. 4.8). In the second case, two or more laser pulses are used, and the time between laser pulses is varied either by a translating retroreflection stage or by an electronic delay component. The first resonant pulse initializes the QD state, and the second pulse probes the QD state after a varying time delay. This technique was used in the DE lifetime measurements (Sect. 4.5).

Polarization-Sensitive Intensity Correlation Measurements

Intensity correlation measurements are a common technique used to establish the quantum nature of light emitted from single photon sources and to characterize radiative cascades in QDs. In these cases,

$$g^{(2)}_{1,2}(\delta t) = \langle I_1(t) I_2(t+\delta t)\rangle / (\langle I_1(t)\rangle \langle I_2(t)\rangle) \quad (4.8)$$

is measured using a two-channel Hanbury Brown-Twiss (HBT) apparatus [66] (Fig. 4.4). Here, $I_i(t)$ is the intensity of light on the ith detector at time t, δt is the time between the detection of a photon in detector 1 and detection of a photon in detector 2, and $\langle\rangle$ means temporal averaging [67]. A radiative cascade [64, 67–69] is characterized by an asymmetric correlation function, due to the temporal order of the emitted photons. Following the detection of the first photon in a cascade, the probability of detecting the second photon is higher than the steady state probability, and bunching $[g^{(2)}_{1,2}(\delta t) > 1]$ is observed [64, 68–71]. Following the detection of the second photon in a cascade, no detection of emission of the first photon is possible and antibunching $[g_{1,2}(\delta t) < 1]$ is observed. Antibunching also results from the single-photon nature of the QD emission.

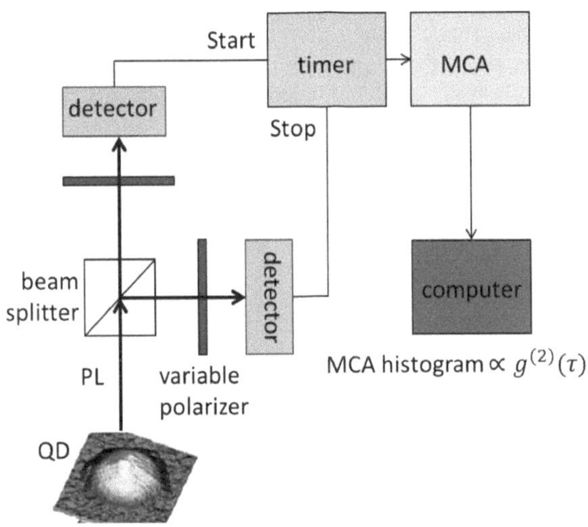

Fig. 4.4 A schematic of a simplified HBT setup. The measured times from the timer are output to a multichannel analyzer (MCA). The histogram from the MCA is proportional to the second order intensity correlation function $g^{(2)}(\delta t)$. In the detailed experimental setup in Fig. 4.3, there are 4 collection channels that can be used for correlation measurements up to $g^{(4)}(\delta t)$

4.4 Probing the Dark Exciton State

We introduce two primary techniques for probing the DE spin state. The results of this section were published as [5, 7].

4.4.1 Optical Observations of the Dark Exciton

Non-resonant excitation of the QD at energies high above the bandgap photogenerates electron-hole pairs in the bulk semiconductor. The photogeneration rate corresponds to the intensity of the exciting laser. Very small numbers of these carriers, uncorrelated and with randomized spin directions [5] arrive at the QD and populate its lower energy levels. For high excitation intensities, four or more carriers can be present in the QD within the radiative recombination time. Thus, the carrier accumulation in the QD exceeds the DE radiative rate by a large margin. In contrast, Fig. 4.5 presents PL measurements of the QD under very weak non-resonant cw excitation. Under these conditions, the carrier accumulation rate is closer to the DE radiative rate, and the probability of finding the QD occupied with a DE is very significant. In this case, if the DE recombines radiatively, its PL emission intensity should be comparable to that of the BE [7].

In Fig. 4.5, an emission line corresponding to optical recombination of the DE (labeled X^0_{DE}) is clearly observed [7]. This spectral line is present 300 μeV lower in energy than the BE spectral lines. This energy difference from the BE corresponds well with previous measurements of the DE-BE separation in a similar sample [13] and in other samples [8, 72]. The line polarization of this emission line matches that of the lowest energy component of the BE (H, along the major axis of the QD). Only one linear polarized transition is observed, in contrast to the BE which has two almost equally strong cross-linearly polarized emission lines.

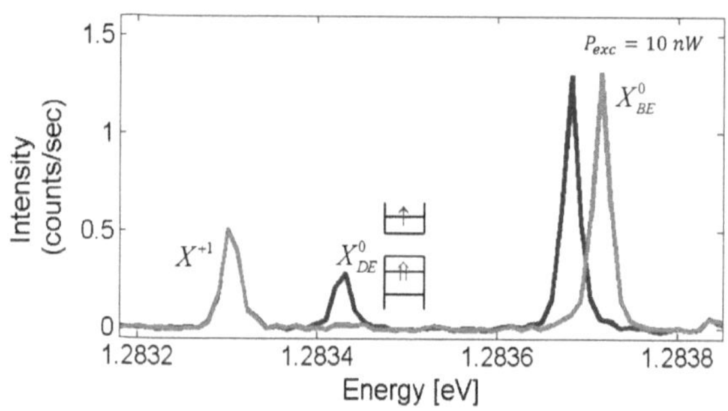

Fig. 4.5 Direct PL emission from the DE. Polarization-sensitive, expanded scale PL spectrum of the QD at low excitation intensity (10 nW). At this excitation level, the DE spectral line is clearly observed 300 μeV below the BE lines. Unlike the BE line, which has two cross-linearly polarized components, the DE has only one horizontal polarization. Image from [7]

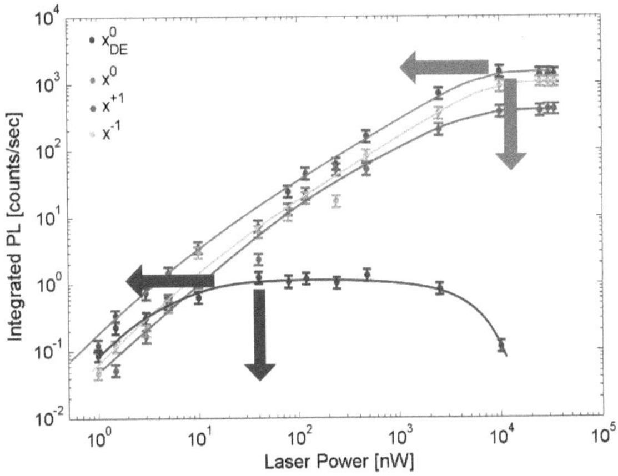

Fig. 4.6 Emission intensity from various exciton lines as a function of laser excitation power. Note that the maximum emission of the DE (*blue arrow*) is three orders of magnitude weaker than that of the BE (*green arrow*) and that the power at which the maximum occurs is also three orders of magnitude weaker than that of the BE. These measurements demonstrate that the DE decays radiatively with an oscillator strength that is three orders of magnitude weaker than that of the BE. Image from [7] (color figure online)

Figure 4.6 shows the measured emission intensities of the DE, BE, and singly charged excitons as a function of the power of the off-resonant excitation [7]. The DE emission saturates at a power 3 orders of magnitude lower than that at which the BE and charged excitons saturate (marked by vertical arrows). The maximum emission intensity of the DE line is also three orders of magnitude weaker (horizontal arrows). This indicates that the DE lifetime is radiative, like that of the BE, and that it is approximately three orders of magnitude longer than that of the BE. In Sect. 4.5.3, the DE radiative lifetime will be directly measured.

Figures 4.5 and 4.6 clearly show that the dipole moment of the DE is substantially stronger than that predicted from heavy hole-light hole mixing alone, and that the direction of polarization is in-plane as opposed to along the QD symmetry axis.

4.4.2 The Spin-Blockaded Biexciton

The $|\pm 3\rangle$ state is a biexcitonic state where the two electrons are paired in a ground state singlet, and the two holes are in the ground state and first excited energy state in a parallel up (+3) or down (−3) spin triplet state. Optical recombination from this state gives rise to the unpolarized $XX^0_{T\pm 3}$ emission line.

This optical recombination also populates the QD with a DE. The $|+3\rangle$ ($|-3\rangle$) biexciton state emits a σ^+ (σ^-) polarized photon, resulting in an excited DE state. In this state, the remaining electron and heavy hole have a spin up (down) parallel spin configuration, and the hole is in the first excited state. This hole is no longer spin blockaded and it quickly (~70 ps) relaxes non-radiatively to the ground state. This fast relaxation is spin preserving [7, 69]. Thus, detection of a σ^+ (σ^-) polarized photon from the $XX^0_{T\pm3}$ emission line heralds the creation of a spin up $|+2\rangle$ (spin down $|-2\rangle$) DE in the QD. These are not eigenstates of the DE (Fig. 4.1). Rather, these states are superpositions of the DE eigenstates $|a\rangle$ and $|s\rangle$, and therefore, the DE spin will precess at a rate given by the energy difference between the eigenstates divided by Planck's constant.

Absorption resonances to the spin-blockaded triplet biexciton states from the various excitonic states were previously identified [1]. One of these resonances is an s-p absorption resonance to the $XX^0_{T\pm3}$ biexciton from the DE states. Here, an electron is added to the electronic ground state and a hole, whose spin is parallel to the DE hole spin, is added to the first excited hole energy level (Fig. 4.10). Now optical recombination of the ground state hole with the added electron is possible, resulting in emission from the $XX^0_{T\pm3}$ at a lower energy.

Figure 4.7a shows the PL spectrum of a neutral QD under nonresonant cw excitation. The ground state $XX^0_{T\pm3}$ emission line is shown by a magenta arrow and the

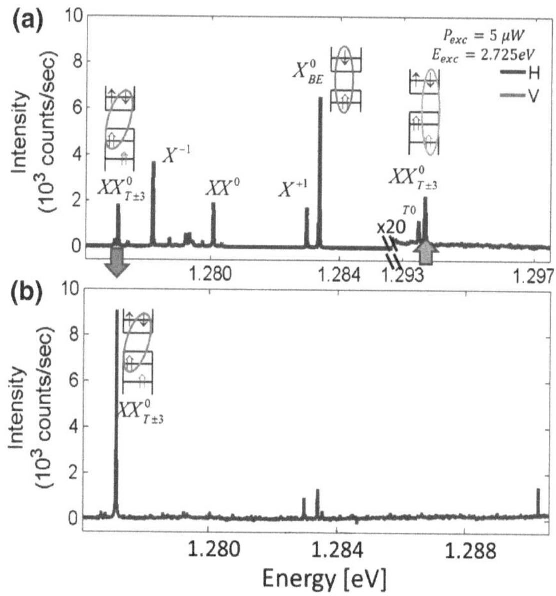

Fig. 4.7 Resonant excitation of the $XX^0_{T\pm3}$. **a** PL spectrum of a neutral QD under non-resonant cw excitation, showing the ground state emission line (*magenta arrow*) and s-p emission line (*green arrow*) corresponding to the absorption resonance used to excite the $XX^0_{T\pm3}$. The relevant electron hole pair is matched by an oval. Upwards (*downwards*) arrows on the energy axis indicate wavelengths used for excitation (detection) in DE experiments. **b** PL spectrum from the same QD under resonant excitation into the higher-energy optical transition of the $XX^0_{T\pm3}$ emission line. Image from [7] (color figure online)

recombining electron-heavy hole pair is indicated with an oval. The s-p absorption resonance is also shown, indicated by a green arrow, and the added electron-heavy hole pair is likewise oval-matched. Figure 4.7b shows the same QD under resonant excitation into the higher-energy optical transition (green arrow in Fig. 4.7a). In this case, the $XX^0_{T\pm3}$ emission line dominates the spectrum.

4.4.3 Probing by Charge Tunneling

Some of the first attempts to optically access the DE relied on the addition of single carriers to the DE state [5]. We will review these experiments here for completeness, though these methods have since been superseded by techniques involving resonant excitation of the spin-blockaded biexciton XX^0_{T3}, which will be discussed in the next section.

After the DE is heralded by detection of polarized emission from the $XX^0_{T\pm3}$, its spin is read by the addition of charge into the QD (Fig. 4.8). While QD charging can be induced externally [6, 48, 50, 51, 58], in this case the measurements were performed using spontaneous charging due to optical deionization of impurities near the QD [73].

When a charge carrier is added to the DE, the DE precession stops. Optical recombination is then possible between an electron and added hole (hole and added electron) in the positively (negatively) charged exciton. The polarization of the emission from this charged exciton state reveals the polarization of the unpaired single carrier and hence of the DE spin projection at the time of charging [5], as shown in Fig. 4.8.

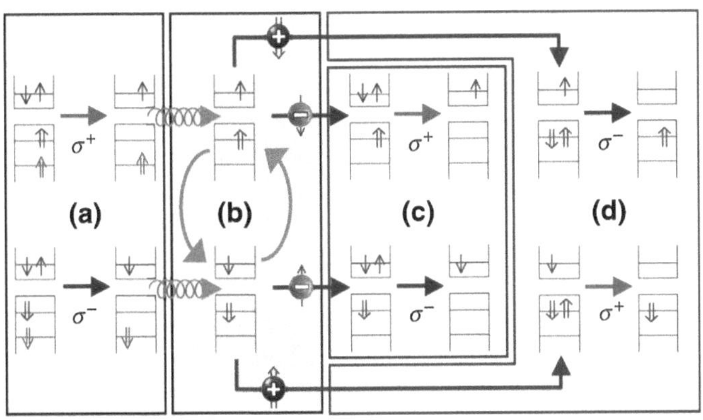

Fig. 4.8 Reading the DE spin state via emission from charged excitons. **a** *Top* (*Bottom*) Preparation of the DE with a spin projection of $|+2\rangle$ ($|-2\rangle$) by detection of a σ^+ (σ^-) polarized photon from the XX^0_{T+3} (XX^0_{T-3}). **b** Fast decay of the hole to its ground state (non-radiative), and precession of the DE spin state. **c** Detection of the DE spin state via charging with an electron. **d** Detection of the DE spin state via charging with a hole. Note that the polarization selection rules are opposite between **c** and **d**. Image from [5]

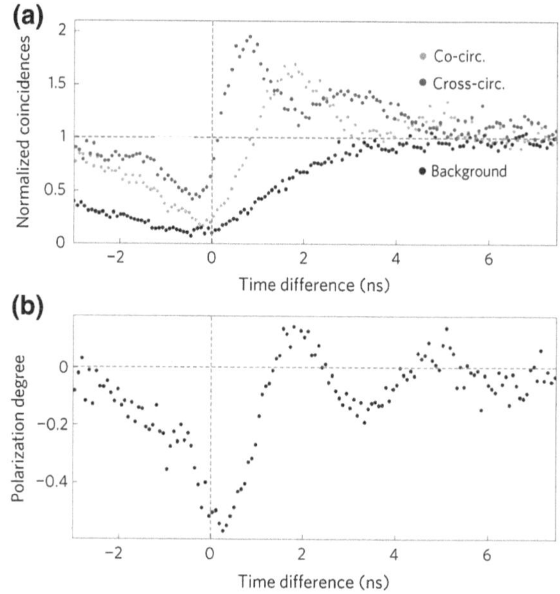

Fig. 4.9 **a** Polarization-sensitive intensity correlation measurements between emission from the $XX^0_{T\pm3}$ and X^{+1} emission lines. *Blue* (*red*) dots represent cross-circular (co-circular) polarization, and the incoherent background, measured by correlation with the X^0 emission line, is shown in *black*. **b** The polarization degree of the second photon (4.9). Image from [5] (color figure online)

This can be demonstrated experimentally through intensity correlation measurements between emission from the $XX^0_{T\pm3}$ and the emission line corresponding to optical recombination from the singly positively charged exciton, X^{+1} (Fig. 4.9). The emission line for recombination from the negatively charged exciton can also be used. The detected polarization of the $XX^0_{T\pm3}$ emission heralds the DE spin state and reveals the spin of the DE at the time of its generation. The polarization of the X^{+1} emission reveals the DE spin at the moment a hole was added to the DE. The measured correlation functions are shown in Fig. 4.9, for co- and cross-circular polarization. For positive time differences, corresponding to detection of emission from the biexciton before emission from the charged exciton, oscillations in the correlation functions are clearly visible (Fig. 4.9a). These oscillations measure the coherent precession of the DE spin as a function of time and provide the first optical measurement of the DE spin projection. The period of the DE oscillation here is measured to be $\sim$3 ns, corresponding to an energy difference between the eigenstates of the DE of $\sim$1.4 μeV.

These initial measurements of the DE spin state depended on spontaneous charging of the QD under off-resonant cw illumination. However, this off-resonant cw illumination also results in other processes in the QD, and spontaneous charging limits the DE lifetime (if the charging rate is high) or the measurement statistics (if the charging rate is low). Measurement by resonant absorption to $|\pm3\rangle$ mitigates these obstacles.

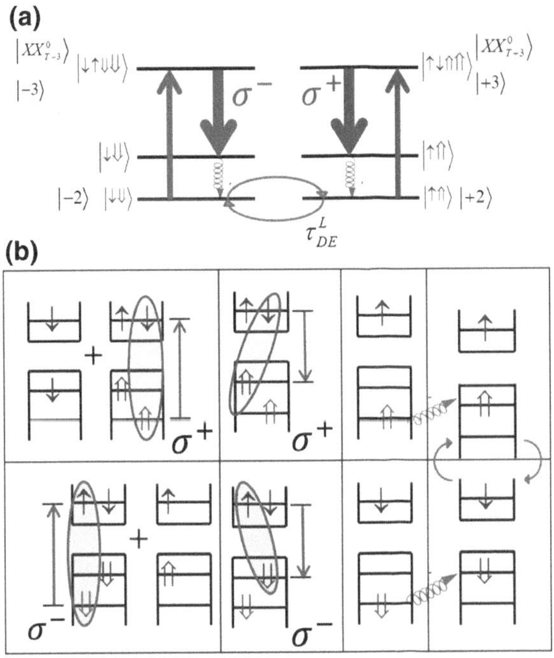

Fig. 4.10 Reading the DE spin state via absorption to the $XX^0_{T\pm3}$. **a** The level system and selection rules for absorption and emission for the $XX^0_{T\pm3}$ states. τ^L_{DE} is the Larmor precession time of the DE. *Grey arrows* represent non-radiative processes. **b** *Top* (*Bottom*) Probing the DE $|+2\rangle$ ($|-2\rangle$) spin state population with resonant absorption to the XX^0_{T+3} (XX^0_{T-3}) state, followed by subsequent optical recombination of ground state carriers. The polarization of the emitted photon matches the polarization of the absorbed photon (color figure online)

4.4.4 Probing by Resonant Absorption

In the previous section, the measurement efficiency and DE lifetime were limited by the spontaneous charging rate in the QD. To enable more efficient measurement, we developed a new technique which uses absorption to the $XX^0_{T\pm3}$ state, and its subsequent emission, to probe the DE spin state.

Since the two hole spins are parallel, the polarization selection rules for excitation of and emission from the $|\pm3\rangle$ biexcitonic states are identical (Fig. 4.10a). Absorption of a σ^+ (σ^-) polarized photon transfers the $|+2\rangle$ ($|-2\rangle$) DE population to the $|+3\rangle$ ($|-3\rangle$) biexciton. The identical emission polarization herald the DE spin state. Thus, it is possible to probe the DE spin state using resonant excitation to the $XX^0_{T\pm3}$, and the emission from the $XX^0_{T\pm3}$ in this case is directly proportional to the DE spin state population.

When measuring the time-dependent PL signal in both the σ^+ and σ^- circular polarizations, the data are often presented in terms of the degree of circular polarization,

$$P(t) = \frac{I_{\sigma^+}(t) - I_{\sigma^-}(t)}{I_{\sigma^+}(t) + I_{\sigma^-}(t)} \tag{4.9}$$

where I_{σ^+} (I_{σ^-}) is the σ^+-polarized (σ^--polarized) PL intensity.

The ground state electrons in the $XX^0_{T\pm3}$ biexciton can recombine optically with either the ground state or the excited state heavy hole, but the biexciton recombination is ~20 times more probable in the lower-energy optical transition due to the better overlap of the ground-level wavefunctions (Fig. 4.7).

Reading the DE spin state via resonant absorption to the $XX^0_{T\pm3}$ is advantageous for several reasons. First, since it does not rely on slow spontaneous charging processes, it increases the measurement efficiency of the DE spin state. Secondly, it can be used with either cw or pulsed laser excitation to measure the DE spin state either as a function of time or at specific times. Finally, the large difference in energy between the absorption resonance and the PL emission (Fig. 4.7) means that filtering the laser light from the PL emission is comparatively simple.

4.5 On-Demand Optical Writing of the Dark Exciton Spin State

The previous section demonstrated how the spin of the DE can be measured using optical techniques and provided a measurement of the coherence time of the DE spin state. Here we demonstrate optical generation of the DE from an empty QD, a direct measurement of the DE radiative lifetime, and ultrafast coherent "writing" of the DE in any desired spin state through the use of excited DE states and polarized picosecond laser pulses. The results in this section were published as [7, 74].

4.5.1 *On-Demand Generation Using Resonant Excitation*

To demonstrate deterministic generation of the DE, we measure the Rabi oscillations of the two-level system formed by the empty QD and the DE $|a\rangle$ state. We consider these oscillations on the Bloch sphere, where the ground state empty QD is at the south pole and the excited state of QD-confined DE is at the north pole. The initial position of the state vector points to the south pole, indicating an empty QD. For a two-level system of given oscillator strength μ, the rotation angle of this state vector on the Bloch sphere is proportional to the pulse area,

$$\Theta = \frac{\mu}{\hbar} \int E(t)dt \tag{4.10}$$

where $E(t)$ is the instantaneous envelope of the electric field of the laser excitation, and the integration is over the pulse duration [75]. The occupancy of the excited state is a function of this pulse area given by

$$P = \sin^2(\Theta/2) \tag{4.11}$$

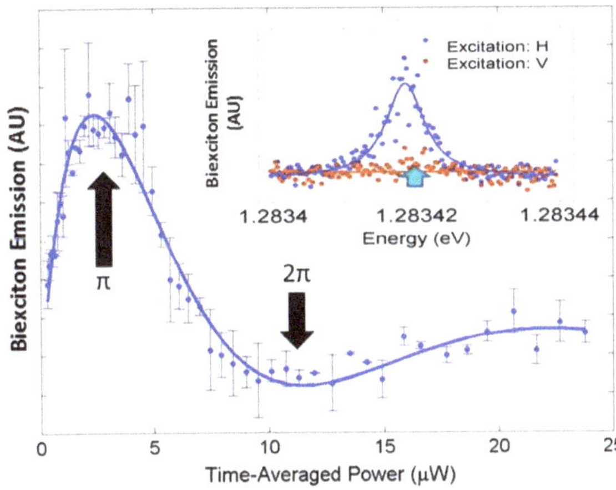

Fig. 4.11 Deterministic optical generation of the DE. $XX^0_{T\pm3}$ emission intensity is monitored as a function of the average resonant excitation power into the DE absorption resonance (*inset* and *emission line* in Fig. 4.5) with a pulse width of 60 ns, in the presence of a cw probe laser tuned to the DE-$XX^0_{T\pm3}$ absorption resonance. The *solid line* represents a model fit of the Rabi oscillations. *Inset* The $XX^0_{T\pm3}$ biexciton emission intensity as a function of the energy of the H (V) polarized excitation into the DE optical transition is depicted by *red* (*blue*) dots. The energy used for resonant, deterministic exciton of the DE is indicated by a *blue arrow*. Image from [7]

and thus for a pulse whose area corresponds to a rotation of $\Theta = \pi$, the excited state is populated with $P = 1$ [75]. Increasing the laser intensity increases the rotation angle until $\Theta = 2\pi$ and the occupation probability of the excited state returns to $P = 0$. Rabi oscillations are identified by an oscillatory dependence of the QD PL emission [76, 77] or photocurrent [78], proportional to the square root of the excitation laser power. Identification of a π pulse in the DE system thus corresponds to a demonstration of deterministic optical DE generation.

We used a pump-probe experiment to demonstrate deterministic generation of the DE. In the first experiment, shown in the inset to Fig. 4.11, one cw probe laser was tuned to the DE-$XX^0_{T\pm3}$ absorption resonance. The energy of a second pump laser was scanned through the energy corresponding to the DE emission line observed in Fig. 4.5. Two rectilinear polarizations of the pump laser were used. In the inset to Fig. 4.11, the increase in biexciton emission indicates an increase in DE population indicating optical generation of the DE. The energy and polarization of this direct DE absorption resonance correspond exactly to the energy and polarization of the DE emission observed in Fig. 4.5. The linewidth of the DE resonance (~5 μeV) reflects the radiative width of the transition to the biexciton, broadened by spectral diffusion caused by the presence of the two laser beams.

In the main panel of Fig. 4.11, the probe laser remains cw but the pump laser is now pulsed using an AOM to generate long laser pulses of 60 ns duration. The pump laser is tuned to the optimal energy for DE generation, indicated by the blue vertical arrow in the inset. The pump laser is H-polarized to match the observed

DE absorption selection rules. The repetition rate of the pump pulse is 1MHz. Rabi oscillations are clearly observed. The intensities corresponding to a π pulse and a 2π pulse are indicated in the Figure with black arrows. The solid line indicates a model fit of the Rabi oscillations.

A comparison between the intensity and temporal pulse width needed to obtain a π pulse to the DE resonance ($\sim$60 ns and 1.9 μW at 1 MHz) with that needed for a π pulse to the BE ($\sim$10 ps and $\sim$0.1 μW at 76 MHz) directly yields the ratio between the oscillator strengths and radiative lifetimes of both excitons. In this way, we verify that the DE oscillator strength is more than three orders of magnitude weaker than that of the BE.[1]

4.5.2 *Coherent On-Demand Generation Using Quasiresonant Excitation*

The optical generation method discussed in Sect. 4.5.1 only generates a DE in the optically active $|a\rangle$ eigenstate, and requires a large laser pulse area. This large pulse area is achieved by using a laser pulse of a few tens of nanoseconds in duration. Consequently, this optical generation process is very slow. An additional drawback of this method is that only one DE spin state can be directly generated, and any other DE spin state would require at least one control pulse (Sect. 4.6). We demonstrate here that, using excited DE states, it is possible to coherently write the DE in any desired spin state with high fidelity using a few-psec long pulse of a given polarization [74]. This coherent writing method is similar to that previously demonstrated for the BE [2].

First, we identified excited state absorption resonances for the DE using PLE similar to that which was performed on the direct DE absorption resonance. Figure 4.12 presents the polarization-sensitive PLE measurements. Figure 4.12a, b presents the PL spectrum of the QD under non-resonant 445 nm excitation. In (b), weak non-resonant excitation is used such that the DE emission line is visible. In (a), the excitation intensity is stronger to enable observation of emission from the $XX^0_{T\pm3}$ biexciton. Figure 4.12c presents the PLE measurements for the DE (bottom, solid lines) and BE (top, dashed lines).

In Fig. 4.12c, two absorption resonances to excited DE states are visible. The absorption resonance at 15 meV is to the $(1e^1)_{\pm1/2}(2h^1)_{\pm3/2}$ DE state and the second absorption resonance at 22 meV is to the $(1e^1)_{\pm1/2}(3h^1)_{\pm3/2}$ DE state. Both of these transitions are a few hundered μeV below corresponding BE transitions [1]. In contrast to the ground state DE absorption resonance, which is H-polarized, these two excited state absorption resonances are unpolarized, indicating equal coupling to H-polarized and V-polarized light [74]. We also observe that excitation of these

---

[1] The pulse area up to constant factors is given by $\sqrt{\frac{P_{avg}\tau}{f}}$ where P_{avg} is the average power as measured on the power meter, f the repetition frequency, and τ the pulse width. Plugging in the numbers above yields a pulse area ratio between the BE and the DE of about 2000.

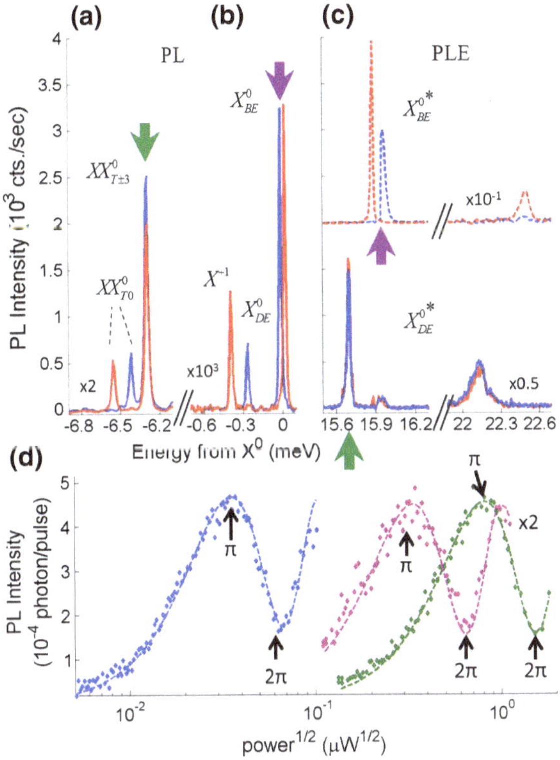

Fig. 4.12 PLE measurements of the DE and BE. **a**, **b** Polarization sensitive spectra of the QD. The **a** biexciton and **b** exciton transitions were excited by 5 μW and 10 nW 445 nm cw laser light, respectively. Observed spectral lines are identified by their initial states. **c** Polarization-sensitive PLE of the BE (*top*, *dashed*) and DE (*bottom*, *solid*). In **a**–**c**, *blue* (*red*) represents H (V) polarization. **d** Rabi oscillations of the X^0_{BE} ground state absorption resonance (*blue*), the X^0_{DE} excited state absorption resonance (*green upward arrow*), measured by monitoring the $XX^0_{T\pm3}$ emission (*green downward arrow*), and the X^0_{BE} excited state absorption resonance (*upwards magenta arrow*), measured by monitoring the X^0_{BE} ground state emission (*downwards magenta arrow*). Points represent the measured PL intensity and the *dashed lines* represent a model fit. Image from [74]

higher-energy DE absorption resonances does not lead to an increase in observed BE emission. During the relaxation of the excited DE state to the DE ground state, spin flip processes are negligible. This agrees well with previous theoretical [79, 80] and experimental [1, 68, 81] studies. Consequently, during the relaxation of the excited DE state to the ground state, the DE predominantly maintains its DE character.

Figure 4.12d presents the PL emission intensity as a function of the square root of the laser power for the absorption resonance to the excited state BE (magenta) and the excited state DE (green) indicated by vertical arrows in Fig. 4.5c. The dashed lines present a model fit to the experimental data. Rabi oscillations are clearly observed, and intensities corresponding to π and 2π pulses are marked.

The ratio between the oscillator strength of the first excited BE, $(1e^1)_{\pm 1/2}(2h^1)_{\mp 3/2}$, resonance to that of the first excited DE, $(1e^1)_{\pm 1/2}(2h^1)_{\pm 3/2}$ is about 6. The ratio between the oscillator strength of the ground BE state oscillator strength to that of the first excited BE state is about 35 [1, 65]. Consequently, the oscillator strength of the excited DE resonance is about 200 times weaker than that of the ground state BE and about an order of magnitude stronger than that of the ground state DE.

The increase in the DE oscillator strength is attributed to increased DE-BE mixing [18] at these elevated energies. That these DE absorption resonances are unpolarized is attributed to the fact that the envelope wavefunction of a higher energy carrier is less restricted to the QD volume and consequently less impacted by any QD deviation from symmetry [27]. The equal polarization distribution of the excited DE states can be straightforwardly obtained from incorporating the DE-BE mixing from [18, 27] to the many body model used in [1].

The increased oscillator strength of the higher-energy DE absorption resonances is further coupled with a broadened absorption resonance compared to the ground state DE absorption resonance. The lifetime of the ground state DE is very long, so the ground state DE absorption resonance is energetically narrow. In contrast, the excited heavy holes relax to their ground state within about 90 ps. This shorter lifetime broadens the corresponding absorption resonance ($\sim$10 μeV), improving the overlap between the laser pulse ($\sim$150 μeV) and the absorption resonance.

Since the DE can be deterministically photogenerated via an excited DE absorption resonance that can be excited using both H and V polarizations, the DE can be written in an arbitrary spin configuration corresponding to the laser polarization in a manner similar to that used to write the BE ground state in [2]. This is demonstrated experimentally using the pulse sequence presented in Fig. 4.13.

In the experiment illustrated in Fig. 4.13, the DE is deterministically generated in a selected coherent spin state using a variably polarized 14 ps pulse (measured by

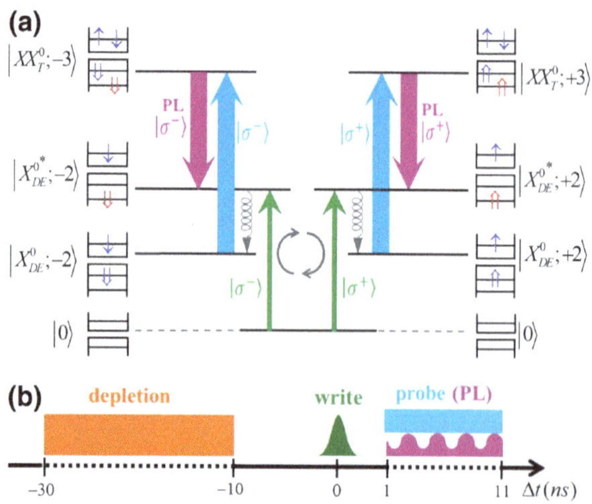

Fig. 4.13 Coherent writing of the DE spin state. **a** The relevant energy levels, spin wavefunctions, and optical transitions involved in the experiment. *Grey arrows* represent non-radiative processes. Other arrows are color-matched to the schematic description of the pulse sequence in **b**. **b** Schematic illustration of the pulse sequence showing the temporal sequence and duration of the optical pulses used in the experiment. Image from [74]

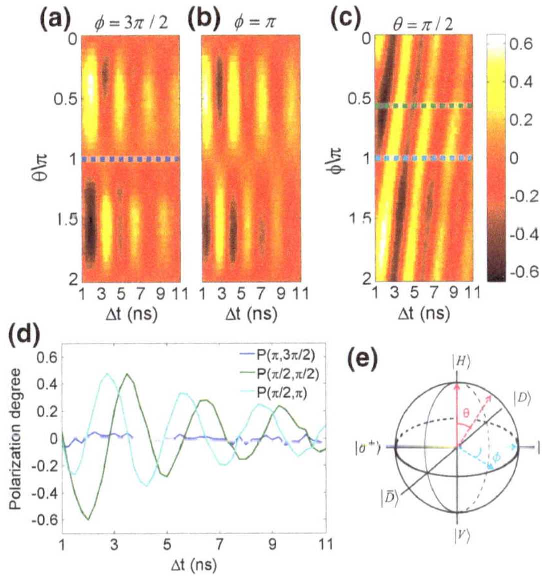

Fig. 4.14 Experimental results indicating coherent writing of the DE. The degree of circular polarization of the $XX^0_{T\pm3}$ as a function of time (*horizontal* axis) and polarization $P(\theta,\phi)$ (*vertical* axis) for **a** $P(\theta, 3\pi/2)$, **b** $P(\theta,\pi)$, and **c** $P(\pi/2,\phi)$. **d** Degree of circular polarization as a function of time for the write pulse polarizations $V = P(\pi, 0)$, $D = P(\pi/2, \pi/2)$ and $\sigma^+ = P(\pi/2, \pi)$. The curve colors match the *dashed lines* in **a**–**c**. **e** The Poincaré sphere for $P(\theta,\phi)$, showing the definition of θ and ϕ. Image from [74]

autocorrelation) tuned to the absorption resonance to the $(1e^1)(2h^1)$ DE absorption resonance indicated by the green arrow in Fig. 4.12c. This excited DE then rapidly relaxes to its ground state non-radiatively [7] by a spin-preserving [79, 81] phonon emission, indicated by the grey curly arrow in Fig. 4.13a, which is faster than the radiative recombination rate of the excited DE state. After relaxation to the ground state, the coherent DE spin state evolves in time. This spin state is then probed via a 10 ns long cw probe laser pulse tuned to the DE-$XX^0_{T\pm3}$ resonance. The temporal dependence of the DE spin state is monitored by the circular polarization of the $XX^0_{T\pm3}$ emission (magenta) as a function of time since the control pulse. The DE is optically depleted from the QD (Sect. 4.8), to enable experimental repetition rates faster than the DE radiative recombination rate and to enhance the fidelity of the "write" operation by ensuring an empty QD.

Figure 4.14 presents the degree of circular polarization of the $XX^0_{T\pm3}$ emission as a function of time from the write pulse and of the write pulse polarization $P(\theta,\phi)$, where the angles θ and ϕ represent the coordinates of the polarization vector on the Poincaré sphere, as shown in Fig. 4.14e. The range of polarizations of the write pulse presented in Fig. 4.14 is (a) $P(0 < \theta < 2\pi, 3\pi/2)$, (b) $P(0 < \theta < 2\pi, 0)$, and (c) $P(\pi/2, 0 < \phi < 2\pi)$. Changes in the polarization angle θ result in corresponding changes in the polarization visibility of the PL. Changes in ϕ result in corresponding changes in the phase of the PL. Figure 4.14 thus demonstrates a one-to-one correspondence between the write pulse polarization $P(\theta,\phi)$ and the amplitude and phase of the oscillations in the degree of circular polarization of the PL emission.

In Fig. 4.14d, the temporal dependence of the degree of circular polarization of the $XX^0_{T\pm3}$ emission is presented for selected write pulse polarizations, indicated by

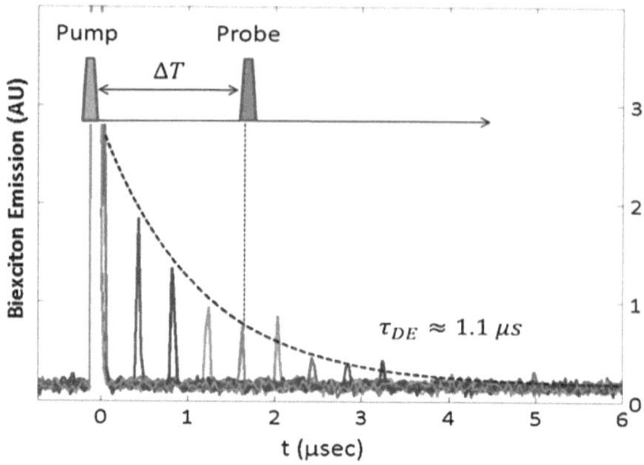

Fig. 4.15 Direct measurement of the DE lifetime. The lifetime is measured by a time-resolved double resonant pump-probe experiment. The *inset* describes the experimental sequence of pulses, where the DE is excited at time $t = 0$ (*blue pulse*) by a resonant π pulse. A π pulse to the DE-$XX^0_{T\pm3}$ absorption resonance probes the DE population at varying time delays from the generation pulse. The *dashed black line* is a fitted exponential decay model with a characteristic lifetime of $\tau_{DE} \approx 1.1\,\mu$s. Image from [7] (color figure online)

color-matched horizontal dashed lines in Fig. 4.14a–c. The maximal degree of polarization observed in the measured data (~0.65) is compatible with the finite temporal resolution of the detectors (~400 ps) [5], indicating a fidelity of initialization for the DE qubit of above 90%.

4.5.3 The Dark Exciton Lifetime

Now that the DE can be deterministically generated with a π pulse to one of the identified DE absorption resonances, it is possible to directly measure the DE lifetime. Figure 4.15 presents an experiment in which the DE is deterministically generated by a resonant π pulse. A time-delayed probe pulse to the DE-$XX^0_{T\pm3}$ absorption resonance follows, and the time delay between the pump pulse and the probe pulse is varied electronically. The experimental pulse sequence is shown as an inset in the image. Since the intensity of the emitted light from the $XX^0_{T\pm3}$ provides a measurement of the DE population, an exponential fit to the PL intensity from the probe pulse as a function of time provides a measurement of the DE lifetime. The dashed black line is a fitted exponential decay model with a characteristic lifetime of $1.1 \pm 0.1\,\mu$s [7]. The fact that the measured lifetime of the DE agrees with its oscillator strength establishes that the DE decays radiatively from its single, H polarized emission line.

4.5.4 The Dark Exciton Coherence Time

Using the $XX^0_{T\pm3}$ biexciton, it is possible to measure the coherence time of the DE [7]. The coherence time of a qubit is an important measure of how well the qubit is decoupled from its environment. For potential applications in QIP, long coherence times are essential.

As we have seen in Fig. 4.9, when the DE is in a coherent superposition of its two eigenstates, $|\psi\rangle = \alpha|a\rangle + \beta|s\rangle$, the phase between the two eigenstates varies periodically with a period of ~3 ns [5, 7], corresponding to precession along the equator of the DE qubit Bloch sphere. The qubit coherence time can be conceptually defined as the characteristic decay time of this precession.

As shown in Fig. 4.10, absorption of a σ^+ (σ^-) photon, resulting in the XX^0_{T+3} (XX^0_{T-3}) state, is directly proportional to the $|+2\rangle$ ($|-2\rangle$) DE spin state population. Detection of emission from the XX^0_{T+3} (XX^0_{T-3}) also heralds the creation of a $|+2\rangle$ ($|-2\rangle$) DE. Consequently, circular-polarization-sensitive intensity autocorrelation measurements of the $XX^0_{T\pm3}$ spectral line in the presence of a laser tuned to the DE-biexciton absorption resonance provide a direct measurement of the temporal evolution of the DE spin state. In this case, detection of a σ^+ polarized "start" photon heralds the creation of the $|+2\rangle$ DE state. Detection of a second photon, co-circularly polarized σ^+ (cross-circularly polarized σ^-) projects the DE onto the $|+2\rangle$ ($|-2\rangle$) spin state at the time of photon absorption.

Figure 4.16a presents intensity autocorrelation measurements of the $XX^0_{T\pm3}$ emission line under resonant cw excitation of the DE-$XX^0_{T\pm3}$ absorption resonance (the probe laser). The degree of circular polarization is calculated from (4.9) (Fig. 4.16b). The oscillations in the measured degree of circular polarization reveal the coherent

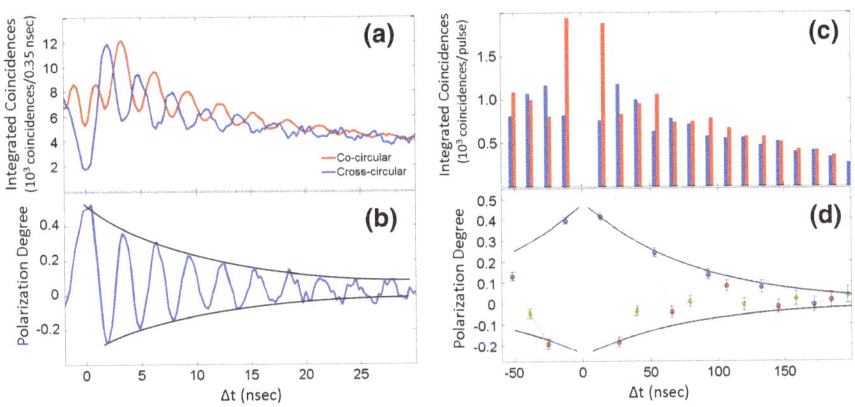

Fig. 4.16 Coherence time of the DE. (**a**) [(**c**)] Polarization-sensitive intensity autocorrelation measurements of the $XX^0_{T\pm3}$ emission line under resonant cw [pulsed] excitation. (**b**) [(**d**)] The degree of circular polarization of the second detected photon as a function of its detection time. The decay of the polarization oscillations is overlaid by an exponential decay curve (*solid black line*) with a characteristic time of ~25 ns [100±20 ns]. Image from [7]

precession of the DE spin state. The characteristic decay time of the oscillations is about 25 ns under this cw excitation. This value is a lower bound on the actual DE coherence time. Under cw probe excitation, there is a constant influx of photons exciting the DE-$XX^0_{T\pm3}$ absorption resonance, and recombination of an electron-hole pair from this biexciton leaves a DE in the QD. The probability that the second detected photon results from absorption of one and only one photon after generation of the DE decreases quickly with time. For longer times, it is likely that the DE was excited to the $XX^0_{T\pm3}$ biexciton state multiple times, but that the spontaneous biexciton recombination was not detected. Indeed, we have observed that the cw polarization decay time depends strongly on the probe laser intensity.

To overcome the impact of the probe laser on the DE coherence time, the measurement was repeated with a pulsed probe excitation, at a repetition rate of 76 MHz. Figure 4.16c presents the intensity autocorrelation measurements under this probe excitation. The measured degree of circular polarization of the second photon as a function of the time difference is given in Fig. 4.16d. Since the DE period is about 3 ns and the laser period is about 13 ns, the periodicity of the measured polarization is 3×13 ns, or about every three laser pulses. This is exactly what is observed in Fig. 4.16d. the maxima (minima) of the oscillations are represented with blue (red) markers and occur every three pulses. The polarization degree decay is indicated by a solid black line, and the characteristic time of this exponential decay is 100 ± 20 ns. This measurement, therefore, sets a lower bound on the coherence time T_2^* of the DE. The actual time is probably longer since, similarly to the case for the cw excitation, the repeated pulsed excitation absorption measurement also shortens the polarization decay time.

4.6 Coherent Control of the Dark Exciton Spin State

In order to use the DE spin state as a qubit, it must be possible to perform operations on the DE spin state, known as gate operations in the language of QIP. One of the simplest gate operations to perform is a rotation of the spin state. In this section, we experimentally demonstrate rotation of the DE spin state using a few-picosecond laser pulse. These results were published as [7].

Rotation of the BE spin has already been demonstrated [3, 4, 55]. In [3], control of the BE spin state was achieved using a circularly polarized pulse detuned slightly from a BE-biexciton absorption resonance. The particular biexciton resonance selected was one where the biexciton electrons were in a $T^e_{\pm1}$ state and the holes were in a $T^h_{\mp3}$ state. This resulted in a Π system with the exciton [54]. In these types of systems, rotation is possible only about one axis [54], essentially rotating the system from eigenstate to superposition of eigenstates. In [3], the circularly polarized control pulse changes the coefficient of one component of the BE wavefunction, affecting the relative amplitude and phase of the eigenstates in the BE wavefunction. This results in an effective rotation of the BE state from the equator of its Bloch sphere towards one of the poles, where the angle is dependent on the detuning of the pulse.

In [4], the two BE states were coupled to a common biexcitonic state. In this case, a Λ system is formed [54], and control is possible about any axis on the Bloch sphere. In this case, a polarized, 2π area control pulse rotates the BE spin projection about the direction of polarization by an angle determined by the detuning of the control pulse from resonance [4]. These measurements have been reproduced by a different research group, using photocurrent measurements [55].

The DE forms a Π system with the XX^0_{T+3} biexciton, as shown in Fig. 4.10 [54]. Thus, control of the DE system is in some respects similar to that in [3]. There is one axis of rotation on the Bloch sphere, and we demonstrate below rotation of the DE from an eigenstate to a coherent superposition of eigenstates by a circularly polarized pulse that couples to one component of the DE state wavefunction. However, due to the Π system's absence of an arbitrary rotation axis [54], full coherent control of the DE requires two control pulses with timed precession in between [7]. This is similar to the case of separate carriers, the electron or the heavy hole [52, 53, 58, 59]. In other respects, the control of the DE system is similar to [4]. The pulse area used is 2π, and the detuning is used to control the angle of rotation about the pulse polarization direction. This control is achieved with a $\sim$10 ps laser pulse, which is more than 5 orders of magnitude shorter than the measured DE lifetime and at least 4 orders of magnitude shorter than the DE spin coherence time.

The experiment is presented schematically in Fig. 4.17. The experiment is shown both via its effect on the DE spin state on the Bloch sphere (Fig. 4.17a) and schematically as a pulse sequence (Fig. 4.17b). The various pulses in the experimental sequence are color-matched to the arrows on the Bloch sphere. The pump pulse deterministically writes the DE in its lower-energy eigenstate, represented by the north pole of the Bloch sphere in Fig. 4.17a. An eigenstate does not evolve in time,

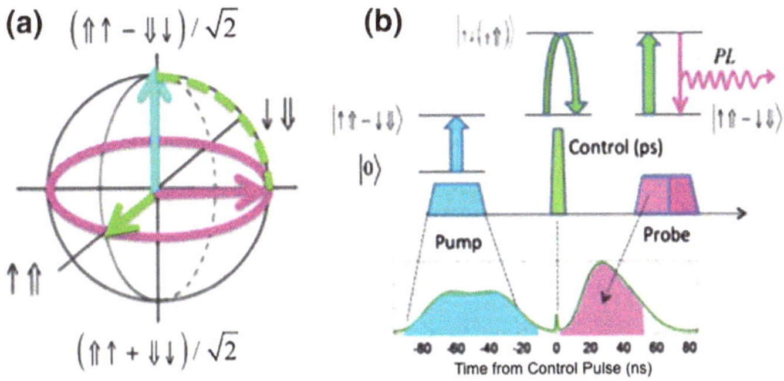

Fig. 4.17 Schematic description of the DE control experiment. **a** The experiment on the Bloch sphere. The pump π pulse deterministically generates the DE in the $|a\rangle$ eigenstate. The detuned, circularly polarized control 2π pulse rotates the DE by an angle of $\pi/2$ about the polarization direction (*green arrow*) and brings the DE to the equatorial plane of the Bloch sphere (*dashed green line*). The DE spin then precesses as described by the magenta trajectory on the Bloch sphere. **b** Schematic description of the pulse sequence used in this experiment. The colors of the pulses match the arrows in **a**. The actual measured temporal sequence of pulses is shown below the schematic description. Image from [7]

and in the absence of a control pulse, the DE will remain in this eigenstate as it radiatively decays. A detuned 2π area optical pulse is used to transfer the DE population through the $XX^0_{T\pm3}$ resonance and back to the DE. The DE state acquires a relative phase difference between the two eigenstates during this pulse, and the relative phase difference is dependent on the detuning from resonance of this control pulse. On the Bloch sphere, this phase accumulation is described as a rotation around the direction of the pulse circular polarization [3, 4, 55]. No detuning results in a π rotation, while negative (positive) detuning results in larger (smaller) rotation angles [4, 55, 82].

Figure 4.17b shows the pulse sequence used in the experiment. After deterministic generation of the DE using a 60 ns H-polarized pulse (blue), a very short (~10 ps) 2π area σ^+ control pulse (green), detuned from the $XX^0_{T\pm3}$ resonance is used to rotate the DE state around the right hand circular polarization direction on the Bloch sphere (indicated by a green arrow). A 60 ns right-hand circularly polarized cw probe (magenta) is then used to re-excite the DE to the $XX^0_{T\pm3}$ biexciton. This probe pulse measures the time evolution of the DE spin state after the control pulse. The entire sequence of pulses lasts ~120 ns and the repetition rate is 1 MHz. In Fig. 4.17b, the measured pulse sequence is also shown below the schematic depiction.

Figure 4.18a presents the measured (points) degree of circular polarization as a function of time after a σ^+-polarized control pulse during the first 35 ns of the probe pulse for a detuning energy of $\Delta/\sigma \approx 0.7$. Here, $\sigma = 100\,\mu\text{eV}$ is the full spectral width of the control laser at half maximum. Δ is the detuning from resonance. Oscillations in the degree of circular polarization are clearly visible, demonstrating that the DE spin state has been rotated from an eigenstate to a coherent superposition of eigenstates by the control pulse. The overlaid solid line presents a best-fit model of an exponentially decaying sinusoidal function

$$f(t) = P_0 \sin(2\pi t/\tau^L_{DE}) \exp[-t/T_{PD}] \tag{4.12}$$

where P_0 is the initial polarization degree (~0.35), τ^L_{DE} is the Larmor precession time of the DE (~3 ns), and T_{PD} is a characteristic polarization decay time. This polarization decay time is relatively short here (~20 ns), due to these conditions of strong cw resonant excitation, similar to the artificial shortening of the polarization decay time observed for cw excitation in Fig. 4.16.

The inset to Fig. 4.18 presents the actual values of the best-fitted P_0 (normalized to 1 at maximum) as a function of the detuning energy. The solid line presents the theoretically expected dependence [82]

$$P_0(\Delta) = \sin(\pi - 2\arctan[\Delta/\sigma]). \tag{4.13}$$

In these experiments, the maximal degree of initial polarization achieved is about 30% for a detuning of $\Delta/\sigma \approx 1.0$. This maximal polarization is limited by two factors: (1) The temporal resolution of the detectors, which limits $|P_0| < 0.6$ [5], and (2) the repetition rate at which the experiments are conducted. At a repetition rate of 1 MHz, the separation between experimental sequences is similar to the DE lifetime. Thus, in ~40% of the cases, a residual DE population is present in the QD and prevents

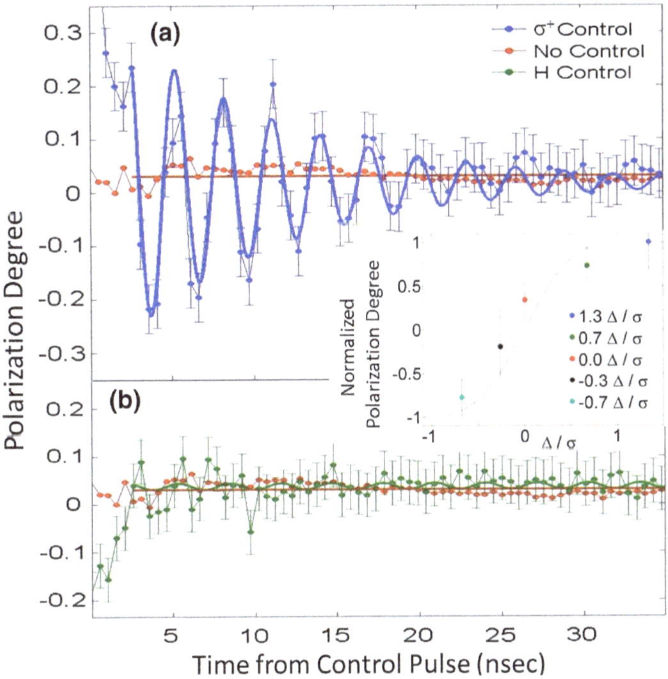

Fig. 4.18 Experimental demonstration of DE spin control. (**a**) [(**b**)] The measured (points) degree of circular polarization of the $XX^0_{T\pm3}$ emission as a function of time after the application of a σ^+-polarized [H-polarized] control pulse (*blue*) [(*green*)], compared to that measured in the absence of a control pulse (*red*). The *solid line* in **a** represents a fit using (4.12). *Inset* Points (*line*) show the measured (model) dependence of the normalized oscillation amplitude on the detuning. Image from [7]

absorption of the pump pulse. For this reason, an optical DE reset technique was developed (Sect. 4.8), to enable experiments on the DE at rates substantially higher than the DE radiative recombination rate.

Figure 4.18b presents the measured degree of circular polarization at the same detuning as in Fig. 4.18a for a linearly (H) polarized control pulse (green). As expected, no polarization oscillations are observed in this case, similar to the case with no control pulse at all (red).

These measurements demonstrate a coherent rotation of the DE spin state with a single picosecond pulse [7]. Consequently, it is possible to perform gate operations on the DE spin state that are fast compared to the DE lifetime and spin coherence time. Having demonstrated many aspects of the DE qubit, we now turn our attention to a necessary feature for high repetition-rate DE experiments, the ability to optically reset the DE from the QD.

4.7 Controlling the Dark Exciton Eigenstates Using an External Magnetic Field

At zero magnetic field, due to the short range e-h exchange interaction, the DE eigenstates are the symmetric $|S_2\rangle = [|+2\rangle + |-2\rangle]/\sqrt{2}$ and anti-symmetric $|A_2\rangle = [|+2\rangle - |-2\rangle]/\sqrt{2}$ coherent superposition of the spin up ($|+2\rangle$) and spin down ($|-2\rangle$) states, where the anti-symmetric state is lower in energy [7]. These states are schematically described in Fig. 4.19. Optical excitation of the DE generates the spin blockaded biexciton $XX^0_{T_3}$ [68]. The lower and higher eigenstates of the $XX^0_{T_3}$ qubit are also the anti-symmetric $|A_3\rangle = [|+3\rangle - |-3\rangle]/\sqrt{2}$ and symmetric $|S_3\rangle = [|+3\rangle + |-3\rangle]/\sqrt{2}$ coherent superpositions of the spin up ($|+3\rangle$) and spin down ($|-3\rangle$).

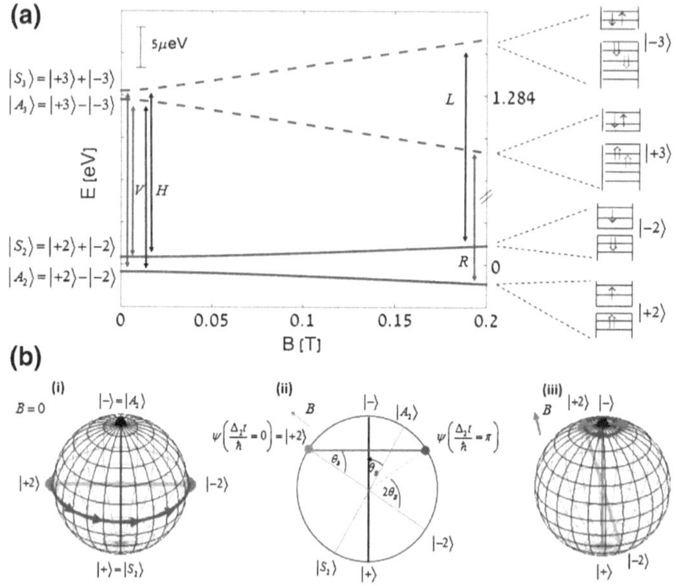

Fig. 4.19 **a** Energy levels and spin wavefunctions of the DE and the $XX^0_{T_3}$ – biexciton as function of an externally applied magnetic field in Faraday configuration. The *blue* and *purple solid* (*dashed*) *lines* represent the energies of the low and high energy eigenstates of the DE (biexciton) respectively. Vertical arrows connecting the DE and biexciton eigenstates mark allowed polarized optical transitions between the eigenstates at zero and high field. **b** Schematic representation of the changes that the external field induces on the Bloch sphere of the DE qubit. Shown are three cases: (i) zero field (ii) cross section of the sphere at arbitrary magnetic field, (iii) strong magnetic field. The eigenstates $|A_2\rangle$, $|S_2\rangle$, at zero field and $|\pm\rangle_2$ at finite field, and the angle θ_B are defined in the text and in (4.15). The eigenstates are always at the poles of the sphere, north pole being the lower energy one. The *pink dot* represents the $|+2\rangle$ state, heralded by detecting a σ^+ polarized biexciton photon. The *blue circle* represents the counter clockwise temporal evolution of the DE state following its heralding. Image from [84] (color figure online)

The DE and $XX^0_{T_3}$ form an optical Π-system with optical transitions between the $|+2\rangle$ ($|-2\rangle$) DE state to and from the $|+3\rangle$ ($|-3\rangle$) biexciton state by right (left) handed circularly polarized light. At zero magnetic field, the DE and $XX^0_{T_3}$ eigenstates are therefore optically connected by linear cross-polarized optical transitions denoted as horizontal (H) and vertical (V). The system is schematically described in Fig. 4.19a.

The time independent Hamiltonian of the DE and the $XX^0_{T_{\pm 3}}$ in the presence of a Faraday configuration magnetic field (field parallel to the optical axis) and expressed in the basis $\{|+2\rangle, |-2\rangle, |+3\rangle, |-3\rangle\}$ is given by:

$$\hat{H} = \frac{1}{2}\begin{pmatrix} -\mu_B(g_e - g_h)B & \hbar\omega_2 & & \\ \hbar\omega_2 & \mu_B(g_e - g_h)B & & \\ & & 2(\Delta + \mu_B g_{2h} B) & \hbar\omega_3 \\ & & \hbar\omega_3 & 2(\Delta - \mu_B g_{2h} B) \end{pmatrix} \tag{4.14}$$

This Hamiltonian represents two decoupled Hamiltonians, one for the DE and one for the $XX^0_{T_{\pm 3}}$, where $\mu_B = e\hbar/2m_e c$ is the Bohr magnetron, B the magnitude of the magnetic field (normal to the sample surface), g_e and g_h are the electron and hole gyromagnetic ratios in the direction of the magnetic field, and g_{2h} is the gyromagnetic ratio of the two heavy holes in triplet configuration. The sign convention for the gyromagnetic factors is such that positive factors mean that electron (heavy hole) with spin parallel (antiparallel) to the magnetic field direction is lower in energy than that with spin antiparallel (parallel) [8]. The triplet state gyromagnetic ratio is not a simple sum of the gyromagnetic ratios of the individual holes [83]. The energy difference between the DE and the $XX^0_{T_{\pm 3}}$ is Δ, and $\hbar\omega_2$ and $\hbar\omega_3$ are the energy differences between the DE and $XX^0_{T_{\pm 3}}$ eigenstates, respectively. All energies are defined at zero magnetic field. Figure 4.19a schematically describes the DE energy level structure, its magnetic field dependence, and the optical transitions between their eigenstates.

The externally applied magnetic field modifies the eigenstates of both qubits: [8]

$$\begin{aligned} |+\rangle_i &= \cos\left(\frac{\pi}{4} + \frac{\theta_B^i}{2}\right)|+i\rangle + \sin\left(\frac{\pi}{4} + \frac{\theta_B^i}{2}\right)|-i\rangle \\ |-\rangle_i &= \cos\left(\frac{\pi}{4} - \frac{\theta_B^i}{2}\right)|+i\rangle - \sin\left(\frac{\pi}{4} - \frac{\theta_B^i}{2}\right)|-i\rangle \end{aligned} \tag{4.15}$$

where $\beta_2 = \mu_B(g_e - g_h)B$ and $\beta_3 = -\mu_B g_{2h} B$ are the magnetic energies. The energy difference between the two eigenstates is given by their Zeeman splitting: $\Delta_i(B) = \sqrt{\beta_i^2 + (\hbar\omega_i)^2}$, and $\tan\theta_B^i = \left(\frac{\beta_i}{\hbar\omega_i}\right)$.

Figure 4.19b presents an intuitive geometrical interpretation for the angle θ_B and the DE Bloch sphere. Since in the Faraday configuration the magnetic field direction is aligned with the direction of the $|+2\rangle$ spin state, it follows that $\pi/2 - \theta_B$ is the angle between the Bloch sphere eigenstate axis and the direction of the magnetic field. Thus, as the magnitude of the external field (B) increases θ_B^i approaches $\pi/2$ and the eigenstates gradually change their nature. Once the Zeeman energies significantly exceed the exchange energies, the eigenstates become the $|\pm 2\rangle$ and $|\pm 3\rangle$ spin states for the DE and the $XX^0_{T_{\pm 3}}$, respectively.

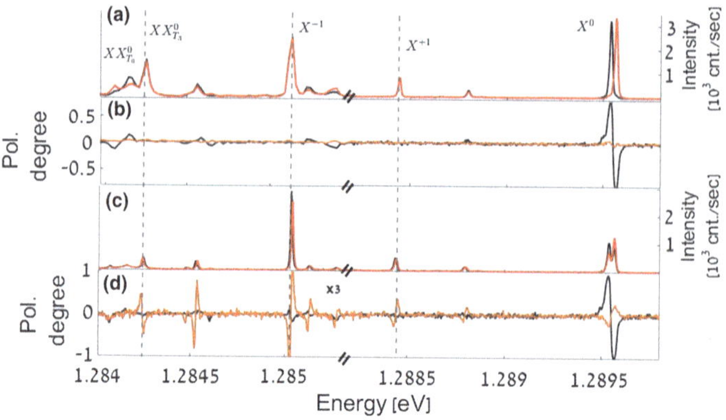

Fig. 4.20 **a** Rectilinear polarization sensitive PL spectra of the QD at zero magnetic field. *Solid black* (*red*) *line* represents horizontal - H (vertical - V) polarization. **b** The degree of rectilinear (*black*) and circular (*orange*) polarizations as a function of the emitted photon energy. **c** Circular polarization sensitive PL spectra at $B = 0.2$T. *Red* (*black*) *line* represents right-σ^+ (left- σ^-) hand circular polarization. **d** The degree of rectilinear (*black*) and circular (*orange*) polarizations as a function of the emitted photon energy at $B = 0.2$T. Note that the Zeeman splitting of the $XX^0_{T_3}$ line is opposite in sign to that of the negative, neutral and positive excitons. Image from [84]

Figure 4.20 shows polarization sensitive PL spectra of the single QD under study. The PL was excited using 445 nm non-resonant cw laser light. Figure 4.20a (c) presents the measured spectra in the two linear (circular) polarizations, in the absence (presence of $B = 0.2$T) external magnetic field. Figure 4.20b (d) presents the obtained degrees of linear (circular) polarizations as a function of the emitted photon energy in the absence (presence of $B = 0.2$T) external magnetic field. In Fig. 4.20a the solid black (red) line represents horizontal - H (vertical - V) polarization and in Fig. 4.20c black (red) line represents left- σ^- (right-σ^+) hand circular polarization. Black (orange) lines in Fig. 4.20b, d represent the degree of linear (circular) polarization. The various exciton and biexciton lines are identified in Fig. 4.20a.

Even in the absence of a magnetic field, one can clearly observe in Fig. 4.20a, b that the BE spectral line is split into two cross linearly polarized components. This split, measured to be $27 \pm 3\,\mu$eV is common to self assembled QDs. It results from the anisotropic e-h exchange interaction, mainly due to the QD deviation from cylindrical symmetry [8, 15]. The DE degeneracy is also removed mainly due to the short range e-h exchange interaction [8, 15]. However, since the splits ω_2 and ω_3 are smaller than the radiative linewidth, the linearly polarized components of the $XX^0_{T_3}$ biexciton line cannot be spectrally resolved. Therefore, only one, unpolarized spectral line is observed. An upper bound for $\omega_3 < 0.2\,\text{ns}^{-1}$ corresponding to split of less than $0.82\,\mu$eV is deduced directly from the degree of circular polarization memory of the $XX^0_{T_3}$ biexciton line at zero magnetic field [7]. At a sufficiently large magnetic field the line splits into two components. The lower energy transition is σ^+-circularly polarized and the upper energy one is σ^--circularly polarized. At a magnitude of

Table 4.1 The measured Zeeman splitting of various spectral lines. The DE splitting was measured from a similar dot from the same sample

Line	Zeeman splitting	Measured at 0.2T in (μeV)
X^0	$\sqrt{(\hbar\omega_0)^2 + [\mu_B\,(g_e + g_h)\,B]^2}$	30 ± 3
X^-/X^{+1}	$\mu_B(g_e + g_h)B$	13.6 ± 3
$XX^0_{T_3}$	$-\mu_B\left(g_{2h} + g_e - g_h^*\right)B$	13.6 ± 3
X^0_D	$\mu_B(g_e - g_h)B$	3.6 ± 1

0.2T, the splitting amounts to 13.6 ± 3 μeV and it exceeds the measured linewidth of 11.4 ± 3 μeV in the absence of external field.

We note that the measured Zeeman splitting of the $XX^0_{T_{\pm3}}$ line is opposite in sign to those of the X^{+1}, the X^{-1}, and the X^0 excitonic lines. It follows from simple considerations that the expected Zeeman splitting of the charged and neutral excitonic spectral lines is proportional to the sum of the hole and electron g-factors ($g_h + g_e$). Therefore the σ^+ polarized part of these spectral lines is expected to be higher in energy than the σ^- polarized part. This is indeed what we experimentally observe. Since the $XX^0_{T_{\pm3}}$ line splits in proportion to $g_{2h} + g_e - g_h^*$ (where g_h^* is the g factor of an excited hole), our experimental observations indicate that the sign of $g_{2h} - g_h^*$ is negative, and its magnitude in this particular QD is larger than that of the electron g-factor. These observations are in agreement with the energy level diagram of Figure. 1a. The dependences of the Zeeman splitting of the various spectral lines on the g-factors are summarized in Table 4.1.

Figure 4.21 shows co-circular polarization sensitive intensity autocorrelation measurements of the emission from the $XX^0_{T_{\pm3}}$ biexciton line under weak non-resonant (P_b) and resonant (P_r) excitation powers, for various externally applied magnetic

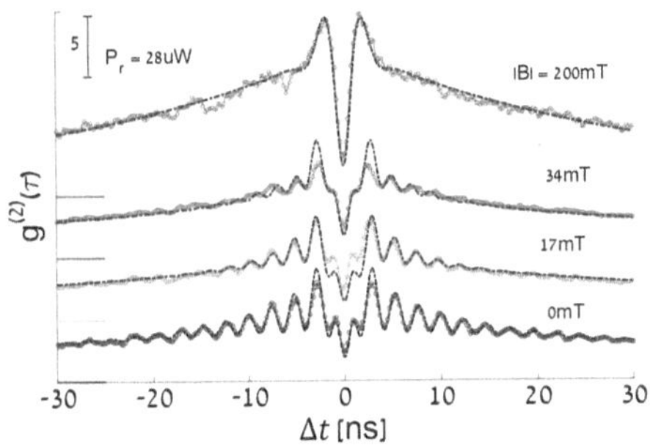

Fig. 4.21 Measured (symbols) and calculated (lines) circularly co-polarized intensity autocorrelation functions ($g^{(2)}\,(\delta t)$) of the emission from the $XX^0_{T_3}$ under weak non-resonant and resonant excitation condition for various externally applied magnetic fields in Faraday configuration. The *solid lines* present best fitted calculations [84] convoluted with the temporal response of the detectors. The *curves* are vertically shifted for clarity and the zero for each measurement is marked by a color-matched *horizontal line*. Image from [84]

fields. The observed reduction in the visibility of the oscillations as the magnetic field increases is observed in both resonant and quasi-resonant excitations [84]. The source of this reduction is explained in Fig. 4.19b as resulting from the field induced changes in the DE qubit eigenstates. For example, at a field of $B = 0.2$T the DE splitting was calculated in Fig. 4.19 to be 4 μeV, which is larger than the measured zero magnetic field splitting of 1.7 μeV. Hence, as expected, no oscillations are observed, and the system can be described as two separated two-level systems.

Consequently, the strength of an applied external field can be used as a tuning knob for varying the dark exciton eigenstates. More details on this experiment and the theoretical model can be found in [84].

4.8 Optical Reset of the Dark Exciton

Demonstrating the coherent properties of the DE is difficult because of its long radiative lifetime. In most of the previous demonstrations of deterministic DE writing and control, the experimental repetition rate had to be substantially slower than the DE radiative recombination rate to ensure that the QD was empty prior to the beginning of the experimental pulse sequence. These low repetition rates substantially increase the time required to collect enough photon events or correlated photons to measure the results of the DE writing and control experiments.

This long lifetime of the DE also impacts the efficiency of QDs as single photon sources. When a QD is excited non-resonantly, the excitonic population generated in the QD is a statistical mix of BEs and DEs. Therefore, sometimes the QD is occupied with a long-lived DE. This prevents the QD from being a true on-demand single photon emitter [70, 85, 86].

In this section, we present an optical method to deplete the long-lived DEs from a QD [87]. We do this via optical pumping of the DE population to a BE population using intermediate excited biexciton states. This technique both substantially reduces the DE population and increases the triggered BE emission. The depletion pulse empties the QD of DEs with near unit probability. This depletion technique was published as [87].

Excited biexciton levels consist of two electron-heavy hole pairs, where at least one of each charge carrier is in an excited energy level. These excited biexciton levels were studied in several previous papers [1, 2, 16, 64, 68, 88]. Their spectra, the spin wavefunctions of carriers in these levels, and their associated dynamics are fairly well-understood. For depleting the DE from a QD, the nine states in which both the electrons and the holes form spin-triplet configurations (T^eT^h) are particularly relevant [1, 64, 68].

In general, the nine T^eT^h excited biexciton states are spin blockaded from relaxation to the ground electron singlet-hole singlet biexciton state ($S_0^eS_0^h$). However, there are efficient spin flip and spin flip-flop processes that permit this relaxation (Sect. 10.4) [69, 88, 89]. Here, due to the enhanced effect of the electron-hole exchange interaction in the presence of a near resonant electron-longitudinal optic

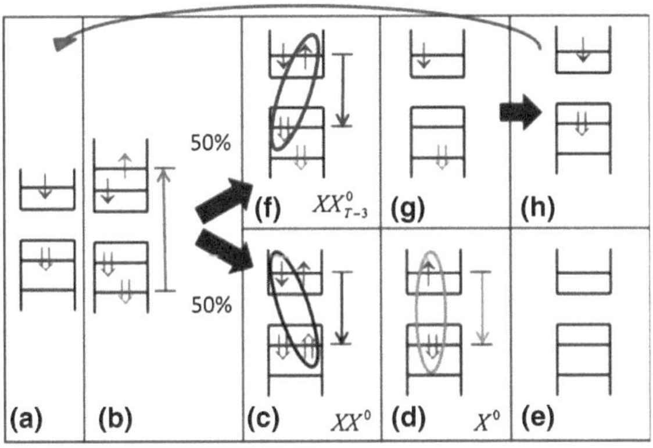

Fig. 4.22 A schematic description of the optical depletion process. *Blue* (*red*) *color* is used to describe a ground (excited) state carrier. *Downward* (*upward*) *arrows* describe optical emission (excitation). **a** The QD is populated with a DE following non-resonant excitation. Only one spin projection is shown for clarity. **b** Optical excitation generates an excited biexciton state, which relaxes nonradiatively with almost equal probabilities either to **c** its ground state XX^0 or **f** to a spin-blockaded biexciton $XX^0_{T\pm3}$. In the first case, the biexciton results in a radiative cascade (**c**–**d**), leaving the (**e**) QD empty. In the second case, **f** one photon is emitted and **g** after hole relaxation, **h** a DE remains in the QD. The process then repeats from **a** for as long as there is a DE in the QD. *Black arrows* represent non-radiative processes. Radiative recombination is indicated by oval-matching the recombining electron-hole pair. The oval colors are matched to the *colors* indicating the corresponding emission lines in Fig. 4.23. Image from [87], reprinted with permission (color figure online)

(LO) phonon Fröhlich interaction, an electron or an electron and a hole flip their spins [88, 89].

Figure 4.22 schematically describes the process of optical depletion. Figure 4.22a shows a QD containing a single ground-state DE. In this case, the DE is generated electrically or via non-resonant optical excitation, where electron-hole pairs with high excess energy are generated in the vicinity of the QD. The carrier spins are randomized during relaxation, yielding a stochastic excitonic population in the QD that is composed on average of approximately 50% DEs and 50% BEs. In our experiments, a weak cw laser pulse (445 nm, few nsec duration) excites the QD, generating a mixed excitonic population. The BE population recombines radiatively within 2–3 ns of the end of the non-resonant pulse, resulting in an empty QD. This occurs in the approximately 50% of the cases corresponding to initial population with a BE. In the remaining cases, the QD is populated with a DE for long after the end of the non-resonant excitation pulse. To deplete the DE, an optical resonant excitation into an excited biexcitonic level is used as described in Fig. 4.22b. These excited biexciton states then quickly relax non-radiatively to one of the lower energy levels of the biexciton (Fig. 4.22c). As shown in Fig. 4.22 and experimentally demonstrated in Fig. 4.23b, the relaxation is into the spin blockaded biexciton $XX^0_{T\pm3}$ states in roughly half the cases. Further relaxation here is prohibited by the spin-parallel configuration of the heavy holes ($T^h_{\pm3}$, Fig. 4.22f). After emission of another photon, a

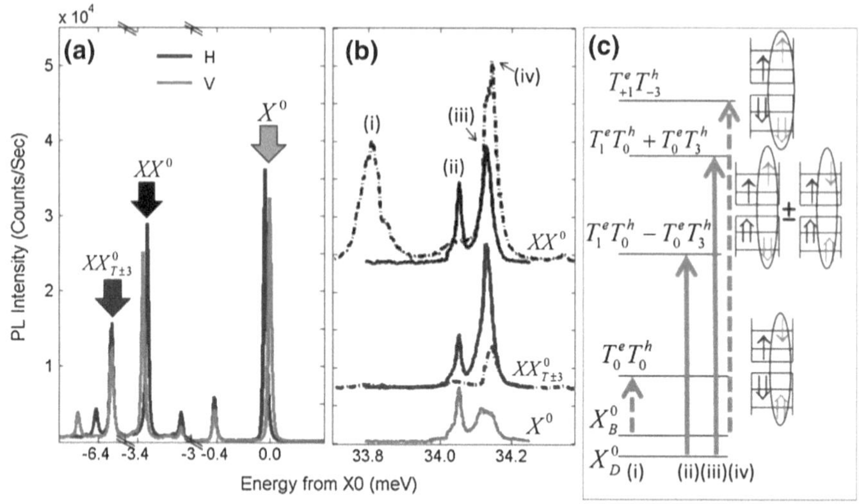

Fig. 4.23 PL and PLE spectra of the depletion resonance. **a** PL spectra showing *H* (*blue*) and *V* (*red*) linearly polarized QD emission. Relevant emission lines are indicated above the spectral line by the initial state of the optical transition. Arrow colors match the PLE spectra in **b**, which were obtained while monitoring the emission from the indicated spectral line. **b** PLE spectra of the indicated lines in **a** showing the T^eT^h biexciton resonances relevant for optical DE depletion. These spectra were obtained using two cw lasers. Spectra indicated by *solid* (*dashed*) lines were obtained with one laser tuned to a *dark* (*bright*) exciton resonance while the energy of the second laser was scanned and the emission from the color matched PL line was monitored. **c** Schematic description of the optical transitions observed in **b**. Optical transitions from the DE (BE) are indicated by *solid* (*dashed*) *arrows* and the added carriers are indicated in the spin state diagram by an oval-matched pair. The spin configuration of the state is provided on the *left*. Roman numerals match observed resonances in **b**. Resonances labeled (ii) and (iii) can be used for the optical depletion process. Image from [87], reprinted with permission (color figure online)

remaining DE is left in the QD. In the other cases, relaxation is to a ground biexciton level. From here, the well-studied two-photon radiative cascade [90–92] from the biexciton leaves the QD eventually empty of charge carriers. Because of the relatively large branching ratio (approximately 0.5) between the two process, the DE can be efficiently depleted from the QD. We estimate that, for a measured exciton (biexciton) radiative lifetime of 470 (330) ps, the QD can be fully emptied using an excitation pulse of several nsec duration.

Figure 4.23 presents (a) the PL and (b) two-laser PLE spectra of the QD. In Fig. 4.23a, relevant emission lines corresponding to optical recombination from the X^0, XX^0, and $XX^0_{T\pm3}$ are indicated by colored arrows. The energy is measured from the BE spectral line at 1.283 eV. Optical transitions from the XX^0 and $m = 0$ spin-blockaded biexciton XX^0_{T0} result predominantly in BEs and lead to sequential emission of a photon due to BE recombination. The presence of a DE in the QD is heralded by the $XX^0_{T\pm3}$ biexciton line. In the PLE data in Fig. 4.23b, the line color corresponds to the arrow color above the monitored emission line.

In the PLE measurements in Fig. 4.23b, one cw laser is used to excite the BE (dashed lines) or DE (solid lines). The PL from the indicated emission line is monitored while the energy of the second laser is varied. Four exciton-biexciton absorption resonances are observed, and indicated by Roman numerals. Resonances indicated (i) and (iv) initiate from the BE. Resonances indicated (ii) and (iii) initiate from the DE. These optical transitions [1, 64] are schematically presented in Fig. 4.23c.

Figure 4.23 clearly demonstrates that the resonances initiating from the BE (resonances labeled (i) and (iv)) contribute mainly to BE emission, but the resonances initiating from the DE (resonances labeled (ii) and (iii)) contribute almost equally to the $XX^0_{T\pm3}$ emission and to the XX^0 emission. From this we infer that in about 50% of these excitations, the DE population is transformed into a BE population and depleted from the QD by subsequent optical recombination. For the optical depletion experiments in the next section, we chose resonance (iii). We note that resonance (ii) performs similarly.

To demonstrate optical depletion, a sequence of three laser pulses was used. This pulse sequence is schematically illustrated above Fig. 4.24. The first pulse was a weak pulse of 445 nm laser light, with a duration of 20 ns. This pulse was generated using an AOM on a cw diode laser. The purpose of this pulse is to generate a stochastic population of BEs and DEs in the QD. A few nsec after the end of this pulse there was a second pulse tuned to resonance (iii) of Fig. 4.23 and of 20 ns duration. This pulse was generated by an EOM on a wavelength-tunable Ti:sapphire laser. The third and final pulse in the pulse sequence was a $\sim$10 ps pulse from a dye laser. This pulse could be tuned either to an X^0 absorption resonance (to verify that the QD is empty) or to the $XX^0_{T\pm3}$ absorption resonance (to measure the DE population remaining in the QD). This pulse occurred 5 ns after the end of the depletion pulse, and its intensity corresponded to a π-pulse for the relevant resonance.

Figure 4.24a presents the PL intensity from the $XX^0_{T\pm3}$ line as given by the color bar, as a function of time (horizontal axis) and the power of the depleting pulse tuned to the (iii) resonance of Fig. 4.23c (vertical axis). In this case, the probe pulse is tuned to a $XX^0_{T\pm3}$ absorption resonance. Figure 4.24 clearly demonstrates that the DE population in the QD decreases as the power of the depletion pulse increases. Comparing the PL intensity during the probe pulse (solid blue line) in Fig. 4.24b, where the depletion power was very weak, to that in Fig. 4.24c, where the power of the depletion pulse is maximal, yields a quantitative measure of the efficiency of the depleting pulse. Under these conditions, the depletion pulse reduces the integrated $XX^0_{T\pm3}$ PL emission intensity during the probe pulse to less than 5% of its value in the absence of the depletion pulse. This suggests a corresponding reduction in the DE population during the depletion pulse.

This verification of the depletion pulse efficiency is complimented by considering the BE PL emission (dashed green line in Figs. 4.24b, c) at low (b) and high (c) depletion pulse powers. In these measurements, the probe pulse is tuned to a BE absorption resonance [1]. The integrated BE emission during the probe pulse increases by almost a factor of 2, indicating that the QD is approximately 50% occupied by DEs in the absence of the depletion pulse and predominantly empty when the depletion pulse is present. Under these conditions, we were able to get more than

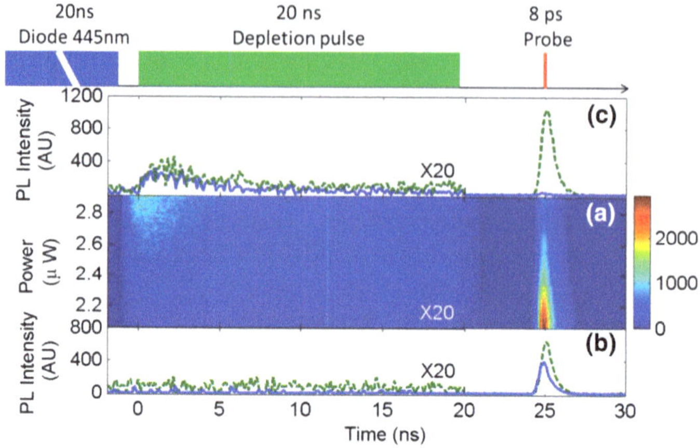

Fig. 4.24 **a** PL emission intensity from the $XX^0_{T\pm3}$ spectral line (color bar) as a function of the power of the depleting pulse to the (iii) resonance (vertical axis) and time (horizontal axis). The temporal sequence of the non-resonant excitation pulse, depletion pulse, and probe pulse tuned to the $XX^0_{T\pm3}$ (X^0) absorption resonance is shown above the Figure. (**b**) [(**c**)] The solid blue line describes the $XX^0_{T\pm3}$ PL emission intensity as a function of time at very low [maximal] depletion pulse power. Close to 95% depletion is clearly observed at high power. The overlaid dashed green line presents the intensity of the PL from the X^0 spectral line while the psec probe pulse is tuned to the BE resonance. The BE PL increases between (b) and (c) by nearly a factor of 2, yet another indication of efficient depletion of the DE population from the QD. Image from [87], reprinted with permission

8500 counts/s on the BE detector during the probe pulse. Taking into account the repetition rate of our experiment (9.5 MHz) and the overall light harvesting efficiency of the experimental setup and sample (approximately 1/1000 [64]), this represents a depletion probability of close to 95%. If the depletion pulse power is increased, the pulse duration can be shortened. Similar depletion probabilities can be achieved with pulses as short as 3 ns.

This demonstration of all-optical depletion of the DE from a semiconductor QD [87], which efficiently converts DEs to BEs and then to an empty QD, on a nanosecond timescale, is essential for using QDs as on-demand sources of single photons, as well as for deterministic writing of the BE [2] and DE [7].

4.9 Outlook

We have discussed and presented here recent experiments demonstrating that the confined dark exciton can be used as a viable matter qubit. Specifically, in the context of the DiVincenzo criteria for a good physical qubit discussed in Sect. 4.2, we have shown the following:

1. The DE is a well-characterized qubit;
2. The DE can be initialized in a known pure state by a single, picosecond long, polarized laser pulse;
3. The DE can be emptied from the QD (reset) by a nanosecond long optical pulse;
4. The DE has long lifetime ($\sim$1 μs) and coherence time ($\sim$100 ns) compared to the time required to perform gate operations on it;
5. The DE state can be easily and efficiently measured optically;

In addition, we showed that the DE qubit's eigenstates can be controlled by application of an external magnetic field.

We have thus demonstrated that the semiconductor quantum dot confined dark exciton fulfills most of the DiVincenzo criteria for a useful physical qubit. These demonstrations and studies make the dark exciton an excellent candidate for future quantum information processing technologies.

These attributes of the DE were recently used in a demonstration [61] of a proposal [93] to generate a photonic cluster state via repeated excitation of a precessing QD-confined spin. These highly-entangled cluster states are a resource for quantum information processing, because quantum computations can be performed by sequential measurement of the qubits in the entangled state [94]. Repeated, timed excitation of the DE spin qubit described in this Chapter results in an entanglement generation "machine" that is capable of deterministically generating a chain of photons in a cluster state. Future efforts in this vein will involve both lengthening the chain of entangled photons and using multiple DE qubits to generate higher-dimensional cluster states via a similar process. The dark exciton is truly a qubit with a bright future.

References

1. Y. Benny, Y. Kodriano, E. Poem, S. Khatsevitch, D. Gershoni, P.M. Petroff, Phys. Rev. B **84**, 075473 (2011)
2. Y. Benny, S. Khatsevich, Y. Kodriano, E. Poem, R. Presman, D. Galushko, P.M. Petroff, D. Gershoni, Phys. Rev. Lett. **106**, 040504 (2011)
3. E. Poem, O. Kenneth, Y. Kodriano, Y. Benny, S. Khatsevich, J.E. Avron, D. Gershoni, Phys. Rev. Lett. **107**, 087401 (2011)
4. Y. Kodriano, I. Schwartz, E. Poem, Y. Benny, R. Presman, T.A. Truong, P.M. Petroff, D. Gershoni, Phys. Rev. B **85**, 241304 (2012)
5. E. Poem, Y. Kodriano, C. Tradonsky, N.H. Lindner, B.D. Gerardot, P.M. Petroff, D. Gershoni, Nat. Phys. **6**, 993 (2010)
6. J. McFarlane, P.A. Dalgarno, B.D. Gerardot, R.H. Hadfield, R.J. Warburton, K. Karrai, A. Badolato, P.M. Petroff, App. Phys. Lett. **94**, 093113 (2009)
7. I. Schwartz, E.R. Schmidgall, L. Gantz, D. Cogan, E. Bordo, Y. Don, M. Zielinski, D. Gershoni, Phys. Rev. X **5**, 011009 (2015)
8. M. Bayer, G. Ortner, O. Stern, A. Kuther, A.A. Gorbunov, A. Forchel, P. Hawrylak, S. Fafard, K. Hinzer, T.L. Reinecke, S.N. Walck, J.P. Reithmaier, F. Klopf, F. Schäfer, Phys. Rev. B **65**, 195315 (2002)
9. Y. Kodriano, E.R. Schmidgall, Y. Benny, D. Gershoni, Semicond. Sci. Technol. **29**, 053001 (2014)

10. E.L. Ivchenko, *Optical Spectroscopy of Semiconductor Nanostructures* (Alpha Science, Harrow, 2005)
11. T. Takagahara, Phys. Rev. B **62**, 16840 (2000)
12. S. Gupalov, E. Ivchenko, A. Kavokin, J. Exp. Theor. Phys. **86**(2), 388 (1998). doi:10.1134/1.558441
13. S. Alon-Braitbart, E. Poem, L. Fradkin, N. Akopian, S. Vilan, E. Lifshitz, E. Ehrenfreund, D. Gershoni, B. Gerardot, A. Badolato, P. Petroff, Physica E **32**(12), 127 (2006)
14. M. Bayer, O. Stern, A. Kuther, A. Forchel, Phys. Rev. B **61**, 7273 (2000). doi:10.1103/PhysRevB.61.7273
15. E.L. Ivchenko, G. Pikus, *Superlattices and Other Heterostructures: Symmetry and Optical Phenomena*, 2nd edn., Springer Series in Solid-State Sciences (Springer, Berlin, 1997)
16. E. Poem, J. Shemesh, I. Marderfeld, D. Galushko, N. Akopian, D. Gershoni, B.D. Gerardot, A. Badolato, P.M. Petroff, Phys. Rev. B **76**, 235304 (2007)
17. M. Korkusinski, P. Hawrylak, Phys. Rev. B **87**, 115310 (2013)
18. M. Zielinski, Y. Don, D. Gershoni, Phys. Rev. B **91**, 085403 (2015)
19. R.J. Young, R.M. Stevenson, A.J. Shields, P. Atkinson, K. Cooper, D.A. Ritchie, K.M. Groom, A.I. Tartakovskii, M.S. Skolnick, Phys. Rev. B **72**, 113305 (2005). doi:10.1103/PhysRevB.72.113305
20. W. Langbein, P. Borri, U. Woggon, V. Stavarache, D. Reuter, A.D. Wieck, Phys. Rev. B **69**, 161301 (2004). doi:10.1103/PhysRevB.69.161301
21. D. Bimberg, M. Grundmann, N. Ledentsov, *Quantum Dot Heterostructures* (Wiley, New York, 1999)
22. G.W. Bryant, M. Zieliński, N. Malkova, J. Sims, W. Jaskólski, J. Aizpurua, Phys. Rev. Lett. **105**, 067404 (2010). doi:10.1103/PhysRevLett.105.067404
23. M.A. Dupertuis, K.F. Karlsson, D.Y. Oberli, E. Pelucchi, A. Rudra, P.O. Holtz, E. Kapon, Phys. Rev. Lett. **107**, 127403 (2011)
24. M. Zielinski, J. Phys. Condens. Matter **25**, 465301 (2013)
25. T. Smolenski, T. Kazimierczuk, M. Goryca, T. Jakubczyk, L. Klopotowski, L. Cywiński, P. Wojnar, A. Golnik, P. Kossacki, Phys. Rev. B **86**, 241305 (2012)
26. B.J. Spencer, J. Tersoff, Phys. Rev. B **87**, 161301 (2013). doi:10.1103/PhysRevB.87.161301
27. Y. Don, M. Zielinski, D. Gershoni. arXiv:1601.05530 (2016)
28. P. Shor, in *35th Annual Symposium on Foundations of Computer Science, 1994 Proceedings* (1994), pp. 124–134
29. L.K. Grover, in *Proceedings of the 28th Annual ACM Symposium on the Theory of Computing* (1996), p. 212
30. M.A. Nielsen, I.L. Chuang, *Quantum Computation and Quantum Information*, Cambridge Series on Information and the Natural Sciences (Cambridge University Press, Cambridge, 2000)
31. D.P. DiVincenzo, Fortschr. Phys. **48**, 771 (2000)
32. T.D. Ladd, F. Jelezko, R. Laflamme, Y. Nakamura, C. Monroe, J.L. O'Brien, Nature **464**, 45 (2010)
33. H.J. Kimble, Nature **453**, 1023 (2008)
34. J.I. Cirac, P. Zoller, Phys. Rev. Lett. **74**, 4091 (1995)
35. T. Monz, P. Schindler, J.T. Barreiro, M. Chwalla, D. Nigg, W.A. Coish, M. Harlander, W. Hänsel, M. Hennrich, R. Blatt, Phys. Rev. Lett. **106**, 130506 (2011)
36. I. Bloch, Nat. Phys. **1**, 23 (2007)
37. M. Anderlini, P.J. Lee, B. Brown, J. Sebby-Strabley, W.D. Phillips, J.V. Porto, Nature **448**, 452 (2007)
38. L.M.K. Vandersypen, M. Steffen, G. Breyta, C.S. Yannoni, M.H. Sherwood, I.L. Chuang, Nature **414**, 883 (2001)
39. C. Negrevergne, T.S. Mahesh, C.A. Ryan, M. Ditty, F. Cyr-Racine, W. Power, N. Boulant, T. Havel, D.G. Cory, R. Laflamme, Phys. Rev. Lett. **96**, 170501 (2006)
40. L. DiCarlo, J.M. Chow, J.M. Gambetta, L.S. Bishop, B.R. Johnson, D.I. Schuster, J. Majer, A. Blais, L. Frunzio, S.M. Girvin, R.J. Schoelkopf, Nature **460**, 240 (2009)

41. E. Lucero, R. Barends, Y. Chen, J. Kelley, M. Mariantoni, A. Megrant, P. O'Malley, D. Sank, A. Vainsencher, J. Wenner, T. White, Y. Yin, A.N. Cleland, J.M. Martinis, Nat. Phys. **8**, 719 (2012)
42. R. Barends, J. Kelley, A. Megrant, A. Veitia, D. Sank, E. Jeffrey, T.C. White, J. Mutus, A.G. Fowler, B. Campbell, Y. Chen, Z. Chen, B. Chiaro, A. Dunsworth, C. Neill, P. O'Malley, P. Roushan, A. Vainsencher, J. Wenner, A.N. Korotkov, A.N. Cleland, J.M. Martinis, Nature **508**, 500 (2014)
43. M.V.G. Dutt, L. Childress, L. Jiang, E. Togan, J. Maze, F. Jelezko, A.S. Zibrov, P.R. Hemmer, M.D. Lukin, Science **316**, 1312 (2007)
44. E. Togan, Y. Chu, A.S. Trifonov, L. Jiang, J. Maze, L. Childress, M.V.G. Dutt, A.S. Sorensen, P.R. Hemmer, A.S. Zibrov, M.D. Lukin, Nature **466**, 730 (2010)
45. J.R. Petta, A.C. Johnson, J.M. Taylor, E.A. Laird, A. Yacoby, M.D. Lukin, C.M. Marcus, M.P. Hanson, A.C. Gossard, Science **309**, 2180 (2005)
46. D. Loss, D.P. DiVincenzo, Phys. Rev. A **57**, 120 (1998)
47. A. Imamoglu, D.D. Awschalom, G. Burkard, D.P. DiVincenzo, D. Loss, M. Sherwin, A. Small, Phys. Rev. Lett. **83**, 4204 (1999)
48. D. Press, T.D. Ladd, B. Zhang, Y. Yamamoto, Nature **456**, 218 (2008)
49. A. Greilich, S.E. Economou, S. Spatzek, D.R. Yakovlev, D. Retuer, A.D. Wieck, T.L. Reinecke, M. Bayer, Nat. Phys. **5**, 262 (2009)
50. J. Berezovsky, M.H. Mikkelsen, N.G. Stoltz, L.A. Coldren, D.D. Awschalom, Science **320**, 349 (2008)
51. B.D. Gerardot, D. Brunner, P.A. Dalgarno, P. Ohberg, S. Seidl, M. Kroner, K. Karrai, N.G. Stoltz, P.M. Petroff, R.J. Warburton, Nature **451**, 441 (2008)
52. K. DeGreve, P.L. McMahon, D. Press, T.D. Ladd, D. Bisping, C. Schneider, M. Kamp, L. Worschech, S. Hofling, A. Forchel, Y. Yamamoto, Nat. Phys. **7**, 872 (2011)
53. T.M. Godden, J.H. Quilter, A.J. Ramsay, Y. Wu, P. Brereton, S.J. Boyle, I.J. Luxmoore, J. Puebla-Nunez, A.M. Fox, M.S. Skolnick, Phys. Rev. Lett. **108**, 017402 (2012)
54. L. Gantz, I. Schwartz, E.R. Schmidgall, G. Bahir, D. Gershoni, Coherent control of the bright exciton using resonant biexciton optical transitions (2015, in preparation)
55. K. Muller, T. Kaldewey, R. Ripszam, J.S. Wildmann, A. Bechtold, M. Bichler, G. Koblmuller, G. Abstreiter, J.J. Finley, Sci. Rep. **3**, 1096 (2013)
56. E.D. Kim, K. Truex, X. Xu, B. Sun, D.G. Steel, A.S. Bracker, D. Gammon, L.J. Sham, Phys. Rev. Lett. **104**, 167401 (2010)
57. D. Press, K.D. Greve, P.L. McMahon, T.D. Ladd, B. Friess, C. Schneider, M. Kamp, S. Hofling, A. Forchel, Y. Yamamoto, Nature Photonics **4**, 367 (2010)
58. A.J. Ramsay, S.J. Boyle, R.S. Kolodka, J.B.B. Oliveira, J. Skiba-Szymanska, H.Y. Liu, M. Hopkinson, A.M. Fox, M.S. Skolnick, Phys. Rev. Lett. **100**, 197401 (2008)
59. A. Greilich, S.G. Carter, D. Kim, A.S. Bracker, D. Gammon, Nat. Photonics **5**, 703 (2011)
60. J. Fischer, W.A. Coish, D.V. Bulaev, D. Loss, Phys. Rev. B **78**, 155329 (2008)
61. I. Schwartz, D. Cogan, E.R. Schmidgall, Y. Don, L. Gantz, O. Kenneth, N.H. Lindner, D. Gershoni. Science **354**, 434–437 (2016)
62. J.M. Garcia, G. Medeiros-Ribeiro, K. Schmidt, T. Ngo, J.L. Feng, A. Lorke, J. Kotthaus, P.M. Petroff, Appl. Phys. Lett. **71**, 2014 (1997)
63. G. Ramon, U. Mizrahi, N. Akopian, S. Braitbart, D. Gershoni, T.L. Reinecke, B. Gerardot, P.M. Petroff, Phys. Rev. B **73**, 205330 (2006)
64. E.R. Schmidgall, I. Schwartz, L. Gantz, D. Cogan, S. Raindel, D. Gershoni, Phys. Rev. B **90**, 241411 (2014)
65. Y. Benny, Y. Kodriano, E. Poem, D. Gershoni, T.A. Truong, P.M. Petroff, Phys. Rev. B **86**, 085306 (2012)
66. R. Hanbury Brown, R.Q. Twiss, Nature **177**, 27 (1956)
67. E. Poem, D. Gershoni, in *Radiative cascades in semiconductor quantum dots*, ed. by Leah Bergman, Jeanne L. McHalei. Handbook of Luminescent Semiconductor Materials, (CRC Press, Boca Raton, FL, USA, 2011)

68. Y. Kodriano, E. Poem, N.H. Lindner, C. Tradonsky, B.D. Gerardot, P.M. Petroff, J.E. Avron, D. Gershoni, Phys. Rev. B **82**, 155329 (2010)
69. E. Poem, Y. Kodriano, C. Tradonsky, B.D. Gerardot, P.M. Petroff, D. Gershoni, Phys. Rev. B **81**, 085306 (2010)
70. D.V. Regelman, U. Mizrahi, D. Gershoni, E. Ehrenfreund, W.V. Schoenfeld, P.M. Petroff, Phys. Rev. Lett. **87**, 257401 (2001)
71. E. Moreau, I. Robert, L. Manin, V. Thierry-Mieg, J.M. Gérard, I. Abram, Phys. Rev. Lett. **87**, 183601 (2001)
72. G. Ortner, M. Bayer, A. Larionov, V. Timofeev, A. Forchel, Y. Lyanda-Geller, T. Reinecke, P. Hawrylak, S. Fafard, Z. Wasilewski, Phys. Rev. Lett. **90**, 086404 (2003)
73. R. Presman, E.R. Schmidgall, I. Schwartz, E. Poem, Y. Benny, D. Gershoni, in *31st International Conference on the Physics of Semiconductors* (Zurich, Switzerland, 2012)
74. I. Schwartz, D. Cogan, E.R. Schmidgall, L. Gantz, Y. Don, M. Zielinski, D. Gershoni, Phys. Rev. B **92**, 201201 (2015)
75. M. Fox, *Quantum Optics: An Introduction* (Oxford University Press, Oxford, 2006)
76. T.H. Stievater, X. Li, D.G. Steel, D. Gammon, D.S. Katzer, D. Park, C. Piermarocchi, L.J. Sham, Phys. Rev. Lett. **87**, 133603 (2001)
77. H. Kamada, H. Gotoh, J. Temmyo, T. Takagahara, H. Ando, Phys. Rev. Lett. **87**, 246401 (2001)
78. A. Zrenner, E. Beham, S. Stufler, F. Findeis, M. Bichler, G. Abstreiter, Nature **418**, 612 (2002)
79. L.M. Woods, T.L. Reinecke, R. Kotlyar, Phys. Rev. B **69**, 125330 (2004)
80. D.V. Bulaev, D. Loss, Phys. Rev. Lett. **95**, 076805 (2005)
81. D. Heiss, S. Schaeck, H. Huebl, M. Bichler, G. Abstreiter, J.J. Finley, D.V. Bulaev, D. Loss, Phys. Rev. B **76**, 241306 (2007)
82. S.E. Economou, T.L. Reinecke, Phys. Rev. Lett. **99**, 217401 (2007)
83. M.P. Nowak, B. Szafran, Phys. Rev. B **83**, 035315 (2011)
84. L. Gantz, E.R. Schmidgall, I. Schwartz, Y. Don, E. Waks, G. Bahir, D. Gershoni, Phys. Rev. B **94** (2016). doi:10.1103/PhysRevB.94.045426
85. A. Muller, E.B. Flagg, P. Bianucci, X.Y. Wang, D.G. Deppe, W. Ma, J. Zhang, G.J. Salamo, M. Xiao, C.K. Shih, Phys. Rev. Lett. **99**, 187402 (2007)
86. Y.M. He, Y. He, Y.J. Wei, D. Wu, M. Atature, C. Schneider, S. Hofling, M. Kamp, C.Y. Lu, J.W. Pan, Nat. Nanotechnol. **8**(3), 213 (2013)
87. E.R. Schmidgall, I. Schwartz, D. Cogan, L. Gantz, T. Heindel, S. Reitzenstein, D. Gershoni, Appl. Phys. Lett. **106**, 193101 (2015)
88. E.R. Schmidgall, Y. Benny, I. Schwartz, L. Gantz, Y. Don, D. Gershoni, Phys. Rev. B **93**, 245437 (2016)
89. Y. Benny, R. Presman, Y. Kodriano, E. Poem, D. Gershoni, T.A. Truong, P.M. Petroff, Phys. Rev. B **89**, 035316 (2014)
90. N. Akopian, N.H. Lindner, E. Poem, Y. Berlatzky, J. Avron, D. Gershoni, B.D. Gerardot, P.M. Petroff, Phys. Rev. Lett. **96**, 130501 (2006)
91. M. Müller, S. Bounouar, K.D. Jöns, M. Glässl, P. Michler, Nat. Photon. **8**, 224 (2014)
92. D. Gammon, E.S. Snow, B.V. Shanabrook, D.S. Katzer, D. Park, Phys. Rev. Lett. **76**, 3005 (1996)
93. N.H. Lindner, T. Rudolph, Phys. Rev. Lett. **103**, 113602 (2009)
94. H.J. Briegel, D.E. Browne, W. Duer, R. Raussendorf, M.V. den Nest, Nat. Phys. **5**, 19 (2009)

Chapter 5
The Mesoscopic Nature of Quantum Dots in Photon Emission

P. Tighineanu, A.S. Sørensen, S. Stobbe and P. Lodahl

Abstract Semiconductor quantum dots share many properties with atoms such as discrete spectrum, which implies the ability to emit high purity single photons. However, they have unique features as well that are unknown to other emitters: they embody tens of thousands of atoms attaining large mesoscopic sizes, and lack the common atomic symmetries. Here we discuss two effects that are mediated by the mesoscopic nature and render quantum dots fundamentally different than atoms. The mesoscopic size and lack of parity symmetry causes the electric-dipole approximation to not be applicable to In(Ga)As quantum dots. As a consequence, the latter do not fulfil the atomic selection rules and thus interact with the electric and magnetic components of light on the same electronic transition. The multi-atomic nature also causes a collective mesoscopic effect in monolayer-fluctuation GaAs quantum dots, namely single-photon superradiance, giving rise to a giant light-matter coupling strength.

Semiconductor quantum dots (QDs) provide the essential link between light and matter and can be integrated monolithically into photonic devices. These nanometer-sized purposefully engineered impurities combine the atomic-like discrete spectra and excellent single-photon purity with the large light-matter interaction strength inherent to solid-state systems [1]. The ability to tailor the photonic environment around QDs has resulted in tremendous progress in manipulating single QD excitations. Strong coupling between a QD and a cavity [2–4] and near-unity coupling to a photonic waveguide [5–10] are a few out of many exciting realizations [1].

The atomic-like properties of QDs are supplemented by a range of new effects owing to their solid-state nature. For instance, vibrations of the underlying crystal lattice, known as phonons, may decohere the light-matter interaction [11–13] or couple non-resonant QD excitations to an optical cavity [14–18]. Similarly, the mesoscopic

P. Tighineanu (✉) · A.S. Sørensen · S. Stobbe · P. Lodahl
The Niels Bohr Institute, University of Copenhagen, Blegdamsvej 17,
2100 Copenhagen, Denmark
e-mail: petrut@nbi.ku.dk

P. Lodahl
e-mail: lodahl@nbi.ku.dk

P. Michler (ed.), *Quantum Dots for Quantum Information Technologies*,
Nano-Optics and Nanophotonics, DOI 10.1007/978-3-319-56378-7_5

ensemble of the nuclei composing the QD can be used to tailor the hyperfine interaction with the electron in spin-based quantum-information science [19]. Recently it was found [20] that QDs may break the dipole approximation, which is often assumed to be valid also in solid-state quantum optics. These realizations unveil the complex nature of QDs, which embody tens to hundred thousand atoms attaining "mesoscopic" sizes that interact relentlessly with the surrounding solid-state environment. In this chapter we present a unified description of the mesoscopic nature of QDs [20–23]. In particular, we discuss two mesoscopic effects that exist solely due to the large physical size of QDs: the breakdown of the dipole theory of In(Ga)As QDs and collective enhancement of light-matter interaction with monolayer-fluctuation GaAs QDs.

The small size L of most quantum emitters compared to the wavelength of light λ has ensured the success of the dipole theory, which states that emitters interact with light as dimensionless entities (point dipoles). Since QDs attain mesoscopic sizes of 10–30 nm [24], the dipole approximation does not necessarily hold because the figure of merit $2\pi nL/\lambda_0 \approx 0.5$ is not negligible. Here, typical values for the wavelength in vacuum $\lambda_0 = 900\,\text{nm}$, refractive index $n = 3.42$ and $L = 20\,\text{nm}$ have been used. This figure of merit may be further enhanced in the vicinity of metal nanostructures, where additional propagating modes (surface plasmons) beyond the light cone arise. It has been observed that the spontaneous-emission dynamics from QDs placed near a metal interface show pronounced deviations from the dipole theory [20]. A theory of light-matter interaction beyond the dipole theory can explain these experimental findings by introducing a single mesoscopic moment to be considered along with the dipole moment in light-matter interactions [21]. Notably, this theory is more general than previously developed models [25–32] because it considers the symmetry of the full quantum-mechanical wavefunction and not only the slowly varying envelope. We show that the discrete atomistic symmetry explains the microscopic origin of the large mesoscopic moment observed experimentally. In particular, the developed theory pinpoints that large structural inhomogeneities at the crystal-lattice level lead to a violation of parity symmetry in In(Ga)As QDs [22]. Quantum dots therefore break the atomic selection rules and probe electric and magnetic fields on the same electronic transition [21]. Moreover, the mesoscopic size of QDs may ease the observation of dipole-forbidden transitions in photonic nanostructures [33]. It has been shown that, in the opposite limit of highest possible (spherical) symmetry present in, e.g., colloidal QDs, a shell theorem is valid, which states that the Purcell enhancement in an arbitrary photonic environment is protected by symmetry and does not depend on the QD size [32].

The second part of this chapter is devoted to presenting another mesoscopic property of QDs, namely collective enhancement of light-matter interaction leading to single-photon superradiance. Quantum dots benefit from their multi-body nature with an enhanced coupling to light compared to atoms, which renders them promising candidates for improving the efficiency of single-photons sources, solar cells and nano-lasers, to name a few important practical applications. Commonly employed QDs have, however, an upper limit for the interaction strength with light, regardless of their size and shape. It has therefore been a long-sought goal in quantum photonics to develop solid-state emitters beyond this upper limit [34–37]. We demonstrate that the fundamental excitation of a monolayer-fluctuation QD [38] is analogous to the phenomenon of single-photon superradiance defined by Dicke for a non-interacting ensemble [39]. This effect leads to an enhanced coupling to light far beyond that of conventional QDs, which may be of interest for fundamental science and technology alike. In particular, such rapid radiative decays will likely exceed relevant dephasing mechanisms resulting in highly coherent flying quantum bits. Furthermore, new and so far largely unexplored solid-state quantum-electrodynamics regimes involving energy non-conserving virtual processes, such as the ultra-strong coupling between light and matter, may become within reach at optical frequencies [23].

5.1 Fundamentals of Light-Matter Interaction with Quantum Dots

In this section we lay the fundamental as well as the experimentally relevant aspects describing the interaction between QDs and light.

5.1.1 Effective-Mass Theory

The commonly employed bandstructure method for QDs is the effective-mass theory. It assumes that the bands, which are exact solutions in the bulk semiconductor, are weakly perturbed by the nanostructure. Formally, a quantized eigenstate within an electronic band can be written as a product of a periodic Bloch function, $u(\boldsymbol{r})$, which captures the properties on the length scale of the crystal unit cell, and a slowly varying envelope, $\psi(\boldsymbol{r})$, that inherits the size and symmetry of the mesoscopic QD potential

$$\Psi_j(\boldsymbol{r}) = \psi_j(\boldsymbol{r})u_j(\boldsymbol{r}), \tag{5.1}$$

where $j = \{\mathrm{e, hh, lh}\}$ labels either of the three relevant bands in zincblende semiconductors: electron, heavy hole, and light hole, respectively. It can be shown [40] that ψ_j is subject to a Schrödinger-type equation

$$E_j\psi_j(\boldsymbol{r}) = -\frac{\hbar^2}{2m_{\mathrm{eff},j}}\Delta\psi_j(\boldsymbol{r}) + V_j(\boldsymbol{r})\psi_j(\boldsymbol{r}), \tag{5.2}$$

where we assume that the effective mass is isotropic. The complicated unit-cell potential is merged into the effective mass, m_{eff}, a parameter that can be inferred from experiments. The potential energy $V_j(\boldsymbol{r})$ contains the smooth mesoscopic potential of the QD as illustrated in Fig. 5.1. This particle-in-a-box problem can be solved either analytically or numerically using the standard techniques of quantum mechanics.

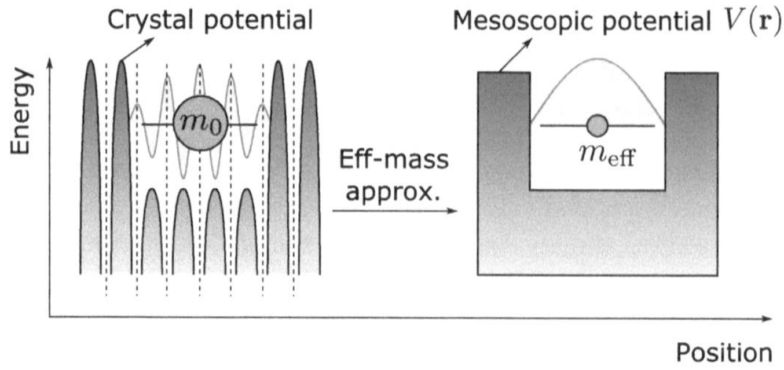

Fig. 5.1 Physical interpretation of the effective-mass approximation. The complicated crystal potential (*left*) is merged into an effective-mass parameter (*right*)

The effective-mass theory describes the properties of QDs remarkably well, since the first valence-band eigenstate is usually heavy-hole-like with a negligible light-hole component. This is related to the presence of compressive strain in In(Ga)As/GaAs QDs, which alters the symmetry of the unit cells and splits the degeneracy of the bands [41]. In strain-free QDs, the ground state is still mostly heavy-hole like due to the small aspect ratio of QDs [24, 36, 42–44].

5.1.2 Excitons

Electrons and holes possess charge and half-integer spin and therefore interact. The electron-hole bound state constitutes a fundamental quasi-particle, the exciton, which governs the optical properties of QDs. Being a two-body system, the exciton wavefunction Ψ_X can be expanded in the single-particle electron and hole wavefunctions [45]

$$\Psi_X(\boldsymbol{r}_e, \boldsymbol{r}_h) = \sum_{n,m} C_{n,m} \Psi_n^e(\boldsymbol{r}_e) \Psi_m^h(\boldsymbol{r}_h) = u_e(\boldsymbol{r}_e) u_h(\boldsymbol{r}_h) \psi_X(\boldsymbol{r}_e, \boldsymbol{r}_h), \tag{5.3}$$

where Ψ_n corresponds to the n-th eigenstate of the QD and $\psi_X(\boldsymbol{r}_e, \boldsymbol{r}_h)$ is the slowly varying envelope of the exciton subject to the two-body effective-mass Schrödinger equation

$$\left(\frac{\boldsymbol{p}_e^2}{2m_e} + \frac{\boldsymbol{p}_h^2}{2m_h} + V_e(\boldsymbol{r}_e) + V_h(\boldsymbol{r}_h) - \frac{e^2}{4\pi\epsilon_0\epsilon \left| \boldsymbol{r}_e - \boldsymbol{r}_h \right|} \right) \psi_X = E\psi_X. \tag{5.4}$$

Here, $\boldsymbol{p}$ is the momentum operator, ϵ_r is the dielectric constant and E the energy of the exciton. In bulk, the electron and hole orbit each other within a distance known as the exciton Bohr radius a_0. Since the Coulomb energy E_C scales inversely with the

QD size, $E_C \propto L^{-1}$, the Coulomb and exchange interactions in a QD are enhanced compared to bulk. On the other hand, the quantum-confinement energy scales as L^{-2}. As a consequence, the exciton motion can be found in two regimes:

(i) The *strong-confinement regime*, in which $L \ll a_0$ [46] and quantum confinement dominates Coulomb confinement. The latter can be neglected and the electron and hole therefore move independently as non-interacting particles $\psi_X(\boldsymbol{r}_e, \boldsymbol{r}_h) = \psi_e(\boldsymbol{r}_e)\psi_h(\boldsymbol{r}_h)$. Most of the studied semiconductor QDs in the literature are in the strong-confinement regime. The dipole moment of an x-polarized exciton is

$$\boldsymbol{\mu} = \frac{e}{m_0}\langle\Psi_h|\hat{p}_x|\Psi_e\rangle\boldsymbol{e}_x \simeq \frac{e}{m_0}p_{cv}\langle\psi_h|\psi_e\rangle\boldsymbol{e}_x, \tag{5.5}$$

where $\boldsymbol{e}_x$ is the Cartesian unit vector, $p_{cv} = V_{\mathrm{UC}}^{-1}\int_{\mathrm{UC}} \mathrm{d}^3\boldsymbol{r} u_x^* \hat{p}_x u_e$ is the interband Bloch matrix element with V_{UC} being the unit-cell volume. In the above equation we have exploited the slow variation of the envelopes ψ over one unit cell. The dipole moment of small QDs has therefore an upper limit of $\mu_{\max} = (e/m_0)p_{cv}$.

(ii) The *weak-confinement regime*, in which $L \gtrsim a_0$ and the electron-hole motion is correlated. Here, (5.4) has to be solved as a two-body problem. Achieving this regime has been a long-sought goal in quantum photonics because such QDs exhibit a giant dipole moment beyond $\mu_{\max}$, cf. Sect. 5.3.

Excitonic effects have a prominent role in determining the QD energy structure. Combining the electron contribution with a spin of $\pm 1/2$ with the heavy-hole projected angular momentum of $\pm 3/2$ yields four possible excitonic configurations: two optically bright with $j_z = \pm 1$ and two optically dark with $j_z = \pm 2$. Bright excitons are higher in energy than dark excitons by several hundred μeV [47]. The splitting between the two bright states is of the order of tens of μeV [47] and is mostly determined by the QD asymmetry [48]. The resulting dipole moments of the bright excitons $j_z = \pm 1$ are orthogonally polarized along the $x = [1, 1, 0]$ and $y = [1, -1, 0]$ crystallographic directions.

5.1.3 Spontaneous Emission

The light-matter interaction strength governs the temporal dynamics of the exciton-to-photon conversion. In the Wigner–Weisskopf approximation, the QD exciton decays exponentially with the rate Γ_{rad} determined by Fermi's Golden Rule

$$\Gamma_{\mathrm{rad}} = \frac{2\pi}{\hbar^2}\sum_f \left|\langle f|\hat{H}_{\mathrm{int}}|i\rangle\right|^2 \delta(\omega - \omega_{if}), \tag{5.6}$$

where $\hat{H}_{\mathrm{int}}$ is the light-matter interaction Hamiltonian triggering a transition from the initial $|i\rangle$ to the final $|f\rangle$ state. In this work we consider the minimal-coupling interaction Hamiltonian [49] between an electron with charge e and mass m_0, and the field described by the vector potential $\hat{A}$

$$\hat{H}_{\text{int}} = -\frac{e}{2m_0}\left(\hat{p}\cdot\hat{A} + \hat{A}\cdot\hat{p} - e\hat{A}\cdot\hat{A}\right). \tag{5.7}$$

Another commonly used Hamiltonian is the multipolar Hamiltonian, which is expressed in terms of electric and magnetic fields. It can be shown [49] that the two Hamiltonians give the same result for energy conserving processes such as spontaneous emission [50].

The nonlinear term $\hat{A}\cdot\hat{A}$ can be neglected for the weak fields studied here. We employ the generalized Coulomb gauge, $\nabla\cdot[\epsilon(\boldsymbol{r})\hat{A}(\boldsymbol{r})] = 0$, yielding

$$\hat{H}_{\text{int}} \simeq -\frac{e}{m_0}\hat{A}(\boldsymbol{r})\cdot\hat{p}, \tag{5.8}$$

where the dielectric constant $\epsilon(\boldsymbol{r})$ is assumed to vary over length scales larger than the QD size [51]. The vector potential can be written in terms of the normal field modes [52]

$$\hat{A}(\boldsymbol{r}) = \sum_l \sqrt{\frac{\hbar}{2\epsilon_0\omega_l}}\left[\hat{a}_l \boldsymbol{f}_l(\boldsymbol{r}) + \hat{a}_l^\dagger \boldsymbol{f}_l^*(\boldsymbol{r})\right], \tag{5.9}$$

where $\hat{a}_l$ ($\hat{a}_l^\dagger$) is the annihilation (creation) operator for the l mode. The QD is approximated as a two-level system with the initial state $|i\rangle = |e\rangle \otimes |0\rangle$ with the exciton in the excited state $|e\rangle$ and the field in the ground state $|0\rangle$, and $|f\rangle = |g\rangle \otimes |1_f\rangle$ the final state with one excitation in the field mode f and the emitter in the ground state $|g\rangle$. Plugging this into (5.6) yields

$$\Gamma_{\text{rad}} = \frac{\pi e^2}{\epsilon_0\hbar m_0^2}\sum_l \frac{1}{\omega_l}\left|\langle g|\boldsymbol{f}_l^*(\boldsymbol{r})\cdot\hat{p}|e\rangle\right|^2 \delta(\omega - \omega_l), \tag{5.10}$$

which is the starting point for the research presented here. The expression is beyond the dipole approximation because the variation of the field over the emitter is taken into account.

5.1.4 *The Dipole Approximation: Oscillator Strength and Density of Optical States*

The standard-textbook approach to evaluating (5.10) is to assume that the field does not vary over the QD, $\boldsymbol{f}_l(\boldsymbol{r}) \approx \boldsymbol{f}_l(\boldsymbol{r}_0)$, where $\boldsymbol{r}_0$ is the QD center. The resulting dipole approximation is excellent for quantum emitters that are much smaller than the wavelength of light. We thus obtain

$$\Gamma_{\text{rad}} = \frac{\pi |\mu|^2}{\epsilon_0 \hbar} \boldsymbol{e}_p \cdot \left[\sum_l \frac{1}{\omega_l} \boldsymbol{f}_l^*(\boldsymbol{r}_0) \boldsymbol{f}_l(\boldsymbol{r}_0) \delta(\omega - \omega_l) \right] \cdot \boldsymbol{e}_p^*, \tag{5.11}$$

where $\boldsymbol{e}_p$ is the unit vector pointing along the direction of the dipole moment. The term in square brackets is proportional to the imaginary part of the Green tensor [25, 52]

$$\text{Im}\underline{\underline{G}}(\boldsymbol{r}, \boldsymbol{r}') = \frac{\pi c_0^2}{2} \sum_l \frac{1}{\omega_l} \boldsymbol{f}_l^*(\boldsymbol{r}) \boldsymbol{f}_l(\boldsymbol{r}') \delta(\omega - \omega_l), \tag{5.12}$$

In the dipole approximation, the light-matter interaction strength is thus governed by two quantities: the dipole moment $\boldsymbol{\mu}$, which is an intrinsic property of the emitter, and the imaginary part of the Green tensor, which is a property of the electromagnetic environment.

It is often useful to recast the emitter and field properties in terms of the oscillator strength f and the projected local density of optical states (LDOS) $\rho(\boldsymbol{r}_0, \omega, \boldsymbol{e}_p)$. The oscillator strength is a dimensionless quantity defined as the ratio between the radiative rate of the QD in a homogeneous environment and the emission rate of a classical harmonic oscillator, and is related to the dipole moment via

$$f = \frac{m_0}{e^2 \hbar \omega} |\mu|^2. \tag{5.13}$$

The oscillator strength of atoms is of the order of 1 and about 10 for QDs [53] because QDs are larger and interact stronger with light. Conventional QDs are in the strong-confinement regime with a maximum oscillator strength of

$$f_{\max} = \frac{E_g}{\hbar \omega}, \tag{5.14}$$

where E_g is the Kane energy, an experimentally accessible quantity. This upper limit for f can be understood from (5.5), where the overlap between the envelopes cannot exceed unity. The LDOS is defined as the number of electromagnetic modes per unit energy and volume that the emitter can decay into

$$\rho(\boldsymbol{r}_0, \omega, \boldsymbol{e}_p) = \frac{2\omega}{\pi c_0^2} \text{Im} \left[\boldsymbol{e}_p \cdot \underline{\underline{G}}(\boldsymbol{r}_0, \boldsymbol{r}_0) \cdot \boldsymbol{e}_p^* \right]. \tag{5.15}$$

Modern fabrication techniques enable accurate tailoring of the LDOS surrounding the QDs in, e.g., photonic-crystal cavities [54] and waveguides [10], micropillar cavities [55], nanowires [7] and plasmonic nano-antennas [56].

In a homogeneous medium, the Green tensor can be evaluated analytically [52] yielding

$$\Gamma_{\text{rad}}^{\text{hom}} = \frac{\mu_0 \omega n}{3\pi \hbar c_0} |\mu|^2. \tag{5.16}$$

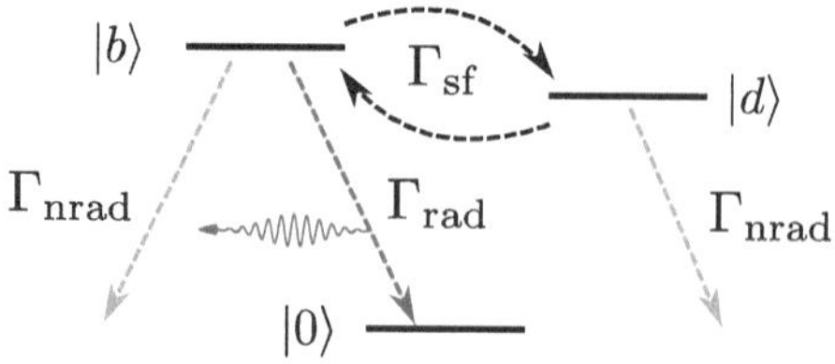

Fig. 5.2 Level scheme describing the population transfer of the exciton in a QD. The bright exciton $|b\rangle$ can decay either radiatively (Γ_{rad}) or nonradiatively (Γ_{nrad}) and can interact with its the dark exciton $|d\rangle$ via the spin-flip rate (Γ_{sf}). Figure reproduced from [42]

Finally we note that only the imaginary part of the Green tensor contributes to the decay rate because energy dissipation is described by the part of the response function, which is $\pi/2$ out of phase with the driving field, as is known for the harmonic oscillator. The real part is proportional to a self-energy term, the Lamb shift, which shifts the frequency of the QD exciton [57].

5.1.5 Decay Dynamics of Quantum Dots

The internal structure of QDs is more complicated than a two-level system: there are 4 excited states comprising 2 bright and 2 dark excitons, which are coupled by spin-flip processes. Furthermore, the omnipresent nonradiative processes, such as defect traps in the vicinity of the QD [58], provide alternative pathways for the recombination of the exciton. The experimentally measured decay rates of QDs therefore depend on radiative, nonradiative, and spin-flip processes. Analyzing the dynamics provides important information about the optical quality of QDs, namely the oscillator strength and the quantum efficiency η. The latter quantifies the probability that the QD exciton recombines radiatively

$$\eta = \frac{\Gamma_{rad}^{hom}}{\Gamma_{rad}^{hom} + \Gamma_{nrad}}, \tag{5.17}$$

where Γ_{nrad} is the nonradiative rate. In the following we present a method that can unambiguously extract these quantities from measurements [23, 42, 53].

Spin-flip processes are inhibited in QDs because it is difficult to simultaneously flip spin and fulfil energy conservation due to the discrete spectrum. The coupling between bright-bright and dark-dark excitons can generally be neglected because it is a second-order process. Only bright-dark excitons are coupled by Γ_{sf} and the level scheme is reduced to the one depicted in Fig. 5.2. Bright-dark and dark-bright spin-flip rates are assumed to be the same because the thermal energy at 4–10 K is larger than the bright-dark energy splitting. Also, the nonradiative rates of bright and dark excitons are the same due to their small energy splitting [59].

The decay dynamics of the bright exciton is governed by the rate equations of the coupled three-level system

$$\begin{bmatrix} \dot{\rho}_B \\ \dot{\rho}_D \end{bmatrix} = \begin{bmatrix} -\Gamma_{\text{rad}} - \Gamma_{\text{nrad}} - \Gamma_{\text{sf}} & \Gamma_{\text{sf}} \\ \Gamma_{\text{sf}} & -\Gamma_{\text{nrad}} - \Gamma_{\text{sf}} \end{bmatrix} \begin{bmatrix} \rho_B \\ \rho_D \end{bmatrix}, \tag{5.18}$$

where ρ denotes the occupation probability. Under the realistic assumption that spin flip-processes are much slower than the radiative rate, i.e., $\Gamma_{\text{sf}} \ll \Gamma_{\text{rad}}$, (5.18) yields for the decay of the bright state

$$\rho_B(t) = \rho_B(0)e^{-(\Gamma_{\text{rad}}+\Gamma_{\text{nrad}})t} + \frac{\Gamma_{\text{sf}}}{\Gamma_{\text{rad}}}\rho_D(0)e^{-(\Gamma_{\text{nrad}}+\Gamma_{\text{sf}})t}. \tag{5.19}$$

The bright exciton exhibits a biexponential decay with the fast rate $\Gamma_F = \Gamma_{\text{rad}} + \Gamma_{\text{nrad}}$ and the slow rate $\Gamma_S = \Gamma_{\text{nrad}} + \Gamma_{\text{sf}}$. Consequently, by fitting the measured decay curves with $f(\tau) = A_F e^{-\Gamma_F \tau} + A_S e^{-\Gamma_S \tau} + C$, where τ is the time delay with respect to the start of the excitation pulse and C is the background level, which is determined by the measured dark-count rate and after-pulsing probability of the detector, the radiative and nonradiative rates can be unambiguously extracted via

$$\Gamma_{\text{rad}} = \Gamma_F - \Gamma_S, \tag{5.20}$$

$$\Gamma_{\text{nrad}} = \Gamma_S - \frac{A_S}{A_F}\frac{\rho_B(0)}{\rho_D(0)}(\Gamma_F - \Gamma_S), \tag{5.21}$$

$$\Gamma_{\text{sf}} = \frac{A_S}{A_F}\frac{\rho_B(0)}{\rho_D(0)}(\Gamma_F - \Gamma_S). \tag{5.22}$$

For non-resonant excitation, QDs trap carriers with random spin, i.e.,

$$\frac{\rho_B(0)}{\rho_D(0)} \simeq 1. \tag{5.23}$$

Notably, the radiative rate Γ_{rad} does not coincide with the homogeneous-medium quantity $\Gamma_{\text{rad}}^{\text{hom}}$ because QDs are often located close to dielectric-dielectric or dielectric-air interfaces, which may modify the LDOS.

5.2 Light-Matter Interaction Beyond the Dipole Approximation with In(Ga)As Quantum Dots

The experimental observation of the dipole-theory breakdown was presented in [20], where QDs placed near a metal interface were probed by time-resolved photoluminescence measurements, cf. Fig. 5.3. In the experiment, the QD spontaneous-emission rate was found to be inhibited relative to the dipole theory for the geometry in

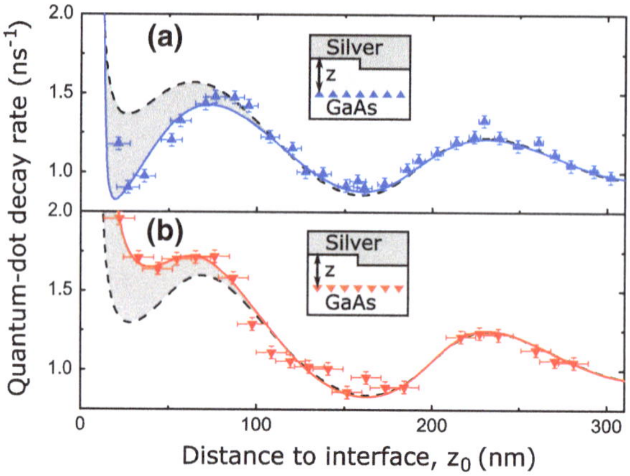

Fig. 5.3 Observation of deviations from the dipole theory for QDs near a metal interface [20]. The decay rate of QDs close to a metal interface was measured for **a** direct and **b** inverted QDs relative to the interface. The *black dashed line* denotes the dipole theory, the *triangles* the data points and the *colored solid lines* the fit. Figure reproduced from [20]

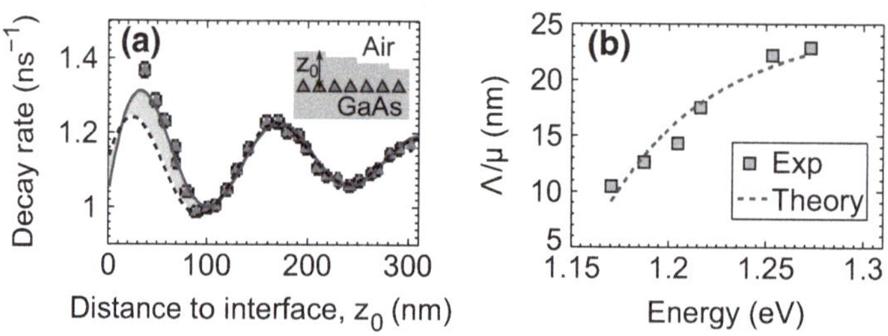

Fig. 5.4 Observation of deviations from the dipole theory for QDs near a dielectric interface. **a** Measured decay rates versus distance z_0 to the GaAs-air interface (data points) at an energy of 1.27 eV. The dipole (multipolar) theory is indicated by the *black dashed* (*blue solid*) *line*. A refractive index $n = 3.5$ of GaAs was used. **b** Extracted mesoscopic strength Λ/μ over the emission spectrum of QDs (*red squares*) along with the prediction of the theoretical model (*blue dashed line*). Figure reproduced from [22] (color figure online)

Fig. 5.3a. In contrast, the inverted structure in Fig. 5.3b showed an increase in the rate; the two structures would exhibit the same rates, if the QDs were dipoles. The observed difference is thus a direct demonstration of the breakdown of the dipole theory. The effects beyond the dipole approximation were merged into a phenomenological QD parameter, the mesoscopic moment $\Lambda = (e/m_0)\langle 0|x\hat{p}_z|\Psi_X\rangle$.

Deviations from the dipole theory have also been observed in the vicinity of an air interface [22, 53, 58], cf. Fig. 5.4, at distances below ~75 nm, which can again be explained by the mesoscopic moment Λ. The extracted mesoscopic strength Λ/μ increases with emission energy and varies from 10 to 23 nm over the inhomogeneously broadened emission spectrum, cf. Fig. 5.4b, and is successfully explained by our microscopic QD theory, which is presented later. The emission rate at the air interface is enhanced while at the metal interface it is suppressed (for the direct structure) due to the phase shift gained upon reflection in the latter case, which flips the sign of the field gradient.

The QD wavefunctions obtained from the effective-mass theory cannot explain the large mesoscopic moment Λ observed experimentally. Sections 5.2.1–5.2.3 present such a theory that is applied to describe the spontaneous-emission process from QDs in Sect. 5.2.5. It is shown that the inhomogeneous quantum-current distribution makes QDs a probe of electric and magnetic fields.

5.2.1 Theory of Light-Matter Interaction Beyond the Dipole Approximation

The starting point for the theory is to account for the variation of the electromagnetic field over the size of the exciton wavefunction in (5.12). The decay rate of the emitter becomes

$$\Gamma(\omega) = \frac{2\mu_0}{\hbar} \int\int \mathrm{d}^3\boldsymbol{r}\mathrm{d}^3\boldsymbol{r}' \mathrm{Im}\left[\boldsymbol{j}(\boldsymbol{r}) \cdot \underline{\underline{G}}(\boldsymbol{r}, \boldsymbol{r}') \cdot \boldsymbol{j}^*(\boldsymbol{r}')\right]. \tag{5.24}$$

Here, we define the quantum-mechanical current density $\boldsymbol{j}(\boldsymbol{r})$ of the QD

$$\boldsymbol{j}(\boldsymbol{r}) = \frac{e}{m_0}\hat{\boldsymbol{p}}\Psi_\mathrm{X}(\boldsymbol{r}, \boldsymbol{r}). \tag{5.25}$$

Unlike the dipole theory, where the effect of the environment on the emitter can be thought of as a self-interaction term at a single point $\boldsymbol{r}_0$, here the self interaction occurs between all possible pairwise points $(\boldsymbol{r}, \boldsymbol{r}')$ within Ψ_X. We perform an expansion in the field modes, $\boldsymbol{f}_l(\boldsymbol{r})$, because the integral formulation offers limited physical insight and is often computationally infeasible.

The transition moment from the ground to the excited state is defined as

$$T_{0\mathrm{X}} = \frac{e}{m_0}\langle 0 | \boldsymbol{f}_l^*(\boldsymbol{r}) \cdot \hat{\boldsymbol{p}} | \Psi_\mathrm{X}\rangle. \tag{5.26}$$

Expanding the normal mode $\boldsymbol{f}_l$ around the QD center $\boldsymbol{r}_0$ yields

$$T_{0\mathrm{X}} = T_{0\mathrm{X}}^{(0)} + T_{0\mathrm{X}}^{(1)} + T_{0\mathrm{X}}^{(2)} + \cdots \tag{5.27}$$

The electric-dipole term neglects the variation of the field over the emitter

$$T_{0\mathrm{X}}^{(0)} = T_{0\mathrm{X}}^{(\mu)} = f_i^*(\boldsymbol{r}_0)\langle 0 | \hat{\mu}_i | \Psi_\mathrm{X}\rangle = f_i^*(\boldsymbol{r}_0)\mu_i, \tag{5.28}$$

where $\mu_i = \langle 0 | \hat{\mu}_i | \Psi_\mathrm{X}\rangle$, $\hat{\mu}_i = (e/m_o)\hat{p}_i$ is the electric-dipole operator, and implicit summation over indices is used. The first-order contribution reads

$$T_{0X}^{(1)} = \partial_j f_i^*(\boldsymbol{r}_0)\Lambda_{ji}, \tag{5.29}$$

where $\Lambda_{ij} = (e/m_0)\langle 0|x_i\hat{p}_j|\Psi_X\rangle$ is the first-order mesoscopic moment. $T_{0X}^{(1)}$ can be written as a sum of the electric-quadrupole, $T_{0X}^{(Q)}$, and magnetic-dipole, $T_{0X}^{(m)}$, contributions

$$T_{0X}^{(1)} = T_{0X}^{(Q)} + T_{0X}^{(m)} = \frac{1}{2}\partial_j f_i^*(\boldsymbol{r}_0)\langle 0|\hat{Q}_{ij}|\Psi_X\rangle + \left[\nabla \times \boldsymbol{f}^*(\boldsymbol{r}_0)\right]\cdot\langle 0|\hat{\boldsymbol{m}}|\Psi_X\rangle, \tag{5.30}$$

where $\hat{Q}_{ij} = (e/m_0)\left(x_i\hat{p}_j + \hat{p}_i x_j\right)$ is the electric-quadrupole and $\boldsymbol{m} = e/(2m_0)\boldsymbol{r} \times \hat{\boldsymbol{p}}$ the magnetic-dipole operator.

The second order correction to the transition moment is

$$T_{0X}^{(2)} = \partial_j\partial_k f_i^*(\boldsymbol{r}_0)\Omega_{kji}, \tag{5.31}$$

where $\Omega_{ijk} = (e/2m_0)\langle 0|x_i x_j\hat{p}_k|\Psi_X\rangle$ is the second-order mesoscopic moment. $T_{0X}^{(2)}$ can be rewritten in terms of electric-octupole, $T_{0X}^{(O)}$, and magnetic-quadrupole, $T_{0X}^{(M)}$, contributions

$$T_{0X}^{(2)} = T_{0X}^{(O)} + T_{0X}^{(M)} = \frac{1}{6}\partial_{jk}^2 f_i^*(\boldsymbol{r}_0)\langle 0|\hat{O}_{ijk}|\Psi_X\rangle + \frac{1}{2}\partial_j\left[\nabla \times \boldsymbol{f}^*(\boldsymbol{r}_0)\right]_i\langle 0|\hat{M}_{ij}|\Psi_X\rangle, \tag{5.32}$$

where $\hat{O}_{ijk} = (e/m_0)(x_k x_j\hat{p}_i + x_k\hat{p}_j x_i + \hat{p}_k x_j x_i)$ is the electric-octupole and $\hat{M}_{ij} = (e/3m_0)\left[x_j\left(\boldsymbol{r}\times\hat{\boldsymbol{p}}\right)_i + \left(\boldsymbol{r}\times\hat{\boldsymbol{p}}\right)_i x_j\right]$ the magnetic-quadrupole operator.

All in all, the multipole expansion to second order results in five contributions

$$T_{0X} = T_{0X}^{(0)} + T_{0X}^{(1)} + T_{0X}^{(2)} = T_{0X}^{(\mu)} + T_{0X}^{(Q)} + T_{0X}^{(m)} + T_{0X}^{(O)} + T_{0X}^{(M)} + \cdots, \tag{5.33}$$

which are summarized in Table 5.1 and sketched in Fig. 5.5. The zeroth order has only electric-dipole contributions, while higher orders include terms of both electric and magnetic nature.

The multipolar expansion depends on the choice of the expansion point $\boldsymbol{r}_0$ [60, 61], which leads to an $\boldsymbol{r}_0$-dependent decay rate. We find that by consistently collecting the expansion orders in the decay rate rather than in the multipolar moments, the rate

Table 5.1 Overview of the different contributions to the multipole expansion of T_{0X} up to second order

Order	Overall	Electric	Magnetic
0	$T_{0X}^{(0)} = \mu_i f_i^*$	$T_{0X}^{(\mu)} = \langle\hat{\mu}_i\rangle f_i^*$	—
1	$T_{0X}^{(1)} = \Lambda_{ji}\partial_j f_i^*$	$T_{0X}^{(Q)} = \frac{1}{2}\langle\hat{Q}_{ij}\rangle\partial_j f_i^*$	$T_{0X}^{(m)} = \langle\boldsymbol{m}\rangle\cdot\left[\nabla\times\boldsymbol{f}^*\right]$
2	$T_{0X}^{(2)} = \Omega_{kji}\partial_j\partial_k f_i^*$	$T_{0X}^{(O)} = \frac{1}{6}\langle\hat{O}_{ijk}\rangle\partial_j\partial_k f_i^*$	$T_{0X}^{(M)} = \frac{1}{2}\langle\hat{M}_{ij}\rangle\partial_j\left[\nabla\times\boldsymbol{f}^*\right]_i$

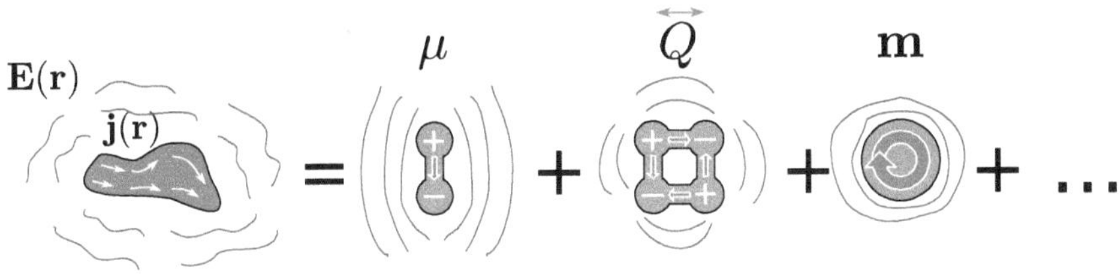

Fig. 5.5 Interpretation of the multipole expansion. The interaction between the current $\boldsymbol{j}(\boldsymbol{r})$ and field $\boldsymbol{E}(\boldsymbol{r})$ is decomposed into a linear superposition of multipoles

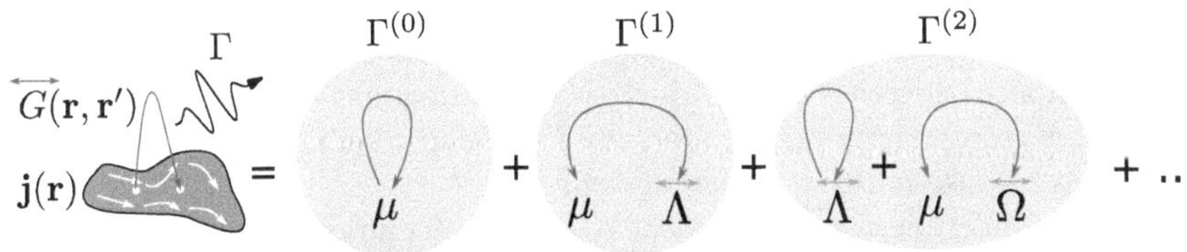

Fig. 5.6 Physical interpretation of the spontaneous-emission rate decomposed into the constituent multipoles. The nonlocal interaction between the points $\boldsymbol{r}$ and $\boldsymbol{r}'$ within the current density of the emitter is converted into an interaction between the different multipoles of the emitter

is remarkably robust against changes in $\boldsymbol{r}_0$. For a detailed derivation the reader is referred to [50]. This aspect is of important for the justification of the multipolar expansion.

We expand Γ up to the second order because the first-order term vanishes in parity-symmetric environments

$$\Gamma \approx \Gamma^{(0)} + \Gamma^{(1)} + \Gamma^{(2)}. \tag{5.34}$$

The zeroth-order term stems from the dipole nature of the emitter (Fig. 5.6),

$$\begin{aligned} \Gamma^{(0)} &= \frac{\pi}{\epsilon_0 \hbar} \sum_l \frac{1}{\omega_l} \left| T_{0X}^{(0)} \right|_l^2 \delta(\omega - \omega_l) \\ &\overset{1.12}{=} \frac{2\mu_0}{\hbar} \mathrm{Im}\left[\mu_i G_{ij}(\boldsymbol{r}_0, \boldsymbol{r}_0) \mu_j^*\right]. \end{aligned} \tag{5.35}$$

The first-order contribution reads

$$\begin{aligned} \Gamma^{(1)} &= \frac{\pi}{\epsilon_0 \hbar} \sum_l \frac{1}{\omega_l} 2\mathrm{Re} T_{0X}^{(1)} T_{0X}^{(0),*}{}_l \delta(\omega - \omega_l) \\ &= \frac{2\mu_0}{\hbar} 2\mathrm{Re}\left[\Lambda_{ki} \mu_j^*\right] \partial_k \mathrm{Im}\left[G_{ij}(\boldsymbol{r}, \boldsymbol{r}_0)\right]\Big|_{\boldsymbol{r}=\boldsymbol{r}_0}, \end{aligned} \tag{5.36}$$

and is proportional to the field gradient at the position of the emitter. $\Gamma^{(1)}$ can be interpreted as an interference between the dipole and mesoscopic quantities. Importantly, $\Gamma^{(1)} = 0$ if either the QD and or environment are parity symmetric [21]. The dipole approximation is therefore protected by parity symmetry to first order.

The second-order contribution to the radiative rate is

$$\begin{aligned}\Gamma^{(2)} &= \frac{\pi}{\epsilon_0 \hbar} \sum_l \frac{1}{\omega_l} \left[2\mathrm{Re} T_{0X}^{(2)} T_{0X}^{(0),*} + T_{0X}^{(1)} T_{0X}^{(1),*} \right]_l \delta(\omega - \omega_l) \\ &= \frac{2\mu_0}{\hbar} \mathrm{Im} \left\{ \left[2\mathrm{Re}\Omega_{lki} \mu_j^* \partial_k \partial_l + \Lambda_{ki} \Lambda_{lj}^* \partial_k \partial_l' \right] G_{ij}(\boldsymbol{r}, \boldsymbol{r}') \big|_{\boldsymbol{r}=\boldsymbol{r}'=\boldsymbol{r}_0} \right\}, \quad (5.37)\end{aligned}$$

and couples to the second-order derivative of the field. Here the first term stems from the interference between $\boldsymbol{\mu}$ and $\underline{\underline{\Omega}}$, which share contributions with the same parity. In contrast, the second term vanishes for parity-symmetric emitters on dipole-allowed transitions because $\boldsymbol{\mu}$ and $\underline{\underline{\Lambda}}$ are orthogonal.

The first-order mesoscopic moment $\underline{\underline{\Lambda}}$ contains 9 entries and the second-order moment $\underline{\underline{\Omega}}$ 27 entries. However, many vanish for symmetry reasons, and only a few capture the essential physics. Motivated by the shape of In(Ga)As QDs [24] we assume the QDs to be lens shaped with in-plane cylindrical symmetry but with no parity symmetry in the growth direction. Notably, this analysis is not bound to this particular QD shape and is also valid for pyramidal or elliptical QDs. The exciton state is in the strong-confinement regime [53] and, using the effective-mass theory, the electron Ψ_e and hole Ψ_h wavefunctions are modeled as

$$\begin{aligned}\Psi_e(\boldsymbol{r}) &= u_e(\boldsymbol{r})\psi_e(\boldsymbol{r}), \\ \Psi_h(\boldsymbol{r}) &= u_h(\boldsymbol{r})\psi_h(\boldsymbol{r}),\end{aligned} \quad (5.38)$$

where u_e (u_h) is the conduction- (valence-) band Bloch function, and $\psi(\boldsymbol{r})$ is the slowly varying envelope. For concreteness we consider the x-polarized exciton but note that the properties of the y-polarized exciton are analogous.

In the following, we investigate the first-order mesoscopic moment

$$\Lambda_{ki} = \frac{e}{m_0} \langle 0 | (x_k - x_{0,k}) \hat{p}_i | \Psi_X \rangle = \frac{e}{m_0} \langle u_x \psi_h | (x_k - x_{0,k}) \hat{p}_i | u_e \psi_e \rangle. \quad (5.39)$$

The choice of x_0 and y_0 is provided naturally by the cylindrical symmetry of the QD. We define z_0 as the z-component of the exciton center-of-mass coordinate [50]. The valence-band Bloch function $u_h = u_x$ inherits the odd symmetry ("-1") of the p_x orbital in the x-direction and even parity ("$+1$") in y and z; u_e inherits the spherical symmetry of the s-orbital. The slowly varying envelopes ψ inherit the symmetry of the QD, cf. Table 5.2. Thus, only Λ_{xz} and Λ_{zx} contain non-zero entries. $\Lambda_{zx} = (e/m_0)\langle \psi_h | z \hat{p}_x | \psi_e \rangle$ can be neglected because it couples to the QD height, which is much smaller than the in-plane size [24], yielding

Table 5.2 Symmetries of the electron and hole wavefunctions for a lens-shaped QD. '1' denotes even parity, '−1' odd parity and '0' no parity

	u_x	ψ_h	Ψ_h	u_e	ψ_e	Ψ_e
x	−1	1	−1	1	1	1
y	1	1	1	1	1	1
z	1	0	0	1	0	0

$$\underline{\underline{\Lambda}} \simeq \begin{bmatrix} 0 & 0 & \Lambda \\ 0 & 0 & 0 \\ 0 & 0 & 0 \end{bmatrix}, \tag{5.40}$$

where $\Lambda \equiv \Lambda_{xz}$. Large values of $\Lambda/\mu \simeq 10$–20 nm were measured for In(Ga)As QDs, cf. Fig. 5.4. With similar arguments it can be shown [50] that all the entries in $\underline{\underline{\Omega}}$ have negligible contributions such that

$$\Gamma = \frac{2\mu_0}{\hbar}\mathrm{Im}\left\{ \left[\mu^*,\ \Lambda^*\right] \left[\begin{matrix} G_{xx}(\boldsymbol{r}_0,\boldsymbol{r}_0), & \partial_x G_{xz}(\boldsymbol{r},\boldsymbol{r}_0) \\ \partial'_x G_{zx}(\boldsymbol{r}_0,\boldsymbol{r}'), & \partial_x \partial'_x G_{zz}(\boldsymbol{r},\boldsymbol{r}') \end{matrix}\right]\Bigg|_{\boldsymbol{r}=\boldsymbol{r}'=\boldsymbol{r}_0} \left[\begin{matrix} \mu \\ \Lambda \end{matrix}\right] \right\}. \tag{5.41}$$

In conclusion, QDs have an additional optical degree of freedom, the mesoscopic moment Λ, which, combined with the dipole moment μ, describes light-matter interaction with QDs.

5.2.2 *Microscopic Model for Mesoscopic Quantum Dots*

The mesoscopic moment Λ has been used as a phenomenological quantity so far with no clear relation to the microscopic origin. The large Λ observed experimentally cannot be reproduced by the effective-mass theory because the size of Λ is governed by the gradient of the QD wavefunction, while, according to the effective-mass theory, gradients can only originate from the envelope functions, and are negligibly small. This can be shown by evaluating Λ with the rules employed for evaluating μ, which is to assume that ψ varies slowly over a unit cell so that $\mu = (e/m_0)\langle\psi_h u_x|\hat{p}_x|u_e\psi_e\rangle \approx (e/m_0)p_{cv}\langle\psi_h|\psi_e\rangle$, where $p_{cv} = V_{\mathrm{UC}}^{-1}\int_{\mathrm{UC}} d^3 r u_x^* \hat{p}_x u_e$ is given by an integral over the unit cell with V_{UC} being the unit-cell volume. A similar calculation yields for Λ

$$\begin{aligned}\Lambda = \frac{e}{m_0}&\left[\langle\psi_h|x|\psi_e\rangle\langle u_x|\hat{p}_z|u_e\rangle_{\mathrm{UC}} + \langle\psi_h|\psi_e\rangle\langle u_x|x\hat{p}_z|u_e\rangle_{\mathrm{UC}}\right. \\ &\left. + \langle\psi_h|x\hat{p}_z|\psi_e\rangle\langle u_x|u_e\rangle_{\mathrm{UC}} + \langle\psi_h|\hat{p}_z|\psi_e\rangle\langle u_x|x|u_e\rangle_{\mathrm{UC}}\right],\end{aligned} \tag{5.42}$$

where $\langle\rangle_{\mathrm{UC}} \equiv V_{\mathrm{UC}}^{-1}\int_{\mathrm{UC}} \mathrm{d}^3\boldsymbol{r}$ denotes integration over a unit cell. The first three contributions vanish for symmetry reasons. The fourth contribution is vanishingly small

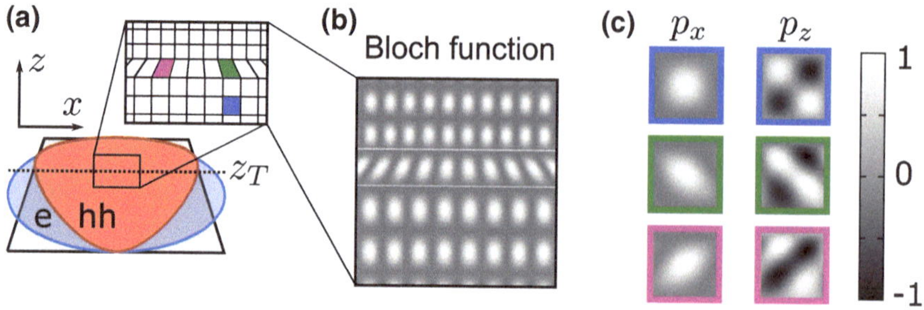

Fig. 5.7 Microscopic model for QDs. **a** The atomic lattice inside the QD is assumed to change periodicity at $z = z_T$. **b** Sketch of the spatial dependence of u_x^2. **c** Illustration of the matrix elements $\langle p_x \rangle \equiv \langle u_x|\hat{p}_x|u_e\rangle$ and $\langle p_z \rangle \equiv \langle u_x|\hat{p}_z|u_e\rangle$ for the three colored unit cells in **a**. The symmetry of the integrand is broken in the transition region around $z = z_T$ giving rise to mesoscopic effects. Figure reproduced from [22]

and does not scale with the QD size: for Gaussian envelopes allowing for realistic mutual displacements of 1–2 nm between the electron and the hole in the growth direction (note that the integral vanishes in the absence of such a displacement) we estimate $\Lambda/\mu \sim 10^{-4}$ nm. In the following we develop a generalization of the effective-mass theory, and find that the mesoscopic moment originates from lattice inhomogeneities at the crystal-lattice level.

Using bulk-material Bloch functions works excellently for quantum wells and lattice-matched QDs, where the structures are strain free and structurally homogeneous. In(Ga)As QDs, on the other hand, are grown by strain relaxation, a violent process that unavoidably leads to the generation of structural gradients. In particular, large lattice-constant shifts are observed in the growth direction of QDs [24, 62]. Motivated by this we assume that the lattice periodicity changes at a certain position $z = z_T$ along the QD height by an amount $\Delta a_l = 110$ pm at a central value $a_l = 605$ pm as measured in [62], see Fig. 5.7a. The Bloch functions change periodicity as well, cf. Fig. 5.7b, and to describe this we expand them in a Fourier series with a position-dependent lattice wavevector $k_l(z)$

$$\begin{aligned} u_x(\boldsymbol{r}) &= \sum_m a_m(y, z) \sin[mk_l(z)x], \\ u_e(\boldsymbol{r}) &= \sum_n b_n(y, z) \cos[nk_l(z)x]. \end{aligned} \tag{5.43}$$

This Ansatz ensures opposite parity of the conduction- and valence-band Bloch functions along x. We assume the shape of the Bloch functions to be the same, and only their periodicity to vary. This yields for the interband matrix element $p_{cv} = \langle u_x | \hat{p}_x | u_e \rangle_{\mathrm{UC}}$

$$p_{cv} = \frac{i\hbar}{V_{\mathrm{UC}}} \sum_n \int_{\mathrm{UC}} \mathrm{d}^3\boldsymbol{r} a_n^*(\boldsymbol{r}) b_n(\boldsymbol{r}) n k_l(z) \sin^2[nk_l(z)x]. \tag{5.44}$$

We now evaluate the mesoscopic moment by separating the slowly and rapidly varying contributions

$$\Lambda = \frac{e}{m_0} \sum_{q=1}^{N} \psi_g^*(\boldsymbol{R}_q) X_q \psi_e(\boldsymbol{R}_q) \int_{\mathrm{UC}} \mathrm{d}^3 \boldsymbol{r} u_x^*(\boldsymbol{r}) \hat{p}_z u_e(\boldsymbol{r}), \tag{5.45}$$

where $\boldsymbol{R}_q$ denotes the position of the q-th unit cell and N is the number of unit cells. In a homogeneous region of the QD (the blue unit cell in Fig. 5.7a) the unit-cell integrand of (5.45) is odd in x- and z-directions, cf. Fig. 5.7c, which leads to a vanishing integral. However, in the transition region around $z = z_T$ strong gradients are present, which destroy the parity of the integrand (see the pink and green unit cells in Fig. 5.7a, c) and generate a contribution to Λ. The z-polarized Bloch matrix element $\langle u_x | \hat{p}_z | u_e \rangle$ is evaluated as

$$\langle u_x | \hat{p}_z | u_e \rangle \simeq i\hbar \sum_{m,n} \int_{\mathrm{UC}} \mathrm{d}^3 \boldsymbol{r} a_m^*(\boldsymbol{r}) b_n(\boldsymbol{r}) \frac{\partial k_l(z)}{\partial z} \sin[m k_l(z) x] n x \sin[n k_l(z) x], \tag{5.46}$$

yielding

$$\Lambda = \sum_{q=1}^{N} \psi_h^*(\boldsymbol{R}_q) X_q \psi_e(\boldsymbol{R}_q) \sum_{m,n} \int_{\mathrm{UC}} \mathrm{d}^3 \boldsymbol{r} a_m^*(\boldsymbol{r}) b_n(\boldsymbol{r}) \frac{\partial k_l}{\partial z} \sin[m k_l x] n (x + X_q) \sin[n k_l x].$$

We assume that $\partial_z k_l$ is slowly varying over an unit cell and pull it in front of the unit-cell integral. The term containing x vanishes because the integral is odd and we are left with

$$\frac{\Lambda}{\mu} = \frac{1}{k_l} \frac{\langle \psi_h(\boldsymbol{r}) | x^2 \left[\partial_z k_l(z)\right] | \psi_e(\boldsymbol{r}) \rangle}{\langle \psi_h(\boldsymbol{r}) | \psi_e(\boldsymbol{r}) \rangle}. \tag{5.47}$$

This expression shows that the mesoscopic strength scales quadratically with the in-plane size of the QD, $\Lambda/\mu \sim L_r^2$, because the term $\langle \psi_g | x^2 \left[\partial_z k_l(z)\right] | \psi_e \rangle$ contains the variance of the exciton wavefunction in the x-direction. Moreover, it increases with decreasing QD height, $\Lambda/\mu \sim L_z^{-1}$, since in shallow QDs the relative importance of the lattice-constant transition region is increased.

For in-plane rotationally symmetric Gaussian envelopes analytic expressions are obtained [22]

$$\frac{\Lambda}{\mu} = -\frac{\Delta a_l}{a_l} \sqrt{\frac{1 + \xi_z}{4\pi} \frac{\sigma_r^2}{\sigma_z}}, \tag{5.48}$$

where σ_z is the height (HWHM) of the electron envelope, σ_r the QD radius, $\Delta a_l / a_l$ the relative lattice-constant shift and $\xi_z \approx 5$ the ratio between the hole and electron effective masses. The largest mesoscopic strengths are achieved in shallow and wide (disk-shaped) QDs, cf. Fig. 5.8.

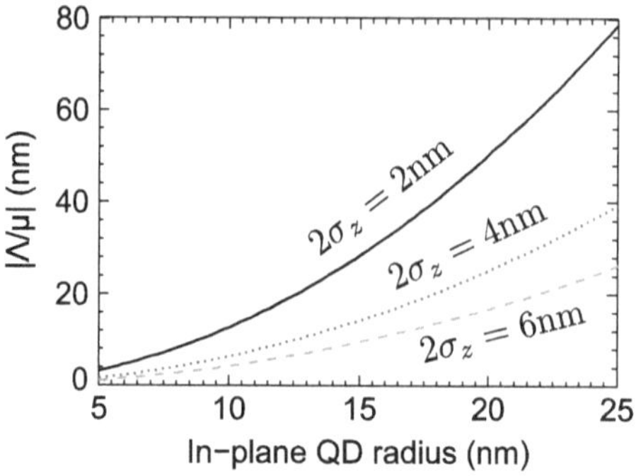

Fig. 5.8 The mesoscopic strength as a function of the in-plane size of the QD for three fixed QD heights. Figure reproduced from [22]

5.2.3 *The Quantum Current Density of Quantum Dots*

To illustrate the physics responsible for the mesoscopic moment we now calculate the quantum-mechanical current density resulting from the above model. The current density $\boldsymbol{j}_{\mathrm{QD}}(\boldsymbol{r})$ flowing through the QD is defined in (5.25) and leads to

$$\boldsymbol{j}_{\mathrm{QD}}(\boldsymbol{r}) = \frac{e}{m_0}\left[\Psi_h(\boldsymbol{r})\hat{p}_x\Psi_e(\boldsymbol{r})\boldsymbol{e}_x + \Psi_h(\boldsymbol{r})\hat{p}_z\Psi_e(\boldsymbol{r})\boldsymbol{e}_z\right]. \tag{5.49}$$

The current density $\boldsymbol{j}_{\mathrm{QD}}(\boldsymbol{r}) = \boldsymbol{J}_{\mathrm{QD}}(\boldsymbol{r})p(\boldsymbol{r})$ is modulated by the Bloch element $p(\boldsymbol{r}) = u_x(\boldsymbol{r})\hat{p}_x u_e(\boldsymbol{r})$. In the following we discuss the slowly varying component $\boldsymbol{J}_{\mathrm{QD}}(\boldsymbol{r})$, which can be written as

$$\boldsymbol{J}_{\mathrm{QD}}(\boldsymbol{r}) = \frac{e}{m_0}\psi_h(\boldsymbol{r})\psi_e(\boldsymbol{r})\left(\boldsymbol{e}_x + x\frac{1}{k_l}\frac{\partial k_l}{\partial z}\boldsymbol{e}_z\right). \tag{5.50}$$

We assume that most of the transition happens over two lattice constants as shown in [24]. In QDs with a homogeneous crystal lattice $\partial k_l/\partial z = 0$ and thus $\Lambda = 0$, the current density flows only along the direction of the dipole moment (see Fig. 5.9a). The presence of lattice inhomogeneities changes the flow due to transverse gradients. The current density flows along a curved path as illustrated in Fig. 5.9b–d. The wider the QD is, the sharper the transverse oscillations of the current are, and the larger Λ/μ is.

We now have the ingredients to provide an intuitive understanding of the experimental data in Fig. 5.3. In the direct (inverted) structure, Fig. 5.3a (Fig. 5.3b), the QD current and the plasmonic field flow along opposite (parallel) curvatures, which leads to enhanced (suppressed) light-matter interaction. We exemplify this for a silver nanowire with a radius of 20 nm and refractive index $n_{\mathrm{Ag}} = 0.2 + 7\mathrm{i}$, see Fig. 5.10. The coupling efficiency to plasmons, β_{pl}, is defined as the coupling rate to plasmons over the coupling rate to all excitations, which include photons, plasmons and ohmic losses. The configuration in Fig. 5.10a exhibits a larger coupling to surface plasmons relative to a dipole (from 75% to 90%), cf. Fig. 5.10c. In contrast, the interaction is diminished if the QD orientation is flipped, cf. Fig. 5.10b, because the QD current

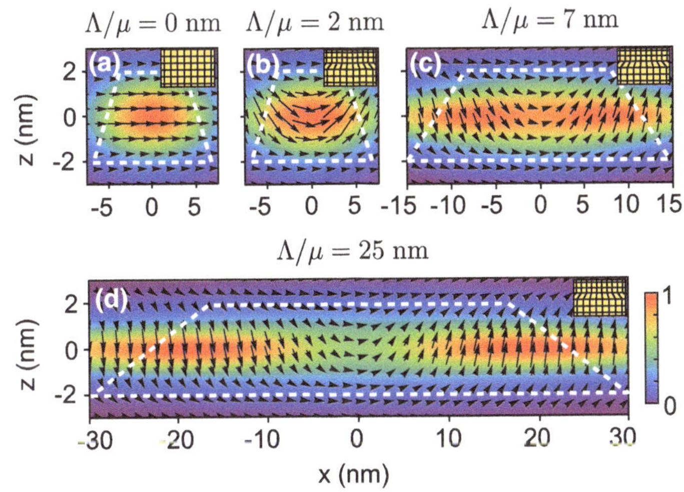

Fig. 5.9 Spatial distribution of the current density $\boldsymbol{J}_{\mathrm{QD}}(\boldsymbol{r})$ in QDs. **a** Homogeneous crystal lattice where the flow is uniform and points along the dipole moment. **b** Inhomogeneous lattice for a QD radius of 5 nm giving rise to a non-uniform flow along a curved path. The QD height is $2\sigma_z = 4$ nm. **c, d** Same as **b** but for QD radii of 10 and 20 nm, respectively. Both the length of the *arrows* and the color scale indicate the magnitude of the flow and the direction of the *arrows* indicates the pointwise direction of the flow. Figure reproduced from [22]

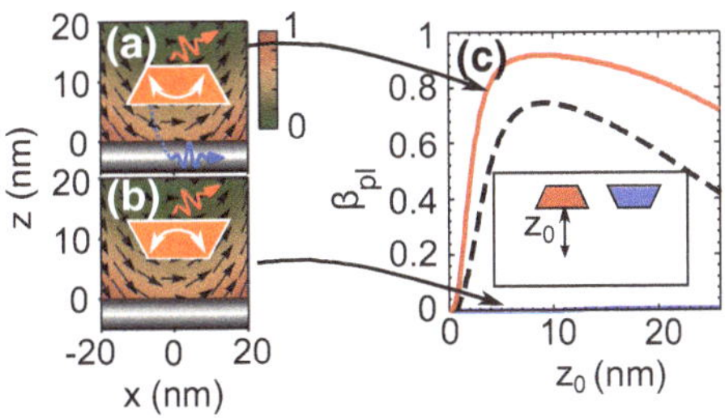

Fig. 5.10 QDs coupled to surface plasmons of a silver nanowire. **a** The field matches the curvature of the QD current and the coupling efficiency to surface plasmons, β_{pl}, is enhanced, *red curve* in **c**, relative to a dipole, *dashed curve* in **c**. **b** If the QD orientation is flipped, the interaction is diminished as shown in **c** by the *blue line*

and of the field oscillate in opposite directions. In other words, μ and Λ interfere constructively in (a) and destructively in (b).

5.2.4 *Lattice-Distortion Effects Beyond the Multipolar Theory*

Knowing the full current distribution according to the microscopic model allows to calculate the decay rate without relying on the multipolar expansion. Inserting (5.50) into (5.24) leads to

$$\Gamma = \frac{2\mu_0 |p_{cv}|^2}{\hbar} \text{Im} \iint d^3r d^3r' \left[J_x(\boldsymbol{r}) \; J_z(\boldsymbol{r}) \right] \begin{bmatrix} G_{xx}(\boldsymbol{r}, \boldsymbol{r}') & G_{xz}(\boldsymbol{r}, \boldsymbol{r}') \\ G_{zx}(\boldsymbol{r}, \boldsymbol{r}') & G_{zz}(\boldsymbol{r}, \boldsymbol{r}') \end{bmatrix} \begin{bmatrix} J_x^*(\boldsymbol{r}') \\ J_z^*(\boldsymbol{r}') \end{bmatrix}, \tag{5.51}$$

where $J_x(\boldsymbol{r}) = (e/m_0)\psi_h(\boldsymbol{r})\psi_e(\boldsymbol{r})$ and $J_z(\boldsymbol{r}) = xk_l^{-1}(\partial k_l/\partial z)J_x$. The zeroth order expansion of G_{xx} contains the electric-dipole contribution $\Gamma^{(0)}$. The other terms are generated by the transverse oscillations of the current density and contain the first $\Gamma^{(1)}$ and second $\Gamma^{(2)}$ order contributions. Equation (5.51) should be preferred over the multipolar theory when the figure of merit $k \times \Lambda/\mu > 1$, i.e., when the multipolar expansion diverges.

5.2.5 *Quantum Dots as Probes for the Magnetic Field of Light*

Electric and magnetic fields play an equally important role in the formation of the light field but interact fundamentally different with matter. The magnetic force acting on a charged particle with velocity v is v/c times smaller than the electric force. Magnetic light-matter interaction is therefore weak. Nevertheless, magnetic-dipole transitions are well known in atomic physics and can be accessed with light despite being weak [63–65], since they have different selection rules than electric-dipole transitions because atoms have parity symmetry [26, 27, 66]. The lack of parity symmetry implies that QDs may be exploited as a probe of electric and magnetic fields on a single electronic transition. We exemplify this by considering again the spontaneous emission of a QD in front of a metal interface [21]. The QD decays into propagating photons with the rate Γ_{RAD}, propagating surface plasmons (Γ_{PL}), or ohmic-lossy modes in the metal (Γ_{LS}) [67]. Γ_{RAD} is negligibly affected by the mesoscopic moment since the radiative modes oscillate weakly, i.e., $\Gamma_{\text{RAD}} \approx \Gamma_{\text{RAD}}^{(0)}$. In contrast, the plasmon field varies strongly in space and Λ plays a major role in the excitation rate of plasmons. The coupling to ohmic losses [68, 69] is normally negligible at distances larger than ~20 nm from the metal and are not discussed further. The three light-matter interaction channels are

$$\begin{aligned} \Gamma^{(0)} &= \frac{2\mu_0}{\hbar}\mu^2 \text{Im} G_{xx}(0,0) = \Gamma_{\text{RAD}} + \Gamma_{\text{PL}}^{(0)}, \\ \Gamma^{(1)} &= \frac{2\mu_0}{\hbar} 2\Lambda\mu \; \partial_x \text{Im} G_{zx}(\boldsymbol{r},0)|_{r=0} \approx \Gamma_{\text{PL}}^{(1)}, \\ \Gamma^{(2)} &= \frac{2\mu_0}{\hbar}\Lambda^2 \; \partial_x \partial_x' \text{Im} G_{zz}(\boldsymbol{r},\boldsymbol{r}')\big|_{r=r'=0} \approx \Gamma_{\text{PL}}^{(2)}, \end{aligned} \tag{5.52}$$

where we assume the QD wavefunctions to be real. Each order has a clear physical meaning as explained in Sect. 5.2.1 and can be visualized in Fig. 5.11.

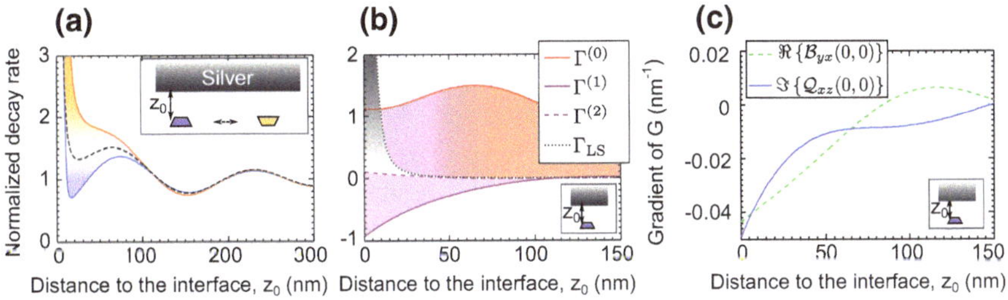

Fig. 5.11 Decay dynamics of QDs near a silver interface. All the rates are normalized to the decay rate in homogeneous GaAs. **a** Decay rate for the direct (inverted) QD orientation marked by *blue* (*orange*) *lines*. The *black dashed line* denotes the dipole theory. **b** Decomposition of the decay rates according to the expansion order. The ohmic losses are indicated by the *dotted black line*. **c** The ED-MD and ED-EQ Green tensor probed by mesoscopic QDs and normalized to $\mathrm{Im}G_{xx}(0,0)$ in homogeneous GaAs. Figure reproduced from [21]

In the following we show that $\Gamma^{(1)}$ probes the magnetic field of light. The mesoscopic moment Λ can be decomposed into multipolar contributions

$$\Lambda_{xz}\partial_x e_{l,z}(0) = \mathrm{i}\omega m_y b_{l,y}(0) + Q_{xz}\left[\partial_x e_{l,z}(0) + \partial_z e_{l,x}(0)\right], \tag{5.53}$$

where $m_y = \Lambda/2$ is the magnetic dipole, $Q_{xz} = \Lambda$ the electric quadrupole of the QD, and $\boldsymbol{e}$ and $\boldsymbol{b}$ are the electric- and magnetic-field modes, respectively. Consequently, $\Gamma^{(1)}$ intertwines the electric dipole, magnetic dipole and electric quadrupole of the QD. The multipolar contribution to $\Gamma^{(1)}$ is

$$\Gamma^{(1)} = \frac{2\hbar}{\mu_0}\left[\omega m_y \mu \mathrm{Re}\mathcal{B}_{yx}(\boldsymbol{r}_0,\boldsymbol{r}_0) + A Q_{xz}\mu \mathrm{Im}\mathcal{Q}_{xz}(\boldsymbol{r}_0,\boldsymbol{r}_0)\right], \tag{5.54}$$

where $\mathcal{B}_{yx}(\boldsymbol{r}_0,\boldsymbol{r}_0) = -\mathrm{i}\omega^{-1}\left[\partial_x G_{zx}(\boldsymbol{r},\boldsymbol{r}_0) - \partial_z G_{xx}(\boldsymbol{r},\boldsymbol{r}_0)\right]_{\boldsymbol{r}=\boldsymbol{r}_0}$ is the magneto-electric Green tensor, and $\mathcal{Q}_{xz}(\boldsymbol{r}_0,\boldsymbol{r}_0) = \left[\partial_x G_{zx}(\boldsymbol{r},\boldsymbol{r}_0) + \partial_z G_{xx}(\boldsymbol{r},\boldsymbol{r}_0)\right]_{\boldsymbol{r}=\boldsymbol{r}_0}$ the electric dipole-quadrupole Green tensor. The probed fields are plotted in Fig. 5.11c for an emitter close to a silver interface, where the two components of the Green tensor vary over length scales of tens of nanometers, which is comparable to the QD size.

The mesoscopic moment can also be used to probe the parity symmetry of nanophotonic environments [21]. If placed in an unknown nanophotonic structure, the spontaneous-emission rate of the QD is generally given by $\Gamma_\blacktriangle \approx \Gamma_\blacktriangle^{(0)} + \Gamma_\blacktriangle^{(1)}$. By flipping the QD orientation, the dipole contribution is unchanged but the first-order term has opposite symmetry and changes sign, i.e., $\Gamma_\blacktriangledown \approx \Gamma_\blacktriangledown^{(0)} + \Gamma_\blacktriangledown^{(1)} = \Gamma_\blacktriangle^{(0)} - \Gamma_\blacktriangle^{(1)}$. As a consequence, both the projected Green tensor $\mathrm{Im}\{G_{xx}(0,0)\}$ and the spatial gradient $\partial_x \mathrm{Im}\{G_{zx}(0,0)\}$ can be extracted using (5.52).

5.3 Single-Photon Superradiance from a Monolayer-Fluctuation Quantum Dot

We address another mesoscopic property of QDs, the collective coupling to light in terms of superradiance. We show that the fundamental optical excitation of a weakly confining QD is a generalization of single-photon superradiance (SPS) [23]. The superradiant state is prepared deterministically with a laser pulse and reaches a five-fold collective light-matter enhancement.

5.3.1 Extending the Concept of Superradiance from Atomic Physics to Solid-State Emitters

In the following we make a formal connection between the proposal of Dicke regarding SPS in an ensemble of atoms [39] and collective light-matter enhancement in a semiconductor QD. We show that the giant oscillator strength of QDs and SPS are two equivalent phenomena.

If N emitters are placed closer than one wavelength apart, the emission dynamics of a shared electronic excitation is strongly enhanced in the symmetric SPS state [39]

$$\left|\Psi_s\right\rangle = \frac{1}{\sqrt{N}} \sum_j \left|g\right\rangle_1 \left|g\right\rangle_2 \ldots \left|e\right\rangle_j \ldots \left|g\right\rangle_N, \tag{5.55}$$

where the j-th emitter is in the excited state $\left|e\right\rangle$ and all others in the ground state $\left|g\right\rangle$. Remarkably, the state $\left|\Psi_s\right\rangle$ decays N times faster to the ground state than a single emitter. This state describes a non-interacting ensemble, where the excitation is bound to either of the emitters, cf. Fig. 5.12a. Harvesting such effects in practice is challenging due to the large size and harmonic spectrum of many ensembles, which decreases the collective enhancement and prohibits deterministic preparation, respectively. These limitations do not apply to QDs, which are small and anharmonic. However, another challenge emerges: in a system of interacting particles the wavefunctions of the underlying atoms overlap leading to delocalized excitations. This causes conventional QDs to be in the strong-confinement regime and thus to have relatively small oscillator strengths of about 10, despite that they consist of tens of thousands of atoms, cf. Fig. 5.12b.

The size of delocalized excitations is a fundamental property of semiconductors and is given by the size of an exciton. Enhancement of light-matter interaction can therefore be achieved only in QDs that confine excitons weakly [70], i.e., that are larger than the exciton radius. We study single GaAs monolayer-fluctuations QDs as sketched in Fig. 5.12c. Bound excitonic states are created by intentionally engineered monolayer fluctuations in a quantum well [38]. Exciton enhancement is achieved only in the plane, where the QD wavefunction is extended beyond the exciton radius.

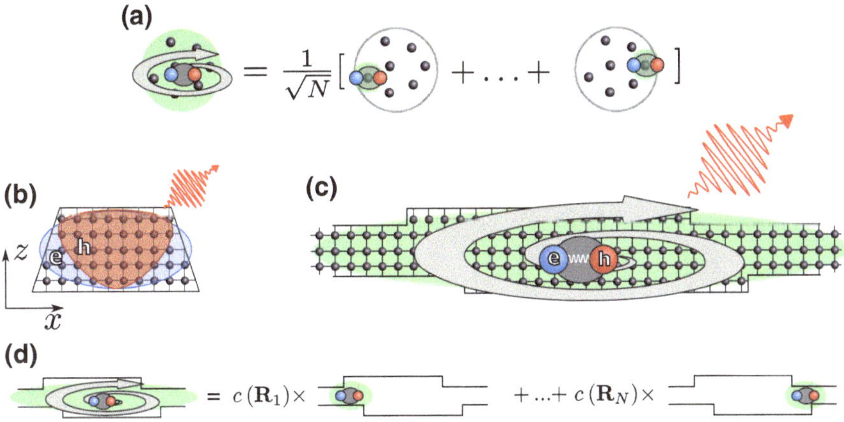

Fig. 5.12 Superradiance with single QDs. **a** SPS is defined in an ensemble of non-interacting emitters as a symmetric superposition of different excitations. **b** In small QDs, such as In(Ga)As QDs, the electrons and holes are strongly confined and uncorrelated, which destroys collective effects. **c** A QD defined by intentional monolayer fluctuations weakly confine electrons (e) and holes (h), which are mutually bound by electrostatic attraction. **d** The excitonic enhancement of light-matter interaction may be regarded as a generalization of SPS: the exciton is in a symmetric superposition of excitations. Figure reproduced from [23]

We assume the QD wavefunctions to be separable into in-plane ψ_X and out-of-plane ϕ components. Due to strong confinement in the z-direction, ϕ is separable as $\phi(z) = \phi_e(z)\phi_h(z)$. We therefore obtain the exciton wavefunction in the effective-mass approximation

$$\Psi_X(\boldsymbol{R}, \boldsymbol{r}, \boldsymbol{r}_e, \boldsymbol{r}_h) = \psi_X(\boldsymbol{R}, \boldsymbol{r})\phi_h(z_h)\phi_e(z_e)u_x(\boldsymbol{r}_h)u_e(\boldsymbol{r}_e), \tag{5.56}$$

where $\boldsymbol{R} = (m_e\boldsymbol{r}_e + m_h\boldsymbol{r}_h)/(m_e + m_h)$ and $\boldsymbol{r} = \boldsymbol{r}_e - \boldsymbol{r}_h$ are the center-of-mass and relative in-plane excitonic coordinates. The superradiant enhancement is therefore governed by the in-plane envelope $\psi_X(\boldsymbol{R}, \boldsymbol{r})$. We consider a symmetric parabolic in-plane confining potential, in which case the excitonic envelope separates into center-of-mass and relative dynamics $\psi_X(\boldsymbol{R}, \boldsymbol{r}) = \chi_{CM}(\boldsymbol{R})\chi_r(\boldsymbol{r})$ with [25]

$$\chi_{CM}(\boldsymbol{R}) = \sqrt{\frac{2}{\pi}\frac{1}{\beta}}e^{-|\boldsymbol{R}|^2/\beta^2}, \tag{5.57}$$

$$\chi_r(\boldsymbol{r}) = \sqrt{\frac{2}{\pi}\frac{1}{a_{QW}}}e^{-|\boldsymbol{r}|/a_{QW}}, \tag{5.58}$$

where a_{QW} is the quantum-well exciton radius with $a_{QW} \simeq a_0/\sqrt{2} \approx 8\,\text{nm}$, and β is the in-plane radius of the exciton wavefunction. The center-of-mass motion can be written as a convolution between a function $c_a(\boldsymbol{R})$ capturing the dynamics on the (uncorrelated) scale a_{QW} and a function $c_s(\boldsymbol{R})$ responsible for the superradiant enhancement

$$\chi_{\rm CM}(\boldsymbol{R}) = c_a(\boldsymbol{R}) * c_s(\boldsymbol{R}) \approx \sum_n c(\boldsymbol{R}_n) c_a(\boldsymbol{R} - \boldsymbol{R}_n). \tag{5.59}$$

Consequently, the slowly varying envelope reads

$$\psi_{\rm X}(\boldsymbol{R}, \boldsymbol{r}) = \sum_n c(\boldsymbol{R}_n) \phi_{\rm X}(\boldsymbol{R} - \boldsymbol{R}_n, \boldsymbol{r}), \tag{5.60}$$

where n runs over the unit cells of the QD. The internal exciton dynamics is governed by $\phi_{\rm X}$, which has a spatial extent of the order of the Bohr radius (~8 nm) and is smaller than the QD. The exciton in (5.60) is therefore in a spatial superposition of excitations corresponding to different positions of $\phi_{\rm X}$ as illustrated in Fig. 5.12d.

The following expression for the oscillator strength is obtained (compare with (5.14))

$$f = \frac{E_g}{\hbar\omega} \chi_r(0) \left| \langle 0 | \chi_{\rm CM}(\boldsymbol{R}) \rangle \right|^2 \left| \langle \psi_h(z) | \psi_e(z) \rangle \right|^2, \tag{5.61}$$

where the first (second) inner product on the right-hand side of the equation denotes a two-dimensional (one-dimensional) integration over $\boldsymbol{R}$ (z). We define the radius of the QD $L = \sqrt{2}\beta$ and, with the help of (5.57) and (5.58), arrive at the superradiant enhancement of the oscillator strength

$$S = \frac{f}{f_{\rm max}} = \left(\frac{\sqrt{2}L}{a_{\rm QW}} \right)^2 \left| \langle \psi_h | \psi_e \rangle \right|^2. \tag{5.62}$$

We calculate $\left| \langle \psi_h | \psi_e \rangle \right|^2 \approx 0.96$ and plot the resulting superradiant enhancement in Fig. 5.13. It scales with the QD area and is a dramatic effect; for realistic QD diameters of 35 nm, the light-matter interaction strength exceeds the upper limit of strongly confined excitons by an order of magnitude.

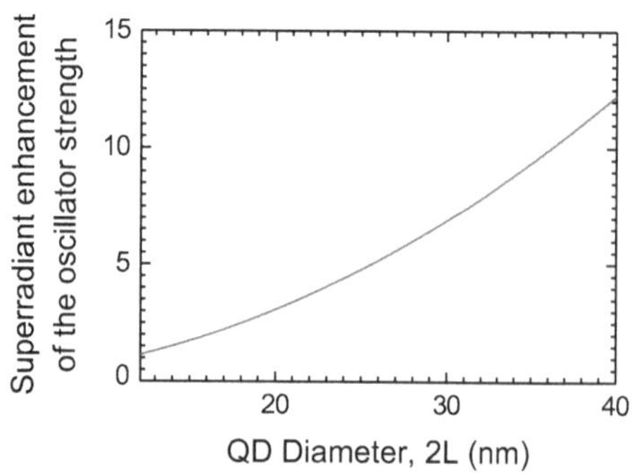

Fig. 5.13 Superradiant enhancement of the oscillator strength, S, for a monolayer-fluctuation QD relative to the strong-confinement limit $f_{\rm max} = 17.4$. Figure reproduced from [23]

5.3.2 *Deterministic Preparation and Impact of Nonradiative Processes*

In the experimental demonstration of SPS, the energy-level structure of the QDs is first probed by photoluminescence-excitation spectroscopy as displayed in Fig. 5.14a, which shows a quasi-continuum band of QD states hybridized with quantum-well resonances followed by the exciton manifold. We identify the 1s, 2s and 3s excitonic states that are denoted according to the two-dimensional hydrogen atom. Note, the recombination of excitons with different symmetry is forbidden. Key features of the spectrum are summarized in Fig. 5.14b. The measurement was carried out using continuous-wave excitation below the saturation power of the 1s exciton.

Deterministic preparation is achieved by exciting the 2s state with a laser pulse having sufficient optical power to saturate the 1s transition. The preparation is deterministic because the decay cascade from 2s to 1s is spin-conserving [71] and spin-dark states are therefore not populated. However, the latter prohibits measurements of the oscillator strength due to the single-exponential character of the exciton decay. We therefore use another excitation scheme (the wavelength labeled "C" in Fig. 5.14) to prepare bright and dark states with equal probability and extract the impact of nonradiative processes using the biexponential model presented in Sect. 5.1.5.

Significantly below the exciton saturation, $P \approx 0.1 P_{\mathrm{sat}}$, only the exciton is prepared, cf. Fig. 5.15a. At saturation, the biexciton line becomes discernible. Above saturation, both the exciton and the biexciton lines are saturated and the spectrum features spectrally continuous multibody emissions. The nature of the exciton and biexciton lines is confirmed by power series measurements as shown in Fig. 5.15c.

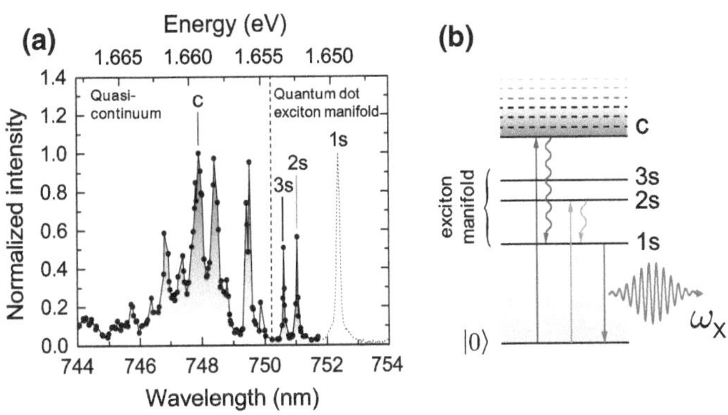

Fig. 5.14 Deterministic preparation of superradiant excitons. **a** Photoluminescence-excitation spectrum obtained by integrating the emission of the 1s transition while scanning the excitation wavelength. It features a quasi-continuum band of states followed by a sequence of QD states. **b** Two excitation schemes are employed. Pumping in the quasi-continuum band at the wavelength "C" results in preparation of carriers with random spin, which is important for extracting the nonradiative processes. For 2s-excitation, the spin is preserved and the bright exciton is prepared deterministically. Figure reproduced from [23]

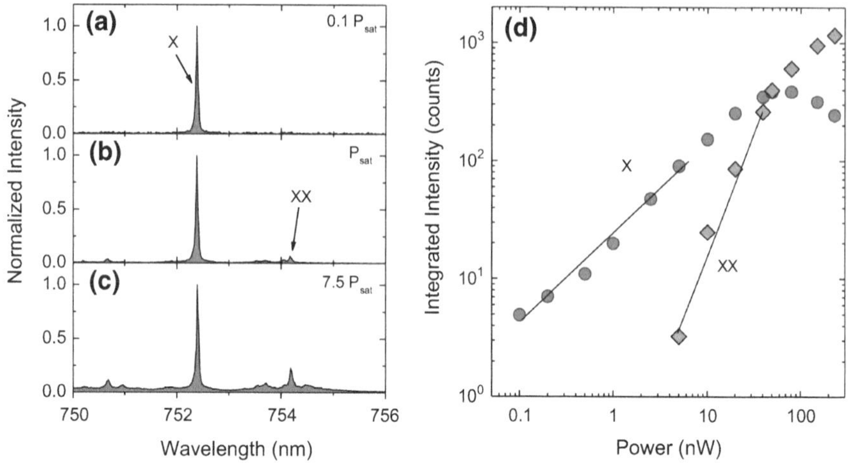

Fig. 5.15 Spectral measurements for "C"-excitation. **a** Measured spectrum at 10% of the exciton saturation power $P_{\mathrm{sat}} = 20\,\mathrm{nW}$. Only the exciton is observed. **b** At saturation of the exciton, the biexciton is visible as a small peak. **c** Significantly above the exciton saturation ($7.5P_{\mathrm{sat}}$), the spectrum acquires further narrow peaks on *top* of a continuous background. **d** The exciton is distinguished from biexcitons by their power-law dependence on excitation power P: the fits yield $P^{0.86}$ and $P^{2.01}$ for the exciton and biexciton, respectively

The spectral broadening of the biexciton line is related to multibody effects between the exciton and the free carriers populating the quantum well [35, 42].

The decay dynamics is recorded by sending the exciton line from Fig. 5.15a to an avalanche photo-diode. The acquired data are fitted by the biexponential model yielding the fast rate $\Gamma_F^C = \Gamma_{\mathrm{rad}}^C + \Gamma_{\mathrm{nrad}} + \Gamma_{\mathrm{sf}}$ and the extracted parameters are outlined in Fig. 5.16. We obtain a nonradiative rate $\Gamma_{\mathrm{nrad}} = 0.19\,\mathrm{ns}{-1}$, and a spin-flip rate $\Gamma_{\mathrm{sf}} = 0.31\,\mathrm{ns}^{-1}$, which are used to extract the oscillator strength in the following.

5.3.3 Demonstration of Single-Photon Superradiance

The experimental signature of SPS is spontaneous emission of single photons with a radiative rate beyond the upper limit for uncorrelated excitons. The 1s bright state is excited deterministically through the 2s shell and a clean emission spectrum below and at saturation is found. The time-resolved measurement is performed at $P = 0.1P_{\mathrm{sat}}$ to ensure that no multi-exciton states are prepared, and the decay is found to be close to single exponential. The radiative rate of the exciton is $\Gamma_{\mathrm{rad}} = \Gamma_F - \Gamma_{\mathrm{nrad}} - \Gamma_{\mathrm{sf}} = 8.4\,\mathrm{ns}^{-1}$, where Γ_F is the fast rate extracted from Fig. 5.17a. The QDs are positioned near an air interface [23] and we calculate an LDOS contribution of 0.95, which is normalized to the LDOS in homogeneous $Al_{0.8}Ga_{0.2}As$, resulting in an oscillator strength of $f = 72.7 \pm 0.8$. The latter is enhanced far beyond the upper

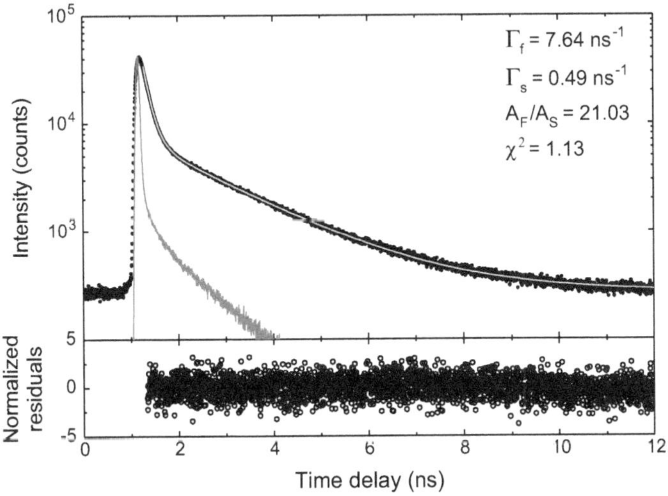

Fig. 5.16 Time-resolved decay of the exciton (*black dots*) under "C"-excitation. The fine-structure model yields an excellent biexponential fit (*yellow line*) with the extracted parameters indicated accordingly. The instrument response of the detector is indicated by the *green line*. Figure reproduced from [23] (color figure online)

limit of $f = 17.4$ for an uncorrelated exciton, cf. Fig. 5.17a. This is a direct signature of exciton superradiance.

To confirm the single-photon nature of the emission, we measure the second-order correlation function $g^{(2)}(\tau) \propto \langle \hat{a}^\dagger(t)\hat{a}^\dagger(t+\tau)\hat{a}(t+\tau)\hat{a}(t)\rangle$ [72], which determines the probability of detecting a photon at time $t = \tau$ given that a photon was detected at $t = 0$. An ideal single-photon source exhibits $g^{(2)}(0) = 0$ but any value below 0.5 is direct evidence of single photons. Figure 5.17b shows the correlation function obtained in an HBT experiment. The data are fitted by a sum of exponentially decaying functions, and $g^{(2)}(0)$ is defined by the ratio between the energy contained in the central peak around $\tau = 0$ and in the adjacent peaks. We find a zero-time correlation of $g^{(2)}(0) = 0.13$, demonstrating the single-photon nature of the emitted light. In conjunction with the measured enhanced oscillator strength for a spatially confined exciton, this is the unequivocal demonstration of SPS in a QD.

Solid-state quantum light sources often suffer from blinking of the emission, in which the QD randomly switches to a dark state and does not emit light [73]. This may happen, if a charge defect in the vicinity of the QD traps the electron or hole composing the exciton, thereby preventing the radiative recombination. This decreases the radiative efficiency of the single-photon source. Blinking normally occurs within nanosecond-to-microsecond time scales with a corresponding bunching in the QD second-order correlation function $g^{(2)}(\tau)$ over such time scales. By numerically integrating each peak in the HBT correlation data we obtain the long-time-scale plot shown in Fig. 5.17c. No bunching effects are observed, which shows that this single-photon source is free from blinking on a time scale of at least 10 μs.

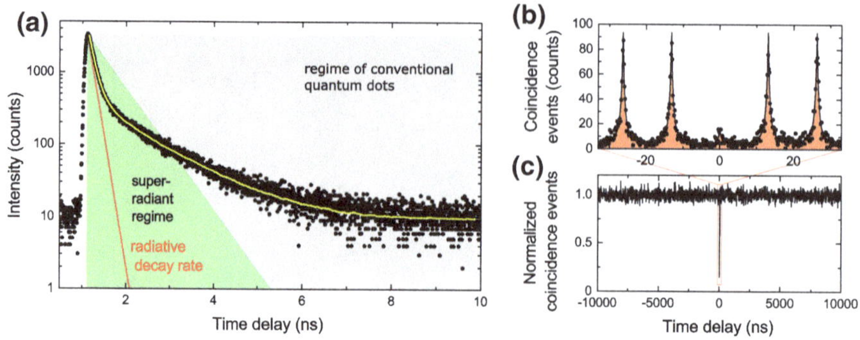

Fig. 5.17 Experimental demonstration of single-photon superradiance from a QD. **a** Time-resolved decay (*black points*) of the 1s exciton obtained under 2s-resonant excitation. The fit to the theoretical model is indicated by the *yellow line*. We take into account the impact of nonradiative processes presented in the previous section and extract a radiative decay rate of $8.4\,\text{ns}^{-1}$ (*red line*), which is deeply in the superradiant regime (*green area*). **b** HBT measurement of the emitted photons showing $g^{(2)}(0) = 0.13$, which demonstrates the single-photon character of the emission. **c** Long-time-scale HBT measurement where each coincidence peak has been numerically integrated. No blinking of the emission is observed. Figure reproduced from [23]

We have measured the oscillator strength of 9 different QDs and found them all to be superradiant with an average oscillator strength of $f = 76 \pm 11$ [23]. Remarkably, an oscillator strength up to $f = 96 \pm 2$ is observed corresponding to an intrinsic radiative rate beyond 10 GHz. Such a highly superradiant QD can deliver a radiative flux of single photons equivalent to more than five conventional QDs.

While the microscopic structure of the out-of-plane wavefunction can be accurately computed because the quantum-well thickness is known with monolayer precision, the in-plane geometry is generally unknown. The information is then inferred from the superradiant enhancement S via (cf. Sect. 5.3.1)

$$L = \frac{a_{\text{QW}}}{\sqrt{2}} \frac{\sqrt{S}}{\left|\langle \phi_h | \phi_e \rangle\right|}. \tag{5.63}$$

From the measured value $S \simeq 4.3$ an in-plane diameter $2L \simeq 24\,\text{nm}$ is obtained. The exciton wavefunction is spread over 90 thousand atoms in a collective superradiant quantum state sharing a single excitation.

5.3.4 *Impact of Thermal Effects on Single-Photon Superradiance*

The quantization energy ΔE_{QD} scales inversely proportional to the QD size squared, $\Delta E_{\text{QD}} \propto L^{-2}$, and, thus, decreases dramatically for large QDs. If ΔE_{QD} is

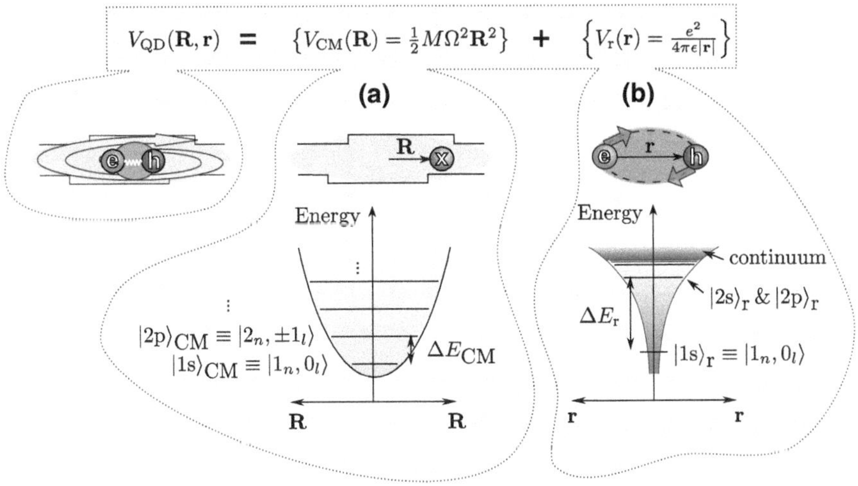

Fig. 5.18 Decomposition of the in-plane exciton dynamics into **a** a center-of-mass and **b** a relative motion. **a** The former describes the motion of the exciton center of mass in a two-dimensional harmonic potential, cf. (5.64). **b** The electron-hole electrostatic attraction is captured by the relative motion. Figure reproduced from [23]

comparable to the thermal energy, $k_B T$, excited states of the exciton manifold become populated and the large 1s oscillator strength is redistributed thus decreasing light-matter interaction. In the following we show that the temperature limits the maximum oscillator strength that can be harvested. In particular, the temperature of the current experiment of 7 K leads to a maximum oscillator strength of ~100.

Since the exciton dynamics can be decomposed into a center-of-mass (CM) and a relative (r) motion, cf. (5.57) and (5.58), the ground state is denoted as $|1s\rangle_{CM}|1s\rangle_r$. The relative motion is equivalent to the two-dimensional Hydrogen problem [29, 74] and is governed by the mutual electron-hole attraction, see Fig. 5.18. In this subspace, the relevant energy difference ΔE_r between the ground $|1s\rangle_r$ and first excited $|2s\rangle_r$ states equals roughly twice the excitonic Rydberg energy and amounts to about 8 meV. At cryogenic temperatures, thermal energies are much smaller (below 1 meV) and thermal population of the relative-motion subspace can be neglected.

The center-of-mass motion is described by a particle in a two-dimensional harmonic potential $V_{CM}(\boldsymbol{R}) = (1/2)M\Omega^2\boldsymbol{R}^2$, cf. Fig. 5.18a, where M is the exciton mass and the spring constant Ω is related to the quantum-dot size L via [25] $\Omega = 4\hbar/ML^2$. The resulting energy eigenstates are given by [29]

$$E_{nl} = (2n - |l| - 1)\hbar\Omega, \tag{5.64}$$

where $n = 1, 2, 3, \ldots$ and $l = 0, \pm 1, \ldots, \pm(n-1)$. The dipole selection rules dictate that states with $l = 0$ are bright (superradiant) and all others are dark (subradiant). For thermal energies $k_B T$ comparable to $\Delta E_{CM} = \hbar\Omega$, the excited states

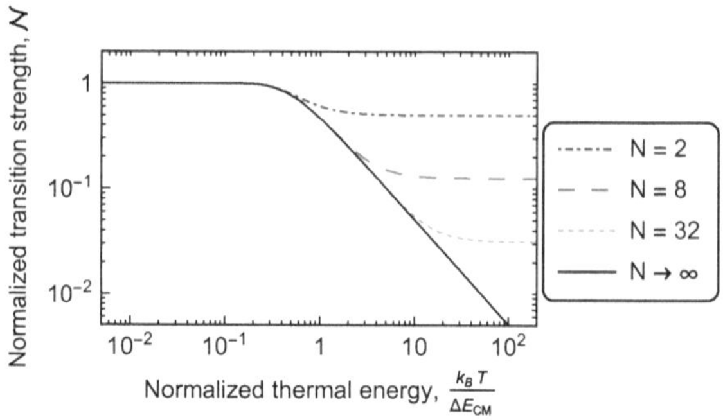

Fig. 5.19 The normalized transition strength $\mathcal{N}$ versus normalized thermal energy $k_B T/\hbar\Omega$ for different numbers of bound states N in a monolayer-fluctuation QD. Figure reproduced from [23]

become populated and the relevant figure of merit for light-matter interaction is the transition strength $F(T) = \mathcal{N}(T) \times f$, which is related to the oscillator strength via a temperature-dependent factor $\mathcal{N}(T)$. The latter describes the distribution of the population within the center-of-mass subspace. For a single excited state, $|e\rangle$, $\mathcal{N}$ is given by

$$\mathcal{N}(T) = \frac{1 + \frac{f_e}{f}\mathcal{B}(T)}{1 + \mathcal{B}(T)}, \tag{5.65}$$

where $\mathcal{B}(T) = \exp(-\Delta E_{\mathrm{CM}}/k_{\mathrm{B}}T)$ is the Boltzmann factor and f_e is the oscillator strength of the excited state. If $|e\rangle$ is dark and the temperature is high, the transition strength is half of f ($\mathcal{N} = 1/2$). We generalize (5.65) with the help of (5.64) for an arbitrarily large subspace

$$\mathcal{N}(T) = \frac{\text{radiative contributions}}{\text{all contributions}} = \frac{\sum_{n=1}^{N} \mathcal{B}^{2(n-1)}}{\sum_{n=1}^{N} \sum_{l=-(n-1)}^{n-1} \mathcal{B}^{2(n-1)-|l|}}, \tag{5.66}$$

where N denotes the number of center-of-mass states. This expression can be evaluated analytically giving

$$\mathcal{N}(T) = \coth\left(\frac{\hbar\Omega}{2k_{\mathrm{B}}T}N\right)\tanh\left(\frac{\hbar\Omega}{2k_{\mathrm{B}}T}\right), \tag{5.67}$$

and is plotted in Fig. 5.19. At small temperatures $4k_{\mathrm{B}}T \lesssim \hbar\Omega$, excited states play a negligible role and $F = f$. This is the regime in which the oscillator strength can be reliably measured. We employ this criterion $\hbar\Omega = 4k_{\mathrm{B}}T$ to estimate the maximum oscillator strength $f_{\mathrm{th,max}}^{\mathrm{sym}}$ that can be resolved at a temperature T and obtain (we consider the out-of-plane overlap $|\langle\psi_h|\psi_e\rangle|^2 \approx 1$ for simplicity)

$$f_{\mathrm{max,th}}^{\mathrm{sym}} = \frac{4\hbar E_P}{M\omega a_0^2}\frac{1}{k_{\mathrm{B}}T}, \tag{5.68}$$

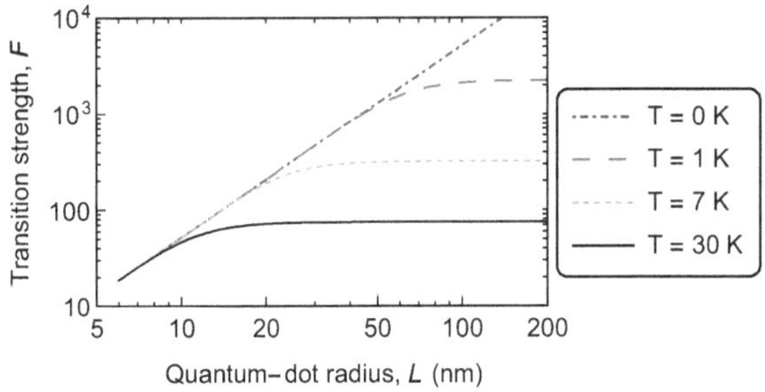

Fig. 5.20 The transition strength F versus QD radius L at different temperatures. For small QDs, F increases quadratically with L due to superradiance but saturates at larger values due to thermally activated excited states. Figure reproduced from [23]

which leads to an oscillator strength of 170 at 7 K. We generalize this expression for a more realistic asymmetric QD with an aspect ratio of 1:ξ with $\xi \geq 1$ and find that the maximum oscillator strength is decreased by ξ. We therefore conclude that oscillator strengths larger than about 100 are unlikely to be resolved at the experimental conditions of the present work. Remarkably, oscillator strengths of 1500 are predicted in monolayer-fluctuation QDs with a radius of about 60 nm. Temperatures below 0.8 K would, however, be required to resolve this effect.

The oscillator strength scales with the QD area $S \propto L^2$ but the normalized transition strength $\mathcal{N}$ scales as L^{-2}, cf. (5.67). Thermal effects therefore saturate the transition strength F, which can be expressed using (5.62)

$$F = f_{\max} \left(\frac{\sqrt{2}L}{a_{\mathrm{QW}}} \right)^2 \tanh \left(\frac{2\hbar^2}{M k_{\mathrm{B}} T L^2} \right), \tag{5.69}$$

which is plotted in Fig. 5.20. The transition strength saturates at

$$\lim_{L \to \infty} F = \frac{8\hbar E_P}{M\omega a_0^2} \frac{1}{k_{\mathrm{B}} T}, \tag{5.70}$$

which is independent of L and, interestingly, happens to equal $2 f_{\mathrm{max,th}}^{\mathrm{sym}}$. Note that for very large $L \gtrsim 100$ nm, deviations from the electric-dipole approximation, which are not accounted for in this study, further reduce the transition strength [25].

5.4 Conclusion and Outlook

In this chapter we discussed two mesoscopic properties of QDs: the breakdown of the dipole theory and single-photon superradiance. Aside from the dipole moment, In(Ga)As QDs have a large mesoscopic moment that contains magnetic-dipole and electric-quadrupole contributions. The impact of the mesoscopic moment is especially pronounced in electromagnetic environments that are not parity symmetric. The developed microscopic theory shows that the mesoscopic moment originates from distortions of the underlying crystal lattice. The resulting current density is curved leading to interaction of both electric and magnetic character. This opens the prospect for designing photonic nanostructures that match the shape of the QD current and thus enhance light-matter interaction efficiency. The simultaneous electric and magnetic nature of the QD current may lead to the vision of designing optical quantum metamaterials made from QDs for tailoring the interaction at the single-electron and single-photon level.

Single-photon superradiance was observed in monolayer-fluctuation QDs by recording the temporal decay dynamics in conjunction with second-order correlation measurements and a theoretical model. This enhanced light-matter coupling is known as the giant oscillator strength and was shown to be equivalent to superradiance. We argued that there is ample room for improving the oscillator strength at lower temperatures with prospects for generating highly coherent photons by outspeeding the noise sources, and for approaching the ultra-strong-coupling regime of cavity quantum electrodynamics with optical photons.

Acknowledgements We gratefully acknowledge the financial support from the Danish Council for Independent Research (natural sciences and technology and production sciences), the European Research Council (ERC consolidator grants "ALLQUANTUM" and "QIOS"), the Lundbeck Foundation and the Carlsberg Foundation.

References

1. P. Lodahl, S. Mahmoodian, S. Stobbe, Rev. Mod. Phys. **87**, 347 (2015)
2. J.P. Reithmaier, A. Löffler, C. Hofmann, S. Kuhn, S. Reitzenstein, L.V. Keldysh, V.D. Kulakovskii, T.L. Reinecke, A. Forchel, Nature **432**, 197 (2004)
3. E. Peter, P. Senellart, D. Martrou, A. Lemaître, J. Hours, J. Gérard, J. Bloch, Phys. Rev. Lett. **95**, 067401 (2005)
4. K. Hennessy, A. Badolato, M. Winger, D. Gerace, M. Atatüre, S. Gulde, S. Fält, E.L. Hu, A. Imamoğlu, Nature **445**, 896 (2007)
5. G. Lecamp, P. Lalanne, J.P. Hugonin, Phys. Rev. Lett. **99**, 023902 (2007)
6. T. Lund-Hansen, S. Stobbe, B. Julsgaard, H. Thyrrestrup, T. Sünner, M. Kamp, A. Forchel, P. Lodahl, Phys. Rev. Lett. **101**, 113903 (2008)
7. J. Claudon, J. Bleuse, N.S. Malik, M. Bazin, P. Jaffrennou, N. Gregersen, C. Sauvan, P. Lalanne, J.-M. Gérard, Nat. Photonics **4**, 174 (2010)
8. A. Schwagmann, S. Kalliakos, I. Farrer, J.P. Griffiths, G.A.C. Jones, D.A. Ritchie, A.J. Shields, Appl. Phys. Lett. **99**, 261108 (2011)
9. A. Laucht, S. Pütz, T. Günthner, N. Hauke, R. Saive, S. Frédérick, M. Bichler, M.-C. Amann, A.W. Holleitner, M. Kaniber, J.J. Finley, Phys. Rev. X **2**, 011014 (2012)
10. M. Arcari, I. Söllner, A. Javadi, S. Lindskov Hansen, S. Mahmoodian, J. Liu, H. Thyrrestrup, E.H. Lee, J.D. Song, S. Stobbe, P. Lodahl, Phys. Rev. Lett. **113**, 093603 (2014)
11. E.A. Muljarov, R. Zimmermann, Phys. Rev. Lett. **93**, 237401 (2004)
12. B. Urbaszek, E.J. McGhee, M. Krüger, R.J. Warburton, K. Karrai, T. Amand, B.D. Gerardot, P.M. Petroff, J.M. Garcia, Phys. Rev. B **69**, 035304 (2004)
13. I. Favero, A. Berthelot, G. Cassabois, C. Voisin, C. Delalande, P. Roussignol, R. Ferreira, J.M. Gérard, Phys. Rev. B **75**, 073308 (2007)
14. I. Wilson-Rae, A. Imamoğlu, Phys. Rev. B **65**, 235311 (2002)

15. U. Hohenester, A. Laucht, M. Kaniber, N. Hauke, A. Neumann, A. Mohtashami, M. Seliger, M. Bichler, J.J. Finley, Phys. Rev. B **80**, 201311 (2009)
16. M. Calic, P. Gallo, M. Felici, K.A. Atlasov, B. Dwir, A. Rudra, G. Biasiol, L. Sorba, G. Tarel, V. Savona, E. Kapon, Phys. Rev. Lett. **106**, 227402 (2011)
17. K.H. Madsen, P. Kaer, A. Kreiner-Møller, S. Stobbe, A. Nysteen, J. Mørk, P. Lodahl, Phys. Rev. B **88**, 045316 (2013)
18. P. Kaer, P. Lodahl, A.-P. Jauho, J. Mork, Phys. Rev. B **87**, 081308 (2013)
19. E. Chekhovich, M. Makhonin, A. Tartakovskii, A. Yacoby, H. Bluhm, K. Nowack, L. Vandersypen, Nat. Mater. **12**, 494 (2013)
20. M.L. Andersen, S. Stobbe, A.S. Sørensen, P. Lodahl, Nat. Phys. **7**, 215 (2011)
21. P. Tighineanu, M.L. Andersen, A.S. Sørensen, S. Stobbe, P. Lodahl, Phys. Rev. Lett. **113**, 043601 (2014)
22. P. Tighineanu, A.S. Sørensen, S. Stobbe, P. Lodahl, Phys. Rev. Lett. **114**, 247401 (2015)
23. P. Tighineanu, R.S. Daveau, T.B. Lehmann, H.E. Beere, D.A. Ritchie, P. Lodahl, S. Stobbe, Phys. Rev. Lett. **116**, 163604 (2016)
24. D. Bruls, J. Vugs, P. Koenraad, H. Salemink, J. Wolter, M. Hopkinson, M. Skolnick, F. Long, S. Gill, Appl. Phys. Lett. **81**, 1708 (2002)
25. S. Stobbe, P.T. Kristensen, J.E. Mortensen, J.M. Hvam, J. Mørk, P. Lodahl, Phys. Rev. B **86**, 085304 (2012)
26. J.R. Zurita-Sánchez, L. Novotny, JOSA B **19**, 1355 (2002)
27. J.R. Zurita-Sánchez, L. Novotny, JOSA B **19**, 2722 (2002)
28. S.V. Goupalov, Phys. Rev. B **68**, 125311 (2003)
29. M. Sugawara, Phys. Rev. B **51**, 10743 (1995)
30. A. Thränhardt, C. Ell, G. Khitrova, H.M. Gibbs, Phys. Rev. B **65**, 035327 (2002)
31. K. Jun Ahn, A. Knorr, Phys. Rev. B **68**, 161307 (2003)
32. P.T. Kristensen, J.E. Mortensen, P. Lodahl, S. Stobbe, Phys. Rev. B **88**, 205308 (2013)
33. M. Cotrufo, A. Fiore, Phys. Rev. B **92**, 125302 (2015)
34. E. Rashba, G. Gurgenishvili, Sov. Phys. Solid State **4**, 759 (1962)
35. J. Hours, P. Senellart, E. Peter, A. Cavanna, J. Bloch, Phys. Rev. B **71**, 161306 (2005)
36. L.C. Andreani, G. Panzarini, J.-M. Gérard, Phys. Rev. B **60**, 13276 (1999)
37. J. Guest, T. Stievater, X. Li, J. Cheng, D. Steel, D. Gammon, D. Katzer, D. Park, C. Ell, A. Thränhardt, G. Khitrova, H.M. Gibbs, Phys. Rev. B **65**, 241310 (2002)
38. D. Gammon, E. Snow, B. Shanabrook, D. Katzer, D. Park, Phys. Rev. Lett. **76**, 3005 (1996)
39. R. Dicke, Phys. Rev. **93**, 99 (1954)
40. D.A.B. Miller, *Quantum Mechanics for Scientists and Engineers* (Cambridge University Press, Cambridge, 2007)
41. P. Harrison, *Quantum Wells, Wires and Dots: Theoretical and Computational Physics of Semiconductor Nanostructures* (Wiley, New York, 2005)
42. P. Tighineanu, R. Daveau, E.H. Lee, J.D. Song, S. Stobbe, P. Lodahl, Phys. Rev. B **88**, 155320 (2013)
43. J. Dreiser, M. Atatüre, C. Galland, T. Müller, A. Badolato, A. Imamoglu, Phys. Rev. B **77**, 075317 (2008)
44. O. Stier, M. Grundmann, D. Bimberg, Phys. Rev. B **59**, 5688 (1999)
45. K. Sivalertporn, L. Mouchliadis, A.L. Ivanov, R. Philp, E.A. Muljarov, Phys. Rev. B **85**, 045207 (2012)
46. S. Schmitt-Rink, D.A.B. Miller, D.S. Chemla, Phys. Rev. B **35**, 8113 (1987)
47. G. Bester, S. Nair, A. Zunger, Phys. Rev. B **67**, 161306 (2003)
48. M. Bayer, G. Ortner, O. Stern, A. Kuther, A.A. Gorbunov, A. Forchel, P. Hawrylak, S. Fafard, K. Hinzer, T.L. Reinecke, S.N. Walck, J.P. Reithmaier, F. Klopf, F. Schäfer, Phys. Rev. B **65**, 195315 (2002)
49. D.P. Craig, T. Thirunamachandran, *Molecular Quantum Electrodynamics: An Introduction to Radiation-Molecule Interactions* (Courier Dover Publications, New York, 1998)
50. P. Tighineanu, Electric and Magnetic Interaction Between Quantum Dots and Light. Ph.D. thesis, The Niels Bohr Institute, Faculty of Science, University of Copenhagen (2015)

51. N. Vats, S. John, K. Busch, Phys. Rev. A **65**, 043808 (2002)
52. L. Novotny, B. Hecht, *Principles of Nano-Optics* (Cambridge University Press, Cambridge, 2012)
53. J. Johansen, S. Stobbe, I.S. Nikolaev, T. Lund-Hansen, P.T. Kristensen, J.M. Hvam, W.L. Vos, P. Lodahl, Phys. Rev. B **77**, 073303 (2008)
54. D. Englund, A. Faraon, I. Fushman, N. Stoltz, P. Petroff, J. Vučković, Nature **450**, 857 (2007)
55. N. Somaschi, V. Giesz, L. De Santis, J. Loredo, M. Almeida, G. Hornecker, S. Portalupi, T. Grange, C. Antón, J. Demory, C. Gómez, I. Sagnes, N. Lanzillotti-Kimura, A. Lemaitre, A. Auffeves, A. White, L. Lanco, P. Senellart, Nat. Photonics **10**, 340 (2016)
56. A.G. Curto, T.H. Taminiau, G. Volpe, M.P. Kreuzer, R. Quidant, N.F. van Hulst, Nat. Commun. **4**, 1750 (2013)
57. P. Yao, V. Manga Rao, S. Hughes, Laser Photonics Rev. **4**, 499 (2010)
58. S. Stobbe, J. Johansen, P.T. Kristensen, J.M. Hvam, P. Lodahl, Phys. Rev. B **80**, 155307 (2009)
59. J. Johansen, B. Julsgaard, S. Stobbe, J.M. Hvam, P. Lodahl, Phys. Rev. B **81**, 081304 (2010)
60. S. DeBeer George, T. Petrenko, F. Neese, Inorganica Chimica Acta **361**, 965 (2008)
61. S. Bernadotte, A.J. Atkins, C.R. Jacob, J. Chem. Phys. **137**, 204106 (2012)
62. H. Eisele, A. Lenz, R. Heitz, R. Timm, M. Dähne, Y. Temko, T. Suzuki, K. Jacobi, J. Appl. Phys. **104**, 124301 (2008)
63. M. Noecker, B. Masterson, C. Wieman, Phys. Rev. Lett. **61**, 310 (1988)
64. I.D. Rukhlenko, D. Handapangoda, M. Premaratne, A.V. Fedorov, A.V. Baranov, C. Jagadish, Opt. Express **17**, 17570 (2009)
65. T.H. Taminiau, S. Karaveli, N.F. van Hulst, R. Zia, Nat. Commun. **3**, 979 (2012)
66. G.Y. Slepyan, A. Magyarov, S.A. Maksimenko, A. Hoffmann, Phys. Rev. B **76**, 195328 (2007)
67. J. Kalkman, H. Gersen, L. Kuipers, A. Polman, Phys. Rev. B **73**, 075317 (2006)
68. D.E. Chang, A.S. Sørensen, P.R. Hemmer, M.D. Lukin, Phys. Rev. B **76**, 035420 (2007)
69. L.D. Landau, E.M. Lifšic, J.B. Sykes, J.S. Bell, M. Kearsley, L.P. Pitaevskii, *Electrodynamics of Continuous Media* (Pergamon Press, Oxford, 1960)
70. E. Hanamura, Phys. Rev. B **37**, 1273 (1988)
71. E. Poem, Y. Kodriano, C. Tradonsky, N.H. Lindner, B.D. Gerardot, P.M. Petroff, D. Gershoni, Nat. Phys. **6**, 993 (2010)
72. C. Gerry, P. Knight, *Introductory Quantum Optics* (Cambridge University Press, Cambridge, 2005)
73. M. Davanço, C.S. Hellberg, S. Ates, A. Badolato, K. Srinivasan, Phys. Rev. B **89**, 161303 (2014)
74. W. Que, Phys. Rev. B **45**, 11036 (1992)

Chapter 6
Single-Photon Sources Based on Deterministic Quantum-Dot Microlenses

T. Heindel, S. Rodt and S. Reitzenstein

Abstract This chapter addresses the design, fabrication and characterization of deterministically fabricated single-photon sources based on quantum dot microlenses. The quantum devices are optimized for future applications in quantum communication systems which require high photon extraction efficiency, strong suppression of multi-photon emission and high indistinguishability of the emitted photons. Highest extraction efficiency is achieved by integrating single quantum dots in the center of monolithic microlenses by means of in-situ electron-beam lithography based on low-temperature cathodoluminescence spectroscopy. Quantum optical studies of deterministic microlenses reveal pure single-photon emission associated with $g^{(2)}(0) < 0.01$ and an indistinguishability exceeding 90% under pulsed p-shell excitation. Mechanisms limiting the indistinguishability are discussed in terms of spectral diffusion at a nanosecond time-scale and phonon-induced dephasing. The chapter closes with an outlook on electrically driven quantum dot microlenses, the application of anti-reflection coatings and future perspectives.

6.1 Introduction

Light sources emitting single photons on demand are key building blocks for quantum communication systems [1–3]. Here, the quantum states of single photons act as information carriers and allow for an unconditionally secure transfer of data. The information can be encoded for instance in the polarization of single quantas of light, often referred to as flying qubits, while quantum mechanical principles such as the no-cloning theorem [4] or entanglement [5] ensure that possible eavesdropping attacks can be detected [6]. In seminal work, the so-called BB84 [7] and the Ekert91 [8] protocol heralded the era of quantum key distribution (QKD), by exploiting single photons or entangled photon pairs to generate and distribute a secret key for data encryption and decryption. Furthermore, long-distance optical quantum communication relies

T. Heindel · S. Rodt · S. Reitzenstein (✉)
Institut für Festkörperphysik, Technische Universität Berlin, Hardenbergstrasse 36,
10623 Berlin, Germany
e-mail: stephan.reitzenstein@physik.tu-berlin.de

P. Michler (ed.), *Quantum Dots for Quantum Information Technologies*,
Nano-Optics and Nanophotonics, DOI 10.1007/978-3-319-56378-7_6

on entanglement distribution via Bell-state measurements, which requires photon sources emitting indistinguishable photons at high flux [2, 9]. In combination, these are stringent physical requirements for quantum light sources which are even complemented by further practical needs such as compactness, robustness, electrical operation and the possibility of direct fiber coupling.

Individual key features of quantum light sources can be addressed by a variety of approaches, including for instance atomic systems [10], non-linear crystals generating entangled photon pairs via parametric down-conversion [11], or vacancy centers in diamond [12, 13]. However, these approaches can hardly meet all requirements in a single implementation. Atomic systems for instance require rather bulky experimental setups, while parametric down-conversion is a probabilistic (i.e. non-deterministic) process. In contrast, semiconductor quantum dots (QDs) are highly promising candidates for the realization of practical on-demand quantum light sources for quantum communication systems [14] and emerging applications such as quantum enhanced sensing [15] or quantum enhanced imaging [16].

Semiconductor QDs realized epitaxially by the Stranski–Krastanov growth mode are particularly appealing for the realization of quantum light sources. This is explained by the discrete nature of their energy levels in combination with excellent optical properties, e.g. in terms of very high quantum efficiency exceeding 90% [17]. Their electronic structure enables high suppression of multi-photon emission events, i.e. $g^{(2)}(0)$-values well below 0.01 [18, 19], and allows for the generation of polarization- [20] or time-bin- [21] entangled photon pairs via the biexciton (XX) - exciton (X) radiative cascade [22] (see this chapter and Chap. 7).

In spite of the superb physical properties, practical issues up till now hindered the application of QDs in commercial products - or at least in proof-of-principle demonstrators. Firstly, total internal reflection limits the photon collection efficiency η_{ext}, which essentially describes the probability of collecting a photon emitted by a QD in an external optical system, such as a lens or an optical fiber with a given numerical aperture. Enhancing η_{ext} from about 2% available in a simple planar structure [23, 24] to values approaching unity requires a precise tailoring of the photonic environment and the introduction of suitable mirror sections. Another issue is related to the statistical character of self-assembled Stranski–Krastanov growth which leads to random positions and random emission energies within the inhomogeneously broadened emission band of respective QDs. This randomness causes significant problems regarding the realization of quantum light sources with well defined properties and with high process yield and has triggered the development of in-situ lithography approaches [25, 26] which allow for the integration of preselected QDs into nanophotonic structures with high alignment accuracy and high process yield. Furthermore, electrical pumping is desirable which requires complex doping and contacting schemes [27–30]. Other important aspects are the posibility of a direct fiber-coupling [31–37] of the source, but also a user-friendly operation in general. This can be achieved by operating QD-based quantum light sources using compact, plug-and-play cryocoolers [38, 39], which make bulky and expensive cooling equipment (such as liquid Helium storage dewars) dispensable.

In this chapter we present and discuss recent advancement in the design, fabrication and characterization of single-photon sources based on self-assembled InGaAs QDs. This involves a brief overview of light-extraction strategies to maximize η_{ext} with a focus on broadband enhancement of emission using monolithically integrated microlenses. The photon extraction efficiency of these nanophotonic structures is performed numerically via solving Maxwell's equations by a finite-element method and will be described in detail. In order to meet the stringent structural requirements and to realize efficient QD-microlenses we introduce in-situ electron-beam lithography based on low-temperature cathodoluminescence spectroscopy. This novel nanotechnology platform allows for a pre-selection of suitable QDs and their integration into photonic nanostructures such as microlenses with about 30 nm alignment accuracy. The high optical quality of QD-microlenses fabricated using this in-situ lithography technique is studied by high resolution optical and quantum optical spectroscopy which demonstrates for instance the excellent suppression of multi-photon emission events of this structures. Beyond that, we study the degree of indistinguishability of emitted photons and limitations of this important parameter in terms of spectral diffusion on the nanosecond timescale and temperature induced dephasing. We close with an outlook on further optimization of QD-microlenses and future perspectives.

6.2 Light Extraction Strategies

With respect to applications of quantum light sources in quantum communication systems it is crucial to maximize their photon extraction efficiency η_{ext}. For a QD embedded in GaAs bulk material this fraction amounts to less than 2% [23, 24], due to total internal reflection of photons impinging at the semiconductor-vacuum interface at angles larger than 17°. However, the semiconductor host matrix, holds the great advantage that light harvesting can be engineered to a large extend by tailoring the photonic environment of the QDs. In this section, the most important strategies for maximizing the photon extraction will be discussed in the context of efficient quantum light sources.

The first report on a QD single-photon source emitting antibunched light was published by Michler et al. [40] based on a microdisk cavity [41, 42]. This type of cavity confines light localized within a circular disk or toroid via so-called whispering gallery modes. These modes are guided at the circumference of the structure via total internal reflection and couple to the vacuum or air via evanescent fields and light scattering due to finite surface roughness of the microdisk. Another type of resonator, which is frequently used in light-matter coupling experiments, is based on a photonic crystal membrane [43, 44]. Here, a modulation of the refractive index on the scale of the wavelength of light, e.g. by etching a periodic grid of holes into a thin air-suspended membrane, introduces a photonic bandgap at which a propagation of light is forbidden. Leaving one or few holes of this photonic crystal unetched, results in the formation of a low mode-volume microcavity capable of localizing light in three dimensions. Both approaches, microdisks and photonic crystal cavities, rely

Fig. 6.1 Illustration of three designs with different strategies for efficient light extraction. The micropillar cavity (**a**), the photonic nanowire (**b**), and the microlens (**c**) (color figure online)

on total internal reflection in growth direction, which limits the extraction efficiency normal to the sample surface. Still, recent experiments using photonic crystal cavities demonstrated out-coupling efficiencies of 46% at 1.3 μm [45] (see Chap. 11).

For efficient coupling of photons into collection optics aligned normal to the sample surface, mainly three different approaches have been established in recent years (see Fig. 6.1): Cylindrical Fabry–Pérot micropillar cavities, photonic wires, and microlenses. In the following sections, we briefly discuss the characteristics of each approach and give an overview to the state-of-the-art.

6.2.1 *Micropillar Cavities*

Micropillar cavities are based on cylindrical Fabry–Pérot resonators and typically consist of a λ-thick cavity sandwiched between a lower and an upper distributed Bragg reflector (DBR) [46, 47]. This geometry allows for a three-dimensional confinement of light, firstly by reflection at the DBRs parallel to the growth direction, and secondly, by mode-guiding via total internal reflection at the micropillar sidewalls in lateral direction. At the same time, highly directional emission of light can be achieved via the top facet of the upper DBR. Due to the low mode-volume and high Q-factors achievable in micropillar resonators with embedded QDs, experiments in the regime of strong light-matter interaction are possible and were first reported in [48]. Operating in the weak coupling regime, a first proof of concept of efficient single-photon extraction using such micropillar cavities was demonstrated in 2002 by Pelton et al. [49]. Additionally, utilizing p-i-n doped micropillar structures, an electrical contacting scheme based on ring-shaped top-contacts was established [30]. Nowadays, state-of-the-art micropillar devices enable the generation of close-to-ideal single-photon states with high efficiency under optical [19] (see Chap. 3) as well as electrical [50, 51] pumping, and can be readily applied for free-space quantum-key distribution experiments [52, 53]. A drawback of micropillar cavities is the narrow-band enhancement of emission typical for microresonator-approaches. Due to the strongly modulated optical density of states, experiments on spectrally separated QD states are laborious and require—if possible at all—complicated cavity designs [54].

6.2.2 Photonic Nanowires

Photonic wires are cylindrical or hexagonal wires of a few μm length and a width on the order of the wavelength of light, which can be fabricated in top-down [55] or bottom-up [56] approaches, the latter also in a deterministic manner [57]. This structure allows to efficiently funnel the emission of an embedded quantum emitter into a single spatial mode propagating along the wires axial direction. An efficient coupling to collection optics is achieved by introducing a mirror-section below the QD in combination with a well-defined tapered section in the upper part of the wire [58], where the latter results in an adiabatic transition from a strongly confined single mode in the near-field to a highly directional Gaussian beam in the far-field [34]. This approach features high photon extraction efficiencies in a broad spectral range of several 10 nm. Exploiting such nanowire designs, QD single-photon sources have been realized with efficiencies of up to 72% [55] using optical pumping. Schemes for electrical current injection have been proposed [59], but remain challenging due to the small dimensions of the nanowire in lateral directions. Another issue is the comparably large spectral linewidth observed for nanowire QD states, which results from the proximity to the nanowire surface and respective charge fluctuations.

6.2.3 Microlenses

The microlens approach is conceptually different to the above discussed cavity- and waveguide-based approaches. Similar to solid immersion lenses (SILs) [60], which have been routinely applied in quantum optics experiments using QDs [61], nitrogen vacancy centers [62] and organic molecules [63], a curved surface is introduced above a quantum emitter in order to increase the fraction of the emission that can leave the semiconductor material. This geometrical effect is not limited to certain spatial or spectral modes and hence allows for a broadband enhancement of the extraction efficiency. While macroscopic SILs, typically made out of ZrO_2 ($n = 2.17$), have a finite refractive index mismatch to the semiconductor sample and need to be glued to its surface (see Chap. 12), microlenses directly integrated into the sample-material itself benefit from further reduced losses. This issue has been previously addressed by fabricating microlenses directly via optical lithography [64] or focused ion-beam milling [65]. However, none of these technologies allowed for a deterministic integration of single quantum emitters.

The remainder of this chapter will introduce a technique which enables the integration of pre-selected QDs within monolithic microlenses. As will be demonstrated, this approach is capable of combining high photon extraction efficiencies with high spectral bandwidth, while a robust and flexible fabrication technique allows even for electrical control of single emitter devices.

6.3 Numerical Optimization of Photon Extraction Efficiency of Quantum Dot Microlenses

For succeeding in the fabrication of nano- and micro-structured optical devices, it is essential to use an optimized structural design. Otherwise, internal reflection, interference, and absorption of light might spoil the functionality and efficiency. Another important aspect concerns the collection optics. Usually a lens system or an optical fiber is used to collect the emitted light. Each collection system has its own angle of acceptance (expressed by the numerical aperture, NA) and the angle of emission should match that NA as good as possible for maximum coupling efficiency.

When the dimensions of the devices are of the same order of magnitude as the wavelength of light, classical ray optics (geometrical optics) fails to give reliable results. Instead Maxwell's equations have to be solved numerically in a rigorous way by sophisticated methods. They have to include realistic material parameters, real geometric properties (including imperfections), and must lead to sufficient numerical convergence.

A number of different simulation techniques have been developed that are well applicable for nanophotonic device simulation [66]. They comprise finite-element methods (FEM), finite-difference time-domain simulations (FDTD), wavelet methods, finite-integration technique (FIT), rigorously-coupled wave-analysis (RCWA), and plane-wave expansion methods (PWE). In the following we will employ FEM simulations which is a general numerical method for solving partial differential equations. Realistic and complex geometries can be exactly treated while the computation times are relatively short. The algorithms are quite robust and convergence is well manageable.

The following calculations are based on the FEM software package JCMsuite from the company JCMwave GmbH and the Zuse Institute Berlin. Besides other setups, time-harmonic resonance problems and scattering problems (by including, e.g., a dipole emitter in the model structure) can be solved for arbitrary 2D and 3D structures. The computations are based on higher-order vector elements, adaptive unstructured grids, and on a rigorous treatment of transparent boundaries [67]. The efficiency for specific setups can further be increased, e.g., by using a subtraction approach for point sources [68], or by exploiting geometrical mirror-symmetries or a cylindrical symmetry, if present.

Figure 6.2 displays a typical 2D grid of meshes as generated by JCMsuite for two steps of grid refinement. The finer the grid is, the more exact becomes the numerical solution. However, a finer grid results in a calculation that consumes more memory and cpu time. So there is always a compromise between exactness and calculation expenses. To overcome this issue, advanced numeric tools apply adaptive mesh refinement which allows for higher accuracy at only moderate increase of calculation expenses [67]. As a further numerical parameter influencing numerical accuracy as well as computational costs, the polynomial order p of the finite-element ansatz functions can be defined. For this parameter also an adaptive setting can be

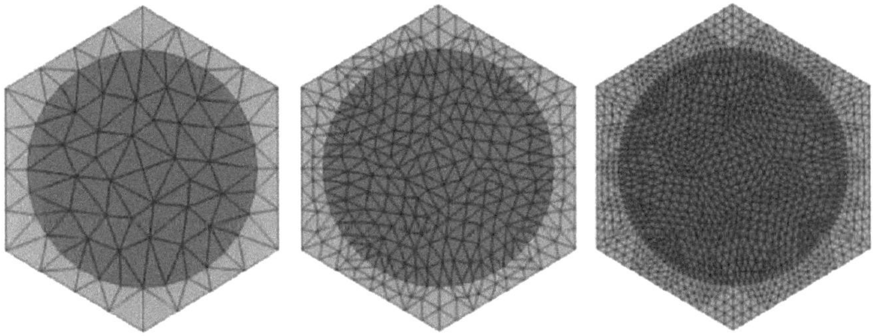

Fig. 6.2 An exemplary FEM grid with increasing refinement from *left* to *right* (color figure online)

used, yielding various orders p on the different patches of the mesh (so-called hp-FEM).

Now we start dealing with the simulations for microlenses above QDs as light emitters. The idea is to find the optimal lens geometry for enhanced light out-coupling into a given spatial angle to match the NA of the collection optics. The scattering-problem solver is best suited to handle this numerical problem. The QD is described by a point source with dipole-like emission of the respective wavelength. Here we investigate the InAs/GaAs material system. Typically such QDs have emission in the 900–1200 nm spectral range. Our target wavelength is 930 nm with respect to the following quantum-optical experiments in Sect. 6.6. To overcome big losses due to light that is emitted to the backside of the sample, a DBR is situated below the QD. It consists of 23 pairs of $Al_{0.9}Ga_{0.1}As$ and GaAs that provide a very high reflectivity close to 100% at normal incidence. The respective model structure with the (mixed, tetrahedral and prismatoidal) 3D FEM grid is depicted in Fig. 6.3a.

The output of the FEM simulation is the distribution of the electro-magnetic field within the computational domain. Derived quantities like power fluxes, far-field distributions, or the Purcell factor are computed from the electro-magnetic near field in post-processes. A typical example is given in Fig. 6.3b. The dipole emitter is located at the position of the red spot that indicates very high local field intensity of the dipole source. Clearly one sees distinct interference fringes that result from the interplay between reflection at the GaAs-air interface on top and at the DBR at the bottom side. The QD is already located at an anti-node position of the electric field which is mandatory for efficient emission into an optical mode of the lens-DBR system.

With respect to applications like coupling to optical fibers or light collection via lens systems it is very important to analyze the fraction of light that will be collected by the respective optics with its given NA in terms of the photon extraction efficiency. This information is acquired by integrating the electro-magnetic-field distribution like that in Fig. 6.3b within the corresponding spatial angle. The resulting photon extraction efficiencies are plotted as a function of NA. Figure 6.4 displays the η_{ext} for a number of lenses with varying base width above a DBR. The lenses have the

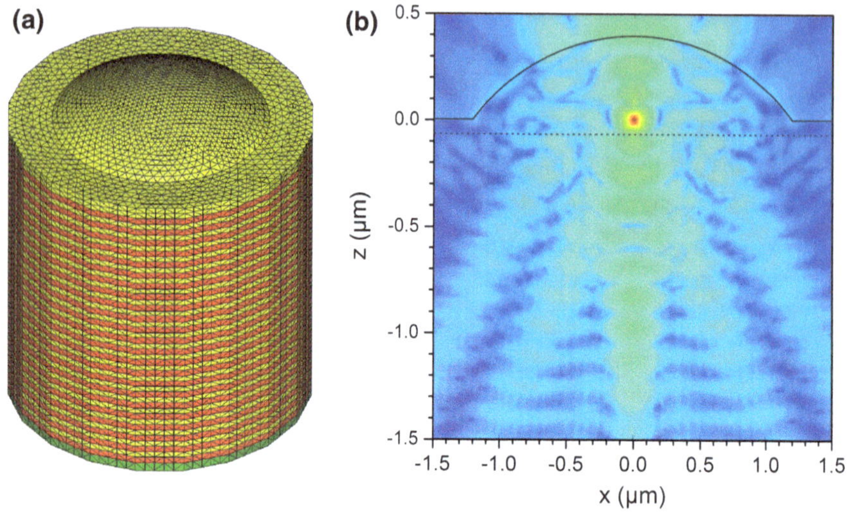

Fig. 6.3 **a** Our model structure consisting of a lower DBR, the active region with the QD (hidden inside the structure), and a hemispheric-section shaped microlens on *top*. *Red* layers correspond to $Al_{0.9}Ga_{0.1}As$ and *yellow* layers to GaAs. **b** Calculated near-field distribution for the hemispheric-section shaped lens with a height of 400 nm and a base width of 2.4 µm. The semiconductor-air interface is marked by the *solid black line*. Below the point source (*bright red spot*), distinct field modulations due to the *bottom* DBR (starting at the *dashed black line*) are visible (the bottom DBR is truncated for a better display format). From [69]

Fig. 6.4 Numerical results for the photon extraction efficiency η_{ext} for a hemispherical-section-shaped lens and a bottom-DBR as a function of the NA of the light-collection optics. The different curves correspond to different base widths of the lenses with a common height of 400 nm. From [69]

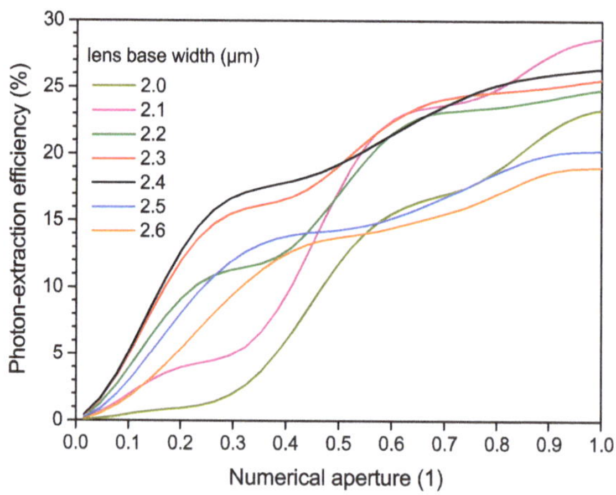

shape of a hemispheric section and a height of 400 nm. There is a strong impact of the base width on η_{ext}. For example, when the collection optics has a NA of 0.4 the optimal base width is 2.4 µm.

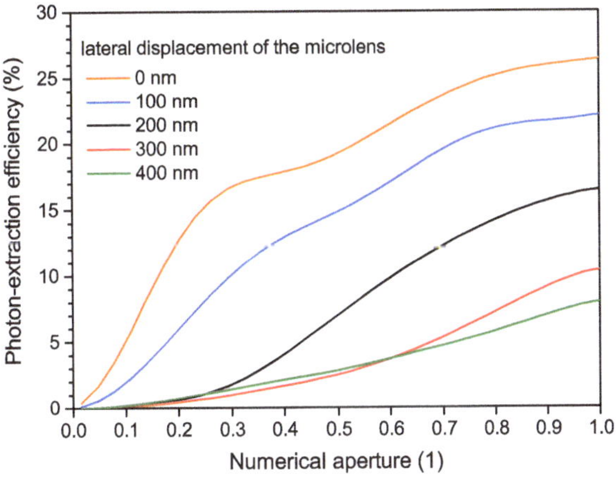

Fig. 6.5 The impact of a lateral displacement of the lens with respect to the QD position is evaluated numerically. The calculations were performed for the 2.4 μm microlens-DBR structure that was discussed before. The x-axis gives the numerical aperture of the external light-collection optics. From [69]

The fabrication process of microlenses (see next section) identifies another important geometrical parameter, namely a possible lateral displacement. If the QD is not located in the center of the lens the interference pattern becomes asymmetric and η_{ext} might be affected. This is investigated in Fig. 6.5. A lateral displacement of 100 nm already leads to a drop in η_{ext} of up to 5% whereas a displacement of 200 nm results in a drastical decrease of up to 15%. This clearly shows that a highly precise fabrication process like a deterministic lithography approach is needed for such nanophotonic structures.

In the previous sections we considered a DBR as back-reflecting mirror. Another possibility would be the use of a gold reflector. Fabricating a lens structure with such a mirror is more demanding as it requires a flip-chip process and well controlled removal of the substrate by volume etching or etching of a sacrificial layer. On the other hand, a gold mirror could be used as electrical contact for electrically driven structures and/or as bonding agent to a piezoelectric actuator for strain and energy tuning [70] (see Chap. 7).

Figure 6.6 shows the comparison between a DBR and a gold mirror in combination with a microlens. Up to a NA of 0.3 there is little difference in η_{ext}. Above NA = 0.3 the gold mirror is calculated to be much more efficient. η_{ext} goes up to 85% for a NA of 1, which makes the gold mirror very promising for future lens structures. The reason for this superior behavior is the dependence of the reflectivity of both mirrors on the angle of incident light. The reflectivity of the DBR starts to drop drastically when the angle is larger than 20°. The gold mirror has almost angle-independent reflectivity that gives rise to the very high η_{ext}.

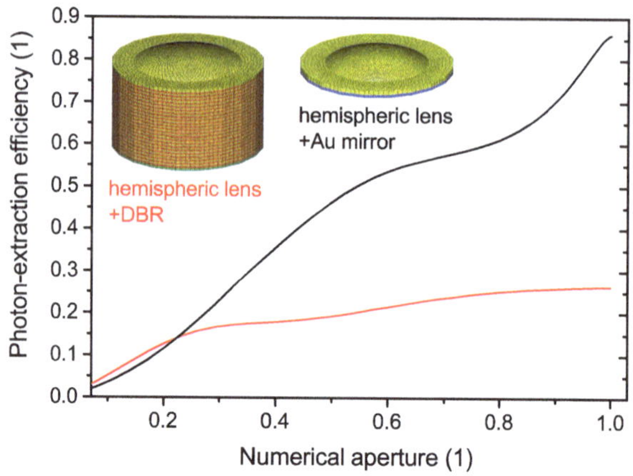

Fig. 6.6 Two different mirror concepts are compared numerically: The first one comprises a DBR as discussed before and the second one is based on a gold layer below the QD acting as mirror. In both cases an optimized lens shape with respect to the given mirror is included in the model structure. *Black* Hemispheric-section lens (height = 400 nm, base width = 2.0 μm) with a gold mirror 150 nm below the QD. Red: Hemispheric-section lens (height = 400 nm, base width = 2.4 μm) with a DBR 65 nm below the QD as described before. The x-axis gives the numerical aperture of the external light-collection optics. The *insets* show the respective model structures (*yellow* = GaAs, *red* = AlGaAs, *blue* = Au). From [69] (color figure online)

6.4 Deterministic Nanophotonic Device Technologies

Many optoelectronic single-QD applications require a QD with specific optoelectronic properties, like precise emission energy and diminishing fine-structure splitting [22]. Such applications might be emitters of entangled photon-pairs for secure quantum cryptography [52] and for quantum repeaters to enable long-distance data transmission via entanglement swapping [2]. Naturally, self-organized semiconductor QDs differ in their optoelectronic properties due to statistical variations in size, geometry, and material composition. This requires the pre-selection of a suitable QD before integrating it into the nanophotonic device. Otherwise the yield of matching devices will be very small or even zero. Deterministic device technologies, as discussed in the following, help to end up with high-performance quantum devices. The key point is the pre-selection as mentioned before. The available techniques can be divided into two groups, depending on the combination of pre-selection and actual device processing. All referenced examples have in common, that the spatial precision for the overall process is better than 100 nm. In the previous section we have seen, that such a precision is indeed prerequisite for proper device efficiency. Most techniques use two different setups to perform the two steps [71–74]. For example, the optical pre-selection is done in a μPL-setup and the sample processing via optical lithography (OL) or electron-beam lithography (EBL) is performed in a dedicated lithography setup. The spatial correspondence between the two setups is realized via marker structures on the sample surface that were fabricated beforehand. The second

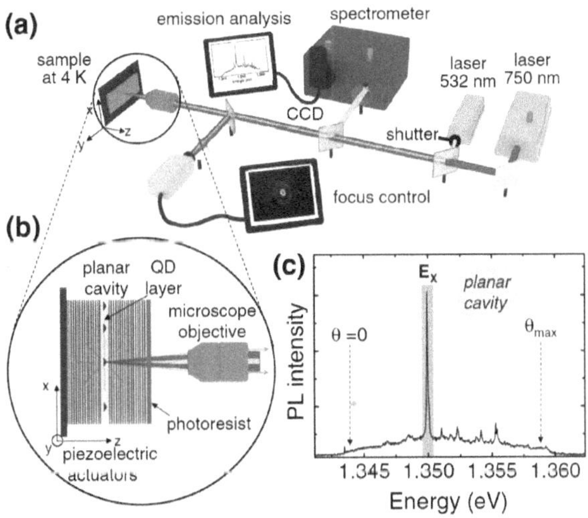

Fig. 6.7 **a** Schematics of an in-situ lithography technique that is based on two lasers with different wavelength. **b** The *red* laser is used to excite the luminescence of the QDs while it does not alter the photo-resist. When a suitable QD is found, the *green* laser is turned on to expose the resist locally. **c** The spectrum shows the emission of a selected QD. From [25] (color figure online)

class of techniques will be called in-situ lithography techniques. They allow for pre-selection *and* lithography in one and the same setup without transferring the sample or heating it up (if the pre-selection was performed at cryogenic temperatures) [75]. A prominent example based on optical lithography is the work by Dousse et. al [25]. This technique is based on a μPL setup with a red and a green laser (Fig. 6.7). The sample is already coated with a photo-sensitive resist when it is initially mounted in the setup. Spectroscopy is done with the red laser that does not alter the resist. When the position of a suitable QD is found, the green laser is turned on. It follows the same optical path as for the red laser and exposes the resist at the given sample position. In this way mesa structures can be fabricated in-situ with a high spatial accuracy.

Our approach utilizes a scanning-electron microscope (SEM) that was extended with spectroscopic attachments and EBL capabilities. As the luminescence of the sample is excited via the SEM's electron beam, that spectroscopy technique is called *cathodoluminescence spectroscopy* (CL) [76]. The overall process is dubbed *cathodoluminescence lithography* (CLL) [26]. Our technique benefits from effectively combining the advantages of high-resolution CL spectrocopy and of high-resolution electron-beam lithography and will be discussed in detail in the next section.

6.5 Fabrication of Deterministic Single Quantum Dot Microlenses

The fabrication of microlenses (cf. Sect. 6.3) is much more demanding than the processing of structures that require only the definition of the respective cross-section, of e.g. pillar or mesa structures, as it involves a high-resolution 3D lithography

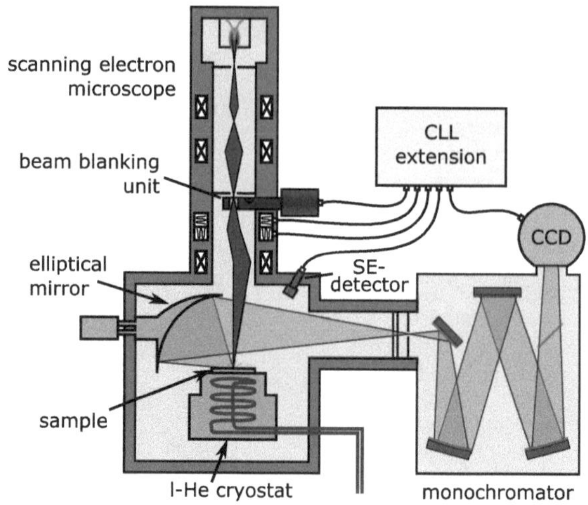

Fig. 6.8 Schematics of the CLL setup. A standard SEM is equipped with luminescence-detection and EBL features. The He-flow cryostat allows for sample temperatures between 4 K and room temperature (color figure online)

process. 3D nano-lithography requires a lithography resist with a low contrast and a high-resolution lithography technique with precise dose application. Here we benefit from the EBL capabilities of our CLL setup and the properties of the EBL resist PMMA (poly-methyl-methacrylate) [77, 78] as described in the following.

Figure 6.8 displays a sketch of the CLL setup. A standard SEM system of type JEOL JSM 840 is equipped with a He-flow cryostat, an elliptical mirror, a spectrometer, and a beam-blanking unit for EBL operation. The sample is mounted on the cryostat's cold finger and is excited via the SEM's electron beam. The sample's luminescence is partly collected by the mirror and focused into the spectrometer that is equipped with a liquid-nitrogen-cooled charge-coupled-device camera.

Before going into details about the CLL process some information on the used resist have to be given. The EBL resist PMMA has the special property that it can be used as positive- and negative-tone resist [77]. A positive-tone EBL resist gets washed away upon development in all regions that were exposed with an electron dose larger than D_C (clearing dose). A negative-tone EBL resist becomes insoluble for doses larger than D_O (onset dose). Figure 6.9 illustrates both regimes by displaying a contrast curve for PMMA. The remaining resist thickness after development is given as a function of initially applied electron dose. If no dose is applied, the resist is almost untouched by the developer agent. When the clearing dose D_C is surpassed, the resist is completely removed. This is the positive tone regime. By introducing doses that are larger than D_O we enter the negative-tone regime and the resist is gradually becoming insoluble in the developer agent - a process which is dubbed as inverting the resist. Crucial for 3D nano-lithography is the resist's contrast in the negative tone regime. Generally, the contrast is defined as $C = -\frac{1}{\log(D_O/D_F)}$ with D_F being the dose for which the resist is fully inverted. A smaller contrast means a smaller slope in the regime of gradual inverting. It helps in the process of 3D

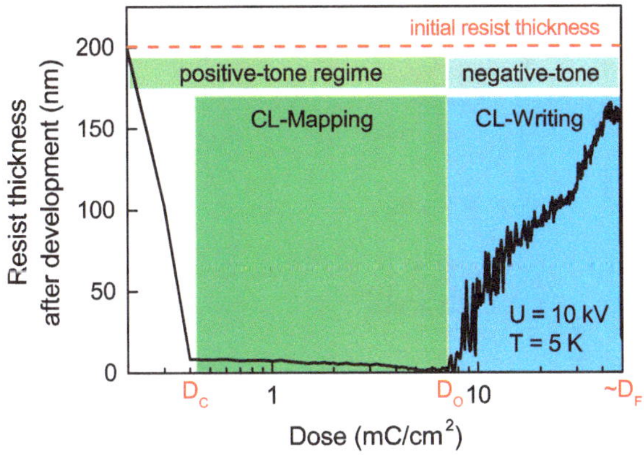

Fig. 6.9 Height profile of 190 nm PMMA after development as a function of applied electron dose. Between 0 and 7 mC·cm^{-2} we observe the characteristics of a positive tone resist. Then cross-linking starts and the remaining height is directly related to the applied dose. D_C: clearing dose, D_O: onset dose, D_F: dose for which the resist if fully inverted

lithography as a targeted remaining resist thickness (= local 3D profile) has a larger tolerance in the corresponding electron dose.

The QD samples are grown by metal-organic chemical vapor deposition (MOCVD) on GaAs(001) substrates. A low-density layer of self-organized InGaAs QDs is deposited above a distributed Bragg reflector (DBR) that consists of 23 alternating $\lambda/4$-thick bi-layers of AlGaAs/GaAs. On top of the QDs, a 400 nm thick GaAs capping layer provides the material for the subsequent microlens fabrication.

The process flow of CLL is as follows (cf. Fig. 6.10a–d): First, the sample is coated with PMMA and mounted onto the cryostat's coldfinger. When the SEM's chamber vacuum is restored, the sample is cooled down to a temperature of 5 K. Next, the spatial and spectral luminescence properties for a chosen sample area are collected. The electron beam scans the sample point-by-point on a regular grid and for each scanning point a full spectrum is taken (Fig. 6.10a). Hereby each scanning point consists of a dense set of sub-pixels to obtain a laterally homogeneous distribution of the electron dose. The electron dose applied along this mapping lies within D_C and D_O. Consequently, the resist in the mapped region is now soluble in a developer agent. Within this luminescence map one or more suitable QDs are chosen and the exact positions of the QDs are determined by a 2D Gaussian fit of the local luminescence distribution. Now the lens patterns are written into the resist at the positions of pre-selected QDs. To end up with a real 3D lens profile, the electron beam is scanned in circular patterns beginning at the center of the desired lens (Fig. 6.10b). The electron dose is determined by the local dwell time of the electron beam. Hence, the dwell time (the dose) decreases from the center of the lens to its outer region. The dwell times (doses) have to be precisely calculated to follow the resist-height to dose relation in the negative-tone regime that was determined from the contrast curve. When the CLL process is finished the sample is removed from

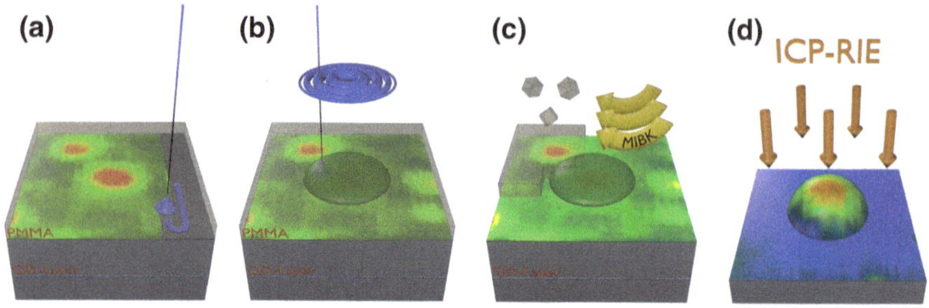

Fig. 6.10 Schematics of the lens fabrication process: **a** The sample's luminescence is mapped by cathodoluminescence spectroscopy. Along this, the resist is exposed in its positive-tone regime and becomes soluble upon development. **b** On *top* of suitable QDs, lens structures are written into the resist by cross-linking the afore cracked resist chains by applying an additional electron dose. The lens shape is defined by writing concentric circles into the resist and by carefully adjusting the respective radial electron doses. **c** Singly exposed resist is removed by applying a resist developer and the lens shape remains in the inverted regions. **d** Upon dry etching the lens profile is transferred from the inverted resist into the semiconductor. A possible lower mirror is omitted for the sake of clarity

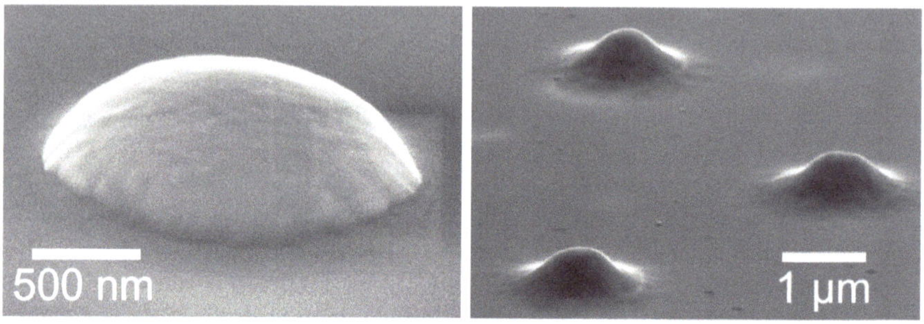

Fig. 6.11 SEM micrographs of fully processed microlenses. The *left* picture shows a hemispheric-section-type lens and on the *right hand side* three Gaussian-shaped lenses are displayed

the CLL setup and transferred into a cleanroom. Here the resist is developed by a standard developing agent, e.g., Methyl-isobutyl-ketone (MIBK) (Fig. 6.10c). For transferring the 3D resist profile into the semiconductor a dry etching process is used (Fig. 6.10d). Inductively-coupled-plasma reactive-ion-etching with a high vertical-to-lateral anisotropy yields good results. The anisotropy is necessary to not have lateral etching of the structures. This would spoil the envisaged lens structure.

Two examples of fully processed microlenses are shown in Fig. 6.11. The shape was chosen to be hemispheric-section-like (a) or Gaussian-like (b). The accuracy of the overall fabrication process was determined to be as good as 34 nm [79]. This includes the determination of the QD's position and the subsequent lithography. So it is very well suited for the precise and deterministic fabrication of nanophotonic structures.

6.6 Optical and Quantum Optical Properties of Deterministic Quantum Dot Microlenses

In this section we present a comprehensive analysis of the optical properties of deterministic QD microlenses fabricated via 3D in-situ electron-beam lithography (cf. Sect. 6.5). Processed microlenses are studied via micro-photoluminescence (μPL) spectroscopy with high spatial and spectral resolution. The sample is mounted to the coldfinger of a liquid Helium-flow cryostat at cryogenic temperatures (4–40 K) and excited optically by a continuous wave (CW) diode-pumped solid-state laser (660 nm) or a mode-locked Ti:Sapphire laser operating in picosecond mode at a repetition rate of 80 MHz (wavelength range: 750–1050 nm). Photoluminescence of single QDs is collected using a 20 × microscope objective (NA = 0.4). The collimated photoluminescence is focused onto the entrance slit of an optical-grating monochromator with attached charge-coupled device camera enabling an overall spectral resolution of 0.017 nm (25 μeV, 6 GHz). Polarization optics (linear polarizer and $\lambda/2$-waveplate) in front of the monochromator enable polarization resolved measurements. Attached to a second output of the monochromator, a fiber-coupled Hanbury-Brown and Twiss (HBT) setup allows for measurements of the second-order photon auto-correlation $g^{(2)}_{\mathrm{HBT}}(\tau)$. The HBT is constituted of a 50:50 multimode beam-splitter and two single-photon counting modules (SPCMs) with a timing resolution of 350 ps in combination with time-correlated single-photon counting electronics.

6.6.1 High Device Yield and High Photon Extraction Efficiency

One major advantage of a deterministic technology is the high yield of suitable devices. Figure 6.12 a shows a SEM image of a representative write-field containing 12 microlenses with a diameter of about 1.5 μm processed onto pre-selected QDs, at which 11 microlenses reveal optical activity of single QDs after fabrication. In fact, a statistical analysis of many write-fields reveals a yield larger than 90% for microlenses containing a QD, which highlights the reliability of our in-situ electron-beam lithography approach [26]. The corresponding chathodoluminescence spectra of each individual microlens, taken at a temperature of 10 K, are depicted in Fig. 6.12b. During lens-processing, QDs emitting at a wavelength around 930 nm have been chosen, which is reflected by the emission centered at an energy of 1.335 eV. Each spectrum shows various emission lines spread over a spectral range of ≈10 meV stemming from the radiative recombination of different charge carrier configurations in a single, integrated QD. The highlighted spectrum (blue line) originates from the microlens marked in Fig. 6.12a, which is chosen for further analysis. Figure 6.12c depicts a spectrum of this microlens under optical above-band ($\lambda = 660$ nm) excitation. Bright and spectrally narrow emission of the neutral exciton (X), biexciton (XX) and singly charged exciton (X^- and X^+) states

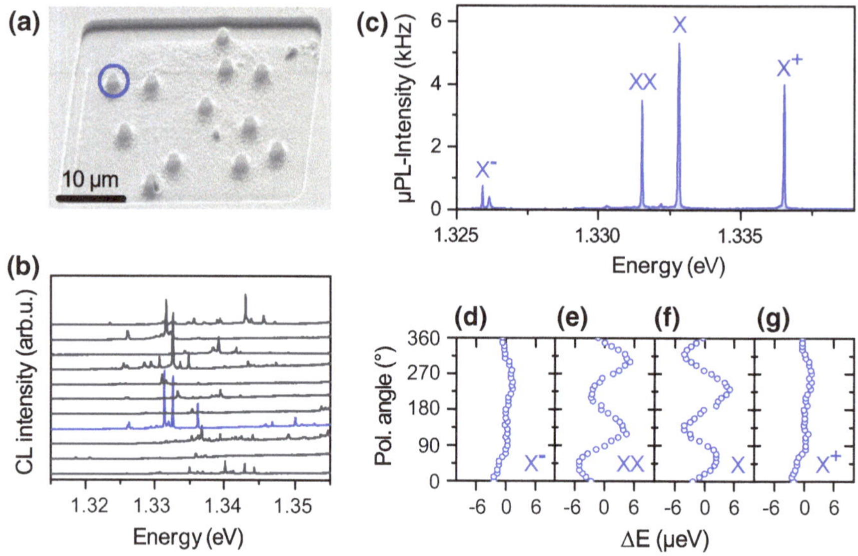

Fig. 6.12 **a** Scanning-electron microscopy image of a write-field processed with 12 QD-microlenses. **b** CL-spectra of the microlenses presented in panel (**a**). The spectrum associated with the marked microlens in panel (**a**) is highlighted in *blue*. **c** Zoom-in of the PL spectrum of the microlens highlighted in panel (**a**), showing different charge carrier configurations in the QD (view main text for details). **d–g** Polarization dependent studies of the four excitonic states indicated in panel (**c**). Evaluating the sinusoidal polarization dependence of X and XX reveals a fine-structure splitting of 7.7 μeV (color figure online)

are observed (see Chap. 1). The assignment to specific charge carrier configurations is thereby confirmed by polarization dependent measurements. The polarization-resolved energetic positions of each excitonic state are shown in Fig. 6.12d–g, which have been extracted by fitting Lorentzian profiles to the μPL spectra. This analysis reveals a counter-phased behavior of the spectral positions of X and XX emission, as expected for emission lines originating from exciton and biexciton of the same QD [80, 81]. Averaging the extracted XX-X splittings for all polarization angles yields a fine-structure splitting of the bright exciton state of $(7.7 \pm 0.1)\,\mu$eV and a binding energy of the biexciton state of (1.266 ± 0.005) meV, both representing typical values for this type of sample. Emission lines marked by X^+ and X^- are identified as the corresponding charged states due to their relative spectral positions, resulting from a smaller effective mass of electrons as compared to those of holes [82].

In order to quantify the photon extraction efficiency of deterministic microlenses, pulsed excitation at a pulse-repetition rate of $f = 80\,$MHz is utilized. The QD was excited at saturation of the X emission. The exciton emission collected via the microscope objective is spectrally filtered by the monochromator and detected with a fiber-coupled SPCM at the second exit of the monochromator. Under such conditions a total countrate of $\dot{n}_X = 120$ kHz is detected by the SPCM. The corresponding single-photon flux into the first lens of the setup can be deduced from a measurement of the setup's transmission, using a tunable diode laser tuned to the wavelength of the X emission line and focused onto a gold mirror inside the cryostat. The laser is attenuated using neutral-density filters in front of the monochromator to achieve count-rates at the SPCMs comparable to those observed for the QD emission. Taking into account the laser power, the reflectance of the gold mirror, the transmission of the cryostat window, the attenuation of the density filters, and the maximal count-rates on the SPCMs we determined a setup efficiency of $\eta_{\text{Setup}} = 0.8\%$. From the detected count-rate $\dot{n}_{\text{X}}$, the setup efficiency η_{Setup}, and the laser repetition rate f, we are are able to deduce a photon extraction efficiency of $\eta_{\text{X}} = \frac{\dot{n}_{\text{X}}}{\eta_{\text{Setup}} f} = (19 \pm 2)\%$ for the X emission. The total photon extraction efficiency of this microlens is finally evaluated, by taking into account the recombination channels of all excitonic states contributing to the emission, which yields broadband enhancement with $\eta_{\text{QD}} = (23 \pm 3)\%$. This value is in quantitative agreement with the simulations presented in Sect. 6.3, where a point light source and an idealized 3D geometry have been assumed for numerical modelling.

The demonstrated high yield of QD-microlenses containing single pre-selected QDs in combination with the achieved enhanced photon extraction efficiency constitutes an excellent basis for further investigations. In the following sections, we address the quantum optical properties of photons emitted by deterministic QD microlenses in terms of the single-photon purity as well as the photon-indistinguishability.

6.6.2 Single-Photon Emission

A typical micro-photoluminescence (μPL) spectrum of a deterministic QD microlens chosen for quantum optics experiments is depicted in Fig. 6.13a. Under pulsed above-bandgap excitation (dashed grey line) a diversity of emission lines stemming from different excitonic states (charge carrier configurations) of the same QD is observed. The charge-neutral exciton and biexciton states (X^0 and XX^0) as well as the singly charged trion states (X^+ and X^-), already discussed in Sect. 6.6.1 (cf. Fig. 6.12), are identified due to their polarization and power dependencies. At longer wavelength the emission of biexcitonic spin-triplet states XX_{T} is observed, whose origin was confirmed via polarization-resolved photon cross-correlation measurements [83]. In contrast to the spin-singlet configuration of the XX^0-state, the XX_{T}-state is constituted of one hole located in the *p*-shell and one hole in the *s*-shell of the QD. This configuration allows for a parallel orientation of both hole-spins - a phenomenon called *spin-blockade*, which is discussed in Chap. 4. Further excitonic states visible in the spectrum are most probably due to recombination of charged biexciton and triexciton states [84]. Switching to quasi-resonant excitation into the QD's *p*-shell at a wavelength of 909 nm (solid blue line), the emission is predominantly generated by the X^0 and XX^0 states while only relatively dim emission remains from the charged excitonic states. This is due to the fact that each laser photon absorbed by the QD's

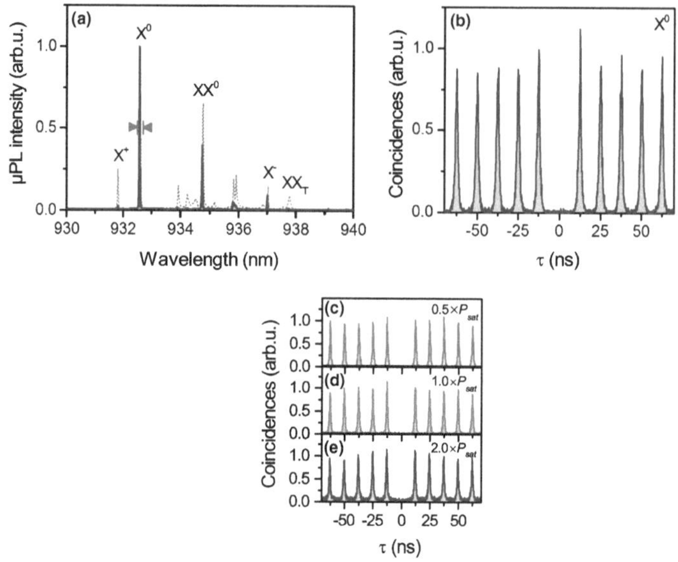

Fig. 6.13 **a** μPL spectrum of a QD microlens under above-band (*dashed grey line*) and *p*-shell (*solid blue line*) excitation. **b** Photon auto-correlation measurement on the neutral exciton state (X^0) from **(a)**. **c–e** Excitation dependent auto-correlation histogramms for excitation powers *P* of 0.5×, 1.0× and 2.0× the saturation power P_{sat} (color figure online)

p-shell generates one electron-hole pair with opposite spins. States which involve parallel spin-configurations hence have negligible occupation, while charged excitonic states require the capture of additional charge carriers from the surrounding semiconductor material. To investigate the photon statistics, the emission of the X^0-state can be spectrally selected via the monochromator (cf. markers in Fig. 6.13a) and coupled to the HBT setup. Figure 6.13b shows the corresponding raw measurement data of the second-order photon auto-correlation $g^{(2)}_{\text{HBT}}(\tau)$. Triggered single-photon emission of the deterministic QD microlens is reflected in the almost absent coincidences at zero time delay ($\tau = 0$). At finite τ, coincidence maxima occur at a periodicity corresponding to the pulse-repetition rate of 80 MHz. To achieve highest photon extraction efficiencies a quantum emitter has to be excited at (or close to) saturation power P_{sat}, e.g. the working point at which the pump rate of a given excitonic state equals its inverse radiative lifetime. Hence, we investigated the influence of the excitation power on the single-photon purity. Figure 6.13 displays the corresponding $g^{(2)}(\tau)$ histograms of the X^0-state for excitation powers P of $0.5 \times P_{\text{sat}}$, $1.0 \times P_{\text{sat}}$ and $2.0 \times P_{\text{sat}}$. We observe almost excitation-power-independent values of $g^{(2)}(0) \leq 0.01$ for up to two times the saturation power. This unique feature of the deterministic microlenses proves that our structures can act as bright quantum light sources with close-to-ideal quantum properties - even at maximum photon flux.

While such excitation-power-independence of $g^{(2)}(0)$ has for instance also been observed in photonic nanowires [55], we go a significant step further in the following section, by demonstrating that our structures not only show a close-to-ideal single-photon emission but at the same time also a high degree of photon indistinguishability. The latter is of crucial importance for advanced quantum communication concepts such as the quantum repeater, which relies on entanglement distribution via Bell-state measurements [85–87].

6.6.3 Generation of Indistinguishable Photons

Advanced quantum communication schemes, such as the quantum repeater, require a high degree of photon indistinguishability to enable entanglement distribution via Bell-state measurements. In an experiment, the photon-indistinguishability, e.g. the mean spatial and spectral photon wave-packet overlap, can be measured by using a Hong-Ou-Mandel (HOM) -type two-photon interference (TPI) setup [88] attached to the second output port of the monochromator (cf. illustration in Fig. 6.14a) (see also Chap. 3). Here, single photons enter an asymmetric Mach-Zehnder interferometer. Within the interferometer the photon stream is equally split into two paths, where one is delayed by the inverse excitation repetition rate of the Ti:sapphire laser. Hence, two consecutively emitted photons interfere at the second beam-splitter. A $\lambda/2$-waveplate allows for rotating the polarization of photons in one arm of the interferometer with respect to the other arm in order to make the photons distinguishable on purpose for reference measurements. In case of ideal indistinguishability, both photons coalesce at the second beam-splitter in a two-photon Fock state $|2\rangle$. This so called HOM-effect leads to a reduced number (ideally zero) of coincidences at zero time delay ($\tau = 0$) in the histogram $g^{(2)}_{\text{HOM}}(\tau)$ obtained from a coincidence measurement at the two interferometer outputs. In the described experiments, we used a Mach-Zehnder interferometer made out of polarization maintaining (PM) single-mode fibers, which, in combination with a variable fiber delay, assures a high spatio-temporal overlapp of two photons at the second beam-splitter.

To investigate the indistinguishability of photons emitted from deterministic QD microlenses, excitation-power-dependent measurements of $g^{(2)}_{\text{HOM}}(\tau)$ were performed on the X^+ emission line of a QD under quasi-resonant excitation. The resulting $g^{(2)}_{\text{HOM}}(\tau)$ histogram recorded at 50% of the saturation excitation power ($P_{\text{sat}} = 30\,\mu\text{W}$) is shown in Fig. 6.14b for co-polarized (red solid curve) and cross-polarized (grey dashed line) measurement configuration. In case of co-polarized photons quantum-mechanical TPI manifests in a strongly reduced number of coincidences at $\tau = 0$, if compared to the measurement in cross-polarized configuration. To quantitatively extract the visibility of TPI, Lorentzian profiles are fitted to the experimental data in co-polarized configuration and evaluated the relative peak areas according to [89]:

$$V = \frac{\bar{A}_S/2 - A_0}{\bar{A}_S/2} = 1 - \frac{A_0}{\bar{A}_S/2}, \tag{6.1}$$

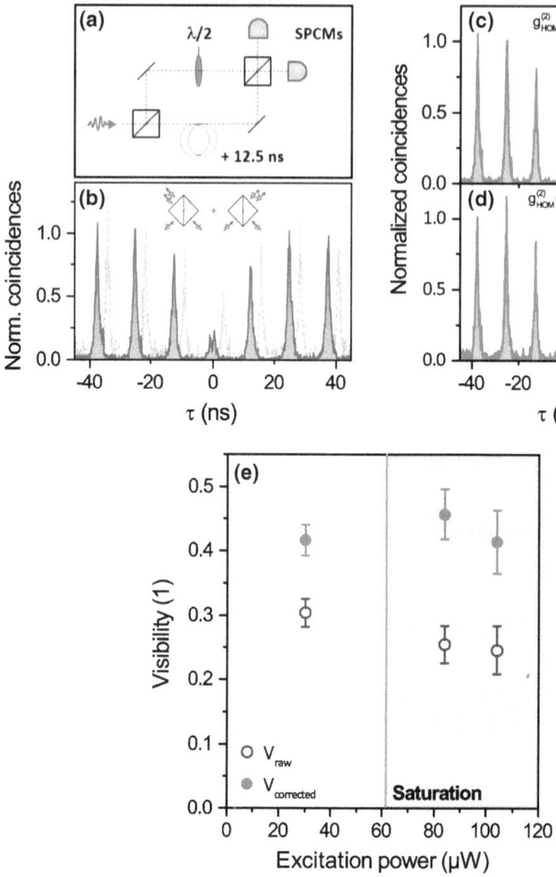

Fig. 6.14 **a** Schematic of a Hong-Ou-Mandel (HOM) -type two-photon interference (TPI) experiment. **b** TPI histogram $g^{(2)}_{\mathrm{HOM}}(\tau)$ measured with a HOM-type setup for co- (*red solid line*) and cross- (*grey dashed line*) polarization of the interfering photons. The cross-polarized data were shifted by +3 ns for clarity. **b–d** TPI histograms for co-polarized photons and increasing optical pump powers P of $0.5 \times P_{\mathrm{sat}}$, $1.0 \times P_{\mathrm{sat}}$ and $2.0 \times P_{\mathrm{sat}}$. **e** Visibility values plotted versus the excitation power P. Shaded in *orange* is the range for which the emission is saturated (color figure online)

where A_0 is the area of the peak at $\tau = 0$ and $\bar{A}_S$ corresponds to the mean value of the side peaks with $|\tau| > 12.5$ ns. A visibility of $V_{12.5\,\mathrm{ns}} = (53 \pm 8)\%$ is extracted at moderate excitation ($P = 0.5 \times P_{\mathrm{sat}}$).

The values of $g^{(2)}_{\mathrm{HOM}}(0)$ remain almost constant with increasing excitation power even beyond saturation. A behavior which was already observed for the bare $g^{(2)}(0)$ value (cf. Fig. 6.13). For a more precise analysis one takes into account the finite probability of multi-photon emission events and the imperfections of the HOM setup, $\mathcal{R}/\mathcal{T} = 0.51/0.49$ and $(1 - \epsilon) = 0.98$, where $\mathcal{R}$ and $\mathcal{T}$ denotes the reflection and transmission coefficients of the fiber-based beam-splitter and $(1 - \epsilon)$ is

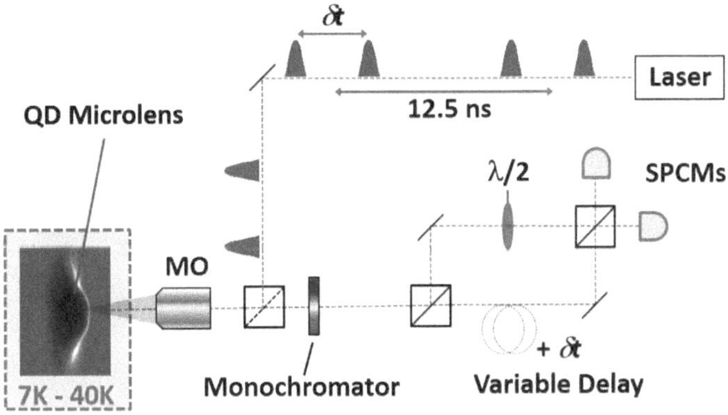

Fig. 6.15 Experimental setup: Hong-Ou-Mandel-type two-photon interference experiments are utilized to probe the indistinguishability of consecutively emitted photons with variable pulse-separation δt (color figure online)

the maximal interference fringe contrast obtained for the balanced interferometer, similar to [89]. A detailed analysis of the data is presented in Fig. 6.14e and reveals the TPI visibility. Well above saturation a corrected (raw) value of $g^{(2)}_{\mathrm{HOM}}(0) = 0.271 \pm 0.019\,(0.373 \pm 0.015)$ is extracted, revealing a corrected (raw) TPI visibility of $V_{\mathrm{HOM}} = (46 \pm 4)\%$ $((26 \pm 3)\%)$.

A possible explanation for the discrepancy to an ideal source of indistinguishable photons with $V = 1$ is the presence of spectral diffusion on a timescale comparable to the excitation pulse separation (12.5 ns) [89]. This might lead to a reduced spectral overlap at the HOM beam-splitter, thus limiting the TPI visibility. To test this interpretation, we conceived an experiment where the delay-time between excitation pulses can be varied, as presented in the following section.

6.6.4 Time- and Temperature Dependent Hong-Ou-Mandel Interferometry

In contrast to the single emitter character described on terms of $g^{(2)}_{\mathrm{HBT}}(0)$, the photon-indistinguishability is particular sensitive to dephasing processes. The dephasing rate of a quantum emitter is described by its coherence time T_2 and the radiative lifetime $T_1 = \Gamma^{-1}$ via $T_2^{-1} = (2T_1)^{-1} + (T_2^*)^{-1}$ [90], where $(T_2^*)^{-1} = \Gamma' + \gamma$ describes pure dephasing due to spectral diffusion (Γ') and phonon coupling (γ). In this section we gain experimental access to both types of pure dephasing independently by means of time- and temperature dependent TPI experiments.

In order to access the timescale of spectral diffusion, the HOM-setup of Fig. 6.14a can be modified by introducing a two-pulse sequence with variable pulse-separation δt as depicted in Fig. 6.15.

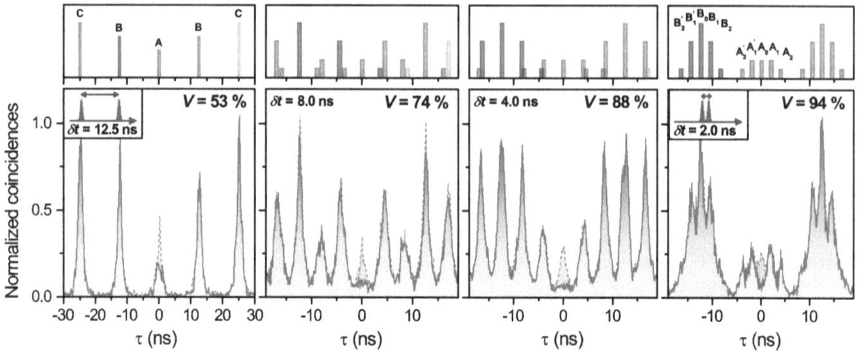

Fig. 6.16 Two-photon interference histograms measured for the X^0-state (cf. Fig. 6.13a) using a two-pulse excitation sequence with variable pulse-separation δt ($T = 7$ K). Experimental data corresponding to co- (cross-) polarized measurement configuration are displayed by *solid red* (*dashed grey*) curves. *Upper panels* show the expected coincidence pulse patterns (in case of distinguishable photons) of a two-pulse sequence with a pulse separation δt repeating every 12.5 ns. The peak area ratios are encoded in the the height of each bar (color figure online)

Here, the periodic excitation pulses of the mode-locked Ti:Sapphire laser (80 MHz repetition rate) are converted to a sequence of double-pulses by utilizing a fiber-based asymmetric Mach-Zehnder interferometer (not shown) in the excitation path. By choosing different fiber-delays within one arm of the interferometer, the pulse-separation δt can be varied from 2.0 to 12.5 ns. This two-pulse sequence is then used to excite the single-QD microlens. For the TPI experiment, a PM-fiber-based HOM setup as introduced in Sect. 6.6.3 is attached to the output port of the spectrometer. Within this HOM interferometer, the variable fiber delay has to be matched precisely (3 ps accuracy) to the respective pulse separation δt. This experimental configuration enables us to probe in the following the mean photon wave-packet overlap, in terms of the TPI visibility of two photons emitted by the QD, as a function of the time elapsed between consecutive emission events.

Figure 6.16 shows HOM-histograms obtained for the QD microlens presented in Fig. 6.13a, where the pulse-separation δt was gradually reduced from 12.5 to 2.0 ns. The respective delay inside the HOM-interferometer was precisely matched for each measurement to assure proper interference of consecutively emitted single photons.

At $\delta t = 12.5$ ns a moderate visibility of $V_{12.5\,\text{ns}} = (53 \pm 8)\%$ is extracted, in agreement with results from Fig. 6.14. For smaller δt complex coincidence-pulse-patterns specific to each δt are observed, which arise from the superposition of five-peak clusters repeating every 12.5 ns according to the pulse separation of the exciting laser [91]. The five-peak cluster in turn arises from the possible pathway-combinations inside the Mach-Zehnder interferometer taken by two photons separated by δt. Hence, the peak area ratios can easily be deduced considering combinatorics, which enables us to extract the TPI visibility quantitatively. The expected peak area ratios are illustrated in the upper panels of Fig. 6.16, where the peak area ratio is represented by the respective bar height. Each cluster yields the peak area ratios 1:4:6:4:1, except

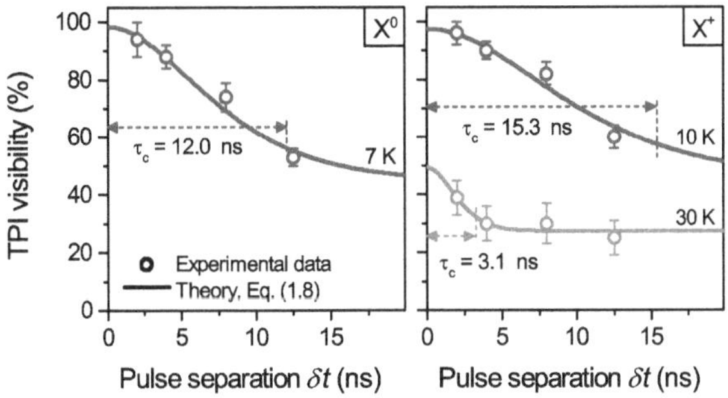

Fig. 6.17 Two-photon interference visibilities of consecutively emitted single photons versus the time δt elapsed between the emission processes. Experimental data for the X^0- (*left*) and the X^+ (*right*)-state are quantitatively described by a theoretical model assuming a non-Markovian noise correlation leading to spectral diffusion at a ns-timescale (see Eqn. 6.8). A characteristic temperature-dependent correlation time τ_c is observed (color figure online)

for the cluster centered at zero-delay ($\tau = 0$). Here, the peak area ratio depends on the photon-indistinguishability. In case of perfect indistinguishability, the coincidences at $\tau = 0$ vanish and the peak area ratios of the cluster become 1:2:0:2:1. Photons which are distinguishable, e.g. due to their polarization, lead to an area ratio of 1:2:2:2:1. In the following, the peak areas of the central clusters are labeled A_2':A_1':A_0:A_1:A_2 and $\bar{A} = (A_1' + A_1)/2$. The corresponding peak areas are extracted from the measurement data by fitting Lorentzian peaks with the expected area ratios to the coincidence histograms. In all fits, the width of the Lorentzian peaks has been fixed to the value obtained from the fit to the data at $\delta t = 12.5$ ns. The TPI visibility for $\delta t = 2$, 4 and 8 ns is then given by

$$V = \frac{\bar{A} - A_0}{\bar{A}} = 1 - \frac{A_0}{\bar{A}}\,. \tag{6.2}$$

In case of $\delta t = 4$ and 8 ns, peaks A_1 and A_1' are overlapping with the adjacent cluster. Hence, the visibility is expressed by

$$V = \frac{2\tilde{A}/3 - A_0}{2\tilde{A}/3} = 1 - \frac{A_0}{2\tilde{A}/3}\,, \tag{6.3}$$

with $\tilde{A}$ being the mean value of A_1 and A_1' and their related overlapping peaks. In case of $\delta t = 4$ ns, A_1 and A_1' overlap with the nearest neighbor cluster B_2 and B_2'. For $\delta t = 8$ ns, the overlapping peaks stem from C_2 and C_2' as seen in Fig. 6.16. To reduce the statistical error of $\bar{A}$ and $\tilde{A}$, instead of taking only A_1, A_1' and their overlapping peak areas into account, we finally averaged over the peak areas for all clusters at $\tau \neq 0$, to infer a more precise normalization of the data.

Figure 6.17 (left panel) summarizes the obtained raw TPI visibilities as a function of the pulse-separation δt for the neutral exciton X^0. At low δt a plateau-like behavior is observed, at which the visibility remains almost constant with values of $V_{2.0\text{ns}} = (94 \pm 6)\%$ and $V_{4.0\text{ns}} = (88 \pm 4)\%$. For pulse-separations larger than 4 ns, a distinct decrease in visibility is observed from $V_{8.0\text{ns}} = (74 \pm 5)\%$ to $V_{12.5\text{ns}} = (53 \pm 3)\%$. The significant decrease in TPI visibility at pulse-separations larger than 8.0 ns indicates the timescale of spectral diffusion. The time-dependent analysis of Γ' has additionally been carried out for the charged exciton state X^+ of the same QD at 10 and 30 K (cf. Fig. 6.17, right panel). Again a characteristic correlation time is observed, which decreases at higher temperature.

In order to gain deeper insight in the underlying dephasing mechanisms,the system is modeled with a Hamiltonian, including the QD, the classical excitation field and the quantized light field:

$$H/\hbar = \underbrace{(\omega_e + F(t))\,\sigma_{ee}}_{\text{Quantum Dot}} + \underbrace{\Omega(t)\left(e^{-i\omega_p t}\sigma_{eg} + e^{+i\omega_p t}\sigma_{ge}\right)}_{\text{Excitation Field}} + \underbrace{\int_0^\infty d\omega\left(\omega\, c_\omega^\dagger c_\omega + g_\omega\, c_\omega^\dagger \sigma_{ge} + g_\omega^* \sigma_{eg} c_\omega\right)}_{\text{Quantized Light Field}} . \tag{6.4}$$

The QD is approximated as a two-level system with the ground state $|g\rangle$ and excited state $|e\rangle$ with respective lowering and raising operators defined by $\sigma_{ij} := |i\rangle\langle j|$. The transition energy between the excited and ground state is denoted by ω_e, where the ground state energy is set to zero. Dephasing is considered by employing the working horse of the phenomenological dephasing description, by including a general stochastic force $F(t)$, which shifts the transition energy of the QD. This stochastic force $F(t)$ will include the dephasing due to spectral diffusion and phonon interaction, as detailed later on.

Given that the classical (pump) field excites the QD fast enough to prevent multiple photon emission processes, the wave function after the two-pulse sequence can be calculated via the Wigner–Weisskopf method:

$$|\Psi(t)\rangle = \int_0^t dt_1 \int_{\delta t}^t dt_2 e^{i(\omega_e + i\Gamma)(t_1+t_2) - i\phi_{\delta t}(t_2) - i\phi_0(t_1)} \times E_2(t_2)E_1(t_1)\,|\text{vac}\rangle \ . \tag{6.5}$$

This wave function includes the two-photon wave packages $E_n(t_n)$ and the time-integrated stochastic forces defined as $\phi_i^X(t) := \int_i^t \mathrm{d}t' X_i(t')$, where $i = 0$ in case the photon was emitted during the first sequence or $i = \delta t$ for photon emission processes due to the second pulse and $X(t)$ denoting the noise. Given this wave function, the observables of the experiment can be calculated.

Considering the interference at the beam-splitter by unitary transformations on the incident electric fields coming from the long and short pathway, the electric fields E_A and E_B at the detector A and B can be calculated. Further, by taking into account the two-photon wave-function and the steady-state limit ($t \longrightarrow \infty$), the unnormalized two-photon correlation $G^{(2)}(t,\tau) = \left\langle E_A^{(-)}(t)E_B^{(-)}(t+\tau)E_B^{(+)}(t+\tau)E_A^{(+)}(t)\right\rangle$

can be deduced (see [92], Supplemental Material for details). This correlation still includes the noise ϕ_i^X and is thus very general in terms of dephasing processes, which can be evaluated for Markovian- and non-Markovian noise correlations. In the following we specify the noise correlations, which is needed to calculated the visibility.

To evaluate the stochastic forces, one needs to average via a Gaussian random number distribution $\langle\langle\cdots\rangle\rangle$. Here, the $\langle\langle\cdot\rangle\rangle$ denotes statistical averaging in terms of a Gaussian random variable, where all higher moments can be expressed by second-order correlation [93]. To include dephasing, we employ the general stochastic force $F(t) = P(t) + D(t)$ with a phonon-induced dephasing (δ-correlated white noise) $P(t)$ and a spectral diffusion $D(t)$ component (colored noise), both shifting the transition energy of the QD. Assuming that the two different noise contributions are independent of each other, we can neglect correlations between $D(t)$ and $P(t)$ in the cumulant expansion. Additionally, we restrict our investigation to the zero-phonon line broadening mechanism [94]. A possible source for such a dephasing mechanism is the quadratic interaction with longitudinal acoustical phonons, which gives rise to a temperature-dependent broadening [95, 96]. Contributions from highly non-Markovian phonon-sidebands are neglected, which also effect the indistinguishability, e.g. in cQED setups [97, 98], and are typically described using the independent Boson model [99] or Feynman path integrals [100]. Here, we employ the simplest possible model for such a dephasing by assuming a Markovian process $\delta-$correlated in time, i.e. as white noise [93, 101]:

$$\left\langle\left\langle \phi_{t_1}^P(t_2)\phi_{t_3}^P(t_4)\right\rangle\right\rangle = \gamma\,(\min[t_2, t_4] - \max[t_1, t_3]) \ . \tag{6.6}$$

The phonon-induced dephasing is highly temperature-dependent and limits the absolute value of the indistinguishability, independent of the temporal distance of the excitation pulses δt.

In contrast to the phonon-induced dephasing, the spectral diffusion reveals a strong dependence on the pulse distance δt, as seen Fig. 6.17. We include this dependence as a finite memory-effect with specific correlation time τ_c:

$$\left\langle\left\langle \phi_{t_1}^D(t_2)\phi_{t_3}^D(t_4)\right\rangle\right\rangle = \Gamma_0'\, e^{-\frac{(t_1-t_3)^2}{\tau_c^2}}\,(\min[t_2, t_4] - \max[t_1, t_3]) \ , \tag{6.7}$$

where Γ_0' describes the maximal amount of pure dephasing induced by spectral diffusion. These kinds of noise correlations stem from a non-Markovian low-frequency noise [102–104] and show plateau-like behavior for temporal pulse distances sufficiently short in comparison to the memory depth (see Chap. 9). Thus, if $\delta t \ll \tau_c$, the effect of spectral diffusion becomes negligible and phonon-induced dephasing limits the absolute value of the visibility.

Using the above noise correlations, the time-integrated normalized two-photon correlation $\bar{g}^{(2)}$ can be calculated, which now depends on the specific dephasing parameters. The two-photon interference visibility V can then be expressed via $V = 1 - \bar{g}^{(2)}$. We are now able to explicitly formulate the dependence of the visibility on the pulse separation δt, the pure dephasing γ, and the diffusion constant Γ':

$$V(\delta t, \tau_c, T) = \frac{\Gamma}{\Gamma_0'(1 - e^{-(\delta t/\tau_c)^2}) + \gamma(T) + \Gamma}, \tag{6.8}$$

where a balanced beam-splitter ($\mathcal{T} = \mathcal{R} = 1/2$) was assumed. Thus, for vanishing phonon-induced dephasing and spectral diffusion, the visibility is 1, i.e. the photons are only Fourier-transform-limited and coalesce at the beam-splitter into a perfect coherent two-photon state. If the phonon-induced dephasing is stronger than other dephasing and relaxation processes $\gamma \gg \Gamma, \Gamma'$, the visibility becomes small, which is typically seen in the high temperature limit. At low temperatures, the phonon-induced dephasing is small and the spectral diffusion with a finite-memory depth dictates the functional form of the visibility for different pulse distances.

Applying the model derived in Eqn. 6.8 to the experimental data of Fig. 6.17, by fixing Γ (measured independently via time-resolved measurements) and assuming $\gamma_{7K,10K} = 0$ (cf. next paragraph), correlation times τ_c as listed in Table 6.1 are deduced. The timescale at which the noise is correlated appears to be close to the fundamental period of the Ti:Sapphire laser for X^0_{7K} and X^+_{10K}, whereas an increase in temperature to 30 K shortens the correlation time of X^+ drastically (cf. Table 6.1). Interestingly, the coherence times T_2^∞ inferred from our model in the limit $\delta t \to \infty$ (see Table 6.1), significantly exceed the values of $T_2 = (291 \pm 6)$ ps for X^0_{7K} and $T_2 = (167 \pm 3)$ ps for X^+_{30K} obtained via measurements using a Michelson-interferometer. A physical origin of the plateau-like behavior of $V(\delta t)$ and the associated non-Markovian decoherence processes are random flips of bistable fluctuators in the vicinity of the QD [103]. Possible candidates for such fluctuators in solid state devices are charge traps or structural dynamic defects [102]. Further evidence for the presence of charge fluctuations is given by the observation of trion states X^+ and X^- under quasi resonant excitation of the QD (cf. Fig. 6.13). To reduce the associated electric field noise, weak optical excitation above-bandgap [105] or a static electric field via gates [106] can be applied (see Chap. 9).

To justify the assumption $\gamma_{7K,10K} = 0$ and to investigate the influence of phonons on the photon-indistinguishability in more detail, we performed complementary temperature dependent TPI experiments. For this purpose, the emission of the trion state X^+ was selected under quasi-resonant excitation and coupled to the HOM-interferometer. The pulse-separation was fixed at $\delta t = 2.0$ ns, while the

Table 6.1 Correlation times τ_c obtained by fitting (6.8) to the experimental data of Fig. 6.17, fixing $\gamma_{7K,10K} = 0$ and Γ. T_2^∞ values have been calculated from the parameters Γ, Γ_0' and γ

	Γ (GHz)	Γ_0' (GHz)	γ (GHz)	τ_c (ns)	T_2^∞ (ps)
X^0_{7K}	0.85	1.02 ± 0.06	0	12.0 ± 1.9	692
X^+_{10K}	0.91	1.03 ± 0.04	0	15.3 ± 2.5	673
X^+_{30K}	0.96	1.55 ± 0.78	$0.29 \pm^{1.10}_{0.29}$	3.1 ± 1.9	431

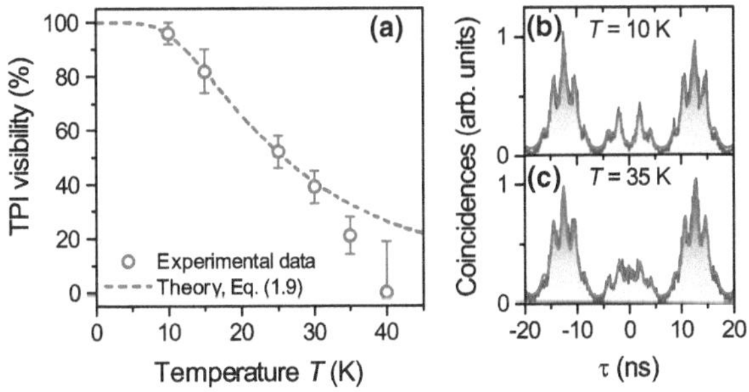

Fig. 6.18 Impact of the temperature on the two-photon interference (TPI) visibiliy ($\delta t = 2\,\text{ns}$). **a** Experimentally obtained TPI visibilities for various temperatures. Qualitative agreement is achieved using a theoretical model assuming dephasing proportional to the square of the phonon number (see Eqn. 6.9) **b** and **c** TPI histograms for co-polarized configuration at 10 and 35 K and corresponding fits (*red curve*) (color figure online)

temperature T was varied. Figure 6.18a depicts the obtained TPI visibilities extracted from the experimental data for temperatures ranging from 10 to 40 K. Exemplary TPI histograms are displayed in Fig. 6.18b, c for 10 and 35 K in co-polarized measurement configuration. A gradual increase in coincidences at $\tau = 0$ is observed, indicating a reduced photon-indistinguishability. At low temperature, close to ideal photon-indistinguishability with $V_{10\text{K}} = (96 \pm 4)\,\%$ is observed. Increasing T results in a distinct decrease of the TPI visibility. Finally, at a temperature of 40 K, V approaches zero within the standard error of our measurement. The observed temperature dependence is further modeled theoretically (dashed red curve).

For this purpose we employed a Markovian approximation for the phonon-induced pure dephasing processes, where the dephasing is proportional to the square of the phonon number [107]: $\gamma(T) = \gamma_0\, \bar{n}(T)\, [\bar{n}(T) + 1]$. Here, we have averaged over the frequency and approximated the expression via an effective phonon number depending on the temperature via the Bose–Einstein distribution for the effective phonon mode. We further employed the formula $\bar{n}(T) = \left[\exp\left[\frac{\alpha}{T}\right] - 1\right]^{-1}$, to underline the experimentally observed behavior qualitatively. To fit the experimental data in Fig. 6.18a, we adjust the parameters γ_0 and α. For the sake of clarity, the remaining dephasing contributions are normalized to one:

$$\begin{aligned} V(T) =& \frac{\Gamma}{\Gamma_0'(1 - \exp[-(\delta t/\tau_c)^2]) + \gamma(T) + \Gamma} \\ \approx& \frac{1}{1 + \gamma_0\, \bar{n}(\alpha, T)\, [\bar{n}(\alpha, T) + 1]} . \end{aligned} \tag{6.9}$$

According to this formula, the fit presented in Fig. 6.18a was performed with $\alpha = \hbar\bar{\omega}/k_B = 44\,\text{K}$ and $\gamma_0 = 3.75$. The model qualitatively describes the experimental observation. This leads to the conclusion that the impact of γ in Eqn. 6.8 is indeed almost negligible at low temperatures ($T \leq 10\,\text{K}$), but has severe impact at elevated

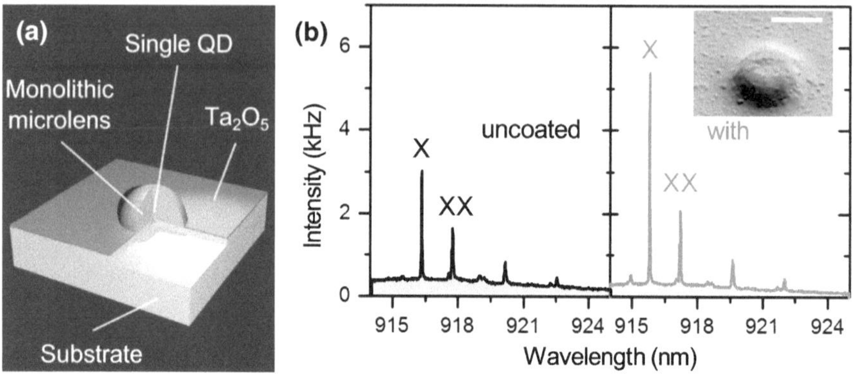

Fig. 6.19 **a** Schematic of a deterministic QD microlens coated with a thin layer of Ta_2O_5 ($n = 2.1$) acting as anti-reflection (AR) coating; **b** μPL spectra measured at saturation of X emission from a QD microlens before and after coating with Ta_2O_5. *Inset* Scanning electron microscopy (SEM) image of a hemisperic microlens coated with Ta_2O_5 (color figure online)

temperatures. For temperatures above 30 K, also in- and out-scattering with wetting layer carriers needs to be included, which explains the deviation between experiment and theory in this temperature range.

6.7 Conclusions and Future Perspectives

The achievements presented in this chapter summarize the advent of a novel and very promising deterministic technology platform based on three-dimensional in-situ electron-beam lithography. We demonstrated the fabrication of monolithic microlenses with pre-selected, integrated QDs. Such microlenses proofed their ability to generate single-photon states with high purity, high photon-indistinguishability combined with an efficient broadband photon extraction. Importantly, these quantum optical properties remain excellent even at highest photon flux, as uncorrelated background emitters are completely eliminated during lens etching. In particular, also the photon-indistinguishability, which is highly sensitive to any additional dephasing, showed no significant degradation even at pump powers beyond saturation. These features make our technology approach highly interesting to future applications in the field of optical quantum information technologies.

To further optimize the photon extraction efficiency of deterministic QD microlenses, an anti-reflection (AR) coating can be applied to the microlens surface (see Fig. 6.19a). Although the microlens geometry proved to largely increase the photon extraction, a fraction of the emitted photons is still lost by partial back-reflection due to the refractive index mismatch at the semiconductor-vacuum interface. In order to reduce the associated losses, a $\lambda/4$-layer of Ta_2O_5 was evaporated onto deterministic QD microlenses via electron beam sputtering. The material Ta_2O_5 fulfills the condition $n_{\text{coating}} \times d = \lambda/4$ with $d = 110\,\text{nm}$ being the thickness of the coating layer and $n_{\text{coating}} = 2.1$ its refractive index at $\lambda = 930\,\text{nm}$, which is close to the optimum optical impedance matching given by $n_{\text{coating}} = (n_{\text{vacuum}} \times n_{\text{GaAs}})^{1/2} = 1.88$ [108]. The impact of the AR-coating on the optical properties of QD-microlenses has been investigated in [109] by performing a statistical spectroscopic analysis on various microlenses, each with a single pre-selected QD. Microlenses were examined within a one-to-one comparison via μPL measurements at saturation pump power of the exciton emission before and after applying the AR-coating. The effect on the photon extraction efficiency can then be quantified by the ratio of the corresponding emission intensity after and before deposition of the Ta_2O_5-layer, respectively. This analysis revealed a mean value for the intensity ratio of 1.57 ± 0.71, in agreement with the value of 1.8 expected from numerical simulations of coated microlenses. This finding confirms, that the observed increase in detected μPL intensity of the QD is due to an effective reduction of reflections by the AR-coating, which enhances η_{ext}. Figure 6.19b exemplarily shows the emission spectrum of a single QD microlens before and after evaporation of the AR-coating onto its surface. Taking a maximum photon extraction efficiency of 29% as found for uncoated microlenses [110] as reference and considering the intensity ratio mentioned above, a photon extraction efficiency of up to 50% can be expected for AR-coated microlenses.

In light of future applications, electrically controlled devices are particular desirable. For instance, combining the efficient broadband photon extraction of microlenses with an electrical current injection in a QD light-emitting diode, holds the great potential to realize ultra-bright, electrically driven sources of polarization entangled photon pairs (see Chap. 7). Up until now, efficient photon-pair extraction and electrical current injection has not been realized in a single device approach [54, 111]. Furthermore, applying a fixed voltage bias to an electrically gated QD microlens, can stabilize the electric field noise in the vicinity of the QD, which could potentially lead to further improved photon-indistinguishabilities [112] (see Chap. 9). To realize an electrically operable microlens device on a scalable technology platform, a novel contacting scheme capable of electrically contacting a curved surface has already been developed. Figure 6.20a schematically illustrates the cross section of our electrically contacted microlens device.

The sample is based on a p-i-n-doped GaAs structure. The QD layer is located in the center of an intrinsic GaAs layer. Single QDs were integrated within microlenses using 3D electron-beam lithography. After lens processing the microlenses are partially planarized via hydrogen silsesquioxane (HSQ), while the sample surface outside the write-field is passivated using a layer of silicon nitride (SiN). Afterwards, a semitransparent metal contact (4 nm Titan, 8 nm Platinum) is evaporated onto the sample. During this step, the free-standing p-doped upper part of each microlens is electrically contacted and allows for efficient current injection. Finally, a 300 nm thick metal contact-pad (Ti/Pt/Au) is defined next to the write-field, which allows for electrically addressing the microlenses via a prober needle or Au wire-bonds. The micro-electroluminescence (μEL) spectrum of a readily processed microlens is shown in Fig. 6.20b at an injection current of 200 nA. Bright emission lines of a single QD can be observed. A current-voltage characteristic of the corresponding device

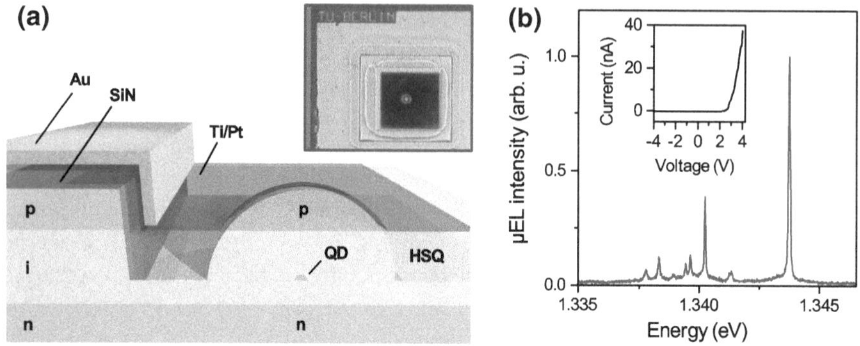

Fig. 6.20 **a** Illustration of an QD-microlens light-emitting diode (QD-µLens LED). The p-i-n-doped microlens is contacted via a planar semitransparent Ti/Pt layer deposited onto a planarization layer of hydrogen silsesquioxane (HSQ). *Inset* Optical microscope image of a microlens LED processed using 3D electron-beam lithography. **b** Electroluminescence spectrum of a QD-microlens at an injection current of 200 nA ($T = 20$ K). *Inset* I–V characteristics of an microlens LED (color figure online)

is depicted in the inset, and reveals the typical behavior of a diode structure. As a first proof-of-principle application, we operated such QD-µLens LEDs in an electro-optically gated mode, where the QD was simultaneously optically excited (using a CW laser) and electrically gated. This enabled triggered single photon emission at a modulation speed of up to 1 GHz [113].

In the near future, the above presented microlens-LED technology can be combined with deterministically fabricated QD microlenses. A step which will result in a powerful and versatile technology platform for advanced quantum information schemes such as the quantum repeater.

Acknowledgements Many people contributed to this work including S. Burger, A. Carmele, L. Gantz, D. Gershoni, M. Gschrey, C. Heine, L. Heindel, C. Hopfmann, A. Knorr, C. Scharfenorth, A. Schlehahn, E. Schmidgall, F. Schmidt, R. Schmidt, P. Schnauber, J.-H. Schulze, M. Seifried, A. Strittmatter, A. Thoma, B. Wohlfeil, U. Woggon, J. Wolters. We are grateful for their enthusiasm and skills.

The research leading to these results has received funding from the German Research Foundation via CRC 787, from the European Research Council under the European Union's Seventh Framework ERC Grant Agreement No. 615613, the German Federal Ministry of Education and Research (BMBF) through the VIP-project QSOURCE (Grant No. 03V0630), and from the German-Israeli-Foundation for Scientific Research and Development, Grant-No.: 1148-77.14/2011.

References

1. N. Sangouard, C. Simon, J. Minář, H. Zbinden, H. de Riedmatten, N. Gisin, Phys. Rev. A **76**, 050301 (2007)
2. N. Gisin, R. Thew, Nat. Photon. **1**(3), 165 (2007)
3. H.J. Kimble, Nature **453**(7198), 1023 (2008)

4. W.K. Wootters, W.H. Zurek, Nature **299**(5886), 802 (1982)
5. A. Einstein, B. Podolsky, N. Rosen, Phys. Rev. **47**, 777 (1935)
6. V. Scarani, H. Bechmann-Pasquinucci, N.J. Cerf, M. Dušek, N. Lütkenhaus, M. Peev, Rev. Mod. Phys. **81**, 1301 (2009)
7. C.H. Bennett, G. Brassard, in *Proceedings of IEEE International Conference on Computers, Systems and Signal Processing* (Bangalore, India, 1984), pp. 175–179
8. A.K. Ekert, Phys. Rev. Lett. **67**(6), 661 (1991)
9. Z.S. Yuan, Y.A. Chen, B. Zhao, S. Chen, J. Schmiedmayer, J.W. Pan, Nature **454**(7208), 1098 (2008)
10. H.J. Kimble, M. Dagenais, L. Mandel, Phys. Rev. Lett. **39**, 691 (1977)
11. P.G. Kwiat, K. Mattle, H. Weinfurter, A. Zeilinger, A.V. Sergienko, Y. Shih, Phys. Rev. Lett. **75**(24), 4337 (1995)
12. A. Gruber, A. Dräbenstedt, C. Tietz, L. Fleurya, J. Wrachtrup, C. von Borczyskowski, Science **276**(5321), 2012 (1997)
13. R. Brouri, A. Beveratos, J.P. Poizat, P. Grangier, Opt. Lett. **25**(17), 1294 (2000)
14. A.J. Shields, Nat. Photon. **1**(4), 215 (2007)
15. V. Giovannetti, S. Lloyd, L. Maccone, Science **306**(5700), 1330 (2004)
16. G.B. Lemos, V. Borish, G.D. Cole, S. Ramelow, R. Lapkiewicz, A. Zeilinger, Nature **512**(7515), 409 (2014)
17. J. Johansen, S. Stobbe, I.S. Nikolaev, T. Lund-Hansen, P.T. Kristensen, J.M. Hvam, W.L. Vos, P. Lodahl, Phys. Rev. B **77**(7) (2008)
18. H. Wang, Z.C. Duan, Y.H. Li, S. Chen, J.P. Li, Y.M. He, M.C. Chen, Y. He, X. Ding, C.Z. Peng et al., Phys. Rev. Lett. **116**(21) (2016)
19. X. Ding, Y. He, Z.C. Duan, N. Gregersen, M.C. Chen, S. Unsleber, S. Maier, C. Schneider, M. Kamp, S. Höfling, et al., Phys. Rev. Lett. **116**(2) (2016)
20. N. Akopian, N.H. Lindner, E. Poem, Y. Berlatzky, J. Avron, D. Gershoni, B.D. Gerardot, P.M. Petroff, Phys. Rev. Lett. **96**(13), 130501 (2006)
21. H. Jayakumar, A. Predojević, T. Kauten, T. Huber, G.S. Solomon, G. Weihs, Nat. Commun. **5** (2014)
22. O. Benson, C. Santori, M. Pelton, Y. Yamamoto, Phys. Rev. Lett. **84**(11), 2513 (2000)
23. W. Barnes, G. Björk, J. Gérard, P. Jonsson, J. Wasey, P. Worthing, V. Zwiller, Eur. Phys. J. D **18**(2), 197 (2002)
24. V. Zwiller, T. Aichele, O. Benson, New J. Phys. **6**, 96 (2004)
25. A. Dousse, L. Lanco, J. Suffczyński, E. Semenova, A. Miard, A. Lemaître, I. Sagnes, C. Roblin, J. Bloch, P. Senellart, Phys. Rev. Lett. **101**, 267404 (2008)
26. M. Gschrey, F. Gericke, A. Schüßler, R. Schmidt, J.H. Schulze, T. Heindel, S. Rodt, A. Strittmatter, S. Reitzenstein, Appl. Phys. Lett. **102**(25), 251113 (2013)
27. Z. Yuan, B.E. Kardynal, R.M. Stevenson, A.J. Shields, C.J. Lobo, K. Cooper, N.S. Beattie, D.A. Ritchie, M. Pepper, Science **295**(5552), 102 (2002)
28. A. Lochmann, E. Stock, O. Schulz, F. Hopfer, D. Bimberg, V. Haisler, A. Toropov, A. Bakarov, A. Kalagin, Electron. Lett. **42**(13), 774 (2006)
29. M. Reischle, G.J. Beirne, W.M. Schulz, M. Eichfelder, R.R. Bach, M. Jetter, P. Michler, Opt. Express **16**(17), 12771 (2008)
30. C. Böckler, S. Reitzenstein, C. Kistner, R. Debusmann, A. Löffler, T. Kida, S. Höfling, A. Forchel, L. Grenouillet, J. Claudon, J.M. Gérard, Appl. Phys. Lett. **92**(9), 091107 (2008)
31. I. Toft, X. Xu, D. Williams, R. Phillips, Microelectron. Eng. **83**, 1352 (2006)
32. X. Xu, I. Toft, R.T. Phillips, J. Mar, K. Hammura, D.A. Williams, Appl. Phys. Lett. **90**(6), 061103 (2007)
33. F. Haupt, S.S.R. Oemrawsingh, S.M. Thon, H. Kim, D. Kleckner, D. Ding, D.J. Suntrup, P.M. Petroff, D. Bouwmeester, Appl. Phys. Lett. **97**(13), 131113 (2010)
34. G. Bulgarini, M.E. Reimer, M. Bouwes Bavinck, K.D. Jns, D. Dalacu, P.J. Poole, E.P.A.M. Bakkers, V. Zwiller, Nano Lett. **14**(7), 4102 (2014)
35. C.M. Lee, H.J. Lim, C. Schneider, S. Maier, S. Höfling, M. Kamp, Y.H. Lee, Sci. Rep. **5**, 14309 (2015)

36. D. Cadeddu, J. Teissier, F.R. Braakman, N. Gregersen, P. Stepanov, J.M. Gérard, J. Claudon, R.J. Warburton, M. Poggio, M. Munsch, Appl. Phys. Lett. **108**(1), 011112 (2016)
37. H. Kumano, T. Harada, I. Suemune, H. Nakajima, T. Kuroda, T. Mano, K. Sakoda, S. Odashima, H. Sasakura, Appl. Phys. Express **9**(3), 032801 (2016)
38. A. Veprik, S. Riabzev, G. Vilenchik, N. Pundak, Cryogenics **45**(2), 117 (2005)
39. A. Schlehahn, L. Krüger, M. Gschrey, J.H. Schulze, S. Rodt, A. Strittmatter, T. Heindel, S. Reitzenstein, Rev. Sci. Instrum. **86**(1), 013113 (2015)
40. P. Michler, A. Kiraz, C. Becher, W.V. Schoenfeld, P.M. Petroff, L. Zhang, E. Hu, A. Imamoğlu, Science **290**(5500), 2282 (2000)
41. S.L. McCall, A.F.J. Levi, R.E. Slusher, S.J. Pearton, R.A. Logan, Appl. Phys. Lett. **60**(3), 289 (1992)
42. K. Srinivasan, O. Painter, Nature **450**(7171), 862 (2007)
43. O. Painter, R.K. Lee, A. Scherer, A. Yariv, J.D. O'Brien, P.D. Dapkus, I. Kim, Science **284**(5421), 1819 (1999)
44. S. Noda, M. Fujita, T. Asano, Nat. Photon. **1**(8), 449 (2007)
45. J.H. Kim, T. Cai, C.J.K. Richardson, R.P. Leavitt, E. Waks, Optica **3**(6), 577 (2016)
46. J.M. Gérard, D. Barrier, J.Y. Marzin, R. Kuszelewicz, L. Manin, E. Costard, V. Thierry-Mieg, T. Rivera, Appl. Phys. Lett. **69**(4), 449 (1996)
47. S. Reitzenstein, A. Forchel, J. Phys. D: Appl. Phys. **43**(3), 033001 (2010). (25pp)
48. J.P. Reithmaier, G. Sęk, A. Löffler, C. Hofmann, S. Kuhn, S. Reitzenstein, L.V. Keldysh, V.D. Kulakovskii, T.L. Reinecke, A. Forchel, Nature **432**(7014), 197 (2004)
49. M. Pelton, C. Santori, J. Vučković, B. Zhang, G.S. Solomon, J. Plant, Y. Yamamoto, Phys. Rev. Lett. **89**(23), 233602 (2002)
50. T. Heindel, C. Schneider, M. Lermer, S.H. Kwon, T. Braun, S. Reitzenstein, S. Höfling, M. Kamp, A. Forchel, Appl. Phys. Lett. **96**(1), 011107 (2010)
51. A. Schlehahn, A. Thoma, P. Munnelly, M. Kamp, S. Höfling, T. Heindel, C. Schneider, S. Reitzenstein, APL Photon. **1**(1), 011301 (2016)
52. T. Heindel, C.A. Kessler, M. Rau, C. Schneider, M. Fürst, F. Hargart, W.M. Schulz, M. Eichfelder, R. Robach, S. Nauerth, M. Lermer, H. Weier, M. Jetter, M. Kamp, S. Reitzenstein, S. Höfling, P. Michler, H. Weinfurter, A. Forchel, New J. Phys. **14**(8), 083001 (2012)
53. M. Rau, T. Heindel, S. Unsleber, T. Braun, J. Fischer, S. Frick, S. Nauerth, C. Schneider, G. Vest, S. Reitzenstein, M. Kamp, A. Forchel, S. Höfling, H. Weinfurter, New J. Phys. **16**(4), 043003 (2014)
54. A. Dousse, J. Suffczyński, A. Beveratos, O. Krebs, A. Lemaître, I. Sagnes, J. Bloch, P. Voisin, P. Senellart, Nature **466**, 217 (2010)
55. J. Claudon, J. Bleuse, N.S. Malik, M. Bazin, P. Jaffrennou, N. Gregersen, C. Sauvan, P. Lalanne, J.M. Grard, Nat. Photon. **4**, 174 (2010)
56. M.E. Reimer, G. Bulgarini, N. Akopian, M. Hocevar, M.B. Bavinck, M.A. Verheijen, E.P. Bakkers, L.P. Kouwenhoven, V. Zwiller, Nat. Commun. **3**, 737 (2012)
57. J. Heinrich, A. Huggenberger, T. Heindel, S. Reitzenstein, S. Höfling, L. Worschech, A. Forchel, Appl. Phys. Lett. **96**(21), 211117 (2010)
58. I. Friedler, C. Sauvan, J.P. Hugonin, P. Lalanne, J. Claudon, J.M. Gérard, Opt. Express **17**(4), 2095 (2009)
59. N. Gregersen, T.R. Nielsen, J. Mørk, J. Claudon, J.M. Gérard, Opt. Express **18**(20), 21204 (2010)
60. S.M. Mansfield, G.S. Kino, Appl. Phys. Lett. **57**(24), 2615 (1990)
61. A. Nick Vamivakas, Y. Zhao, C.Y. Lu, M. Atatüre, Nat. Phys. **5**(3), 198 (2009)
62. T. Schröder, F. Gdeke, M.J. Banholzer, O. Benson, New J. Phys. **13**(5), 055017 (2011)
63. Y.L.A. Rezus, S.G. Walt, R. Lettow, A. Renn, G. Zumofen, S. Götzinger, V. Sandoghdar, Phys. Rev. Lett. **108**(9) (2012)
64. F.W. Ostermayer, Appl. Phys. Lett. **43**(7), 642 (1983)
65. Y.S. Kim, J. Kim, J.S. Choe, Y.G. Rob, H. Jeon, J. Woo, I.E.E.E. Photon, Technol. Lett. **12**(5), 507 (2000)

66. A.V. Lavrinenko, J. Laegsgaard, N. Gregersen, F. Schmidt, T. Soendergaard, *Numerical Methods in Photonics* (CRC Press, Bocca Raton, 2014)
67. J. Pomplun, S. Burger, L. Zschiedrich, F. Schmidt, Phys. Stat. Sol. (b) **244**, 3419 (2007)
68. L. Zschiedrich, H.J. Greiner, S. Burger, F. Schmidt, Proc. SPIE **8641**, 86410B (2013)
69. M. Gschrey, A. Thoma, P. Schnauber, M. Seifried, R. Schmidt, B. Wohlfeil, L. Krüger, J.H. Schulze, T. Heindel, S. Burger et al., Nat. Commun. **6**, 7662 (2015)
70. R. Trotta, P. Atkinson, J.D. Plumhof, E. Zallo, R.O. Rezaev, S. Kumar, S. Baunack, J.R. Schröter, A. Rastelli, O.G. Schmidt, Adv. Mater. **24**(20), 2668 (2012)
71. G. Nogues, Q. Merotto, G. Bachelier, E. Hye Lee, J. Dong Song, Appl. Phys. Lett. **102**(23), 231112 (2013)
72. T. Kojima, K. Kojima, T. Asano, S. Noda, Appl. Phys. Lett. **102**(1), 011110 (2013)
73. M. Pfeiffer, K. Lindfors, H. Zhang, B. Fenk, F. Phillipp, P. Atkinson, A. Rastelli, O.G. Schmidt, H. Giessen, M. Lippitz, Nano Lett. **14**(1), 197 (2014)
74. L. Sapienza, M. Davanço, A. Badolato, K. Srinivasan, Nat. Commun. **6**, 7833 (2015)
75. F. Donatini, L.S. Dang, Nanotechnology **21**(37), 375303 (2010)
76. B.G. Yacobi, D.B. Holt, J. Appl. Phys. **59**(4), R1 (1986)
77. M. Gschrey, R. Schmidt, A. Kaganskiy, S. Rodt, S. Reitzenstein, J. Vac. Sci. Technol. B **32**(6), 061601 (2014)
78. P. Schnauber, R. Schmidt, A. Kaganskiy, T. Heuser, M. Gschrey, S. Rodt, S. Reitzenstein, Nanotechnology **27**(19), 195301 (2016)
79. M. Gschrey, R. Schmidt, J.H. Schulze, A. Strittmatter, S. Rodt, S. Reitzenstein, J. Vac. Sci. Technol. B **33**(2), 021603 (2015)
80. R.J. Young, R.M. Stevenson, A.J. Shields, P. Atkinson, K. Cooper, D.A. Ritchie, K.M. Groom, A.I. Tartakovskii, M.S. Skolnick, Phys. Rev. B **72**(11), 113305 (2005)
81. D. Sarkar, H.P. van der Meulen, J.M. Calleja, J.M. Becker, R.J. Haug, K. Pierz, J. Appl. Phys. **100**(2), 023109 (2006)
82. S. Rodt, A. Schliwa, K. Pötschke, F. Guffarth, D. Bimberg, Phys. Rev. B **71**, 155325 (2005)
83. Y. Kodriano, E. Poem, N.H. Lindner, C. Tradonsky, B.D. Gerardot, P.M. Petroff, J.E. Avron, D. Gershoni, Phys. Rev. B **82**(15) (2010)
84. Y. Benny, Y. Kodriano, E. Poem, D. Gershoni, T.A. Truong, P.M. Petroff, Phys. Rev. B **86**, 085306 (2012)
85. H.J. Briegel, W. Dür, J.I. Cirac, P. Zoller, Phys. Rev. Lett. **81**, 5932 (1998)
86. J. Hofmann, M. Krug, N. Ortegel, L. Gérard, M. Weber, W. Rosenfeld, H. Weinfurter, Science **337**(6090), 72 (2012)
87. Z.L. Yuan, A.R. Dixon, J.F. Dynes, A.W. Sharpe, A.J. Shields, Appl. Phys. Lett. **92**(20), 201104 (2008)
88. C.K. Hong, Z.Y. Ou, L. Mandel, Phys. Rev. Lett. **59**, 2044 (1987)
89. C. Santori, D. Fattal, J. Vučković, G.S. Solomon, Y. Yamamoto, Nature **419**(6907), 594 (2002)
90. J. Bylander, I. Robert-Philip, I. Abram, Eur. Phys. J. D - Atom. Mol. Opt. Plasma Phys. **22**(2), 295 (2003)
91. M. Müller, S. Bounouar, K.D. Jöns, M. Glässl, P. Michler, Nat. Photon **8**(3), 224 (2014)
92. A. Thoma, P. Schnauber, M. Gschrey, M. Seifried, J. Wolters, J.H. Schulze, A. Strittmatter, S. Rodt, A. Carmele, A. Knorr, T. Heindel, S. Reitzenstein, Phys. Rev. Lett. **116**(3) (2016)
93. C. Gardiner, P. Zoller, *Quantum Noise: A Handbook of Markovian and Non-Markovian Quantum Stochastic Methods with Applications to Quantum Optics*, Springer Series in Synergetics (Springer, Berlin, 2004)
94. P. Borri, W. Langbein, S. Schneider, U. Woggon, R.L. Sellin, D. Ouyang, D. Bimberg, Phys. Rev. Lett. **87**, 157401 (2001)
95. J. Förstner, C. Weber, J. Danckwerts, A. Knorr, Phys. Status Solidi (b) **238**(3), 419 (2003)
96. E.A. Muljarov, R. Zimmermann, Phys. Rev. Lett. **93**, 237401 (2004)
97. P. Kaer, N. Gregersen, J. (Mørk), New J. Phys. **15**(3), 035027 (2013)
98. P. Kaer, P. Lodahl, A.P. Jauho, J. (Mørk), Phys. Rev. B **87**, 081308 (2013)
99. E. Stock, M.R. Dachner, T. Warming, A. Schliwa, A. Lochmann, A. Hoffmann, A.I. Toropov, A.K. Bakarov, I.A. Derebezov, M. Richter, V.A. Haisler, A. Knorr, D. Bimberg, Phys. Rev. B **83**, 041304 (2011)

100. A. Vagov, M.D. Croitoru, M. Glässl, V.M. Axt, T. Kuhn, Phys. Rev. B **83**, 094303 (2011)
101. C. Santori, D. Fattal, K.M.C. Fu, P.E. Barclay, R.G. Beausoleil, New J. Phys. **11**(12), 123009 (2009)
102. Y.M. Galperin, B.L. Altshuler, J. Bergli, D.V. Shantsev, Phys. Rev. Lett. **96**(9), 097009 (2006)
103. B.D. Laikhtman, Phys. Rev. B **31**(1), 490 (1985)
104. J.H. Eberly, K. Wdkiewicz, B.W. Shore, Phys. Rev. A **30**(5), 2381 (1984)
105. O. Gazzano, S. Michaelis de Vasconcellos, C. Arnold, A. Nowak, E. Galopin, I. Sagnes, L. Lanco, A. Lemaître, P. Senellart, Nat. Commun. **4**, 1425 (2013)
106. M.J. Stanley, C. Matthiesen, J. Hansom, C. Le Gall, C.H.H. Schulte, E. Clarke, M. Atatüre, Phys. Rev. B **90**(19), 195305 (2014)
107. H. Carmichael, *Statistical Methods in Quantum Optics 1 - Master Equation and Fokker-Planck Equations* (Springer, Berlin, 1999)
108. H.K. Raut, V.A. Ganesh, A.S. Nair, S. Ramakrishna, Energy Environ. Sci. **4**, 3779 (2011)
109. P. Schnauber, A. Thoma, C.V. Heine, A. Schlehahn, L. Gantz, M. Gschrey, R. Schmidt, C. Hopfmann, B. Wohlfeil, J.H. Schulze, A. Strittmatter, T. Heindel, S. Rodt, U. Woggon, D. Gershoni, S. Reitzenstein, Technologies **4**(1), 1 (2016)
110. A. Schlehahn, M. Gaafar, M. Vaupel, M. Gschrey, P. Schnauber, J.H. Schulze, S. Rodt, A. Strittmatter, W. Stolz, A. Rahimi-Iman, T. Heindel, M. Koch, S. Reitzenstein, Appl. Phys. Lett. **107**(4), 041105 (2015)
111. C.L. Salter, R.M. Stevenson, I. Farrer, C.A. Nicoll, D.A. Ritchie, A.J. Shields, Nature **465**(7298), 594 (2010)
112. G. Moody, C. McDonald, A. Feldman, T. Harvey, R.P. Mirin, K.L. Silverman, Phys. Rev. Lett. **116**, 037402 (2016)
113. A. Schlehahn, R. Schmidt, C. Hopfmann, J.H. Schulze, A. Strittmatter, T. Heindel, L. Gantz, E.R. Schmidgall, D. Gershoni, S. Reitzenstein, Appl. Phys. Lett. **108**(2), 021104 (2016)

Part III
Biexcitons and Entangled Photon Emission

Chapter 7
Polarization Entangled Photons from Semiconductor Quantum Dots

Fei Ding and Oliver G. Schmidt

Abstract Semiconductor quantum dots are among the most promising candidates for the deterministic generation of polarization entangled photon pairs. In this chapter we review the most recent progress on this topic. First, we recall the basic concepts of polarization entangled photon qubits, the biexciton cascade process and exciton fine structure splitting in single quantum dots. The experimental techniques for controlling the fine structure splitting, which are critical for the generation of photon pairs with a high fidelity to the symmetric Bell state, are briefly discussed. A main focus is given to the recently developed anisotropic strain engineering technique, which has been used to fabricate an entangled light-emitting-diode with high yield and fast triggering speed, and an entangled photon source with wavelength tunability. Experimental progress on improving the collection efficiency of the entangled photon sources are also mentioned. We envision that, with the remarkable achievements in the field, the entanglement distribution, the hybrid interfacing with atoms, the telecom band emissions, and the on-chip integration will be realized soon with quantum dot based polarization entangled photon sources.

7.1 Introduction

The introduction of *lichtquanten* by Einstein, together with the heroic endeavours of Bohr, Heisenberg, Schrödinger and many others, gave birth to quantum mechanics in 1925 [1]. Though initially driven by curiosity, quantum mechanics has now fuelled many revolutionary applications in our society. A particularly important one is quantum information processing (QIP), which exploits the superposition states (qubits) in quantum mechanics to store and manipulate information. Stemmed from

F. Ding (✉) · O.G. Schmidt
Institute for Integrative Nanosciences, IFW Dresden, Helmholtzstraße 20, 01069 Dresden, Germany
e-mail: f.ding@fkp.uni-hannover.de

F. Ding
Institute for Solid State Physics, Leibniz University of Hannover, Appelstraße 2, 30167 Hannover, Germany

P. Michler (ed.), *Quantum Dots for Quantum Information Technologies*, Nano-Optics and Nanophotonics, DOI 10.1007/978-3-319-56378-7_7

the 1935 paper by Einstein, Podolsky and Rosen on the completeness of quantum mechanics [2], flourished in the 1980 s with the theoretical proposals on quantum computation (Feynman [3] and Deutsch [4]) and on quantum cryptography (Bennett and Brassard [5]), QIP has great potential for practical applications with recent breakthroughs in physical concepts, material sciences, and in electrical and photonic engineering techniques.

A number of important QIP protocols rely on the *spooky* behavior of quantum entanglement between spatially separated qubits. Prominent examples are quantum teleportation [5], quantum dense coding [6], quantum cryptography [7], cluster-state quantum computing [8], and quantum sensing, metrology and imaging [9]. We witnessed considerable theoretical and experimental progress in recent years, and entangled qubits have been demonstrated with superconducting circuits [10], trapped ions [11] and even macroscopic atomic ensemble [12] and diamonds [13]. Photon qubits are also attractive, since they travel fastest in nature, interact weakly with the environment and have long coherence times. In many QIP applications (e.g. quantum key distribution [7] and nondestructive quantum CNOT gate [14]) entangled photon pairs can be used as ideal flying qubits to distribute quantum information among individual nodes/systems.

The generation of entangled photon pairs is a prime theme in quantum information science, and several candidate systems can be employed for this purpose. The aim of this chapter is to discuss the the generation of polarization entangled photon pairs with self-assembled semiconductor quantum dots (QDs) [15]. Compared to the most commonly used spontaneous parametric down conversion (SPDC) sources, semiconductor QDs have a significant advantage of being able to emit entangled photons with sub-Poissonian statistics, an important requirement for large scale QIP applications. An excellent review on the QD growth was given by Rastelli, Kiravittaya and Schmidt in 2009, see [16]. The early researches on QD-based polarization entangled photon sources have been reviewed by Shields, Stevenson and Young in the same book [17], and since then much progress has been made by the community. Therefore we shall focus mainly on the most recent experimental results in the past few years. Being compatible with mature semiconductor technologies, QD-based entangled photon sources can be fabricated on chip [18] and triggered optically [19–27] with high brightness (up to 0.12 pair per excitation pulse)[25] and high indistinguishability (0.86 ± 0.03) [28]. Electrical triggering of QD entangled photon sources [29, 30] with repetition rates of up to several hundreds MHz [30] have also been demonstrated. Armed with these powerful techniques, semiconductor QDs have the potential to fulfill the wish-list of a perfect entangled photon source for many exciting QIP applications [31].

7.1.1 Photon Qubits

We begin by describing the basic concepts of single photon qubits. The term *photon* was firstly introduced in 1926 [32] and it represents a single quantum of light. A

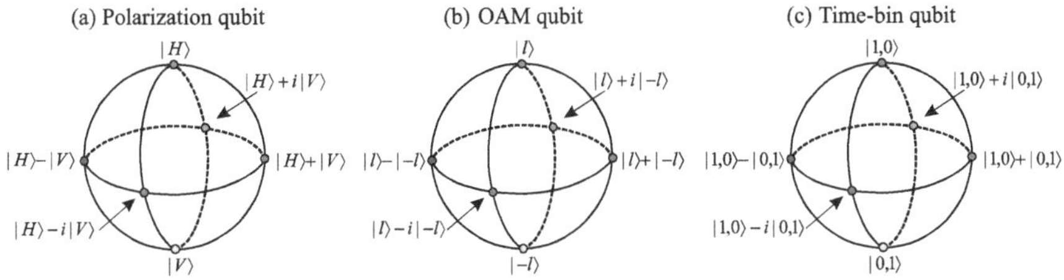

Fig. 7.1 Poincaré (Bloch) sphere representations of **a** polarization qubit, **b** orbital angular momentum qubit and **c** time-bin qubit

single photon can possess simultaneously many degrees of freedom, for example, energy (frequency), polarization, spin and orbital angular momentum, and arrival time. Therefore a coherent superposition of possible quantum states (qubit) can be encoded by using different degree of freedom of a single photon. Figure 7.1 shows the Poincaré sphere representations of three most commonly studied qubit encoding schemes.

Polarization Photon Qubits

Due to its wave-particle duality, a photon may be described as a travelling electromagnetic wave. The photon polarization describes the oscillation direction of the electric field. Therefore, a superposition state (qubit) can be represented by the polarization projection onto any orthonormal basis. For example, we denote the horizontal and vertical polarization states as $|H\rangle$ and $|V\rangle$, respectively. And then, a polarization qubit in a pure state can be written as:

$$|\psi\rangle = \alpha|H\rangle + \beta|V\rangle, with|\alpha|^2 + |\beta|^2 = 1 \tag{7.1}$$

Measuring $|\psi\rangle$ in the $\{|H\rangle, |V\rangle\}$ basis yields polarization H with probability $|\alpha|^2$, and polarization V with probability $|\beta|^2$. We can also choose the orthogonal basis $\{|R\rangle, |L\rangle\}$ or $\{|D\rangle, |A\rangle\}$, where R, L, D, A denote the right-hand circular ($|R\rangle = (|H\rangle - i|V\rangle)/\sqrt{2}$), left-hand circular ($|L\rangle = (|H\rangle + i|V\rangle)/\sqrt{2}$), diagonal ($|D\rangle = (|H\rangle + |V\rangle)/\sqrt{2}$), anti-diagonal ($|A\rangle = (|H\rangle - |V\rangle)/\sqrt{2}$) polarization states, respectively. Measuring the photon qubit in these bases yields different probabilities. Therefore, more generally, we can express any polarization state as a point on a Poincaré sphere, as shown in Fig. 7.1a. The two linear polarized states lie at the north and south poles, and the two circularly polarized states lie on the equator. The states lie in between them represent elliptical polarized states. Each pair of antipodal points correspond to orthogonal states. The points on the Poincaré surface correspond to the pure states, whereas the interior points correspond to the mixed states.

Orbital Angular Momentum Photon Qubits

The polarization encoding of photon qubits has been extensively studied in the past decade, mostly due to the ease of manipulation and detection of the spin angular momentum (SAM) of light. However the Hilbert space associated with the

SAM degree of freedom is restricted to two dimensions, which prevents the high-dimensional encoding of quantum states. Besides SAM, orbital angular momentum (OAM) degree of freedom of photons can be also used to encode quantum information [33]. OAM encoding, due to its almost unlimited dimensions, can enhance both the information carrying capacity and tolerance of noise in quantum communication channel.

The OAM is associated with the spatial distribution of light. The phase front of z-direction propagated light wave with well-defined OAM possess an azimuthal dependence of $e^{il\phi}$, where l represents the integer number of azimuthal phase rotations in one full cycle from 0 to 2π [33]:

$$\psi(r, \phi, z) = \psi_0(r, z)e^{il\phi} \tag{7.2}$$

where ψ_0 is the amplitude distribution, ϕ is the azimuthal angle. There are a number of ways of generating OAM states with arbitrary l. It is worth mentioning that a controlled quantum state transfer between polarization and OAM of a single photon have been demonstrated, allowing the generation and manipulation of hyper-entangled quantum states [34, 35]. An analogous Bloch sphere for OAM qubit is shown in Fig. 7.1b, where the north and south poles represent the states $|l\rangle$ and $|-l\rangle$.

Time-Bin Photon Qubits

The polarization and OAM encoded photon qubits are susceptible to environmental turbulence (for example, due to the birefringence effect in free space, fiber, and in on-chip waveguides). Therefore the qubit encoding by using the single photon arrival time was proposed [36]. A detailed introduction to time-bit qubit and entangled time-bin photon pairs has been given in the Chap. 8 by G. Weihs et al., and here we include only a brief discussion for completeness.

To generate a time-bin qubit, a single photon pulse is sent into a phase stabilized (the variation in path length difference is much less than the wavelength of light) and unbalanced Mach–Zehnder interferometer. The output of the interferometer consists of two temporal modes, due to the different length of the two interferometer arms. A single photon has a non-zero probability of taking either the short or the long arm, then a time-bin qubit can be defined as a superposed state of these two temporal modes. The quantum state in which a photon is in the first (second) temporal position can be expressed as $|1, 0\rangle$ ($|0, 1\rangle$), see Fig. 7.1c for the Bloch sphere representation.

$$|\psi\rangle = \alpha|1, 0\rangle + \beta|0, 1\rangle, with|\alpha|^2 + |\beta|^2 = 1 \tag{7.3}$$

While the time-bin encoding technique is very robust against environmental induced decoherence, it does not allow easy single qubit unitary operations and interaction between the different qubits. Therefore, various schemes have been proposed by converting a time-bin qubit to a polarization qubit for qubit operation purposes and then converting it back for qubit transferring purposes.

Experimental QIP with photon qubits has progressed rapidly in the past years. So far, most photonic QIP experiments (especially, those sophisticated ones) were performed with polarization qubits by using standard optical elements. In the following sections we focus mainly on the two-qubit entanglement based on the polarization degree of photons.

7.1.2 Entangled Photon Qubits

Schrödinger coined the term "entanglement" (in German, *Verschränkung*) in 1935. [37] Einstein later famously derided entanglement as "*spooky action at a distance*" (in German, *spukhafte Fernwirkung*). It refers to particles whose individual states cannot be written without reference to the state of other particles. Mathematically these states are said to be non-separable, and measurements of the state of one particle will change the measurement outcome of all other particles. Consider two systems A and B with respective Hilbert spaces H_A and H_B, and the combined state ρ_{AB} is called separable (not entangled) iff it can be written as a convex combination of pure product states:

$$\rho_{AB} = \sum_i \omega_i |A_i\rangle\langle A_i| \otimes |B_i\rangle\langle B_i| = \sum_i \omega_i \rho_A^i \otimes \rho_B^i, \sum_i \omega_i = 1 \tag{7.4}$$

where ω_i are positively valued probabilities, and $|A_i\rangle$, $|B_i\rangle$ are state vectors on the Hilbert spaces H_A and H_B, respectively. Per this definition, a separable state can be considered as a probability distribution over uncorrelated states.

But how do we define an entangled state? If ρ_{AB} is non-separable, then it is called an entangled state. To decide the separability of ρ_{AB}, Peres–Horodecki criterion was proposed [38, 39]. The partial transpose of ρ_{AB} with respect to the subsystem A is written as:

$$\rho_{AB}^{T_A} = \sum_i \omega_i (|A_i\rangle\langle A_i|)^T \otimes |B_i\rangle\langle B_i| = \sum_i \omega_i (\rho_A^i)^T \otimes \rho_B^i, \tag{7.5}$$

where $(\rho^A)^T = (\rho^A)^*$ are non-negative matrices with unit trace. The criterion states that if ρ_{AB} is separable, $\rho_{AB}^{T_A}$ has non-negative eigenvalues. In other words, if $\rho_{AB}^{T_A}$ has a negative eigenvalue, ρ_{AB} is said to be entangled. This is valid for 2×2 and 2×3 composite systems. As an example let us consider the two-photon state:

$$|\phi^+\rangle = \frac{1}{\sqrt{2}}(|HH\rangle_{AB} + |VV\rangle_{AB}) \in H_A \otimes H_B \tag{7.6}$$

where H, V denote the polarization. It can be shown that the density matrix:

$$\rho = \begin{bmatrix} \frac{1}{2} & 0 & 0 & \frac{1}{2} \\ 0 & 0 & 0 & 0 \\ 0 & 0 & 0 & 0 \\ \frac{1}{2} & 0 & 0 & \frac{1}{2} \end{bmatrix}, \rho^{T_A} = \begin{bmatrix} \frac{1}{2} & 0 & 0 & 0 \\ 0 & 0 & \frac{1}{2} & 0 \\ 0 & \frac{1}{2} & 0 & 0 \\ 0 & 0 & 0 & \frac{1}{2} \end{bmatrix} \tag{7.7}$$

where the partial transpose ρ^{T_A} gives a negative eigenvalue. Therefore $|\phi^+\rangle$ is said to be an entangled two-photon state. While Peres–Horodecki criterion is a strong measure for entanglement, Bell's inequalities [40] and its generalization, Clauser–Horne–Shimony–Holt (CHSH) inequalities [41] were the first quantitative criteria for determining the separability of ρ_{AB}. And there are also several other good measures of the degree of entanglement, such as concurrence, entropy of entanglement and tangle. A detailed discussion is beyond the scope of this chapter, and the reader is referred to the bibliography for more information.

Two independent photon qubits can be entangled in many degree of freedoms, for example, in polarization [42], spin angular momentum [43], orbital angular momentum [44] or in frequency [45] and photon arrival time (time bins) [36], as we briefly discussed above. The phenomenon of hyperentanglement, which is the simultaneous quantum entanglement in multiple and different degrees of freedom, is also being studied extensively. In this chapter we restrict our discussion to the polarization entangled photon pairs for the sake of simplicity. It is also worth mentioning that the quantum non-locality holds for particles entangled not only in space but also in time, as shown very recently by the polarization entanglement between two photons that never coexisted in time [46, 47].

In fact the state in (7.6) is a maximally entangled state. The polarization state measurement on system A in any basis will yield a completely random result of either H or V (with equal probability of $\frac{1}{2}$). But there is a perfect correlation between the measurements on systems A and B. If we measure H in one system then we will measure also H in the other system, and vice versa. There are four maximally entangled Bell states:

$$|\psi^+\rangle = \frac{1}{\sqrt{2}}(|HV\rangle_{AB} + |VH\rangle_{AB}) \tag{7.8}$$

$$|\psi^-\rangle = \frac{1}{\sqrt{2}}(|HV\rangle_{AB} - |VH\rangle_{AB}) \tag{7.9}$$

$$|\phi^+\rangle = \frac{1}{\sqrt{2}}(|HH\rangle_{AB} + |VV\rangle_{AB}) \tag{7.10}$$

$$|\phi^-\rangle = \frac{1}{\sqrt{2}}(|HH\rangle_{AB} - |VV\rangle_{AB}) \tag{7.11}$$

7.2 Semiconductor Quantum Dots Based Entangled Photon Sources

The first experiments to test Bell's inequality were performed by using the radiative cascade of single calcium atoms [48, 49]. The energy level scheme is shown in Fig. 7.2a. The excited state ($4p^2\ {}^1S_0$) and the final state ($4s^2\ {}^1S_0$) both have total angular momentum $J = 0$. The intermediate state ($4p4s\ {}^1P_1$) is a triply degenerated state with the orbital magnetic quantum numbers $m = \pm 1, 0$. The cascade process

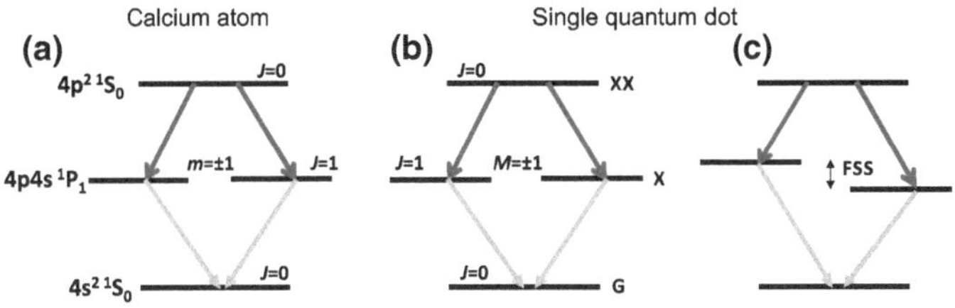

Fig. 7.2 Energy level scheme of **a** a single calcium atom, **b** an ideal semiconductor QD with zero FSS and **c** with finite FSS

occurs as $J = 0 \rightarrow J = 1 \rightarrow J = 0$ and produces a pair of photons at 551.3 nm and 422.7 nm, involving only the $m = \pm 1$ states. By performing polarization correlation experiments, Aspect et al. proved that the generated photon pairs are in the one of the maximally entangled Bell states [49]:

$$|\phi^+\rangle = \frac{1}{\sqrt{2}}(|L_{551.3}L_{422.7}\rangle + |R_{551.3}R_{422.7}\rangle) \equiv \frac{1}{\sqrt{2}}(|LL\rangle + |RR\rangle), \qquad (7.12)$$

However, demonstrating scalable QIP applications with single atoms is seemingly a technology challenge. In 1970 Burnham and Weinberg demonstrated photon pairs emission based on the nonlinear effect of spontaneous parametric down conversion (SPDC) [50], and then in 1988 Shih and Alley demonstrated that the photon pairs generated from SPDC are entangled in their polarization states and are able to violate Bell's inequality [51], which opened the door for various polarization-entanglement based QIP experiments. SPDC has served as the main workhorse for generating polarization entangled photons in the past decade. However the generated photons are characterized by Poissonian statistics, i.e. one usually does not know when an entangled photon pair is emitted. This fundamentally limits their applications in complex quantum protocols.

The intrinsic limitations of SPDC call for next generation entangled photon sources. Single self-assembled QDs based on III-V semiconductor materials (e.g. InAs QDs embedded in GaAs matrix, GaAs QDs embedded in $Al_xGa_{1-x}As$ matrix) are often referred to as *artificial atoms*, due to the fact that they possess discrete energy levels. They are among the leading candidates for the deterministic generation of single photons [52] and polarization entangled photons [15]. As proposed by Benson et al. single QDs can generate polarization entangled photon pairs via its biexciton (XX) cascade decay through the intermediate exciton states X. The biexciton state (with a total angular momentum $J = 0$) consists of two electrons (with $S_{h,z} = \pm\frac{1}{2}$) and two holes (with $J_{h,z} = \pm\frac{3}{2}$) and can be written as: $|\uparrow\downarrow\Uparrow\Downarrow\rangle$. The exciton state, on the other hand, consists of one electron and one hole, and there are four possible configurations characterized by their angular momentum projections $M = S_{e,z} + J_{h,z}$. The states with $|M| = 2$ are the so called dark excitons, and cannot couple with the light field. The states with $|M| = 1$ are the bright excitons (with a total angular momentum $J = 1$) and can be written as: $|\uparrow\Downarrow\rangle$ or $|\downarrow\Uparrow\rangle$. Similar to the cascaded emission from single calcium atoms, the emission from biexciton to the ground state via the intermediate exciton states also produce polarization entangled photon pairs (Fig. 7.2b):

$$|\psi^+\rangle = \frac{1}{\sqrt{2}}(|L_{XX}R_X\rangle + |R_{XX}L_X\rangle) \equiv \frac{1}{\sqrt{2}}(|LR\rangle + |RL\rangle), \quad (7.13)$$

or, in the linear and diagonal basis:

$$\equiv \frac{1}{\sqrt{2}}(|HH\rangle + |VV\rangle) \equiv \frac{1}{\sqrt{2}}(|DD\rangle + |AA\rangle) \quad (7.14)$$

7.2.1 Biexciton Cascade and Fine Structure Splitting

In real III-V semiconductor QDs, however, the anisotropy in strain, composition and shape reduces the QD symmetry to C_{2v} (e.g., in a pure lens shaped InAs/GaAs QD with circular or elliptical base elongated along the [110] or $[1\bar{1}0]$ directions [53]) or the even lower C_1 (in general alloyed InGaAs QDs) [54]. The anisotropy mixes the circularly polarized bright exciton states $|M| = 1$, and the new states are:

$$\frac{1}{\sqrt{2}}(|+1\rangle \pm |-1\rangle) \equiv \frac{1}{\sqrt{2}}(|\uparrow\Downarrow\rangle \pm |\downarrow\Uparrow\rangle) \quad (7.15)$$

The photons emitted from the radiative recombination of these states are linearly polarized along the in-plane anisotropy axes, and have different energies due to exchange interaction [54]. The energetic splitting between these two states is called fine structure splitting (FSS), see Fig. 7.2c. More details on the origin of FSS can be found in the seminar work by Bayer et al. [54].

Although the classical polarization correlation still holds true for the linearly polarized photon emitted from the cascaded process in Fig. 7.2c, the transformation into entangled states in the diagonal and circular bases are not possible due to the existence of FSS. The two-photon state is therefore given by [17, 55]:

$$|\psi\rangle = \frac{1}{\sqrt{2}}(|HH\rangle + e^{i\tau s/\hbar}|VV\rangle) \quad (7.16)$$

where τ is the radiative lifetime of the exciton and s is the FSS.

In time-integrated polarization correlation experiments, high fidelity to the maximally entangled Bell state $|\phi^+\rangle = \frac{1}{\sqrt{2}}(|H_{XX}H_X\rangle + |V_{XX}V_X\rangle)$ can be observed only with a vanishing FSS (typically, smaller than the radiative linewidth of $\hbar/\tau$, which is about $\sim 1\mu$eV in InGaAs/GaAs QDs). This was firstly demonstrated by the Cambridge group [19] and the Technion group [20] in 2006. The former work used in-plane magnetic fields to eliminate the FSS, while the latter used a narrow bandpass filter to select photons with energy differences below $\hbar/\tau$. For related works before 2009, the readers may refer to the previous review by Shields, Stevenson and Young [17]. In time-resolved experiments, Stevenson et al. demonstrated that a time-dependent oscillation of the entangled states can be observed in a single QD with finite FSS [55], and the technique was then used to reveal the time-evolving Bell state in telecom QDs [56].

To obtain entangled photons without discarding most of the photons, it is therefore necessary to have near zero FSS. However, the probability of finding such QDs in an as-grown sample is $<10^{-2}$. This is a great challenge for fabricating a practical entangled photon source based on single QDs. A good example is that, since the first demonstration of a QD-based entangled light-emitting-diode (LED) in 2010 [29], no second group was able to realize a similar device due to the very low probability of finding QDs with zero FSS.

7.2.2 Manipulation of Fine Structure Splitting

In the past years we witnessed considerable progress in the field, and the elimination of FSS can be achieved by applying rapid thermal annealing [25, 57–60], magnetic fields [19, 61], optical Stark effect [26],vertical [23, 62, 63] and lateral electric fields [64, 65]. Here we review the recent works based on two new strategies developed since 2009. The first is to use novel growth methods to obtain single QDs with ultrahigh symmetry and therefore intrinsically small FSS. And the second is to use anisotropic strain fields (and their combinations with other fields) to control the FSS precisely in each single QDs [18, 27, 30, 66–68].

Vanishing FSS in Highly Symmetric QDs

The degeneracy of the bright excitonic states is lifted mainly by the anisotropic in-plane electron-hole exchange interaction. Therefore it is intuitive to to recover the symmetry in single QDs, in order to eliminate the FSS. Although various post-growth tuning methods can be used for the tuning of FSS in each individual QDs, it is desirable to develop novel growth methods to create ensembles of QDs with sufficiently small FSS.

In 2009 there were two theoretical works suggesting that In(Ga)As/GaAs QDs grown on GaAs(111) surfaces, instead of the conventional GaAs(100) surfaces, could have intrinsically small FSS and can be used as ideal sources for entangled photons [70, 71]. The main idea of using GaAs(111) surfaces is to avoid the C_{2v} symmetry of the atomic alignment on GaAs(100) surfaces. Both (111)A and (111)B surfaces of GaAs have a higher C_{3v} symmetry [72, 73] and there will be no preferential growth axis for the QDs.

Unfortunately, it is known that the self-assembled QD growth via Stranski–Krastanov method is prohibited on (111) surfaces. In recent years, several different growth methods were developed to solve this problem. One solution is to grow InGaAs/GaAs QDs in inverted tetrahedral recesses on GaAs(111)B surfaces [22, 74, 75] by using metalorganic vapour phase epitaxy machines, see Fig. 7.3a. The second

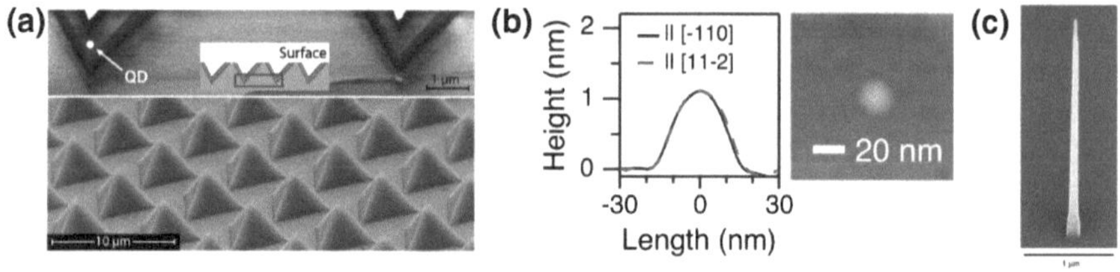

Fig. 7.3 Highly symmetric semiconductor QDs grown on (111) surfaces. **a** Pyramidal $In_{0.25}Ga_{0.75}As_{1-\delta}N_{\delta}$ QDs formed in the tetrahedral recesses on GaAs(111)B. The *upper figures* shows the AFM image of a cleaved QD sample, and the *lower figures* shows a HRSEM image of the sample after the substrate removal procedure. Figures adapted from the work by Juska et al. [22]. **b** AFM linescan and image of a single GaAs QD grown by droplet epitaxy. Figures adapted from the work by Kuroda et al. [24]. **c** A single QD-containing InP nanowire grown on InP (111)B substrate. Figure adapted from the work by Versteegh et al. [69]

solution is to use droplet epitaxy technique with standard molecular beam epitaxy machines. For example, in the work by Kuroda et al. [24], a layer of $Al_{0.3}Ga_{0.7}As$ was deposited on the Ga-rich surface of a GaAs (111)A substrate to form Ga droplets. Then As_4 was then supplied to crystallize the droplets into GaAs QDs, see see Fig. 7.3b. From the AFM linescan, it is shown that a single QD exhibits no lateral elongation. With a similar technique InGaAs QDs were grown on (111)B GaAs surfaces and a mean FSS of $5.6 \pm 0.6\,\mu eV$ averaging over 22 quantum dots was reported [76]. Also, telecom wavelength QDs with vanishing FSS were reported [77]. The third solution employs a chemical beam epitaxy machine to grow InP nanowires containing single InAsP QDs on SiO_2 patterned InP (111)B substrates, see Fig. 7.3c. Polarization entangled photon emissions have been demonstrated with all three systems. In the work by Juska et al., up to 15% QDs show entangled photon emissions. In the work by Kuroda et al., about 5% QDs show no detectable FSS and therefore entangled photon emissions.

The situation on GaAs (100) surfaces is more difficult, due to the different surface adatom mobilities along [110] and [$1\bar{1}0$] directions. Nevertheless, highly symmetric, strain-free GaAs/AlGaAs QDs on GaAs (100) substrates have been successfully demonstrated in a recent work [78] by infilling GaAs into AlGaAs droplet-etched nanoholes [79, 80]. The samples were grown by molecular beam epitaxy, see Fig. 7.4a. A thin $Al_xGa_{1-x}As$ layer was grown on the GaAs substrate, and Al droplets were formed by depositing 0.5 monolayer (ML) excess Al at a growth rate of 0.05 ML/s. Low-density nanoholes were etched into the surface by dissoluting the Al droplets. QDs were formed by diffusing 2 nm GaAs into the nanohole template and by capping the filled holes with AlGaAs as top barrier [78]. Because of the weak intermixing between GaAs and AlGaAs, the GaAs QDs assume the shape of the predefined holes. Figure 7.4b, c show the representative AFM image and linescans of a single nanohole prior to GaAs infilling. The circular symmetry of these QDs can be clearly identified. Micro-photoluminescence was then performed to investigate the FSS of neutral excitons. The linear polarization of the emitted light was analyzed by a rotatable achromatic $\lambda/2$ retarder followed by a fixed polarizer in the

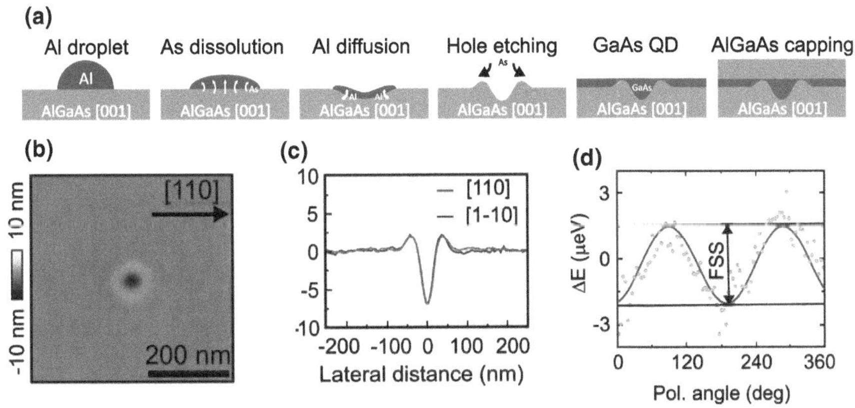

Fig. 7.4 Highly symmetric semiconductor QDs grown on (100) surfaces. **a** Schematic representaion of the growth procedures. **b** Representative AFM image and **c** corresponding linescans along [110] and [1$\bar{1}$0] crystal directions of a single nanohole prior to GaAs infilling. **d** Neutral exciton emission of a single QD as a function of polarization detection angles. FSS is obtained from the amplitude of a sine fit. **b–d** are adapted from the work by Huo et al. [78]

optical path. The sample was aligned in such a way that the polarization angle of 0° corresponds to the [110] crystal direction within an error of about 5° [78]. The FSS is then defined as the absolute value of the energy difference between the two bright excitonic extremes, see Fig. 7.4d. A mean FSS of 3.9 ± 1.8 μeV with random polarization angles was revealed by measuring 20 QDs. Since the radiative recombination lifetime of these QDs are around 0.2 ns, the low FSS is comparable to the Fourier limited excitonic linewidths. Therefore, these QDs are promising candidates for entangled photon sources. It is also worth mentioning that the Fourier limited linewidth can be achieved in this type of QDs with both resonant and non-resonant excitations [81], which may lead to the emissions of indistinguishable entangle photon pairs without the need of complex two photon pumping schemes [28]. With further improvements in growth techniques, it might be possible to finally achieve ensembles of high quality entangled photon emitters on single substrates.

Eliminating FSS with Giant Quantum Stark Effect

Before presenting the relative works on strain tuning of FSS, we first introduce the work on tuning FSS with vertical electric fields. It is a good starting point for understanding the effects of strain fields on FSS, since Gong et al. pointed out that applying an electric field along the [001] direction has the same symmetry as applying in-plane stress along the [110] and [1$\bar{1}$0] directions [82].

Bennett et al. designed a QD heterostructure (Fig. 7.5a) that allows very larger electric fields to be applied before the tunneling of carriers out of QDs [62]. At large FSS the two exciton eigenstates can be considered as radiating dipoles aligned along the [110] and [1$\bar{1}$0] directions, and their energies follow the quantum confined Stark shift $E = E_0 + pF + \beta F^2$, where p is the permanent dipole moment and β

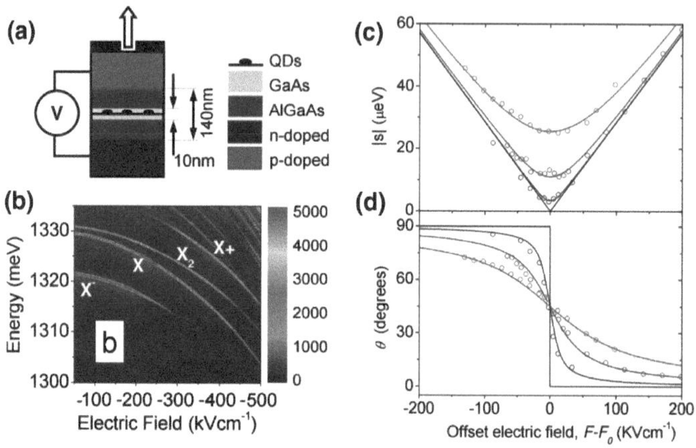

Fig. 7.5 **a** Device design for giant Stark shift of exciton emissions. **b** Photoluminescence versus electric field for a single QD. **c** FSS (denoted as $|s|$) and polarization angle θ as a function of the applied electric field for three QDs. The *black lines* indicate the ideal behaviors of a QD with vanishing FSS during tuning. Figures adapted from the work by Bennett et al. [62]

the polarizability and F the applied field. Since the polarizability does not depend on the in-plane anisotropy of the QD, it does not contribute to the *relative* change of the two excitonic energies. Due to the different confinement potentials in the two directions, the two excitonic eigenstates shift differently with applied electric field and FSS is given by $|s| \propto (p_H - p_V)F$ [62, 67]. Therefore the FSS decreases monotonically with applied vertical electric field before the two excitonic eigenstates reach a hybridization.

At a certain field the FSS reaches a minimum value, which indicates an anti-crossing in the energies of the two excitonic states, see Fig. 7.5c. The polarization angle θ, which is defined as the angle between the orientation of the eigenstates relative to the crystal axis [110], also rotates during the tuning of FSS. In Fig. 7.5c, d it is interesting to see that, the QD exhibiting a sharper rotation of θ has a smaller minimum FSS value at the anti-crossing point. With a simple model it was predicted that the polarization angle θ of a QD with vanishing minimum FSS at the anti-crossing point will show an abrupt change of exactly 90° during the tuning.

7.2.2.1 Eliminating FSS with Anisotropic Strain Fields

The behaviors of anti-crossing in the exciton states and the rotation of polarization angle were firstly predicted for QDs under uniaxial stresses [53, 82, 83]. In the seminal work by Seidl et al. [66] a QD sample (4×3×0.5 mm^3) was glued tightly onto a piezoelectric lead zirconic titanate (PZT) ceramic stack, and the FSS showed a monotonic change with increasing voltages on the PZT. However, the FSS was not tuned to zero due to the limited tuning range of exciton energy (~0.5 meV).

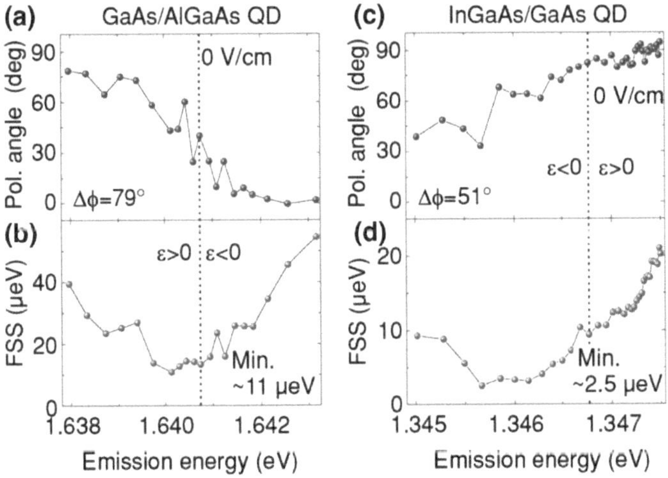

Fig. 7.6 Experimental studies of the influence of anisotropic strain fields on the QD emissions. **a**, **b** Polarization angle and FSS behavior of a GaAs/AlGaAs QD and **c**, **d** of an In GaAs/GaAs QD. The polarization angle is defined as the angle of the higher energy components of the exciton emission with respect to the x direction of the PMN-PT substrate. For both QDs the x direction roughly corresponds to the $[1\bar{1}0]$ direction of the QD nanomembrane. Figures adapted from the work by Plumhof et al. [86]

In 2009 Zander et al. used the $Pb(Mg_{1/3}Nb_{2/3})O_3$-$PbTiO_3$ (PMN-PT) piezoelectric actuators to tune the emission energies of QDs embedded in microrings. In 2010 Ding et al. used PMN-PT to investigate the excitonic binding energies of QDs embedded in a thin nanomembrane (with a thickness of a few hundred nanometers) [84]. In these two works the use of high performance PMN-PT together with the nanomembrane designs facilitated a much larger tuning range of exciton energies ($\sim$10 meV), leading to the observations of many strain-related physics in single QDs [85–87]. More details on the technique can be found in the review by Rastelli et al. [88]. However, due to the use of PMN-PT substrates providing biaxial strain fields, there was no significant changes in the FSS.

Plumhof et al. developed a technique to apply anisotropic strain fields on the QD nanomembranes [86]. The relation between the strain components is $\varepsilon_\perp \approx -0.7 \times \varepsilon_\parallel$, where $\varepsilon_\perp$ and $\varepsilon_\parallel$ are the strain parallel to the pseudo-cubic cut directions (x and y) of the PMN-PT. 150–200 nm thick QD-containing nanomembranes were created by selective etching techniques and transferred onto the piezo, with their edges carefully aligned along the x and y directions of the PMN-PT. Two different types of QD were studied in their work, GaAs/AlGaAs QDs and InGaAs/GaAs QDs, and the results are shown in Fig. 7.6. When moving from low to high emission energies, that is, from tensile to compressive strain along x direction ($[1\bar{1}0]$), the FSS goes to a non-zero minimum value before increasing again, accompanied by a gradual rotation in the the polarization angle. These results are quite similar to the ones obtained by vertical electric field tuning shown in Fig. 7.5.

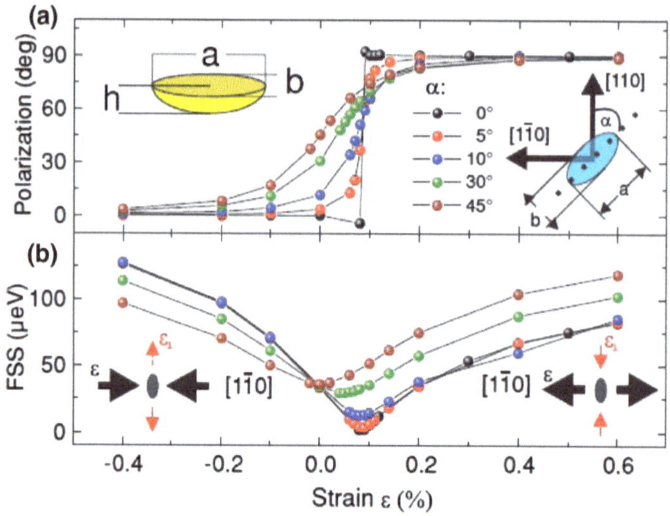

Fig. 7.7 Theoretical studies of the influence of anisotropic strain fields on the QD emissions. **a** Polarization of the high energy excitonic component with respect to the x direction ([1$\bar{1}$0]) for different values of α. The *left inset* shows the shape of the model GaAs/AlGaAs QD. The *right inset* defines the angle of the QD elongation with respect to the [110] crystal direction. **b** Corresponding values of the FSS. Figures adapted from the work by Plumhof et al. [86]

In order to gather more insight on the strain induced anti-crossing of the bright neutral exciton states, the excitonic FSS of a model GaAs/AlGaAs QD was calculated by combining the eight band $k \cdot p$ model and the configuration interaction method [86]. The results are shown in Fig. 7.7a, b for different angles α between the QD elongation axis and the [110] crystal direction. In the ideal situation of $\alpha = 0$, the polarization angle remains almost unchanged while the FSS shows a monotonic change (increase or decrease) in a broad strain tuning range. At a certain strain field (0.086%), the FSS reaches its minimum at below 0.4 μeV and the polarization angle changes abruptly by 90°. When the QD elongation axis is not aligned along [110] (that is, $\alpha \neq 0$), the minimum reachable FSS increases with increasing α and the polarization angle changes smoothly. By inspecting the single-particle states it was found that small stain fields produce relevant changes on the ground-state hole wave function, while the effect on the electron wave function is much weaker. The enhanced mixing of the heave hole band with the light hole band modifies the effective mass, causing a pronounced anisotropy along the principal stress axes [110] and [1$\bar{1}$0]. The joint influence of the structural anisotropy and the anisotropy of the effective mass leads to the elongation of the wave functions along the [110] or [1$\bar{1}$0] directions, depending on the sign of the applied strain. The smooth rotation of the hole wave function for finite α accounts for the polarization rotation observed in Fig. 7.6.

Similar results can be obtained when the direction of the applied strain is changed and the QD elongation axis is kept fixed at [110] direction. This indicates that it would be even possible to reduce the FSS in every single QDs to a minimum value (below the lifetime limited linewidth of exciton) by carefully aligning the stress direction to the QD principal axis. This "universal tuning" of FSS is very appealing for the practical applications of QD based entangled photon sources. In practice, however, controlling the stress directions locally for each single QDs is very challenging. Trotta et al. proposed a different and elegant approach to achieve the universal tuning of FSS in InGaAs QDs without a structure symmetry. A dual-knob device was used their works, where in-plane strain fields and vertical electric fields can be applied to the QD nanomembrane simultaneously [68, 89]. The relevant theory was inspired by the work of Gong et al. [82] and the main conclusion is that [68, 90]:

$$s = |(\eta + \sigma p + \beta F)^2 + (\kappa + \gamma p)^2| \tag{7.17}$$

$$tan(\theta_{\pm}) = \frac{\kappa + \gamma p}{\eta + \sigma p + \beta F \pm s} \tag{7.18}$$

where s is the FSS, p the anisotropic stress, F the electric field, η and κ account for the QD structural asymmetry, σ and γ are related to the applied anisotropic stress, and β is related to the vertical electric field. It is clear that s equals to zero when:

$$p_{crit} = -\frac{\kappa}{\gamma}; \; F_{crit} = \frac{\sigma\kappa}{\gamma\beta} - \frac{\eta}{\beta} \tag{7.19}$$

In the situation of only anisotropic stress field applied to the QDs, $s = 0 \rightarrow \frac{\kappa}{\eta} = -\frac{2\gamma}{\sigma}$. Since η and κ are different for every QDs, this equation implies that $s = 0$ can be always achieved if one has active control over the stress related parameters γ and σ [90]. This corresponds to the ideal situation of $\alpha = 0$ in Fig. 7.7. With the dual-knob device, it is clear from (7.19) that the FSS of every single QDs can be erased regardless of their structural symmetry. The tuning results are presented in Fig. 7.8. Due to this unprecedented control on FSS, the violation of Bell's inequality was demonstrated without discarding any emitted photons from the QDs [27].

The discussions above prove that the strain field is a powerful tuning knob for engineering the optical properties (including the FSS) of single QDs. It is also possible to combine the strain field with other tuning knobs, such as the electric field, magnetic field or even another strain field [18, 91, 92]. A combination of electrical injection of carriers into the QDs and local stress control of the FSS would allow the fabrication of an all-electrically operated QD-based entangled LED for practical applications. This goal has been demonstrated recently and will be discussed in the following section.

7.2.3 Electrical Injection of the Sources

III-V semiconductor QDs possess an important advantage of being compatible with mature semiconductor technology. For practical quantum applications, electrical triggering would allow the realization of compact and deterministic sources of entangled

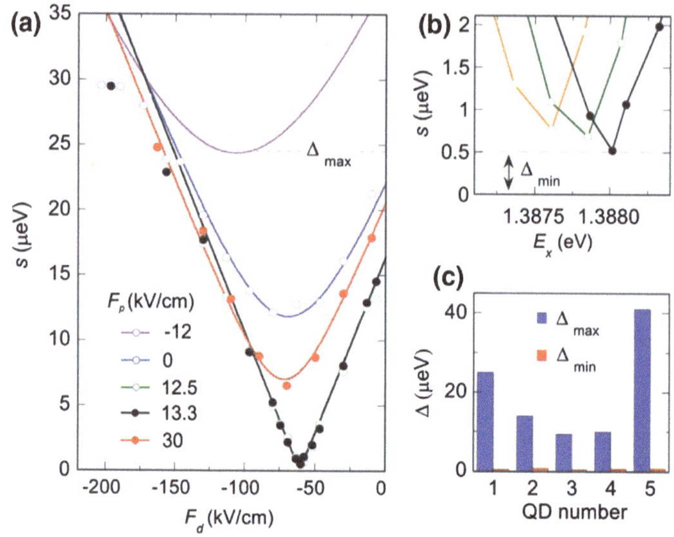

Fig. 7.8 **a** Behavior of the FSS as a function of the vertical electric field F_d. The different curves correspond to different strain fields F_p. The solid lines are theoretical fits. **b** FSS as a function of the exciton energy. Δ_{min} indicates the minimum value of Δ (lower bound for the FSS versus F_d). **c** Histogram of Δ_{max} (the maximum value of Δ) and Δ_{min} for the different measured QDs. Figures adapted from the work by Trotta et al. [68]

photons. In recent years the entangled photon pairs generation at room temperature in an AlGaAs semiconductor waveguide has received considerable attention. The source is based on SPDC with a counter-propagating phase-matching scheme [93]. Thanks to the direct bandgap of AlGaAs, an electrically injected AlGaAs waveguide device was demonstrated in late 2013. However, due to the intrinsic nature of SPDC process, the generated photons from these devices are characterized by Poissonian statistics, that is, one usually does not know when an entangled photon pair is emitted. This results in the generation of zero or even multiple entangled-photon pairs in most excitation cycles and unavoidably limits the success of realizing deterministic photonic quantum technologies.

Entangled light-emitting-diodes (LEDs) based on semiconductor QDs can potentially address this task. It was firstly realized by the Cambridge group in 2010 [29], see Fig. 7.9. The design of the device is based on a single layer of QDs embedded in a doped planar microcavity. Carrier injection into the quantum dots was achieved by biasing the diode beyond its turn-on voltage. To facilitate the emission of entangled photons, a single QD with vanishing FSS was pre-selected for the experiment. Since the two-photon wavefunction can be expressed as the superposition of co-linearly, co-diagonally or cross-circularly polarized photon pairs (see (7.13) and (7.14)), the polarization correlation experiment shown in Fig. 7.9c demonstrated clearly the emission of entangled photon pairs. The device has been used by the same group to

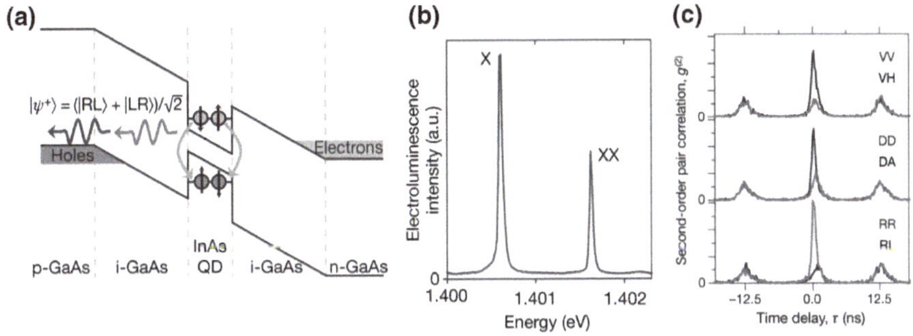

Fig. 7.9 **a** Schematic of the active region of a QD based entangled LED. Polarization entangled photon pairs are emitted from the biexciton cascade. **b** Electroluminescence from a single QD under d.c. electrical injection, and the FSS of this pre-selected QD is $0.4 \pm 0.1 \mu eV$. **c** Polarization correlation experiment in the rectilinear, diagonal and circular bases with a time resolution of 0.2 ns. The device was pulsed with an alternating current at a repetition rate of 80 MHz. Figures adapted from the work by Salter et al. [29]

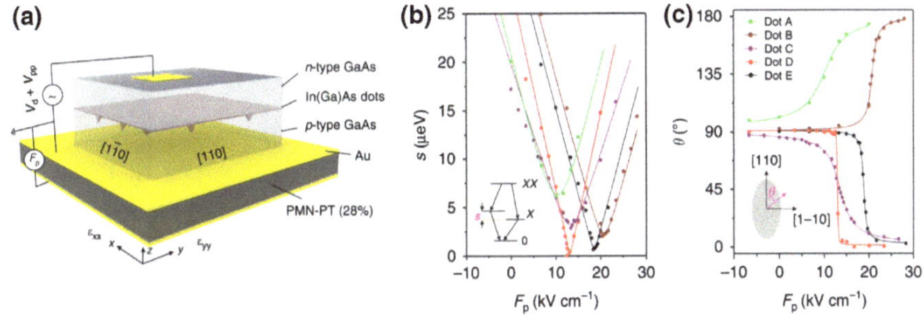

Fig. 7.10 **a** Sketch of the diode structure. Different from the previous works [84, 86, 89], the PMN-PT top surface has (011) orientation, which imposes large anisotropic strain fields with well-defined orientation onto the overlying ELED. **b**, **c** Representative variation of the FSS s and the polarization direction θ of the high-energy component of the exciton as a function of F_p for five QDs. The definition of θ is shown in the *inset* of **c**. Figures adapted from the work by Zhang et al. [30]

demonstrate several important components in quantum communication, e.g. a quantum relay over 1 km [94] and quantum teleportation [95].

However, such devices rely on the pre-selected "hero" QDs with zero FSS, and the extremely low yield ($<10^{-2}$) limits the further studies on entangled LED based on QDs. In-plane magnetic fields can be used to tune the FSS of single QDs embedded in a LED structure, but it requires bulky setups. Tuning the FSS with vertical electric fields, on the other hand, does not compatible with electric injections of carriers. Therefore, piezoelectric induced strain fields hold strong promise to realize a high yield entangled LED based on QDs.

A strain tunable entangled LED was introduced by Zhang et al. in 2015 [30]. The device is schematically shown in Fig. 7.10a. Arrays of 440 nm *n-i-p* GaAs nanomembranes (each with a size of $120 \times 160\,\mu m^2$) containing InGaAs QDs were integrated onto a PMN-PT single crystal substrate. The details on the nanomembrane fabrications can be found in the literatures [30, 88, 89]. The long edge of the GaAs nanomembranes was processed along the [110] crystal directions of GaAs and carefully aligned along the [110] pseudo-cubic cut direction of the PMN-PT substrate, see Fig. 7.10a. When the PMN-PT is poled along the z axis, in-plane strain fields with normal components ε_{xx} along the x axis and ε_{yy} along the y axis with opposite sign can be transferred to the nanomembrane. Accounting for its relevant piezoelectric coefficients the in-plane anisotropy is estimated to be $\varepsilon_{xx} \approx -0.37\varepsilon_{yy}$. Under this condition the anisotropic strain fields are applied to the QDs with $\alpha = 0$ as discussed in Fig. 7.7a.

Electroluminescence from the QDs can be observed when the diode is biased with a DC voltage (V_d) of above -1.7 V. For the triggered generation of photons, an ultrafast electrical pulse stream V_{pp} was superimposed onto a -1.6 V DC bias. The strain provided by the PMN-PT crystal can be controlled by the electric field F_p, see Fig. 7.10a. The FSS and polarization angle θ are shown as a function of F_p in Fig. 7.10b, c. When F_p was varied from -6.7 to 28 kVcm^{-1}, all studied QDs exhibit the anti-crossing behavior as discussed above. In the meanwhile, the rotations of θ was also observed. At the largest available tensile (compressive) strain, θ tends to be aligned along the $[1\bar{1}0]$ ([110]) direction. The minimum achievable FSS was, however, different for each QDs. Gong et al. pointed out that the lower bound of FSS under external stress can be predicted by the polarization angle θ_0 and FSS s_0 at zero strain fields. The experiments showed excellent agreement with this theoretical prediction. For QD D and E, their FSS can be reduced well below 1 μeV due to the exact alignment between θ_0 and the strain axis at zero strain fields, and an abrupt change in θ can be also observed.

Statistical studies on 82 randomly selected QDs in the device showed that the majority of QDs have their polarization angle θ_0 aligned along the $[1\bar{1}0]$ crystal direction, see the supplementary materials in Zhang's work [30]. With strain engineering, as high as 33% QDs can be tuned to entangled photon emitters. Compared to the only work on entangled LED with QDs [29], the yield is more than an order of magnitude higher (a factor of 30). This probability is higher than what was reported for highly symmetric pyramidal QDs [22] where, however, electrical injection has not been realized yet.

The ability to tube the FSS allows the generation of entangled photon pairs without the need of post-filtering. Figure 7.11a shows the polarization resolved polarization correlation between the XX and X photons. The FSS of the QD has been tuned to $0.6 \pm 0.2\,\mu$eV. Strong polarization correlations were observed for the co-polarized photons in $\{|H\rangle, |V\rangle\}$ and $\{|D\rangle, |A\rangle\}$ bases, while strong polarization anti-correlation were observed for the co-polarized photons in $\{|R\rangle, |L\rangle\}$ basis. Two-photon density matrix was then reconstructed by performing 16 polarization correlation measurements [96], see Fig. 7.11b, c. The fidelity f^+ to the maximally entangled Bell State $|\phi^+\rangle = \frac{1}{\sqrt{2}}(|H_{XX}H_X\rangle + |V_{XX}V_X\rangle)$ was 0.766 ± 0.051. With ultrafast electric pulses, it is possible to achieve fast generate rate of the entan-

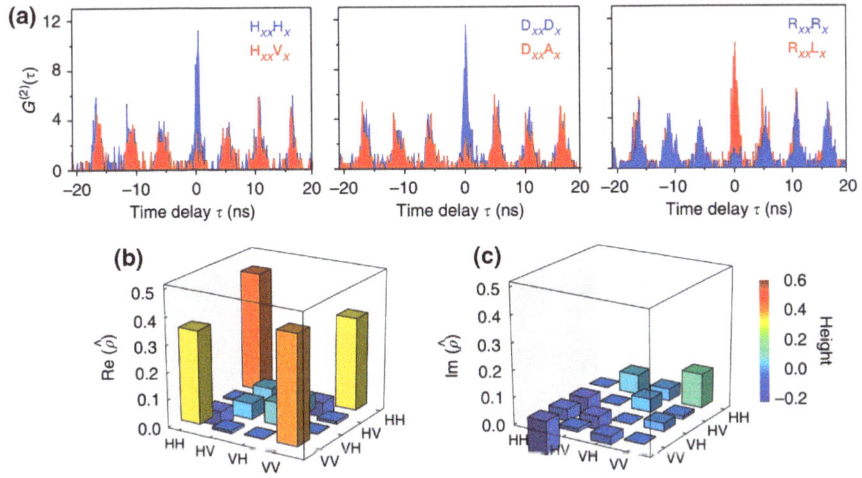

Fig. 7.11 **a** Polarization correlation experiment on a QD in a strain tunable entangled LED. **b** Real part and **c** imaginary part of the density matrix, which were reconstructured with 16 coincidence counts integrated in a 1.8 ns temporal window centred at zero delay time. Figures adapted from the work by Zhang et al. [30]

gled photon pairs, which is highly desirable for high data-rate quantum information processing. An electrical excitation rate of up to 400 MHz has been tested with the strain tunable entangled LED. A small decrease in the entanglement fidelity was observed, which was most likely due to the contributions of a small amount uncorrelated photon pairs generated during a time-dependent re-excitation process [30].

7.2.4 Scalability of the Sources

Many of the quantum information applications rely on indistinguishable sources of polarization-entangled photons. The wavelength tunability has therefore become a fundamental requirement for a number of envisioned applications, for example, nesting different dots via the entanglement swapping [97] and interfacing dots with cavities/atoms [98]. However, with single tuning knobs, the FSS can only be eliminated under particular tuning parameters. Therefore, any attempt to manipulate the emission wavelength increases the FSS and spoils the entanglement. This fact undoubtedly restricts the entangled photon emissions at arbitrary wavelengths. The inability to tune the emission wavelength without restoring the FSS, which is unfortunately the common disadvantage associated with most of the FSS tuning technologies, has become a major stumbling block to the QIP applications based on scalable QD sources. Several recent works have solved this problem successfully by combining two or three FSS tuning knobs.

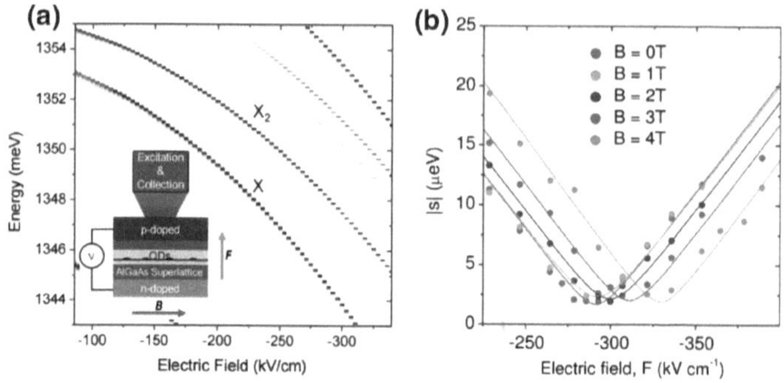

Fig. 7.12 **a** Photoluminescence as a function of electric field, F, without magnetic field. (*Inset*) Schematic diagram of a device structure showing the simultaneous applications of the electric and magnetic fields. **b** FSS as a function of electric field F, plotted at five different magnetic fields B. Figures adapted from the work by Pooley et al. [99]

Combination of Electric and Magnetic Fields

A p-i-n diode, which allows the application of giant quantum confined Stark shifts of exciton emissions, was used in the work by Pooley et al. [99]. A magnetic field was applied simultaneously to the diode structure in Voigt geometry, see Fig. 7.12a. As discussed in Fig. 7.5, the FSS can be minimized by the vertical electric field F and the minimum FSS s_0 can be achieved at a certain field F_0. The studied QD has a s_0 of around $2.0 \pm 0.2\,\mu$eV. By changing the magnetic field B, the electric field F_0 also changes, see Fig. 7.12b. This was the first experimental proof that wavelength-tunable entangled photons may be generated by semiconductor QDs. However, the QDs with sufficiently s_0 must be pre-selected for this purpose.

Combination of Three Strain Fields

Tuning the FSS in single QDs with magnetic fields in Voigt geometry has been a successful approach in the last years, but the bulky setup renders a practical implementation very inconvenient. Wang et al. suggested that the FSS can be eliminated in wavelength-tunable InGaAs/GaAs QDs by using three combined strain fields [100]. However, the realization of the proposed 3D stressor is experimentally challenging.

Trotta et al. proposed a device to fully control the in-plane strain tensor by applying three independent uniaxial stresses in the QD nanomembrane plane [101]. Figure 7.13a shows the sketch of the device. It consists of a micromachined single crystal PMN-PT substrate featuring six legs and capable of deforming in any direction a nanomembrane containing QDs. Quasiuniaxial stresses, and therefore, the deformation of nanomembrane along different directions, are achieved by applying three independent voltages ($V_{1,2,3}$) on the legs. The same voltage is applied to opposite legs to limit displacements of the central nanomembrane, see more details in [101].

The device operation principle was theoretical investigated in [92, 101] and the main results are shown in Fig. 7.13b, c. The polar plots show ΔE (the half of the

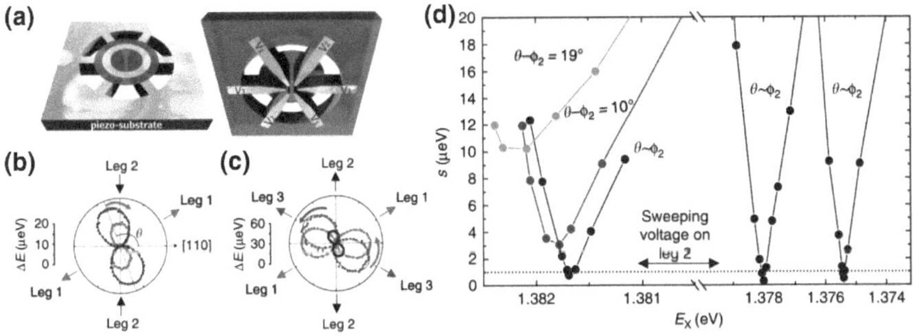

Fig. 7.13 **a** Sketch of the proposed device. The *top* and *bottom* views are depicted on the *left* and *right panel*, respectively. Three independent voltages ($V_{1,2,3}$) applied across pairs of legs and the *top* (grounded) contact allow the in-plane stress in the QD membrane to be controlled. Reference [101] **b**, **c** Polar plots of ΔE under the situation of two and three legs, respectively. **d** FSS as a function of the exciton energy E_x at different stress conditions. Figures adapted from the work by Trotta et al. [92]

energy difference between the XX and X energies minus its minimum value, and max(ΔE) equals to the FSS) as a function of the angle the linear polarization analyser forms with the [110] crystal axis. Leg 1 is used to rotate the polarization direction θ of the exciton until it aligns along the direction Leg 2, see Fig. 7.13b. Then it is similar to the situation of $\alpha = 0$ as discussed in Fig. 7.7, and the (quasi-)uniaxial stress provided by Leg 2 will be able to eliminate the FSS (reduce ΔE to zero). Note that, ϕ_2 in Fig. 7.13 is the angle between the direction of Leg 2 and the [110] crystal axis, which is 90° in this specific example. With a third uniaxial stress, the strain condition is modified and the combination of the first two stresses to achieve $\Delta E = 0$ will be different, see Fig. 7.13c. With this elegant design, wavelength tunable entangled photon emissions from single QDs have been successfully demonstrated.

Combination of Two Strain Fields

Another approach to the generation of wavelength tunable entangled photons from QDs was introduced by Chen et al. [18] and published back-to-back together with the work from Trotta et al. [92]. The newly developed concept of hyperactive PMN-PT MEMS on /Si [102] was successfully applied in the study of III-V quantum photonics. Unlike the piezo substrate used in all previous works on strain tuning of QDs, a 15 μm PMN-PT thin-film bonded on a silicon substrate was employed to realize a novel MEMS devices with sophisticated functionalities on chip, see Fig. 7.14. Arrays of QD-containing GaAs nanomembranes, as described above, were then transferred onto the PMN-PT MEMS with four actuation legs. Owing to the small footprint and the compatibility with mature semiconductor technologies, large scale on-chip integration is feasible. The crystal axes $[1\bar{1}0]$ and [110] of the GaAs nanomembrane were carefully aligned along the designed stress axes of the actuators, see Fig. 7.14c. When applying negative (positive) voltages to the electric contacts, the PMN-PT legs

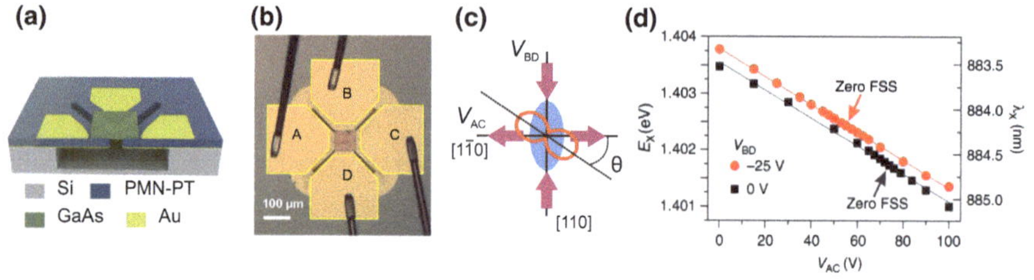

Fig. 7.14 **a** MEMS device for anisotropic strain engineering of IIICV QD-based quantum light sources. Focused ion beam (FIB) cut was used to define trenches on the PMN-PT thin film, and then wet-chemical undercut was used to form four suspended actuation legs. A thin GaAs nanomembrane containing In(Ga)As QDs was transferred onto the suspended region between the four legs. **b** Micrograph showing the zoom-in of a completed device. Electrical contacts are made on the four legs A-D. The centre region is a bonded QD-containing nanomembrane. **c** Definition of the polarization angle *theta* is the same to that in Fig. 7.7. **d** Exciton wavelength is plotted as a function of V_{AC} for the two different V_{BD}. The *solid lines* are linear fits. The *arrows* indicate exciton wavelengths at which the FSS are erased. Figures adapted from the work by Chen et al. [18]

expand (contract) in-plane and therefore exert quasi-uniaxial compressive (tensile) stresses to the QDs.

Specific QDs with initial polarization angle $\theta_0 = 0$ were chosen for the experiments. According to the statistical studies performed by Zhang et al. [30], the probability of finding such QDs are rather high due to the elongation of QDs along [110] during MBE growth. With negative (positive) voltages applied, the legs exert uniaxial compressive (tensile) stresses to the QD which causes a blue (red) shift in exciton emission. This is confirmed by sweeping voltage on the legs A&C V_{AC} from 0 to 100 V while keeping the voltage on the legs B&D V_{BD} at 0 V. A red shift of the emission was observed during the tuning, see Fig. 7.14d. The device performs remarkably well and there was no hysteresis in the wavelength tuning, as indicated by the linear fit. From the discussions in Figs. 7.7, 7.10 and 7.13, we know that the FSS can be eliminated during this tuning. The arrows in Fig. 7.14d indicate the exciton wavelengths at which the FSS reaches zero. The effect is similar when changing V_{BD} from 0 to −25 V. Therefore a high degree of control on the exciton wavelength and the FSS can be achieved, for the QDs with $\theta_0 = 0$, by using two pairs of orthogonal actuation legs.

Two dimensional scanning on the two pairs of legs by sweeping both V_{AC} and V_{BD} was performed, see the three-dimensional plot in Fig. 7.15a. The astonishing result is that, with this four-legged device providing orthogonal uniaxial stresses, multiple zero FSS points with different exciton wavelength can be achieved. At different V_{BD}, the electronic symmetry of QD can be always recovered by sweeping V_{AC} and the FSS is erased. The dashed line on the bottom plane of the plot indicates the combinations of (V_{AC} and V_{BD}) at which the FSS reaches its minimum. A linear relationship was found for the ratio of voltage changes $\Delta V_{AC}/\Delta V_{BD}$. This result was confirmed theoretically by an effective two-level model, see more details in [18]. Figure 7.15b shows the polarization correlation experiment when the FSS is tuned to near zero

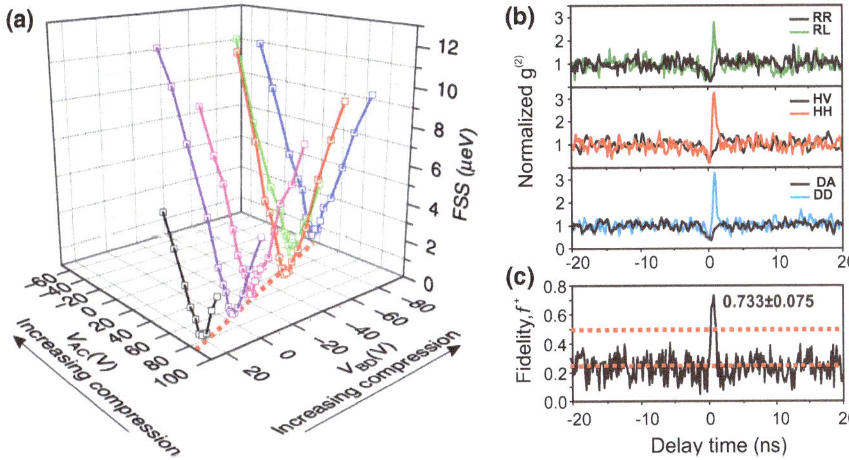

Fig. 7.15 **a** The changes in FSS when both V_{AC} and V_{BD} were scanned. The *dashed line* on the *bottom plane* indicates a linear shift of the voltage combination (V_{AC} and V_{BD}) at which FSS reaches the minimum values. **b** Polarization correlation spectroscopy was performed in different bases, when the QD FSS is tuned to $0.21 \pm 0.20 \mu eV$. The normalized coincident counts were given in the plot. **c** The peak near the zero time delay yields a fidelity f^+ of 0.733 ± 0.075 without any background subtraction. The two *dashed lines* indicate the threshold of 0.5 for the classically correlated light, and the threshold of 0.25 for the uncorrelated light. Figures adapted from the work by Chen et al. [18]

($0.21 \pm 0.20\,\mu$eV). The entanglement fidelity f^+ to the maximally entangled Bell state can be determined from the measurements in Fig. 7.15b. The peak near the zero time delay yields a fidelity f^+ of 0.733 ± 0.075 without any background subtraction, which exceeds the threshold of 0.5 for a classically correlated state by >3 s.d.

Combination of Electric and Strain Fields

Although wavelength tunable entangled photon emissions from QDs can be achieve by applying two or three in-plane strain fields (as shown above), the active feed-back stabilization of the strain control [88] has to be implemented for practical applications due to the unavoidable hysteresis and creep of the piezoelectric materials. Very recently, Zhang et al. reported a device which combines the advantages of vertical electric field (giant Stark shift, see Fig. 7.5) and in-plane uniaxial strain field tuning knobs [103]. The device consists of a n-i-p diode nanomembrane integrated onto a conventional PMN-PT substrate. Different from the strain tunable entangled LED shown in Fig. 7.10, the QD nanomembrane contains a layer of QDs in the middle of a 150 nm-thick intrinsic GaAs/AlGaAs quantum well. The quantum well was used in [62] to enable a giant quantum confined Stark shift of QD emissions. Although a similar device was used [68], extending the work to achieve wavelength tunable entangled photons is not straightforward as the anisotropic biaxial strain field cannot eliminate the FSS solely [103]. A PMN-PT substrate with pseudo-cubic direction [110], [0$\bar{1}$1] and [011] providing quasi-uniaxial strain fields was used in Zhang's

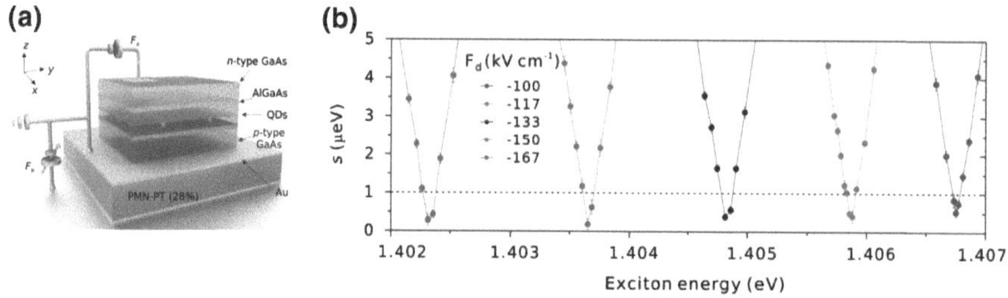

Fig. 7.16 **a** Sketch of the device combining giant Stark shift and in-plane strain tuning. **b** FSS as a function of exciton energy E_x (tuned by strain fields), for five different values of F_d applied to the diode. Figures adapted from the work by Zhang et al. [103]

work. By choosing the QDs with initial polarizations aligned along $[1\bar{1}0]$ (or $[110]$), the FSS tuning effect of combined vertical electric field and in-plane uniaxial strain field is similar to that of two orthogonal in-plane uniaxial strain fields. Therefore, the independent control on the FSS and the exciton emission energy can be also achieved, see Fig. 7.16b. More details on the theoretical simulations and on the triggered generation of wavelength tunable entangled photons can be found in [103].

7.2.5 Photon Collection Efficiency

A major challenge in the practical applications of QD-based polarization entangled photon sources is the photon extraction efficiency. In bulk GaAs only less than 2% of the photon pairs can be collected and most of the light are confined in the high refractive index material. In recent years there has been considerable progress in improving the extraction efficiency of QD-based single photon sources. The most common approach is to design an optical microcavity, where the QD emissions are funnelled into a cavity mode. When the QD is both spatially and spectrally coupled to a cavity mode, the emission rate Γ_{cav} is enhanced by the so called Purcell effect:

$$\Gamma_{cav} = \frac{3Q\lambda^3}{4\pi^2 V}\Gamma_{free} \tag{7.20}$$

where Γ_{free} is the QD emission rate in free space, Q the quality factor, V the mode volume, λ the wavelength on resonance. $\Gamma_{cav}/\Gamma_{free}$ is the Purcell factor F_p, and the fraction $F_p/(F_p+1)$ of the total QD emission can be funnelled into the cavity mode. The spectral matching of QD emission and cavity mode can be achieved by tunings with temperature, electric, magnetic fields and the strain fields discussed above. The spatial matching of the QD position and the field maximum of cavity can be achieved by careful positioning of QDs before the cavity fabrication [104, 105], or by the newly developed in situ lithography techniques [106, 107]. From (7.20) it

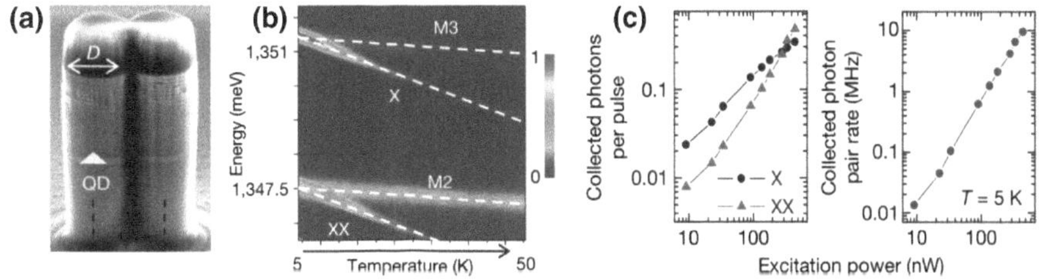

Fig. 7.17 Principle of entangled photon extraction using a photonic molecule. **a** Two identical pillar microcavities with diameter *D* are coupled. A single QD is inserted in one of the pillars. **b** Emission intensity as a function of energy and temperature. **c** *Left* number of X and XX collected photons for each excitation pulse as a function of the excitation power. *Right* collected entangled photon pair rate (MHz) as a function of the excitation power. Figures adapted from the work by Dousse et al. [25]

is clear that a high F_p requires a high Q. Since $Q = \lambda/\Delta\lambda$, it is only possible to use a low Q cavity to enhance both X and XX photons due their large energy difference.

Dousse et al., however, proposed an interesting design based on high Q cavities to realize an ultrabright entangled photon source. A single QD is deterministically coupled to a photonic molecule, see Fig. 7.17a. To obtain two cavity resonances for both XX and X photons, the QD is inserted in a micropillar cavity coupled to a second identical but empty micropillar. By choosing pillar diameter and distance, it is possible to independently tune the energies of the photonic molecule modes and the energy differences between them to match both X and XX energies [25]. By temperature tuning, the X emission is resonant to mode 3 and the XX emission is resonant to mode 2 at 5 K, see Fig. 7.17b. The Purcell effect leads to a strong increase in the QD emissions and reduces the radiative lifetime to 200–300 ps for each line, corresponding to a expected $F_p = 3 - 5$. To generate entangled photon emissions, single QDs with FSS of 1–3 μeV were chosen after the thermal annealing of the sample. Figure 7.17c presents the rate of collected entangled photon pairs for an excitation rate of 82 MHz and a rate of photon pairs collected in the first lens of about 10 MHz was achieved [25].

Versteegh et al. demonstrated a bright entangled photon source from a position controlled nanowire QD [69]. A single nanowire with an embedded QD is shown in Fig. 7.3c, and it has the key features of vanishing FSS (due to the growth on (111) substrate) and bright emissions (due to the tapered waveguide geometry). An extraction efficiency of $18 \pm 3\%$ was achieved. But the two-photon state emitted by the QD is modified by the birefringence effect during propagation along the nanowire, therefore certain polarization compensations must be implemented for the source. A similar work was reported by Huber et al. by using very similar nanowire QD samples [108]. This tapered nanowire design is in principle compatible with the III-V QDs grown by MBE, and a strain tunable nanowire antenna with embedded InAs/GaAs QDs has been demonstrated by Kremer et al. [109].

7.3 Outlook

SPDC based single and entangled photon sources have been serving as the main workhorse for optical QIP experiments in the last years. Do we really need a QD based entangled photon source, if we have the easily operated SPDC sources in hand? The answer to this question, in our opinion, depends highly on the targeted applications. For example the applicability of *many photon*-based QIP schemes is limited due to the stochastic character of the photon sources. Therefore, a worldwide effort has focused on overcoming the limitation of probabilistic emission by generating two-photon entangled states "on demand" [110]. Although heralded generation of polarization entangled photons has been realized with SPDC sources, the experimental implementation requires complex setups and the performance of the sources still need to be improved. A source with deterministic and sub-Poissonian emissions is therefore very attractive. Here we outline several possible applications, where the QD-based entangled photon sources could play an important role.

7.3.1 Entanglement Distribution

Building a quantum network is among the most exciting applications of QIP. An ideal quantum network could, in principle, take the advantages of both stationary qubits and flying qubits. Information can be stored and processed with stationary qubits in each individually addressable node, and the faithful transfer of information can be done via flying qubits between distant nodes. Spins are natural two-level systems which fulfil the quantum superposition principle, and they are ideal candidates for stationary qubits. Photons travel fastest in nature and have long coherence times, thus they are considered as ideal flying qubits to carry quantum information over long distances and to nest separated quantum nodes.

Semiconductor QDs are among the most promising candidates for both tasks. Entangling the stationary (spin) and flying (photon) qubits in QDs has been demonstrated by several groups internationally [111–114]. To distribute the entanglement in a quantum network, a proven technique is to use the so called entanglement swapping [97, 115, 116]. In Fig. 7.18a, the FSS of two spatially separated QDs are tuned to zero and therefore polarization entangled photon pairs 1 and 2, 3 and 4 are emitted from the two QDs. A Bell state measurement on photons 2 and 3, if they are indistinguishable, heralds the entanglement of photons 1 and 4 despite the fact that they never interacted and may be far away from each other. With the recently developed wavelength tunable entangled photon sources [18, 92, 99, 103], it is even possible to perform "event-ready" entanglement swapping [115] with QDs. Figure 7.18b shows one of our recent results, where the energy (wavelength) coincidence of two QDs emitting entangled photons has been achieved. However the brightness and the photon indistinguishability of the studied source need to be improved for a successful entanglement swapping experiment.

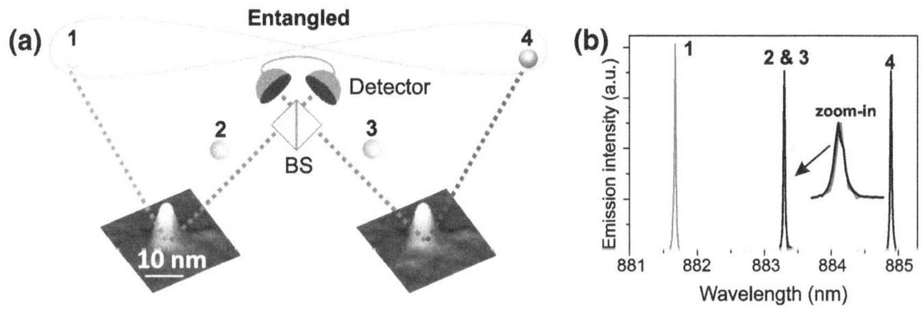

Fig. 7.18 **a** Principle of entanglement swapping with the polarization entangled photons from two semiconductor QDs. **b** Normalized photoluminescence spectrum showing the exciton (1, 3) and biexciton (2, 4) emissions from from two spatially separated QDs with FSS of zero. The charged excitonic line are omitted for simplicity. The energy (wavelength) coincidence of photons 2 and 3 has been achieved

7.3.2 Hybrid Interfacing with Atoms

Entanglement swapping and quantum memory are two most critical elements in building a quantum repeater. One may use atomic systems (either atomic ensemble or single atoms) to store quantum information carried by photons [117, 118]. The hybrid interfacing between entangled photons and quantum memories have applications not only in long-distance quantum communication, but they also provide a route to a more efficient multiphoton entanglement or linear-optics quantum computing, see more details in the review by Pan et al. [1]. A deterministic entangled photon source (e.g. single QDs) will be quite useful in this scheme, as it will enable an efficient synchronization of multiple nodes in a network.

The coupling between semiconductor QDs and atomic systems was firstly realized by Akopian et al. in 2011 [98]. The single photon emissions from GaAs/AlGaAs QDs were tuned into resonance with the ^{87}Rb D2 transitions at 780 nm and then stored in the slow-light medium for 15 times of the single photon temporal width. With different choices of semiconductor materials, single QDs can emit single photons in a broad wavelength range, from the ultraviolet-visible region to the near-infrared region. Together with the wavelength fine tunings provided by external perturbations (e.g., magnetic, electric and strain fields), the hybrid interfacing has been realized, for example, with QD-Rubidium atoms (780 nm) [81, 98], QD-Cesium atoms (895 nm) [92, 119], QD-Yb^+ ion (935 nm) [120], QD-Nd^{3+}:YVO_4 crystals (880 nm) [121]. Thanks to the tunability as discussed in Sect. 7.2.4, the hybrid interfacing between the QD based entangled photon sources and the atomic systems can be also achieved now [92], which may lead to the implementations of a high efficiency solid-state quantum repeater, see detailed discussions in [121–123].

7.3.3 Telecom Band Sources

For long distance quantum communication, there is strong motivation to develop entangled photon sources emitting at telecom wavelengths. To date most experiments have been performed with SPDC sources. Until now the studies on the telecom QD based entangled photon sources (e.g. at O band:1260–1360 nm and C band:1530–1565 nm) are rather primitive. This is mainly due to the difficulty to grow high quality telecom QDs with vanishing FSS. The first attempt was made by Ward et al. in 2014 [56], where, however, the entangled emissions was observed only in a short time window. The QD used in their work has a large FSS of ∼6μeV, therefore, a phase difference of $e^{i\tau s/\hbar}$ is developed between the two-photon states (see (7.16)). This fact leads to a time-evolving Bell state and a fidelity of >0.5 in ∼5 ns (7 periods of evolution of the exciton state). In an effort to reduce the FSS in telecom QDs, Liu et al. reported the growth of highly symmetric InAs/InAlAs QDs on C_{3v} symmetric InP(111)A substrates [77]. The distribution of the FSS (from 3 to 70 μeV) was considerably smaller than those reported in previous works on telecom QDs. The FSS tuning techniques discussed in Sect. 7.2.2 can be in principle applied to the telecom QDs. To explore the possibility of on-demand generation of indistinguishable photons, the resonant pumping technique has been also tested with telecom QDs [124]. A hybrid source was recently reported, where the single photons from QDs emitting at 775 nm were converted by SPDC process into single pairs of entangled photons at 1.55 μm [125]. With further improvements on the photon qualities (coherence and brightness), this hybrid source could be an interesting alternative to the telecom QD based sources. It is also worth mentioning that, the interfacing between polarization entangled photons and quantum memories is also feasible at the telecom wavelengths [126].

7.3.4 On-Chip Integration

Quantum information processing with on-chip integrated architectures has received considerable attention due to the improved performance, miniaturization and scability of the devices. Integrated circuits for the generation and manipulation of superconducting qubits can be fabricated with hundreds to thousands of elements on a chip [129]. In terms of QIP with photonic qubits, Harris et al. demonstrated that the generation of time-correlated photon pairs with non-linear processes can be combined with the spectral filtering on a single silicon chip. Together with the on-chip detection techniques [130], these results open an exciting possibility for large-scale QIP with photonic qubits [131]. There has been also significant progress in the development of on-chip integrated platforms with semiconductor QDs, and a detailed review is given in the Chap. 13 by Rengstl et al. Here we mention several most recent examples. Reithmaier et al. demonstrated a device which combines the on-chip generation, routing and detection of QD resonance fluorescence all on the

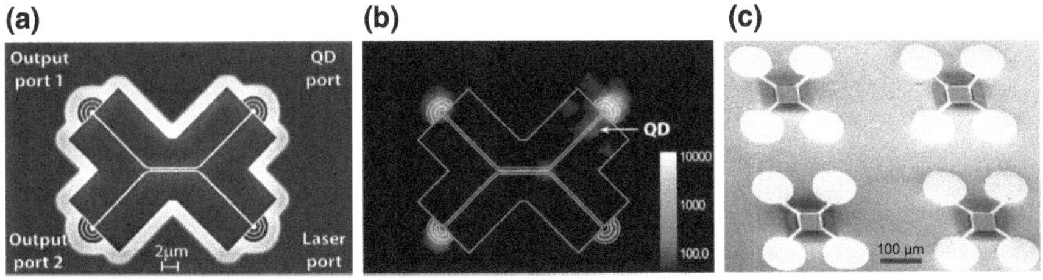

Fig. 7.19 **a** SEM image of a four port directional coupler. The input ports for the QD and a weak laser are indicated. The quantum interference occurs at the central 2 × 2 beamsplitter and are detected by the two output ports. **b** Photoluminescence map of the tested device with an overlaid device contour. The emission of a single QD was used as an internal source illuminating the device. **a** and **b** are adapted from the work by Prtljaga et al. [127]. **c** Individually addressable and strain tunable QD sources on a PMN-PT/Si substrate. The false color coded areas are the gold contacts (*yellow*) and the QD nanomembranes (*cran*), respectively. Adapted from the work by Zhang et al. [128] (color figure online)

same chip [132]. Prtljaga et al. [127] realized the on-chip two photon interference of single photons from a QD and an external laser, see Fig. 7.19a, b. However, the on-chip quantum interference of multiple QDs, which may find interesting applications in Boson sampling and several other QIP experiments, is very challenging, due to the difficulty to fine tune the QD emissions on chip. By using the in-plane anisotropic strain engineering technique, it would be possible to fabricate arrays of scalable single and entangled photon sources on chip. Figure 7.19c shows the SEM image of our recently fabricated device, where four individually addressable and strain tunable QD sources are integrated on a same chip [128]. The fabricate process is similar to that described in Sect. 7.2.4.

7.4 Conclusion

As said in the 2009 review by Shields, Stevenson and Young: "Semiconductor quantum dots are perhaps the world's youngest proven technology for entangled light generation. Since their first successful operation in 2006, there has been continuous improvements to the quality of the entangled light generation, and to the understanding of the fundamental concepts that both enable and degrade entanglement." [17]. Another 7 years has passed since that review, and the quest for a "perfect" entangled photon sources still continues. Polarization entangled photon pairs can be now triggered optically from QDs with much higher brightness and indistinguishability. Electrically triggered sources can be fabricated with high yield and operated at high repetition rates. The stumbling block of non-zero FSS, which has plagued the generation of entangled photon sources for years, has been also removed with the recent progress. However, the indistinguishability and brightness of the sources need to be further improved, before they can surpass their SPDC counterparts. Very recently the excellent work by Somaschi and Senellart et al. showed that the performance of QD based single photon sources is on par with that of the state-of-the-art SPDC sources [133]. We believe that, with the continuous efforts of the community, the "***Eureka!***" moment for QD based entangled photon sources will also come very soon.

Acknowledgements This work was financially supported by the ERC Starting Grant No. 715770 (QD-NOMS), the BMBF Q.Com-H (16KIS0106) and the European Union Seventh Framework Programme 209 (FP7/2007–2013) under Grant Agreement No. 601126 210 (HANAS).

References

1. J.W. Pan et al., Rev. Mod. Phys. **84**, 777 (2012)
2. A. Einstein, B. Podolsky, N. Rosen, Phys. Rev. **47**, 777 (1935)
3. R. Feynman, Int. J. Theor. Phys. **21**, 467 (1982)
4. D. Deutsch, Proceedings of the Royal Society of London A: Mathematical. Phys. Eng. Sci. **400**, 97 (1985)
5. C.H. Bennett et al., Phys. Rev. Lett. **70**, 1895 (1993)
6. C.H. Bennett, S.J. Wiesner, Phys. Rev. Lett. **69**, 2881 (1992)
7. A.K. Ekert, Phys. Rev. Lett. **67**, 661 (1991)
8. R. Raussendorf, D.E. Browne, H.J. Briegel, Phys. Rev. A **68**, 022312 (2003)
9. J.P. Dowling, K.P. Seshadreesan, Journal of Lightwave Technology **33**, 2359 (2015)
10. N. Roch et al., Phys. Rev. Lett. **112**, 170501 (2014)
11. R. Blatt, D. Wineland, Nature **453**, 1008 (2008)
12. R. McConnell et al., Nature **519**, 439 (2015)
13. K. Lee et al., Science **334**, 1253 (2011)
14. M. Koashi, T. Yamamoto, N. Imoto, Phys. Rev. A **63**, 030301 (2001)
15. O. Benson, C. Santori, M. Pelton, Y. Yamamoto, Phys. Rev. Lett. **84**, 2513 (2000)
16. A. Rastelli, S. Kiravittaya, O.G. Schmidt, *Single Semiconductor Quantum Dots* (Springer, Berlin, 2009), pp. 31–69
17. A.J. Shields, R.M. Stevenson, R.J. Young, *Single Semiconductor Quantum Dots* (Springer, Berlin, 2009)
18. Y. Chen et al., Nat. Commun. **7**, 10387 (2016)
19. R.M. Stevenson et al., Nature **439**, 179 (2006)
20. N. Akopian et al., Phys. Rev. Lett. **96**, 130501 (2006)
21. R. Hafenbrak et al., New J. Phys. **9**, 315 (2007)
22. G. Juska et al., Nat. Photon. **7**, 527 (2013)
23. M. Ghali, K. Ohtani, Y. Ohno, H. Ohno, Nat. Commun. **3**, 661 (2012)
24. T. Kuroda et al., Phys. Rev. B **88**, 041306 (2013)
25. A. Dousse et al., Nature **466**, 217 (2010)
26. A. Muller, W. Fang, J. Lawall, G.S. Solomon, Phys. Rev. Lett. **103**, 217402 (2009)
27. R. Trotta et al., Nano Lett. **14**, 3439 (2014)
28. M. Muller et al., Nat. Photon. **8**, 224 (2014)
29. C.L. Salter et al., Nature **465**, 594 (2010)
30. J. Zhang et al., Nat. Commun. **6**, 10067 (2015)
31. C.-Y. Lu, J.-W. Pan, Nat. Photon. **8**, 174 (2014)
32. G.N. Lewis, Nature **118**, 874 (1926)
33. L. Allen, M.W. Beijersbergen, R. Spreeuw, J. Woerdman, Phys. Rev. A **45**, 8185 (1992)
34. E. Nagali et al., Phys. Rev. Lett. **103**, 013601 (2009)
35. X.L. Wang et al., Nature **518**, 516 (2015)

36. J. Brendel, N. Gisin, W. Tittel, H. Zbinden, Phys. Rev. Lett. **82**, 2594 (1999)
37. E. Schrödinger, in *Mathematical Proceedings of the Cambridge Philosophical Society*, vol. 04 (Cambridge University Press, Cambridge, 1935), pp. 555–563
38. A. Peres, Phys. Rev. Lett. **77**, 1413 (1996)
39. M. Horodecki, P. Horodecki, R. Horodecki, Phys. Lett. A **223**, 1 (1996)
40. J.S. Bell, On the Einstein Podolsky Rosen Paradox (1964)
41. J.F. Clauser, M.A. Horne, A. Shimony, R.A. Holt, Phys. Rev. Lett. **23**, 880 (1969)
42. P.G. Kwiat et al., Phys. Rev. A **60**, R773 (1999)
43. P.G. Kwiat et al., Phys. Rev. Lett. **75**, 4337 (1995)
44. A. Mair, A. Vaziri, G. Weihs, A. Zeilinger, Nature **412**, 313 (2001)
45. C. Law, I. Walmsley, J. Eberly, Phys. Rev. Lett. **84**, 5304 (2000)
46. E. Megidish et al., Phys. Rev. Lett. **110**, 210403 (2013)
47. X.S. Ma et al., Opt. Express **20**, 23126 (2012)
48. S.J. Freedman, J.F. Clauser, Phys. Rev. Lett. **28**, 938 (1972)
49. A. Aspect, P. Grangier, G. Roger, Phys. Rev. Lett. **47**, 460 (1981)
50. D.C. Burnham, D.L. Weinberg, Phys. Rev. Lett. **25**, 84 (1970)
51. Y. Shih, C.O. Alley, Phys. Rev. Lett. **61**, 2921 (1988)
52. P. Michler et al., Science **290**, 2282 (2000)
53. R. Singh, G. Bester, Phys. Rev. Lett. **104**, 196803 (2010)
54. M. Bayer et al., Phys. Rev. B **65**, 195315 (2002)
55. R.M. Stevenson et al., Phys. Rev. Lett. **101**, 170501 (2008)
56. M. Ward et al., Nat. Commun. **5**, 4316 (2014)
57. W. Langbein et al., Phys. Rev. B **69**, 161301 (2004)
58. A. Tartakovskii et al., Phys. Rev. B **70**, 193303 (2004)
59. D.J.P. Ellis et al., Appl. Phys. Lett. **90**, 011907 (2007)
60. R. Seguin et al., Appl. Phys. Lett. **89**, 263109 (2006)
61. P. Mrowiński et al., Appl. Phys. Lett. **106**, 053114 (2015)
62. A.J. Bennett et al., Nat. Phys. **6**, 947 (2010)
63. J. Mar et al., Phys. Rev. B **93**, 045316 (2016)
64. B.D. Gerardot et al., Appl. Phys. Lett. **90**, 041101 (2007)
65. M. Vogel et al., Appl. Phys. Lett. **91**, 051904 (2007)
66. S. Seidl et al., Appl. Phys. Lett. **88**, 203113 (2006)
67. J.D. Plumhof, R. Trotta, A. Rastelli, O.G. Schmidt, Nanoscale Res. Lett. **7**, 336 (2012)
68. R. Trotta et al., Phys. Rev. Lett. **109**, 147401 (2012)
69. M.A. Versteegh et al., Nat. Commun. **5**, 6298 (2014)
70. R. Singh, G. Bester, Phys. Rev. Lett. **103**, 063601 (2009)
71. A. Schliwa et al., Phys. Rev. B **80**, 161307 (2009)
72. V. Troncale et al., J. Appl. Phys. **101**, 081703 (2007)
73. D. Oberli et al., Phys. Rev. B **80**, 165312 (2009)
74. Y. Sugiyama, Y. Sakuma, S. Muto, N. Yokoyama, Appl. Phys. Lett. **67**, 256 (1995)
75. A. Mohan et al., Nat. Photon. **4**, 302 (2010)
76. J. Treu et al., Appl. Phys. Lett. **101**, 022102 (2012)
77. X. Liu et al., Phys. Rev. B **90**, 081301 (2014)
78. Y.H. Huo, A. Rastelli, O.G. Schmidt, Appl. Phys. Lett. **102**, 152105 (2013)
79. A. Rastelli et al., Phys. Rev. Lett. **92**, 166104 (2004)
80. P. Atkinson, O.G. Schmidt, S.P. Bremner, D.A. Ritchie, Comptes Rendus Physique **9**, 788 (2008)
81. J.-P. Jahn et al., Phys. Rev. B **92**, 245439 (2015)
82. M. Gong, W. Zhang, G.-C. Guo, L. He, Phys. Rev. Lett. **106**, 227401 (2011)
83. G.W. Bryant et al., Phys. Rev. Lett. **105**, 067404 (2010)
84. F. Ding et al., Phys. Rev. Lett. **104**, 067405 (2010)
85. K.D. Jons et al., Phys. Rev. Lett. **107**, 217402 (2011)
86. J.D. Plumhof et al., Phys. Rev. B **83**, 121302 (2011)
87. Y.H. Huo et al., Nat. Phys. **10**, 46 (2014)

88. A. Rastelli et al., Physica status solidi (b) **249**, 687 (2012)
89. R. Trotta et al., Adv. Mater. **24**, 2668 (2012)
90. R. Trotta, A. Rastelli, *Engineering the Atom-Photon Interaction* (Springer, Berlin, 2015)
91. J. Wang, M. Gong, G.-C. Guo, L. He, Appl. Phys. Lett. **101**, 063114 (2012)
92. R. Trotta et al., Nat. Commun. **7**, 10375 (2016)
93. A. Orieux et al., Phys. Rev. Lett. **110**, 160502 (2013)
94. C. Varnava et al., npj Quantum Information **2**, 16006 (2016)
95. J. Nilsson et al., Nat. Photon. **7**, 311 (2013)
96. D.F. James, P.G. Kwiat, W.J. Munro, A.G. White, Phys. Rev. A **64**, 052312 (2001)
97. J.W. Pan, D. Bouwmeester, H. Weinfurter, A. Zeilinger, Phys. Rev. Lett. **80**, 3891 (1998)
98. N. Akopian et al., Nat. Photon. **5**, 230 (2011)
99. M.A. Pooley et al., Phys. Rev. Appl. **1**, 024002 (2014)
100. J. Wang, M. Gong, G.C. Guo, L. He, Phys. Rev. Lett. **115**, 067401 (2015)
101. R. Trotta et al., Phys. Rev. Lett. **114**, 150502 (2015)
102. S. Baek et al., Science **334**, 958 (2011)
103. J. Zhang et al., Nao Lett. **17**, 501 (2017)
104. K. Hennessy et al., Nature **445**, 896 (2007)
105. L. Sapienza, M. Davanco, A. Badolato, K. Srinivasan, Nat. Commun. **6**, 7833 (2015)
106. A. Dousse et al., Phys. Rev. Lett. **101**, 267404 (2008)
107. M. Gschrey et al., Nat. Commun. **6**, 7662 (2015)
108. T. Huber et al., Nano Lett. **14**, 7107 (2014)
109. P. Kremer et al., Phys. Rev. B **90**, 201408 (2014)
110. S. Barz, G. Cronenberg, A. Zeilinger, P. Walther, Nat. Photon. **4**, 553 (2010)
111. W.B. Gao et al., Nature **491**, 426 (2012)
112. A. Delteil et al., Nat. Phys. **12**, 218 (2015)
113. K. De Greve et al., Nature **491**, 421 (2012)
114. J.R. Schaibley et al., Phys. Rev. Lett. **110**, 167401 (2013)
115. M. Zukowski, A. Zeilinger, M.A. Horne, A.K. Ekert, Phys. Rev. Lett. **71**, 4287 (1993)
116. H.J. Briegel, W. Dr, J.I. Cirac, P. Zoller, Phys. Rev. Lett. **81**, 5932 (1998)
117. N. Gisin, R. Thew, Nat. Photon. **1**, 165 (2007)
118. H. de Riedmatten, M. Afzelius, *Engineering the Atom-Photon Interaction* (Springer, Berlin, 2015)
119. S. Ulrich et al., Phys. Rev. B **90**, 125310 (2014)
120. H. Meyer et al., Phys. Rev. Lett. **114**, 123001 (2015)
121. J.-S. Tang et al., Nat. Commun. **6**, 8652 (2015)
122. N. Sinclair et al., Phys. Rev. Lett. **113**, 053603 (2014)
123. S. Guha et al., Phys. Rev. A **92**, 022357 (2015)
124. R. Al-Khuzheyri et al., Appl. Phys. Lett. **109**, 163104 (2016)
125. G. Chen et al. Sci. Rep. **6** (2016)
126. C. Clausen et al., New J. Phys. **16**, 093058 (2014)
127. N. Prtljaga et al., Appl. Phys. Lett. **108**, 251101 (2016)
128. Y. Zhang et al., Nano Lett. **16**, 5785 (2016)
129. M.H. Devoret, R.J. Schoelkopf, Science **339**, 1169 (2013)
130. F. Najafi et al., Nat. Commun. **6**, 6873 (2015)
131. N.C. Harris et al., Phys. Rev. X **4**, 041047 (2014)
132. G. Reithmaier et al., Nano Lett. **15**, 5208 (2015)
133. N. Somaschi et al., Nat. Photon. **10**, 340 (2016)

Chapter 8
Time-Bin Entanglement from Quantum Dots

Gregor Weihs, Tobias Huber and Ana Predojević

Abstract The desire to have a source of single entangled photon pairs can be satisfied using single quantum dots as emitters. However, we are not bound to pursue only polarization entanglement, but can also exploit other degrees of freedom. In this chapter we focus on the time degree of freedom, to achieve so-called time-bin entanglement. This requires that we prepare the quantum dot coherently into the biexciton state and also build special interferometers for analysis. Finally this technique can be extended to achieve time-bin and polarization hyper-entanglement from a suitable quantum dot.

8.1 Introduction

While the realization of a large, universal quantum computer appears to be some time away, small-scale and special purpose quantum computing devices have been realized or are under construction. Quantum cryptographic protocols, in particular quantum key distribution (QKD), which lets us distribute a secure cryptographic key between two parties, are already commercial to some degree. Yet the distance over which the key exchange can be realized is limited to a few hundred kilometers of optical fiber, due to the inevitable exponentially growing losses and the noise floor or background level of any realistic detector. The practical limits are even shorter because for long distances the key rates will be extremely low.

G. Weihs (✉) · T. Huber
Institut für Experimentalphysik, Universität Innbruck, Technikerstr. 25, 6020 Innsbruck, Austria
e-mail: gregor.weihs@uibk.ac.at

T. Huber (*Present address*)
Joint Quantum Institute, National Institute of Standards and Technology and University of Maryland, Gaithersburg, MD 20849, USA

A. Predojević
Institute for Quantum Optics, University Ulm, Albert-Einstein-Allee 11, 89081 Ulm, Germany
e-mail: ana.predojevic@uni-ulm.de

P. Michler (ed.), *Quantum Dots for Quantum Information Technologies*, Nano-Optics and Nanophotonics, DOI 10.1007/978-3-319-56378-7_8

While for QKD one may resort to classical, trusted repeaters, thus sacrificing the absolute physical security of the key exchange, for connecting quantum information processing devices we will have to implement so-called quantum repeaters [1]. Quantum repeaters break a long distance connection into smaller links over which entanglement is established. Via Bell-state measurements (BSM) at the intermediate nodes the entanglement over the smaller links is then converted into entanglement between the endpoints. What sounds simple in this very abstract description is much more difficult in practice, because we must not assume perfect quantum channels even for the smaller links. While protocols [1] have been devised to cope with the errors, the resulting overhead in resources appears to be forbidding. Only if we start with a high degree of entanglement in the small links will it be feasible to establish the end-to-end quantum channel.

The most frequently used source of entanglement is the spontaneous parametric down-conversion (SPDC) source [2, 3], which produces pairs of entangled photons through a nonlinear optical effect from a pump laser. Unfortunately, SPDC does not create a single entangled pair at a time, but is rather probabilistic, so that a (very) small fraction of the pump laser's photons are converted resulting in a random number of pairs per output pulse or time window. This limits their applicability in quantum repeaters, because there is a fundamental trade-off between a high pair emission rate and the error rate that is caused by multi-pair emissions. This error rate dramatically reduces the achievable distance in a multi-link repeater scenario, even at a two-pair emission probability of only 1% [4].

This is the ultimate reason why quantum communication will eventually need sources of single entangled photon pairs. For now these are limited to single quantum emitters with cascaded optical transitions. Atoms can serve as entangled photon pair sources [5–7], but they require complex atomic beam or trap setups and their overall emission rate is limited. To our knowledge no entangled photon pairs have been produced from single molecules or color centers in solids, which otherwise seem to work well as single photon sources. This leaves semiconductor quantum dots as the only viable solid-state single quantum emitter of entangled photon pairs.

Proposed initially in [8] the biexciton-exciton cascade may emit polarization entangled photon pairs, if the two spin configurations of the intermediate exciton state are degenerate and thus no which-path information is available. The status of polarization entanglement from quantum dots is discussed in detail in Chap. 7. In this chapter we would like to point out that once we have an emitter of photon pairs, there may be other degrees of freedom available to us for realizing photon entanglement. Further we will discuss our and others' results on time-bin entanglement from quantum dots with an outlook on improvements and the possibility of generating hyperentanglement of photons in two degrees of freedom.

The chapter will start by discussing the degrees of freedom of a photon and their measurement, followed by a more detailed discussion of the related phenomena of energy-time and time-bin entanglement. We will show that a coherent excitation mechanism is required for obtaining time-bin entanglement from a quantum dot and will discuss the optimal conditions. Finally we will present the results on time-bin entanglement and an outlook.

8.2 Photon Degrees of Freedom

Without resorting to a particular interpretation we may define a photon to be an elementary excitation of a quantized mode of the electromagnetic radiation field. A mode is a solution to the wave equation under particular boundary conditions and in particular we can always resort to monochromatic solutions so that the modes are harmonic solutions with a particular frequency ω and the spatial part is the corresponding solution of the Helmholtz equation. In a box-like quantization volume with fixed or periodic boundary conditions these will be plane waves. From these plane, monochromatic waves we may build other monochromatic modes by unitary transformations. In experiments these unitary transformations between different sets of modes or wavevectors are effected by beam-splitters or other, similar couplers. We may further resort to non-monochromatic or spatio-temporal modes, like wavepackets, which obviously will not necessarily be orthogonal, but in most practical cases may be constructed close to orthogonal [9].

A plane wave is characterized by a wavevector **k** with magnitude $k = \omega n(\omega)/c$ and a polarization unit vector **e**, which is orthogonal to **k**. Because there are always two orthogonally polarized modes for any **k**, a photon may have any state of polarization that can be described as a superposition of the two, i.e. any state on the Poincaré sphere.

A single photon with a given frequency and wavevector is thus a perfect two-state system, or qubit, with degenerate energy levels. The polarization of a photon can be manipulated easily using retarders (wave-plates) and measured using polarizers, which effectively project any incoming polarization to the one transmitted. Polarizing beam-splitters (PBS), also called two-channel polarizers, are devices that couple polarization and spatial mode (wavevector) by transmitting light that is polarized parallel to the plane of incidence (p) and reflecting light that is polarized perpendicular to the same (s).

In contrast to the polarization, which is discrete, the continuous degrees of freedom frequency and wavevector allow storing more information in one photon [10]. In most practical cases we will strive to define a discrete but not necessarily binary set of modes for transmitting and manipulating photonic quantum information, because the analysis in the presence of noise and distortion through a channel will become difficult for continuous encoding. Several schemes have been put forward and demonstrated for wavevector-spatial coding: the dual-rail qubit [11] and its multi-rail extension, transverse paraxial mode coding, in particular orbital angular momentum (OAM) [12] and similar rotationally invariant coding [13]. In the frequency/energy-time dimension time-bin [14] and multi-time-bin coding have been used as well as generalized temporal mode [15, 16] and frequency mode coding [17]. A more thorough discussion of photonic quantum information encoding is given in Chap. 7.

Some of these encodings promise good stability of the quantum state under propagation either in free space or in optical waveguides. On the other hand their manipulation and analysis (decoding) present more difficulties than in the simple case of polarization. In all cases one has to deal with some kind of interferometer. As an example, for OAM decoding only recently efficient methods were found [18]. In practice one will thus choose an encoding that is robust for the chosen channel. There seems to be some general agreement that the most robust encoding for long-distance transmission of quantum information in optical fibers is time-bin encoding or some variant of it, e.g. differential phase shift keying (DPSK).

8.3 Time-Bin Encoding and Entanglement

In classical communication a large variety of modulation schemes is known, both in incoherent and coherent communication, analog and digital. For photons, any classical scheme can be used or adapted in principle. The only thing that changes are the fundamental noise limits given by the uncertainty principle for amplitude and phase.

In this sense time-energy wavefunctions and time-bin qubits are particular quantum variants of classical phase-shift-keying (PSK), even though in the quantum realm we rarely use continuous-wave carriers. A time-bin qubit is defined via two usually pulse-like quasi-orthogonal temporal wavepackets as shown in Fig. 8.1. A general pure state is thus $|\psi\rangle = \alpha\,|\mathrm{E}\rangle + \beta\,|\mathrm{L}\rangle$. The superposition bases $|\mathrm{E}\rangle \pm |\mathrm{L}\rangle$ are sometimes called energy bases, even though this terminology is only accurate if we are talking about energy (frequency) eigenstates, i.e. plane waves, which are complementary to a time basis with temporal δ-distributed wavefunctions. The particular wavepacket shape will either be determined by the generating optical (laser) pulse or the decay properties of the generating quantum emitter.

Before discussing advantages and disadvantages of the time-bin encoding we would like to look at the historical perspective. Temporal superpositions of photons were first proposed by Franson in [19] in the context of entanglement and Bell's inequalities. To our knowledge, his original proposal of using cascaded transitions in atoms was never realized to generate energy-time entangled photon pairs or demonstrate a violation of Bell's inequality. Most experiments [20–22] used SPDC as the photon pair source in which the coherence of the pump laser provides the coherent superposition of the early and late times. The requirement for using discrete time

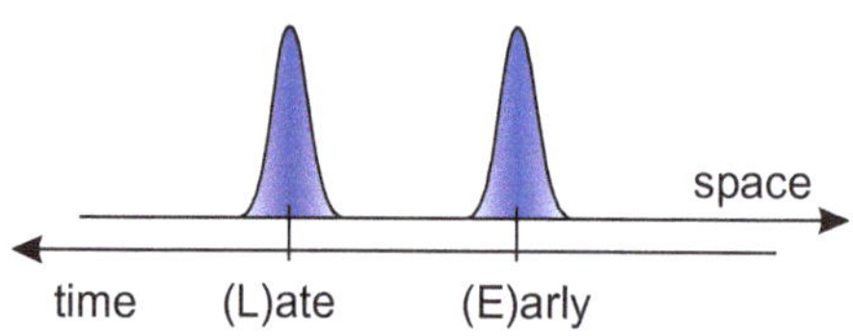

Fig. 8.1 Time bins are quasi-orthogonal wavepacket envelopes, which form the two quantum states of a photonic qubit when occupied by a single photon. In this picture we imagine the wavepackets propagating to the right in real space. The photon can then be in either the early (E) or late (L) state or any superposition thereof

bins through pulsed pumping of the SPDC instead of the continuous variant derived from the desire to use the so-encoded qubits in protocols that require interferometric Bell-state analysis on photons from different sources such as quantum teleportation [23] and entanglement swapping [24].

For long distance quantum communication through optical fibers time-bin encoding has a decisive advantage. The relative phase between two pulses that are only a few nanoseconds apart in time is only altered by changes in the environment that are faster than their temporal separation, i.e. in the GHz range. This kind of encoding can also be seen as a temporally multiplexed version of the dual-rail qubit (see Fig. 8.2). Nevertheless, chromatic dispersion can play a role both through the induced pulse spreading and inside the imbalanced interferometers that are required for time-bin analysis.

In order to measure a time-bin qubit in a superposition basis we have to delay the early time bin and interfere it with the late one. For this purpose we use an imbalanced Mach-Zehnder or Michelson interferometer (see Fig. 8.2). Obviously this analysis is lossy, because only 50% of the photons will experience the correct delay. Half of the early time bin will not be delayed and half of the late time bin will be delayed even further. This results in a temporal pulse pattern as shown in Fig. 8.3. If the beam splitters are symmetric, then only bases on the equator of the Bloch sphere (with E and L at the poles) can be analyzed. For $\phi = 0$ the interferometer outputs correspond

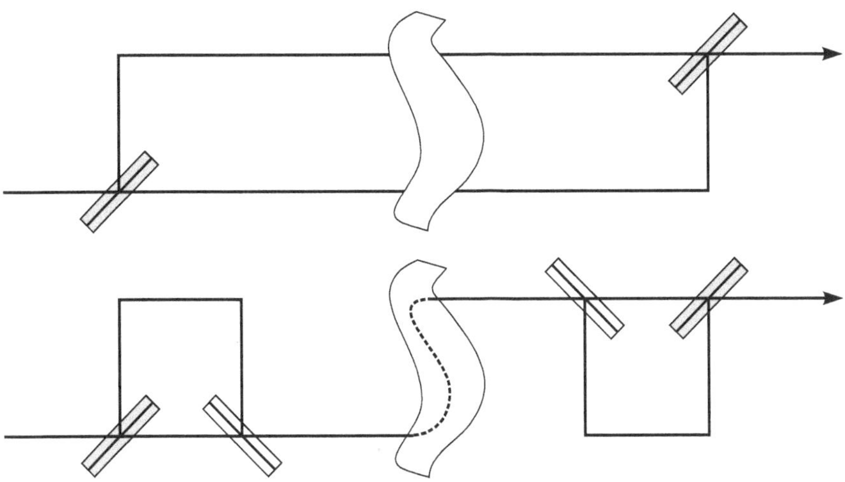

Fig. 8.2 *Top* the first beamsplitter creates a superposition of a photon being either in the *upper* mode (rail) or the *lower* one. The second one analyzes the superposition depending on the phase difference accrued between the two paths along their entire length, or in other words, it converts the superposition back to the photon going to either of its outputs. *Bottom* additional mirrors (not shown) and beamsplitters multiplex the *upper* mode onto the *lower* one with some delay. The second imbalanced interferometer undoes the delay for measurement. This works only probabilistically, with 50% efficiency in each interferometer, i.e. 25% overall for analysis in superposition bases. Better performance could be achieved by using a switch

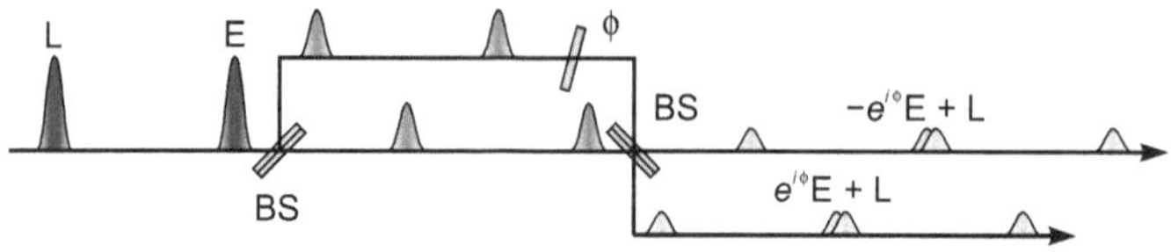

Fig. 8.3 Time-bin analysis occurs through imbalanced interferometers built from two beamsplitters (BS) and mirrors (not shown). The output probability of a photon is distributed into three pulses, where the middle ones are the two complementary superpositions. However, only half of the total probability (photons) will be in the middle pulses. Depending on the phase of the initial state and the interferometer phase ϕ this probability will be distributed between the two interferometer outputs. The phase shift allows us to analyze with regard to a particular basis, i.e. X, Y or any other equal-amplitude superposition of the E and L states. The first and third pulse contain the other half of the early and late pulses, i.e. when detected, they give projections to the E/L (Z) basis

to the qubit states $\pm X$, for $\phi = i$ to $\pm Y$. More general, universal time-bin analyzers require beamsplitters with adjustable splitting ratios [25] to allow arbitrary amplitude superpositions.

The ultimate time-bin analyzer (and encoder) uses a switch (switchable mirror) instead of the first beamsplitter. The switch would have to route the early time bin along the long arm and the late time bin along the short arm. The splitting ratio of the second beamsplitter then defines the amplitude ratio of the superposition states that are to be analyzed.

While time-bin encoding is very stable under propagation through optical communication channels, the stability of an imbalanced interferometer may be a concern. They have been realized in free-space and fiber versions and in both cases one needs to add a phase stabilization laser and ensure the best possible mechanical and thermal stability. Because of the large imbalance the stabilization laser not only needs to have a long enough coherence length but is also required to be locked in its absolute wavelength, which is almost always chosen different from the wavelength of the single photons to avoid stray laser light reaching the sensitive single photon detectors.

So far the discussion concerned a single time-bin qubit. Things get somewhat more complicated for two qubits, which may be entangled or not. The situation is visualized in Fig. 8.4, where a source emits pairs of time-bin encoded photons. Both photons are analyzed in identical interferometers and detectors, whose detection times are recorded as t_1 and t_2. A simple start-stop measurement between the two sides is not sufficient, as it would lump the superposition basis events in with other simultaneous detection events. Instead one needs to either record (time-tag) the arriving photons in absolute time or at least determine the time difference to a synchronization signal from the source on each side.

A perfect source of time-bin entanglement would produce the maximally entangled state

$$|\Phi\rangle(\phi_l) = \frac{1}{\sqrt{2}}\left(|\mathrm{E}_1\mathrm{E}_2\rangle + e^{i\phi_l}\,|\mathrm{L}_1\mathrm{L}_2\rangle\right), \tag{8.1}$$

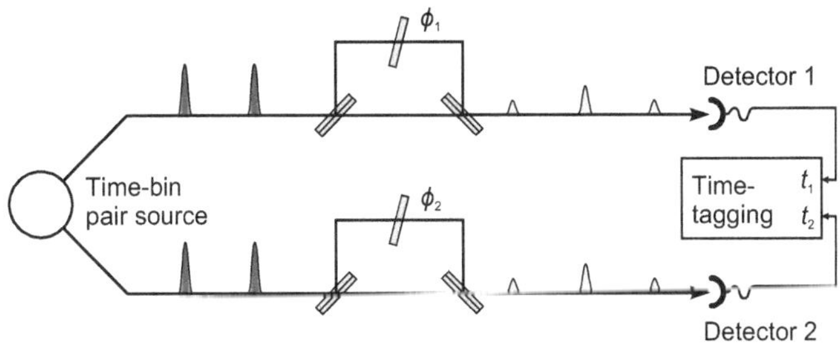

Fig. 8.4 Analyzing time-bin encoded photon pairs requires time-bin analyzers with phase settings ϕ_1 and ϕ_2 to set the basis on both sides and time-correlated detection. The amount of imbalance is kept at the minimum allowed by the time resolution of the detectors or the minimum possible wavepacket duration for the source

where the phase ϕ_l is internal to the source. It may originate, for example, from the superposition of pump pulses in SPDC. By setting $\phi_l = 0, \pi$ one thus obtains the $\Phi^{+,-}$ Bell states, respectively. The Bell states are defined and discussed in Sect. 7.1.2. For the Φ^+ state the coincidence count rate for the two middle pulses in a pairing of two equivalent outputs of the analyzers will then vary as $\frac{1}{16}(1 + \cos(\phi_1 + \phi_2))$, i.e. a coincidence probability of $1/8$ for $\phi_1 + \phi_2 = 0$. The same is true for the second equivalent pairing and both are complementary to the coincidence count rates for the inequivalent output pairings. Therefore, in total only for one quarter of all emitted pairs both photons are detected in the superposition basis, for another quarter both photons are detected in the time-bin basis and for the remaining half, one photon each is detected in the superposition and time-bin basis, respectively.

8.4 Time-Bin Entanglement from Single Quantum Emitters

In the original proposal by Franson [19] a long-lived upper level in an atom provided the required coherence between the early and late cascade emission of a photon pair. In SPDC the coherence of an earlier or later produced photon pair is provided by the coherence time of the pump laser, whose phase will be the sum phase of the paired photons. This can either happen with a continuous-wave laser for Franson-type entanglement or with a coherent superposition of an early and late pump pulse for time-bin entanglement. The laser pulses can be produced by an imbalanced interferometer or directly from a mode-locked laser.

For most quantum dots we do not know of really long-lived levels that decay in a cascade, so time-bin entanglement appears to be the only option, which also happens to be more relevant for quantum communication purposes. To achieve time-bin entanglement from a quantum dot the phase difference between the two pump laser pulses has to be carried over to the phase difference between the emitted photon pairs, and thus also intermediately to the phase difference between the two possible

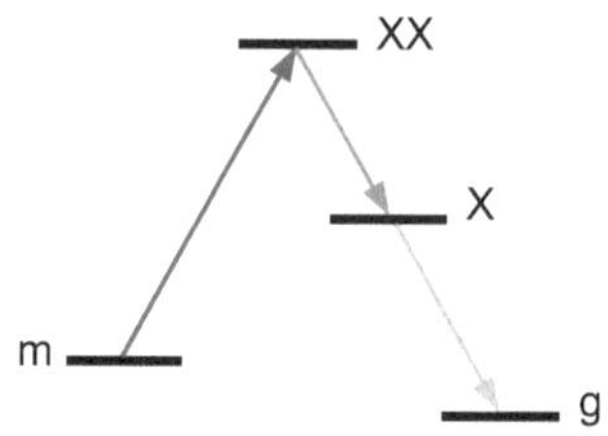

Fig. 8.5 Bringing the quantum dot to a metastable level before exciting it to the biexciton state ensures that it will not be excited a second time by the second excitation pulse. This prevents the emission of two pairs, which constitutes an error

excitations of the upper level that will decay in a two-photon cascade. It was originally proposed [26] that the quantum dot be brought to a metastable state and then further excited to the topmost level of the cascade, the biexciton level as shown in Fig. 8.5. Dark excitons were proposed in the same reference as potential metastable states and are discussed in Chap. 4. Because of the difficulty of exciting quantum dots into dark exciton states a simpler version is to go directly from the ground state to the biexciton, with the drawback of a possible second excitation by the late pump pulse. Yet, such an experiment can prove the general possibility of time-bin entanglement.

In either case it is good to keep in mind that in order for the two possible emission cascades to be indistinguishable it is necessary that no trace be left in the pump field or the environment of the quantum dot. This however, does not mean that the biexciton level has to have a coherence time that is long enough to span the gap between the two excitation pulses, but only that its dephasing is considerably shorter than the emitted wavepacket. The fact that the cascade itself is not always the same, i.e. that the exciton state has a finite lifetime, does not degrade the achievable time-bin entanglement. As in SPDC the phase difference between the exciton and biexciton photon is irrelevant for the entanglement, which depends only on the sum phase. However, as discussed in [26] the uncertainty stemming from the exciton lifetime does lead to an entanglement between the biexciton and exciton photons of a pair. Because this would limit their usefulness for multi-photon protocols such as quantum repeaters, the authors of [26] proposed to employ microcavities to modify the lifetimes such that this unwanted entanglement would be eliminated.

8.5 Two-Photon Coherent Excitation of a Quantum Dot

The central goal of photon pair generation from quantum dot systems is to get exactly one photon at the biexciton and one photon at the exciton frequency that are produced within a short time interval and with a well defined sum phase. This is possible and the exciton and biexciton transition frequencies are well separated due to the existence of the biexciton binding energy. Nonetheless, to accomplish the generation process coherently, the quantum dot needs to be excited resonantly. A thorough review of the various optical excitation and control methods is given in Chap. 3.

This task is, despite the favorable energetic structure, not trivial to achieve in epitaxial semiconductor quantum dots. The first, and most important reason is the excess scattered laser light that can easily be much stronger than the single photon signal emitted by the quantum dot. Therefore, the traditional way to excite quantum dots is above-band excitation. Here, one uses a laser with an energy higher than any transition in the quantum dot. This laser creates a multitude of carriers in the vicinity of the quantum dot that can be probabilistically trapped in the quantum dot potential. While it is possible to achieve very high single photon count rates with this method, the probabilistic nature of this process reduces the suitability of such a source for quantum information protocols. Another negative feature of the above-band excitation is related to how exactly the quantum dot levels are populated. Namely, biexciton excitations will only be created once the exciton level has been filled and therefore a high rate of biexciton photons requires a very large number of carriers in the quantum dot vicinity. This, however, is very unfavorable because it promotes dephasing of the quantum dot levels due to the electric field fluctuations and causes poor photon statistics properties due to processes like carrier re-capture.

A way to overcome these issues is to exploit the biexciton binding energy, which sets the emission lines of exciton and biexciton photons far apart. When in such a system the excitation laser light is tuned to an energy in between these two energies it produces a resonant two-photon coupling between the ground and biexciton states as shown in Fig. 8.6. The two-photon approach to excite quantum dots was initially shown in [27] on II–VI quantum dots, but it is quite a bit more difficult to apply it to III–V quantum dots. II–VI quantum dots typically have a much larger biexciton binding energy (the difference between the exciton and the biexciton line wavelengths can be more than 10 nm), but exhibit otherwise unfavorable optical properties; II–VI quantum dots emit photons in the blue and green spectral range that are, due to losses in the optical fibers, not very suitable for quantum communication. The values for the energy difference between biexciton and exciton lines in III–V quantum dots are in the range of 1–2 nm. Therefore, these systems demand a more thoughtful approach to reduce the laser scattering. Earlier work on III–V quantum dots [28] showed the signatures of resonant excitation, like Rabi oscillations, but only in photo-current measurements and not in the optical signal. The first optical measurements under two-photon resonant excitation on III–V quantum dots were shown in [29]. It turns out that this type of excitation also enables and improves several other emission properties compared to the traditional above-band excitation [30, 31] but does not completely remove the blinking due to the random occurrence of charged quantum dot states. This blinking behavior can be improved to some degree by photo-neutralization [32].

The coherence of the excitation process enables coherent manipulation of the phase of the ground-biexciton state superposition, which is crucial for obtaining time-bin entanglement as pointed out in the previous section. The traditional way to characterize the coherence between energy levels is to perform a Ramsey interference measurement in which the investigated system is excited using a sequence of two

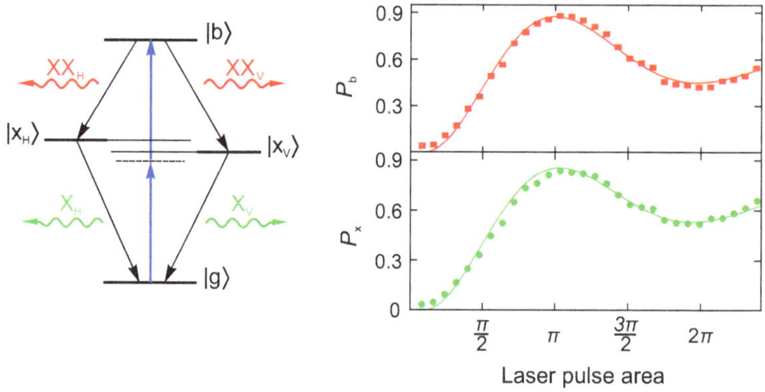

Fig. 8.6 The quantum dot energy scheme on the *left* shows the exciton fine structure splitting as the energy difference between two exciton levels $|x_H\rangle$ and $|x_V\rangle$. In the process of two-photon resonant excitation a pulsed laser (shown as arrows pointing *upwards*) with half the energy of the biexiton state $|b\rangle$ coherently couples the ground ($|g\rangle$) and biexciton states through a virtual level (*dashed line*). The biexciton recombination takes place through the intermediate exciton states ($|x_H\rangle$ or $|x_V\rangle$) emitting biexciton ($XX_{H,V}$) and exciton ($X_{H,V}$) photons, respectively. On the *right* the measured biexciton emission probability, P_b, and exciton emission probability, P_x, as functions of the laser pulse area are compared to a simulation (*solid line*) that includes linearly intensity-dependent dephasing. The experimental error bars are smaller than the symbols

consecutive $\pi/2$ pulses, Fig. 8.7a. The first of these pulses brings the quantum dot into a superposition of the ground and biexciton states. Upon this pulse, the system is allowed to evolve freely for a time defined by the variable delay between the pulses, Fig. 8.7. During the free evolution the excitation pseudo-spin precesses along the equator of the Bloch sphere. The second pulse will map the population either back to the ground state or flip it further to the biexciton state, depending on the evolution of the pseudo-spin and the relative phase between the two pulses. A very thorough review of the coherent manipulation of excitons and spins in quantum dot systems is given in [33].

When such an experiment is performed in two-photon excitation it results in Ramsey interference fringes in both the exciton and the biexciton emission [27]. It is important to note here that in the case of the biexciton emission these fringes are a direct result of the laser driving the transition. The interference observed in the exciton channel closely follows the behavior of the biexciton but comes as a consequence of the cascade decay of the system. The Ramsey interference measurement characterizes the coherence of the ground-biexciton state superposition and by varying the delay between the two Ramsey pulses one can measure the coherence decay of this pseudo-spin. An example of the decay of the Ramsey interference fringe visibility is shown in Fig. 8.7c.

Decoherence caused by low frequency noise can be eliminated by applying a refocusing pulse. Such a measurement is commonly called spin echo (also Hahn echo) and requires a sequence of three consecutive pulses of different intensities

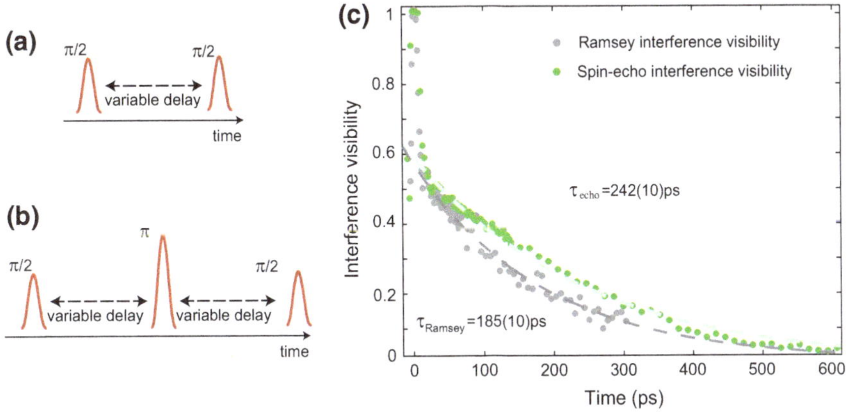

Fig. 8.7 **a** Pulse sequence consisting of two $\pi/2$ pulses applied with variable delay. **b** The spin-echo pulse sequence. **c** The Ramsey interference visibility decay experiment as monitored by the emitted biexciton photons is shown in *gray*. The data shown in *green* are from a spin-echo measurement performed on the same emitter

$(\pi/2, \pi, \pi/2)$ as illustrated in Fig. 8.7b. Due to their lifetime quantum dots are usually excited using laser pulses that are not longer than a few picoseconds. Therefore the simplest way to obtain the sequence of Ramsey pulses is by feeding pulsed laser light into a variable-length Michelson interferometer. Concerning the spin echo measurements, it is quite straightforward to implement such a measurement in systems that have long lifetimes and coherence times. For example, for a trapped ions system where the coherence times are of the order of a millisecond one can use light derived from a continuous-wave laser and create the pulse sequence using an acousto-optical modulator. Driving the ground-biexciton state superposition of a quantum dot, however, requires pulse durations of a few picoseconds. In [29] it was shown that the echo sequence with such pulses can be constructed by using a Michelson interferometer in double-pass configuration. Such an implementation is capable of delivering the three consecutive pulses necessary for the spin-echo sequence with the middle pulse being a result of the interference between the light passing once through the interferometer with the light passing twice. Figure 8.7c shows two sets of data, one taken in a Ramsey and the other in a spin-echo interference experiment.

The creation of time-bin entanglement requires a phase stable generation of subsequent photon pairs, which can be hampered by the phase uncertainty in the biexciton generation. To predominantly generate single pairs of photons through the biexciton decay, one needs to avoid populating the single exciton state as well as the re-excitation of the biexciton state after a decay within the same laser pulse. This creates conflicting requirements for the excitation pulse length. Namely, short pulses suppress dephasing and decay within the pulse duration, but have large bandwidth and high peak intensity, which increases the off-resonant generation of single

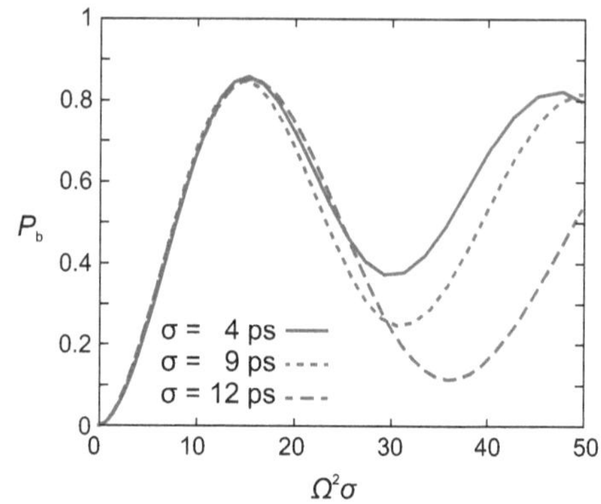

Fig. 8.8 Simulated emission probability for the biexciton, P_b for constant dephasing as a function of $\Omega^2\sigma$, where Ω is the Rabi frequency and σ the pulse length. The damping of the Rabi oscillations strongly depends on the length of the excitation pulse

excitons and power induced phase shifts. Longer pulses make the system more vulnerable to background dephasing, decay during the pulse and thus multiple excitations.

Nevertheless, one can find an optimized operation regime for the parameters of the system under consideration. The interaction of the quantum dot with the semiconductor environment does not seem to influence this optimization. In [31] we showed that one can choose an excitation pulse length that favors the creation of biexcitons while suppressing the creation of unpaired excitons. This dependence is illustrated in Fig. 8.8 showing the results of a theoretical simulation conducted in [31].

This result has an important consequence. It indicates an existence of a tradeoff between the excitation-pulse length and the biexciton binding energy. In particular it favors the use of quantum dots with large biexciton binding energy that in return allow using short excitation pulses. In addition, such excitation pulses reduce the excitation jitter and are therefore more favorable for quantum information applications.

8.6 Time-Bin Entangled Photon Pairs from a Quantum Dot

Written in terms of the biexciton (XX) and exciton (X) photon modes, the state given in (8.1) reads

$$|\Phi\rangle = \frac{1}{\sqrt{2}}\left(|\mathrm{E_{XX}E_X}\rangle + e^{i\phi_l}\,|\mathrm{L_{XX}L_X}\rangle\right), \tag{8.2}$$

where E(L) denotes the early (late) time bin, XX (X) the biexciton (exciton) recombination photon and ϕ_l is the phase between the two pump pulses. In the previous section, we explained how the quantum dot can be excited resonantly. The phase ϕ_l in (8.2) is the reason why a resonant pumping scheme is necessary. If the pump process is not phase preserving, like above-band excitation, ϕ_l will not be the phase between the two pump pulses but some random phase in each emission event, resulting in an overall mixed state. The coherent excitation of the biexciton directly from the ground state, enables the possibility to transfer the phase from the laser onto the quantum dot system thereby creating an output of the desired form given in (8.2).

While still superior to SPDC sources, which emit thermal pair distributions, one drawback of the presented scheme is the rare but inherent creation of four-photon events, even with a perfect quantum dot with zero multi-photon emission. This comes from the fact that the entanglement generation depends on a probabilistic generation of one photon cascade either in the E or in the L time bin. A photon cascade in both of the time bins is a four-photon event, outside the single-pair Hilbert space, and therefore the excitation probability has to be kept at a reasonably low level. The same problem occurs with time-bin entangled photon pairs from SPDC. Contrary to the case of SPDC, two solutions to this problem are known. As discussed above, the first one was proposed by Simon and Poizat [26], which is using a metastable state as the initial state. Thereby, a deterministic creation of the time-bin entangled state is possible, without the loss of other degrees of freedom. This idea has not yet been demonstrated experimentally. The second one, which was already demonstrated experimentally, is to create the entanglement in a different degree of freedom, e.g. polarization, and convert this entanglement to time bin [34]. This however requires the availability of suitable quantum dots with zero fine-structure splitting, which may have other disadvantages. The conversion requires that either fast polarization switches are used or an extra 75% combined loss for the pairs is accepted. A further drawback of this solution is that the simultaneous creation of entanglement in the polarization and time-bin degrees of freedom, so called hyper-entanglement is not possible.

Let us come back to the analysis of the time-bin entanglement, which was already discussed in Sect. 8.3. As shown in Fig. 8.4, the middle pulses coming out of each analyzing interferometer yield the superposition bases measurements that are important to demonstrate entanglement. As the phases ϕ_1 and ϕ_2 are varied, entanglement manifests itself in a variation of the rate of coincidence counts between pairings of two output pulses, one of each interferometer. For a maximally entangled state like (8.1) the individual, single count rate would remain constant, independent of the phases, because either photon is individually in a mixed state of the early and late time bins. In coincidence, however, the time bins are interfering, because it is not possible, not even in principle, to tell in which time bin the photon cascade was created and which paths the photons took in the analyzing interferometers. For an imperfect state the coincidence rate will oscillate with an interference visibility that depends on the indistinguishability of the early and late cascades.

In our experimental realization a pulsed laser (80 MHz repetition rate, 12 ps pulse duration) coherently drove the ground-biexciton transition with a probability of 6%. To create the two pump pulses, we sent the laser light through an imbalanced Michelson interferometer with a fixed length difference of 1 m. This interferometer plus the resonantly pumped quantum dot is the time-bin pair source in Fig. 8.4. After frequency and polarization selection of the XX and X photon, the photons were sent through additional beam paths inside the same physical interferometer (see Fig. 8.9). This ensured the same path length difference for all the three interferometers. Furthermore, any global phase drift would affect all three interferometers equally, thus

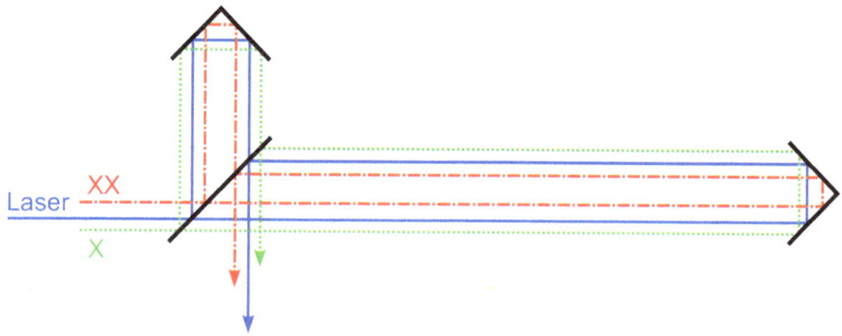

Fig. 8.9 Three different paths in one physical realization of one interferometer were used as the pump interferometer and the three analyzing interferometers

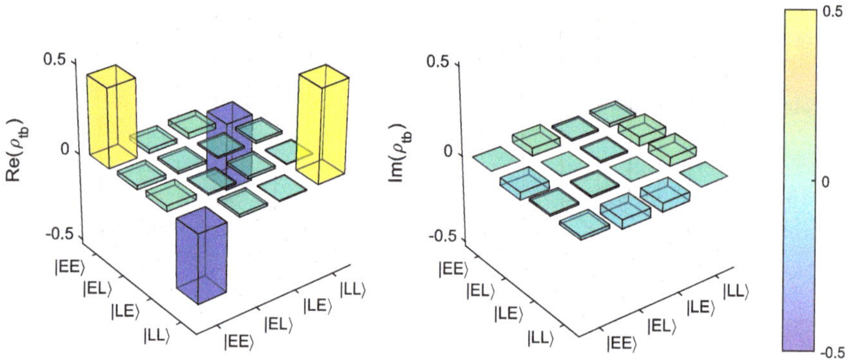

Fig. 8.10 Real and imaginary part of a reconstructed density matrix. This matrix was measured with a 12 ps excitation pulse and 6% excitation probability. $|\mathrm{EE}\rangle$ - $|\mathrm{LL}\rangle$ denote the measurement basis states

no stabilization is required. If one wants to use the time-bin entangled photons for any real-world quantum protocol, this interferometer has to be unfolded and stabilized.

For the analysis of the created time-bin entangled state, we used the method of tomographic reconstruction, which needs measurements in a variety of bases, i.e. phase settings. Details on the reconstruction can be found in [35]. The resulting two-photon density matrix ρ can be seen in Fig. 8.10. In general it will not be a pure state, but exhibit some degree of mixedness. The diagonal of the matrix from $|\mathrm{EE}\rangle\langle\mathrm{EE}|$ to $|\mathrm{LL}\rangle\langle\mathrm{LL}|$ represents classical correlations in the E/L basis. The off-diagonal elements are also called the coherences of the state and quantify the entanglement present in the output.

If a source is reasonably close to the desired ideal state it makes sense to quantify the overlap with that state, the so-called fidelity as an elementary measure of the achieved quality. The fidelity $\mathcal{F}$ of an arbitrary mixed state ρ with a pure target state $|\psi\rangle$ is defined as $\mathcal{F} = \langle\psi|\,\rho\,|\psi\rangle$. For the density matrix shown in Fig. 8.10 the fidelity towards the state $\left|\Phi^-\right\rangle$ (see (7.11)) is $\mathcal{F} = 0.88(3)$.

Unfortunately, the quantum dot community has long been using the fidelity as a substitute for a proper entanglement measure. This is not a good practice and it is better to calculate the concurrence, which is defined as

$$C(\rho) = \max(0, \lambda_1 - \lambda_2 - \lambda_3 - \lambda_4), \tag{8.3}$$

where λ_1, ..., λ_4 are the eigenvalues, in decreasing order, of the matrix

$$R = (\sqrt{\rho}\,\widetilde{\rho}\sqrt{\rho})^{1/2}, \tag{8.4}$$

where $\widetilde{\rho} = (\sigma_y \otimes \sigma_y)\rho^*(\sigma_y \otimes \sigma_y)$. σ_y is the Pauli matrix $\begin{pmatrix} 0 & -i \\ i & 0 \end{pmatrix}$ for a spin flip and ρ^* is the complex conjugate of ρ. The concurrence is $C = 0$ if no entanglement is present and $C = 1$ for a maximally entangled state. For the density matrix given in Fig. 8.10 the concurrence is 0.78(6).

These values compare well with earlier achievements in polarization entanglement from quantum dots and allow, at least in principle, a violation of Bell's inequality. For practical applications we would still like to see some improvements. Looking at the density matrix (Fig. 8.10) we can identify some shortcomings. First, there is a small imbalance between $|\mathrm{EE}\rangle$ and $|\mathrm{LL}\rangle$, which is either due to a slightly different pump pulse energy or different transmissivity of the long and short analyzing interferometer arms. Second, we notice that the magnitude of the coherences is smaller than that of the diagonal elements. This is a result of several effects that limit the indistinguishability of the early and late cascade, including dephasing during the excitation process and during the lifetime of the biexciton state. The former effects were discussed in detail in Sect. 8.5. The dephasing during the lifetime of the biexciton is most likely due to the phonon environment remaining at temperatures around 5 K and also due to the fast components (comparable to the biexciton lifetime) of spectral diffusion, which in turn is usually attributed to the fluctuations in the charge environment around the quantum dot. The impact of both these detrimental effects could be reduced most by a lifetime reductions, e.g. using a microcavity and its Purcell effect, but so far no results on resonant two-photon excitation of quantum dots in microcavities have been reported.

8.7 Outlook

The level of time-bin entanglement that has been achieved with quantum dots to date is quite remarkable. Direct single-pair emission, however has not yet been achieved. At this point one needs to work at rather low excitation probabilities and the quantum dot structures that have been used exhibit rather low outcoupling and collection efficiencies. This results in an overall low pair count rate even though the rate of actual excitation events is in the MHz range. This also means that it is difficult to optimize all the relevant parameters for a given quantum dot. For this reason one should apply the same technique to new structures that promise much higher count rates such as nanowire quantum dots [36] or quantum dot microlenses [37]. The two-photon resonant excitation of these structures may be more difficult, but should be achievable with stronger spectral filtering of the luminescence. Having higher pair count rates will allow investigating the conditions that are required for even better time-bin entanglement. To reduce any unwanted entanglement within a time bin, it would be interesting to also try this out with micropillar microcavities [38]. The cavity would be tuned to resonance with the biexciton transition so that the biexciton lifetime is shortened by the Purcell effect as proposed in [26]. At the same time the increased collection efficiency might make this the perfect time-bin entanglement source.

For creating single time-bin entangled pairs it will be most interesting to investigate the dark exciton preparation (see Chap. 4 and [39]) and how to coherently transfer from the dark exciton to the biexciton. Single entangled pairs enable quantum communication protocols with much higher efficiency than entangled pairs from SPDC. Another idea that could readily be demonstrated is the creation of hyperentanglement in the polarization and time-bin degrees of freedom, i.e. a state of the form

$$|\psi\rangle = (|\mathrm{HH}\rangle + |\mathrm{VV}\rangle) \otimes (|\mathrm{EE}\rangle + |\mathrm{LL}\rangle). \tag{8.5}$$

This is useful for certain linear optical quantum information protocols. For example, it is possible to exploit the extra entangled degree of freedom to perform perfect Bell-state analysis, the central process of quantum teleportation and entanglement swapping. Another direction lies in the observation that time-bin encoding and entanglement is not limited to two-dimensional (qubit) configurations. In other systems high-dimensional time encoding has been investigated [40] but not yet for any single quantum emitter. Finally, if we consider multi-photon entanglement through multi-level cascades in quantum dots or quantum dot molecules, time-bin entanglement may be the only possible way to establish multipartite entangled states such as the GHZ or W states directly from the source.

In summary, the temporal degree of freedom of the photon can be a valuable resource, which has not yet been sufficiently explored for single quantum emitters. It is versatile, because it can apply to any cascaded transition without any particular requirements on energy or spin structure. It does, however require the possibility of coherent control of the topmost energy level of the cascade. In our opinion this is something that ties in with other developments in single emitters, where all properties of single photon sources improve when dedicated coherent interactions are used rather than the primitive above-band pumping. Admittedly, the coherent control increases the complexity of the optical setup, but barring any massive breakthroughs there seems no other way to go.

References

1. H.J. Briegel, W. Dür, J.I. Cirac, P. Zoller, Phys. Rev. Lett. **81**, 5932 (1998). doi:10.1103/PhysRevLett.81.5932
2. P.G. Kwiat, K. Mattle, H. Weinfurter, A. Zeilinger, A. Sergienko, Y. Shih, Phys. Rev. Lett. **75**, 4337 (1995). doi:10.1103/PhysRevLett.75.4337
3. A. Predojević, S. Grabher, G. Weihs, Opt. Express **20**(22), 25022 (2012). doi:10.1364/OE.20.025022, http://www.opticsexpress.org/abstract.cfm?URI=oe-20-22-25022
4. S. Guha, H. Krovi, C.A. Fuchs, Z. Dutton, J.A. Slater, C. Simon, W. Tittel, Phys. Rev. A **92**, 022357 (2015). doi:10.1103/PhysRevA.92.022357
5. S.J. Freedman, J.F. Clauser, Phys. Rev. Lett. **28**(14), 938 (1972). doi:10.1103/PhysRevLett.28.938
6. A. Aspect, P. Grangier, G. Roger, Phys. Rev. Lett. **47**(7), 460 (1981). doi:10.1103/PhysRevLett.47.460
7. B. Weber, H.P. Specht, T. Müller, J. Bochmann, M. Mücke, D.L. Moehring, G. Rempe, Phys. Rev. Lett. **102**, 030501 (2009). doi:10.1103/PhysRevLett.102.030501
8. O. Benson, C. Santori, M. Pelton, Y. Yamamoto, Phys. Rev. Lett. **84**(11), 2513 (2000). doi:10.1103/PhysRevLett.84.2513, http://link.aps.org/abstract/PRL/v84/p2513
9. W. Vogel, D.G. Welsch, *Quantum Optics* (Wiley-VCH, Weinheim, 2006)
10. W. Weihs, Quantum Inf. Comput. **1**, 3 (2001), http://www.rintonpress.com/journals/qic-1-2/mainfile.zip
11. T.C. Ralph, G.J. Pryde, *Progress in Optics*, vol. 54 (Elsevier, 2010), pp. 209–269. doi:10.1016/S0079-6638(10)05409-0, http://www.sciencedirect.com/science/article/pii/S0079663810054090
12. A.M. Yao, M.J. Padgett, Adv. Opt. Photonics **3**(2), 161 (2011). doi:10.1364/AOP.3.000161, http://aop.osa.org/abstract.cfm?URI=aop-3-2-161
13. V. D'Ambrosio, E. Nagali, S.P. Walborn, L. Aolita, S. Slussarenko, L. Marrucci, F. Sciarrino, Nat. Commun. **3**, 961 (2012). doi:10.1038/ncomms1951
14. I. Marcikic, H. de Riedmatten, W. Tittel, V. Scarani, H. Zbinden, N. Gisin, Phys. Rev. A **66**, 062308 (2002). doi:10.1103/PhysRevA.66.062308
15. X. Xing, A. Feizpour, A. Hayat, A.M. Steinberg, Opt. Express **22**(21), 25128 (2014). doi:10.1364/OE.22.025128, http://www.opticsexpress.org/abstract.cfm?URI=oe-22-21-25128
16. B. Brecht, D.V. Reddy, C. Silberhorn, M.G. Raymer, Phys. Rev. X **5**, 041017 (2015). doi:10.1103/PhysRevX.5.041017
17. A. Zavatta, M. Artoni, D. Viscor, G. La Rocca, Sci. Rep. **4**, 3941 (2014). doi:10.1038/srep03941
18. M. Mirhosseini, M. Malik, Z. Shi, R.W. Boyd, Nat. Commun. **4**, 2781 (2013). doi:10.1038/ncomms3781
19. J.D. Franson, Phys. Rev. Lett. **62**(19), 2205 (1989). doi:10.1103/PhysRevLett.62.2205
20. J. Brendel, E. Mohler, W. Martienssen, Europhy. Lett. **20**(7), 575 (1992)
21. P.G. Kwiat, A.M. Steinberg, R.Y. Chiao, Phys. Rev. A **47**(4), 2472 (1993). doi:10.1103/PhysRevA.47.2472
22. J. Brendel, N. Gisin, W. Tittel, H. Zbinden, Phys. Rev. Lett. **82**, 2594 (1999). doi:10.1103/PhysRevLett.82.2594
23. D. Bouwmeester, J.W. Pan, K. Mattle, M. Eibl, H. Weinfurter, A. Zeilinger, Nature **390**, 575 (1997)
24. T. Jennewein, G. Weihs, J.W. Pan, A. Zeilinger, Phys. Rev. Lett. **88**, 017903 (2002). doi:10.1103/PhysRevLett.88.017903
25. F. Bussières, J.A. Slater, J. Jin, N. Godbout, W. Tittel, Phys. Rev. A **81**, 052106 (2010). doi:10.1103/PhysRevA.81.052106
26. C. Simon, J.P. Poizat, Phys. Rev. Lett. **94**(3), 030502 (2005). doi:10.1103/PhysRevLett.94.030502, http://link.aps.org/abstract/PRL/v94/e030502
27. T. Flissikowski, A. Betke, I.A. Akimov, F. Henneberger, Phys. Rev. Lett. **92**, 227401 (2004). doi:10.1103/PhysRevLett.92.227401

28. S. Stufler, P. Machnikowski, P. Ester, M. Bichler, V.M. Axt, T. Kuhn, A. Zrenner, Phys. Rev. B **73**, 125304 (2006). doi:10.1103/PhysRevB.73.125304
29. H. Jayakumar, A. Predojević, T. Huber, T. Kauten, G.S. Solomon, G. Weihs, Phys. Rev. Lett. **110**, 135505 (2013). doi:10.1103/PhysRevLett.110.135505
30. A. Predojević, M. Ježek, T. Huber, H. Jayakumar, T. Kauten, G.S. Solomon, R. Filip, G. Weihs, Opt. Express **22**(4), 4789 (2014). doi:10.1364/OE.22.004789, http://www.opticsexpress.org/abstract.cfm?URI=oe-22-4-4789
31. T. Huber, L. Ostermann, M. Prilmüller, G.S. Solomon, H. Ritsch, G. Weihs, A. Predojević, Phys. Rev. B **93**(20), 201301 (2016). doi:10.1103/physrevb.93.201301
32. T. Huber, A. Predojević, G.S. Solomon, G. Weihs, Opt. Express (2016). arXiv:1603.09085
33. A.J. Ramsay, Semicond. Sci. Technol. **25**, 103001 (2010). doi:10.1088/0268-1242/25/10/103001
34. M.A.M. Versteegh, M.E. Reimer, A.A. van den Berg, G. Juska, V. Dimastrodonato, A. Gocalinska, E. Pelucchi, V. Zwiller, Phys. Rev. A **92**, 033802 (2015). doi:10.1103/PhysRevA.92.033802
35. H. Jayakumar, A. Predojević, T. Kauten, T. Huber, G.S. Solomon, G. Weihs, Nat. Commun. **5**, 4251 (2014). doi:10.1038/ncomms5251
36. T. Huber, A. Predojević, M. Khoshnegar, D. Dalacu, P.J. Poole, H. Majedi, G. Weihs, Nano Lett. **14**(12), 7107 (2014). doi:10.1021/Nl503581d
37. M. Gschrey, A. Thoma, P. Schnauber, M. Seifried, R. Schmidt, B. Wohlfeil, L. Kruger, J.H. Schulze, T. Heindel, S. Burger, F. Schmidt, A. Strittmatter, S. Rodt, S. Reitzenstein, Nat. Commun. **6**, 7662 (2015). doi:10.1038/ncomms8662
38. C. Schneider, T. Heindel, A. Huggenberger, P. Weinmann, C. Kistner, M. Kamp, S. Reitzenstein, S. Höfling, A. Forchel, Appl. Phys. Lett. **94**(11), 111111 (2009). doi:10.1063/1.3097016
39. E. Poem, Y. Kodriano, C. Tradonsky, N.H. Lindner, B.D. Gerardot, P.M. Petroff, D. Gershoni, Nat. Phys. **6**(12), 993 (2010). doi:10.1038/nphys1812
40. H. de Riedmatten, I. Marcikic, H. Zbinden, N. Gisin, Quantum Inf. Comput. **2**(6), 425 (2002), http://www.rintonpress.com/xqic1/qic-2-6/425-433.pdf

Part IV
Spin Properties and Integrated Systems

Chapter 9
A Self-assembled Quantum Dot as Single Photon Source and Spin Qubit: Charge Noise and Spin Noise

Richard J. Warburton

Abstract A self-assembled quantum dot confines both electrons and holes to a nano-sized region inside a semiconductor. An exciton in a single self-assembled quantum dot is a potentially excellent source of single photons. A quantum dot also acts as a host for a spin qubit with the advantage that the spin can be initialized, manipulated and read-out with optical techniques. However, the exciton and spins couple strongly not just to an external optical probe but also to internal excitations of the host semiconductor: the semiconductor is a source of noise resulting in exciton and spin dephasing. The noise can be suppressed in some cases, circumvented in others, leading to an improvement in quantum dot performance. In particular, resonant excitation at low temperature using high quality material results in a small level of charge noise. A heavy hole spin in an in-plane magnetic field is decoupled from the spin noise arising from fluctuations in the nuclear spin bath. Presented here is a series of experiments which probe the noise in advanced quantum dot devices: single quantum dot resonance fluorescence as a sensor of both charge noise and spin noise; nuclear magnetic resonance on the quantum dot nuclear spins to probe the electron spin hyperfine interaction; and coherent population trapping to probe the hole spin hyperfine interaction.

9.1 A Self-assembled Quantum Dot for Quantum Technology

A self-assembled quantum dot in GaAs has a nano-size, typically 20 nm in base diameter, 5 nm in height, resulting in discrete, atom-like quantum states. This small size, a size which is difficult to access with top-down nano-fabrication, confers an immediate advantage: both the confinement energies and on-site Coulomb interactions are large, tens of meV. The net result is that Coulomb blockade is well established: ultra-low temperatures are not required. Also, the self-assembly process along with

R.J. Warburton (✉)
Department of Physics, University of Basel,
Klingelbergstrasse 82, 4056 Basel, Switzerland
e-mail: richard.warburton@unibas.ch

P. Michler (ed.), *Quantum Dots for Quantum Information Technologies*,
Nano-Optics and Nanophotonics, DOI 10.1007/978-3-319-56378-7_9

the direct gap of the host semiconductor results in a strong optical transition across the fundamental gap of the semiconductor: the optical dipole moment is large. This, together with the large number of photon modes in a bulk structure, leads to short radiative lifetimes, $\sim$0.8 ns.

A self-assembled quantum dot is potentially useful in quantum technology, as a single photon source and as a spin qubit. On the one hand, self-assembled quantum dots can be embedded in semiconductor heterostructures, and devices can be made from the wafer material by semiconductor nano-fabrication for which a large and established tool-box is available. On the other hand, electrons and holes in the quantum dot are not just strongly coupled to external probes such as the light field: they are also strongly coupled to excitations in the host semiconductor. This means that dephasing, the interaction of quantum dot-based quantum states with the semiconductor environment, is strong and it is challenging to design structures where the dephasing is kept to a manageable level. In fact much of the interesting physics lies in the dephasing which can be complex [1].

This chapter describes a series of experiments on single self-assembled quantum dots. The aim of these experiments is to investigate the noise which limits the performance of the quantum dot as a single photon emitter and as a spin qubit. An underlying feature is the use of high quality material at low temperature, and resonant optical excitation with resonance fluorescence detection.

9.2 Photonics of a Self-assembled Quantum Dot

9.2.1 The Optical Transition

An InGaAs/GaAs quantum dot, Fig. 9.1a, has a strong optical transition between the highest valence state and the lowest conduction state [2]. In semiconductor language, optical excitation creates an exciton, an electron-hole pair. For as-grown InAs/GaAs quantum dots, the optical transition lies at an inconvenient wavelength, $\sim$1,200 nm at low temperature, but can be shifted to the more convenient $\sim$950 nm either during growth [3] or via post-growth annealing [4]. The radiative lifetime is short, typically $\sim$0.8 ns [5], corresponding to a dipole moment of 0.6e nm·C [6]. The level structure is explained in Fig. 9.1b–d. Note that the two bright excitons with spin ± 1 are coupled by the so-called fine structure: this lifts the degeneracy of the two exciton states even at zero magnetic field.

The main optical transition forms the basis of a quantum dot single photon source. The problem of extracting the photons out of the high-index host material – GaAs has a refractive index of 3.5 – can be solved by engineering also the photonic states. For instance, by embedding the quantum dots in a GaAs waveguide, and using a taper as an out-coupler, high fidelity single photon emission with a quantum efficiency as high as $\sim$70% has been achieved [7].

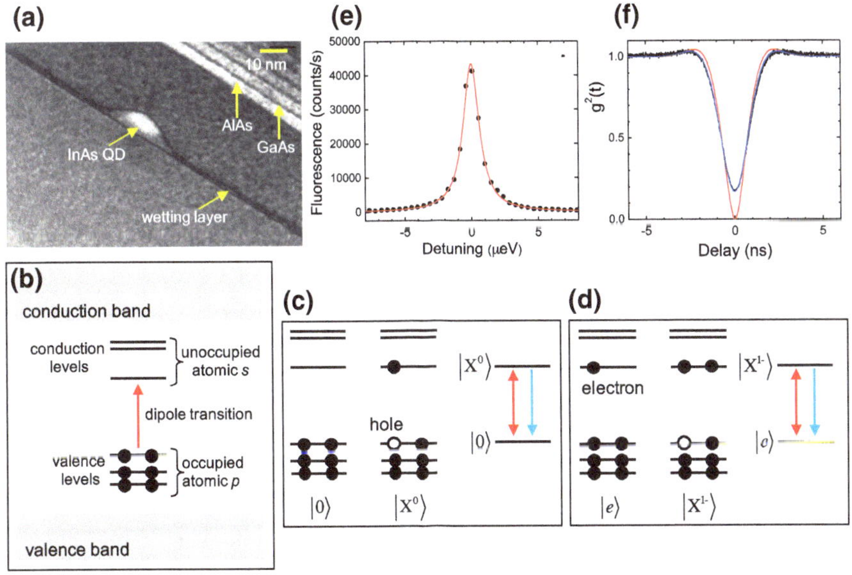

Fig. 9.1 Photonics of a single self-assembled quantum dot at zero magnetic field. a Transmission electron microscopy image of a sample grown by molecular-beam-epitaxy showing an InAs quantum dot with its associated wetting layer, embedded in GaAs and capped with an AlAs/GaAs superlattice. Image provided courtesy of Jean-Michel Chauveau, Arne Ludwig, Dirk Reuter and Andreas Wieck. **b** Schematic energy level structure of a self-assembled quantum dot. On account of quantization, there are discrete, atom-like conduction and valence levels with wave functions localized in the quantum dot. At higher energies, there are the conduction and valence energy bands associated with the two-dimensional wetting layer (and bulk GaAs at higher energy still). At low temperature and in the absence of doping, the valence levels are occupied and the conduction levels are unoccupied. A strong optical dipole transition connects the highest energy valence level with the lowest energy conduction level, the transition taking place across the fundamental gap of the semiconductor. **c** The vacuum state $|0\rangle$ and the optically excited state $|X^0\rangle$ are represented as the ground and excited states of a 2-level atom (*red arrow* represents the optical coupling, *blue arrow* spontaneous emission). In $|X^0\rangle$, a valence electron has been promoted to a conduction level creating a so-called exciton, an electron-hole pair. **d** A quantum dot can be loaded with a single excess electron (see Fig. 9.2). In this case, the ground state is $|e\rangle$ and the optically excited state $|X^{1-}\rangle$ consisting of two electrons (in a singlet) and a hole. **e** Laser spectroscopy on a single quantum dot at a wavelength close to 950 nm at temperature 4.2 K. The resonance fluorescence is plotted as a function of laser detuning. The linewidth is 1.6 μeV (400 MHz). The signal corresponds to the count rate on a silicon avalanche photodiode. The transform-limited linewidth is $\Gamma_0 = 0.8$ μeV, equivalently 200 MHz (radiative lifetime 0.8 ns). **f** The intensity correlation coefficient $g^{(2)}$ of the resonance fluorescence measured with a Hanbury Brown-Twiss interferometer (*black line*). The dip at zero delay shows clear photon antibunching. The signal at zero delay is dominated by the jitter of the detector (0.5 ns); the slight overshoot at delay ~1 ns is the first hint of a Rabi oscillation which becomes marked at higher laser power. The *blue line* is a convolution of the two-level atom $g^{(2)}$ with the response of the detectors; the *red line* is the two-level atom $g^{(2)}$ alone. **e**, **f** data courtesy of Andreas Kuhlmann, Julien Houel and Arne Ludwig

9.2.2 *Vertical Tunneling Structures*

Implementing the concept of an electron spin qubit [8] with a self-assembled quantum dot involves trapping a single electron. This can be achieved by including a δ-doped layer close to the quantum dot layer in the growth such that some of the quantum dots are permanently occupied with an excess electron [9] or hole [10] at low temperature. A more flexible technique allowing for considerable in situ tuning is to embed the quantum dot layer in a vertical tunnelling device [2, 11, 12], Fig. 9.2a. The device operates in the Coulomb blockade regime which at 4 K is highly pronounced based on the huge on-site Coulomb energy to thermal energy ratio ($\sim$25 meV : 0.4 meV). The quantum dots are in tunnel contact with a Fermi sea; the quantum dot potential with respect to the Fermi energy is controlled by applying a gate voltage to the top contact. The top contact is often a Schottky gate on the sample surface, an n-i-Schottky structure with the quantum dots in the intrinsic "i" region. Alternatively, the Schottky contact can be replaced with a p-type layer, an n-i-p structure where the p-type layer acts as an "epitaxial gate" [13].

Structures for trapping a single hole swap n-doping for p-doping, i.e. p-i-Schottky and p-i-n. However, quantum dots grown after the p-type layer have relatively poor optical properties [13, 15]. The solution is the n-i-p structure with the quantum dots in tunnel contact with the p-layer as in this case the p-type layer is grown last, Fig. 9.3.

In these structures, the Coulomb blockade is revealed most immediately in the single quantum dot photoluminescence by clear steps in the photoluminescence energy [12], Fig. 9.2b. A single electron is trapped in the quantum dot over the Coulomb blockade plateau, Fig. 9.2b: a voltage chosen within this region allows access to single spin physics [16]. In the Coulomb blockade regime, tunnelling is suppressed to first order but the second order process, co-tunnelling, survives [17]. Co-tunnelling represents a spin relaxation mechanism: the quantum dot electron spin is swapped by a two-electron process with the spin of an electron close to the Fermi energy in the Fermi sea of the n^+ layer. A convenient property of the vertical tunnelling structure is that the co-tunnelling rate is large at the edge of the plateau but is suppressed by several orders of magnitude in the plateau centre, providing useful in situ control [17–21].

The tunnel barrier thickness is important. For very small values, the quantum states hybridize strongly with the states at the Fermi energy in the back contact [22]; at very large values, the tunneling time becomes much longer than the recombination time. An important point of detail in these structures concerns the thickness of the capping layer, the i-GaAs grown between the quantum dot layer and the blocking barrier (an AlAs/GaAs superlattice). At intermediate thicknesses, fluctuating minority carriers at the GaAs/blocking barrier interface result in unwanted charge noise [23]. This noise can be suppressed with a thin $\sim$10 nm capping layer pushing the minority charge states well above the quantum dot levels such that the states are unlikely to be occupied at low temperature. A thin capping layer also prevents the optically-excited minority carrier (the hole in the case of an electron spin device; the electron in the case of a hole spin device) from tunneling out of the quantum dot.

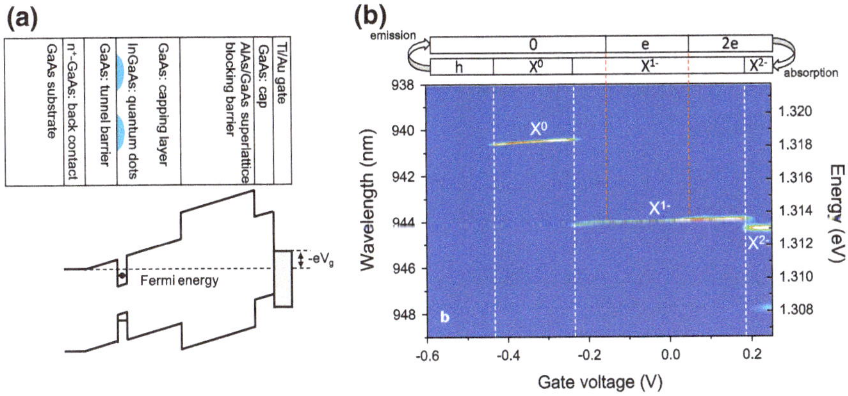

Fig. 9.2 Coulomb blockade of a single quantum dot. a Layer structure of a typical heterostructure for experiments controlling the charge state of a quantum dot. A layer of self-assembled quantum dots is embedded in a vertical tunnelling structure. The quantum dots are in tunnel contact with the Fermi sea in the n^+ layer; the blocking barrier prevents current flow to the surface; a Schottky gate on the surface allows control of the vertical electric field. Typically, the tunnel barrier (gallium arsenide, GaAs) is 12–40 nm thick, the capping layer 10–150 nm, blocking barrier > 100 nm, and the Schottky gate is a semi-transparent metal layer, e.g. 5–10 nm Ti/Au. A voltage applied to the gate, V_g, tunes the energy of the first confined electron level relative to the Fermi energy as shown schematically in the band diagram. At low temperature and large electric field, the quantum dot conduction level lies above the Fermi energy and is therefore unoccupied; when the conduction level lies below the Fermi energy but close to it (the case shown), the conduction level is singly occupied (the case shown); at more positive V_g it is doubly occupied. **b** The photoluminescence (PL) from a single quantum dot in a vertical tunnelling structure is shown as a function of V_g at a temperature of 4.2 K. The steps in the photoluminescence energy correspond to charging events. X^0 refers to the neutral exciton (an electron-hole pair); X^{1-} to the negatively charged trion (a two electron-one hole complex); X^{2-} the doubly charged exciton (a three electron-one hole complex). Note that the charging event without a hole, $|0\rangle \rightarrow |e\rangle$, takes place at slightly more positive V_g than the charging event with a hole, $|X^0\rangle \rightarrow |X^{1-}\rangle$, on account of the Coulomb energies: the electron-hole on-site Coulomb energy is larger than the electron-electron on-site Coulomb energy. Conversely, the $|e\rangle \rightarrow |2e\rangle$ charging event takes place at more negative V_g than the $|X^{1-}\rangle \rightarrow |X^{2-}\rangle$ charging event as the $|X^{2-}\rangle$ state has a total of three electrons, the "third" forced to occupy the first excited conduction level by the Pauli principle. The main features in the PL characterization correspond to charging events in the initial state, the exciton state (*white dashed lines*). However, charging events in the final state are revealed by hybridization effects in the X^{1-} plateau (*red dashed lines*) [14]. Probing a single spin with resonant laser excitation involves working in the V_g window defined by the two dashed red lines. Experimental data provided courtesy of Paul Dalgarno

9.2.3 Resonance Fluorescence Detection

Many optical experiments in this field rely on non-resonant excitation in which a high-energy continuum is occupied with electron-hole pairs. The exciton level in the quantum dot is populated by typically fast relaxation, and the spontaneous emission on exciton decay (the photoluminescence) can be detected. An example is shown in Fig. 9.2b. This is a relatively simple experiment but lacks the power of true laser

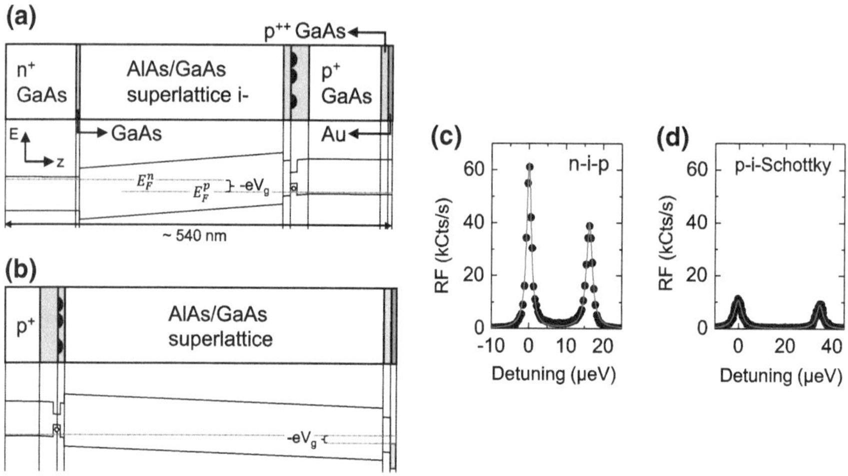

Fig. 9.3 Devices for loading a quantum dot with a single hole. **a** n-i-p device; **b** p-i-Schottky device: heterostructure design and schematic band diagram. **c** Resonance fluorescence spectrum at 4.2 K on an empty quantum dot in the p-i-n device. The two lines (split by the exciton fine structure) have a linewidth of 1.5 μeV, equal to the linewidth in the very best n-type devices, demonstrating the extremely low level of charge noise. **d** Resonance fluorescence spectrum at 4.2 K on an empty quantum dot in the p-i-Schottky device. The larger linewidths of 3.6 μeV and also the smaller signals signify problems related to the p-doping: increased charge noise and non-radiative decay

spectroscopy methods which involve driving the optical resonance with a coherent laser. The interaction of a single self-assembled quantum dot with a coherent laser tuned to the optical resonance was initially detected via a change in the transmission coefficient, a "ΔT-experiment" [24]. Meanwhile, the resonance fluorescence can also be detected (see also Chap. 3), discriminating resonance fluorescence from scattered/reflected laser light with a dark field technique based either on the propagation direction [25] or on the polarization [26–29], Fig. 9.1e. Remarkably, all the features of a driven two-level system known from atomic physics have been observed on a single quantum dot. These include a Lorentzian absorption lineshape [24], Fig. 9.1e; power broadening and power-induced transparency [30]; the ac Stark effect [31]; Rabi oscillations [25], the Mollow triplet, and antibunching of resonance fluorescence [25], Fig. 9.1f. Routinely, close-to-transform limited optical transitions are observed in laser spectroscopy experiments on single InGaAs/GaAs quantum dots [23]. Phonon-induced exciton decoherence is significant either at low temperature with large Rabi couplings [32, 33], or at elevated temperatures [30, 34–36].

The capability to detect resonance fluorescence has given the field a significant boost: the resonance fluorescence itself represents single photon output; in spectroscopy terms it is a low noise, low background technique.

9.3 Exciton Dephasing

A key goal in quantum communication is to create highly indistinguishable photons which are separated in space by more than 100 Km for device-independent quantum key distribution and for a quantum repeater [37]. This is potentially possible using a semiconductor quantum dot: single photons are generated either by spontaneous emission from the upper level [38–40], Fig. 9.1, or by coherent scattering of a resonant laser [41–43]. Optimizing performance demands an understanding of noise and a strategy to circumvent its deleterious effects [1]. There are two main sources of noise in a semiconductor. Charge noise arises from occupation fluctuations of the available states and leads to fluctuations in the local electric field. This results in shifts in the optical transition energy of a quantum dot via the dc Stark effect and is one mechanism by which the optical linewidth of a self-assembled quantum dot can be significantly increased above the transform limit [18, 23, 24]. Charge noise can also result in spin dephasing via the spin-orbit interaction, and, in particular for hole spins, via the electric field dependence of the g-factor [44, 45]. The second source of noise, spin noise, arises typically from fluctuations in the nuclear spins of the host material and, on account of the hyperfine interaction, results in a fluctuating magnetic field (the Overhauser field) experienced by an electron spin [46, 47]. Spin noise from noisy nuclei results in rapid spin dephasing in an InGaAs quantum dot [9, 48, 49].

Strategies for reducing noise involve working with ultra-clean materials to minimize charge noise and the use of dynamic decoupling, schemes which employ complex echo-like sequences to "protect" the qubit from environmental fluctuations [50–52]. In this case, it is absolutely crucial that the noise power decreases with increasing frequency.

For quantum dot-based single photon sources, the linewidths are in the best case (high quality material with resonant excitation) typically about a factor of two larger than the transform limit in which the linewidth is determined only by the radiative decay time [18, 23, 24], Fig. 9.1. This is a poor state of affairs for applications which rely on photon indistinguishability, the resource underpinning a quantum repeater for instance. On the positive side, there is evidence that with low power, resonant excitation, there is no significant upper level dephasing apart from radiative recombination [41, 42] such that over short timescales indistinguishable photons are emitted [39]. It has been surmised that the increase in linewidth above the ideal limit arises from a spectral wandering [23, 24].

The particular challenge posed by charge noise in self-assembled quantum dot devices is illustrated in Fig. 9.4. In this experiment, resonant laser spectroscopy on single quantum dots was carried out as a function of the power of an additional non-resonant laser [23]. The non-resonant laser is weak such that the induced photoluminescence is negligible but it introduces a small number of holes in the device. A steady state population of holes builds up at the capping layer/blocking barrier interface, in this case 35 nm above the quantum dot layer. The quantum dot transition frequency increases step-wise, Fig. 9.4. The interpretation is that there are localiza-

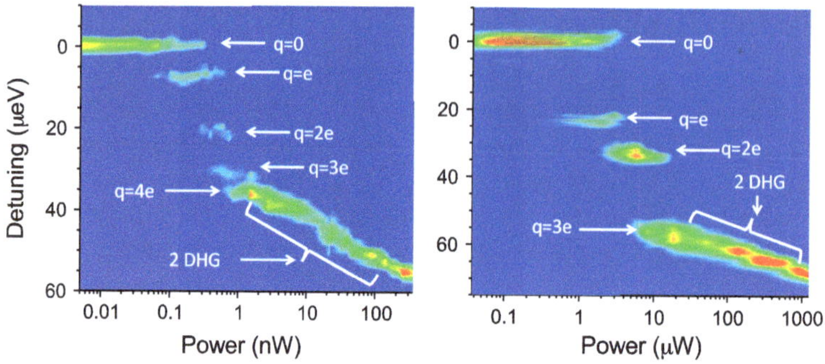

Fig. 9.4 Resonant laser spectroscopy on single quantum dots with close-by charge localization centres. Colour-scale plot (linear scale, *blue*: 0.061%; *red*: 0.61%) of differential reflectivity versus non-resonant laser power on two different quantum dots. The non-resonant laser power controls the steady state occupation of defects, hole localization centres located 30 nm above the quantum dot at the capping layer/blocking barrier interface. The charge is labelled in each case. At large non-resonant laser power, a two-dimensional hole gas forms (2DHG)

tion centres for the holes directly above the quantum dot. Occupation of one of these localization centres by a single hole shifts the quantum dot frequency by several linewidths, suppressing the resonance scattering. On the one hand, the sensitivity of the quantum dot to the small levels of charge noise reflects the potential of quantum dots as ultra-sensitive electrometers [23, 53, 54]. On the other hand, the same sensitivity makes the generation of transform-limited single photons challenging. The mechanism is the dc Stark effect [55].

9.3.1 The Charged Exciton

The resonance fluorescence signal itself can be used to investigate the noise [28, 56]. The linewidth determines the spectral purity of the single photons. Measured on second time-scales, the single quantum dot linewidths are 1.6 μeV in Fig. 9.5 for both neutral and charged excitons, X^0 and X^{1-}, respectively. The linewidth can be recorded on smaller time-scales: Fig. 9.5 shows the linewidths versus scanning frequency. The linewidths are constant up to about 1 kHz and then decrease, reaching constant values above about 50 kHz. Remarkably, the constant values at the highest frequencies correspond closely to the transform limit. (The difference in X^0 and X^{1-} transform-limited linewidths reflects the slightly different radiative decay times [5].) The implication is two-fold. First, any pure exciton dephasing mechanisms must be much weaker than radiative recombination in this experiment. Secondly, the quantum dot produces a stream of identical photons over time-scales of about 1–10 μs: over this time the noise is "frozen". At longer times, there are fluctuations in the quantum dot's optical frequency, a spectral wandering.

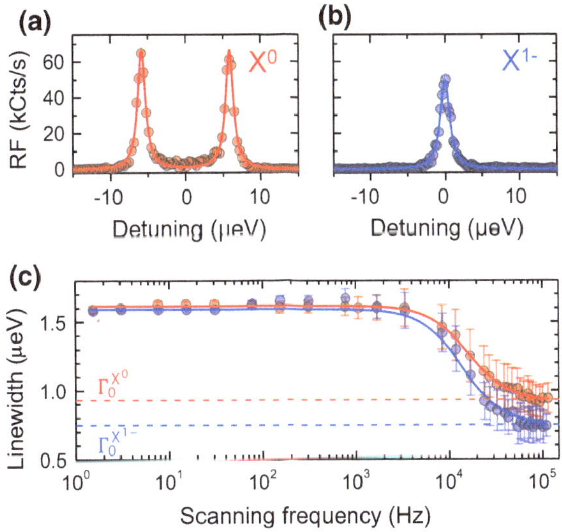

Fig. 9.5 Linewidth versus scanning frequency. **a, b** X^0, X^{1-} resonance fluorescence versus detuning δ at 4.2 K, $B = 0.0$ mT with 100 ms integration time per point. The *solid lines* are Lorentzian fits to the data. The linewidths are $\Gamma^{X^0} = 1.29\,\mu$eV, $\Gamma^{X^{1-}} = 1.49\,\mu$eV; the Rabi energies $\Omega/\Gamma_0 = 0.5$ (X^0), 0.4 (X^{1-}); and transform-limits $\Gamma_0^{X^0} = 0.92 \pm 0.10\,\mu$eV, $\Gamma_0^{X^{1-}} = 0.75 \pm 0.10$ μeV. **c** RF linewidth against scanning frequency $d\delta/dt/\Gamma_0$. Γ approaches Γ_0 for scanning frequencies above 50 kHz. For each scanning frequency, the error bar represents the standard deviation of several hundred linewidth scans. *Solid lines* represent a Lorentzian fit of the data with linewidth 30 ± 3 kHz

An immediate question concerns the origin of the noise which leads to the spectral wanderings in the quantum dot's optical frequency. Insight comes from a noise spectrum of the resonance fluorescence: a time-trace is recorded at a particular detuning, $\delta = 0$ or $\delta = \Gamma/2$, Fig. 9.6a; a Fourier transform yields a noise spectrum [28, 56]. A typical time trace of the RF is shown in Fig. 9.6b with binning time 10 ms. The main contribution to the noise comes from shot noise. However, the level of extrinsic noise is highly reproducible: its spectrum, $N_{\text{EXP}}(f)$, is recorded carefully and then subtracted from the total noise to determine the intrinsic noise, the noise power of the normalized RF signal, $N_{\text{QD}}(f)$, Fig. 9.7a. Specifically, the fast Fourier transform (FFT) of the normalized RF signal $S(t)/\langle S(t)\rangle$ provides a noise spectrum:

$$N_{\text{RF}}(f) = |\text{FFT}[S(t)/\langle S(t)\rangle]|^2 (t_{\text{bin}})^2/T, \tag{9.1}$$

where t_{bin} is the binning time and T the total integration time. $N_{\text{QD}}(f)$ is determined by

$$N_{\text{QD}}(f) = N_{\text{RF}}(f) - N_{\text{EXP}}(f). \tag{9.2}$$

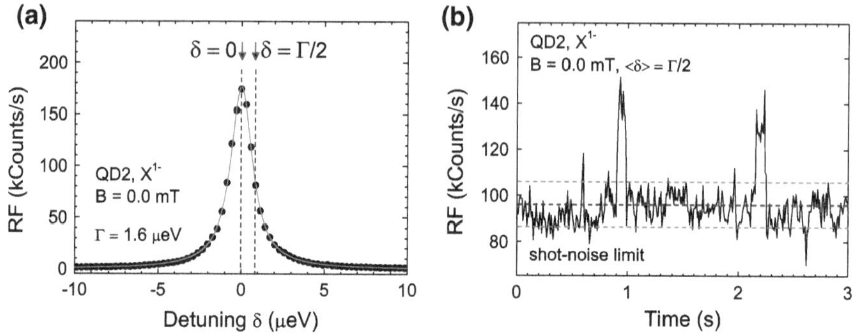

Fig. 9.6 Resonance fluorescence (RF) on a single quantum dot. a RF recorded on a single InGaAs quantum dot at wavelength 950.61 nm at a power corresponding to a Rabi energy of 0.55 μeV at a temperature of 4.2 K without external magnetic field. The RF was detected with a silicon avalanche photodiode operating in single photon mode; the detuning was achieved by sweeping the gate voltage with respect to the laser using the dc Stark effect. In this case, the integration time per point was 100 ms. The *solid line* is a Lorentzian fit to the data with linewidth $\Gamma = 1.6\,\mu\text{eV}$ (390 MHz). **b** A time-trace of the RF recorded with detuning set to half the linewidth, $\langle\delta\rangle = \Gamma/2$. The arrival time of each detected photon is stored allowing a time trace to be constructed post-experiment with an arbitrary binning time. An example is shown using a binning time of 10 ms

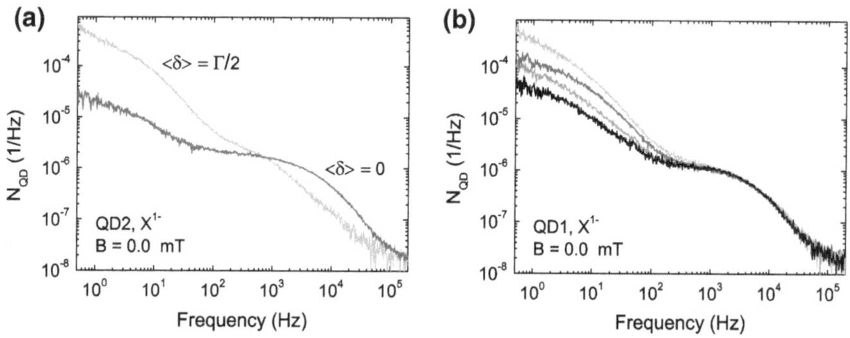

Fig. 9.7 Resonance fluorescence noise. a RF noise spectra recorded on a quantum dot occupied with a single electron, the trion X^{1-}, for average detuning equal to zero, $\langle\delta\rangle = 0$ (*blue*), and for $\langle\delta\rangle = \Gamma/2$ (*red*) at 4.2 K and $B = 0.0$ mT. Following the scheme in Fig. 9.8, the noise at low frequencies is shown to originate from charge noise, that at high frequencies from spin noise. Plotted is the noise power spectrum of the normalized RF, $S(t)/\langle S(t)\rangle$, where $S(t)$ is the RF signal, $\langle S(t)\rangle$ the average RF signal, corrected for external sources of noise. **b** RF noise spectra recorded on X^{1-} with $\langle\delta\rangle = 0$ under identical experimental conditions (4.2 K, $B = 0.0$ mT) in the course of the experiment. The charge noise at low frequency depends on the sample history; the spin noise at high frequency does not (color figure online)

Figure 9.7a shows noise spectra over six decades of resolution in the noise power over six decades of frequency, from 0.1 Hz to 100 kHz, Fig. 9.7a, b. The noise falls very rapidly above 10 kHz: this is consistent with the observation of transform-limited linewidths in the spectroscopic measurement on exactly the same quantum

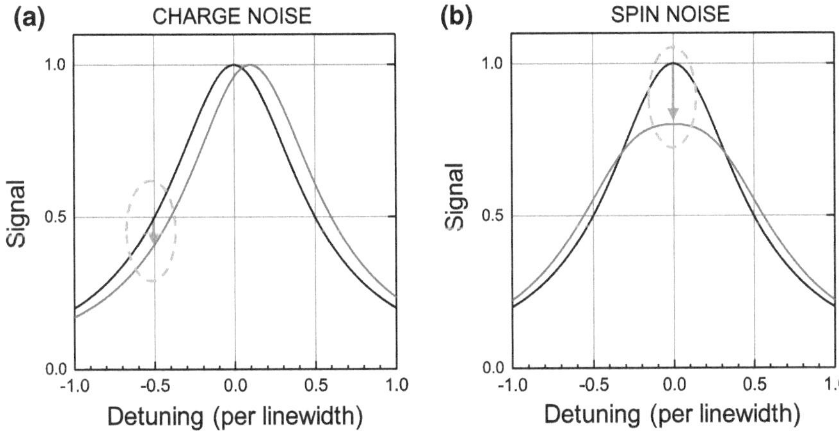

Fig. 9.8 Distinguishing between charge noise and spin noise. Schematic showing the effect of charge noise and spin noise on the charged, X^{1-}, exciton (applied magnetic field zero). **a** Charge noise (noise in the local electric field) results in a "rigid" shift of the optical resonance leading to a small change in resonance fluorescence (RF) for zero detuning $\delta = 0$ and a large change in RF at $\delta = \Gamma/2$. **b** Without an external magnetic field, spin noise induces a Zeeman splitting in the resonance resulting in a large change in RF at $\delta = 0$ and a small change in RF at $\delta = \Gamma/2$ (zero for $\delta = \Gamma/2\sqrt{3}$), opposite to charge noise. This difference, a "rigid" shift of the X^{1-} resonance from charge noise, a "breathing motion" in the X^{1-} resonance from spin noise, allows charge noise and spin noise to be identified

dot, Fig. 9.8c, d. The entire noise spectrum can be described by two Lorentzian features along with a $1/f$-like component. Significantly, there is a spectroscopic technique to assign these noise sources to charge noise or spin noise, Fig. 9.8, based on the different response of X^{1-} to charge noise and spin noise.

As the local electric field F fluctuates, the detuning δ of the quantum dot optical resonance with respect to the constant laser frequency fluctuates on account of the dc Stark effect. For small electric field fluctuations, the Stark shift is linear: the optical resonance shifts rigidly backwards and forwards on the detuning axis, as shown in Fig. 9.8a. The response in the RF to charge noise has a first order component in electric field for $\delta = \Gamma/2$ giving rise to large changes in the RF. Conversely, for $\delta = 0$ the first order component vanishes. Sensitivity to charge noise in the RF is therefore weak for $\langle\delta\rangle = 0$ yet strong for $\langle\delta\rangle = \Gamma/2$. Spin noise results in a complementary behaviour in the absence of an external magnetic field, $B = 0$. Fluctuations in the local magnetic field B_N arising from spin noise do not shift the X^{1-} resonance backwards and forwards. Instead, a typical B_N fluctuation induces a sub-linewidth Zeeman splitting of the X^{1-} resonance, as shown in Fig. 9.8b. Sensitivity to spin noise in the RF is therefore strong for $\langle\delta\rangle = 0$, weak for $\langle\delta\rangle = \Gamma/2$. The crucial point is that, for X^{1-} at $B = 0$, the dependence of the RF noise on $\langle\delta\rangle$ is *opposite* for charge noise and spin noise.

The detuning dependence, Fig. 9.7a, therefore identifies the main noise at low frequencies (Lorentzian spectrum and $1/f$-like component) as charge noise, the main noise at high frequencies spin noise. The charge noise gives large noise powers but only at low frequencies. The spin noise gives much weaker noise powers but over a much larger bandwidth. It is striking that the resonance fluorescence reveals the full spectrum of the fluctuating nuclear spin ensemble.

The simple rules connecting RF intensity with the local electric field F (charge noise) and with the local magnetic field B_N (spin noise) allow quantitative statements on the noise to be made. The charge noise has root-mean-square (rms) electric field noise $F_{rms} = 0.46$ Vcm^{-1} (bandwidth starting at 0.1 Hz). It is striking that, first, the charge noise is very small: the rms noise in the local potential is just 1.2 μV; the charge noise contribution to the low-scanning-frequency X^{1-} linewidths are <0.05 μeV. The low charge noise is a consequence of both the ultra-pure material and also the carefully controlled experimental conditions. Secondly, it is striking that the charge noise is concentrated at such low frequencies. The rms noise in the Overhauser field measured on X^{1-} amounts to $B_{N,rms} = 9$ mT with correlation time 100 μs. It is this noise which makes the dominant contribution to the X^{1-} linewidth at low scanning frequency. The random fluctuations of N nuclear spins lead to a $B_{N,rms}$ which scales as $1/\sqrt{N}$ [46, 47]; applied to an InGaAs quantum dot with $N \sim 10^5$, the expectation is $B_{N,rms} \sim 20$ mT [57, 58], reasonably close to the value measured here. The timescale is characteristic of the nuclear spin dipole-dipole interaction [46].

Figure 9.8b shows $N_{QD}(f)$ curves measured on the same quantum dot over the course of the experiment (several months) under nominally identical conditions. There are changes in the low frequency noise power (up to a factor of 10) but the high frequency noise remains exactly the same. The charge noise therefore depends on the sample's history. Conversely, the spin noise arises from the host nuclear spins of the quantum dot which remain the same and retain their properties: this results in the unchanging spin noise at high frequency.

A Lorentzian noise spectrum is characteristic of a two-level fluctuator [59]. The associated on-off behaviour, equivalently telegraph noise, is however not observed here. Instead, the Lorentzian noise arises from fluctuations in an *ensemble* of two-level fluctuators, each with approximately the same transition rates, $0 \rightarrow 1$, $1 \rightarrow 0$. For the charge noise, the fluctuators are hole localization centres at an interface 150 nm above the quantum dot. Electrostatic noise arises via fluctuations in the exact configuration of occupied (state 0) and unoccupied (state 1) localization sites in the ensemble. For the spin noise, each nuclear spin acts as a fictitious two-level fluctuator. A Monte Carlo simulation enables both sources of noise to be treated on an equal footing [28]. The $1/f$-like noise varies from quantum dot to quantum dot and its exact origin is an open question.

9.3.2 *The Neutral Exciton*

There is evidence that at low temperature, there is negligible pure upper level decoherence of the neutral exciton [35, 41–43]. Photons emitted subsequently are close to indistinguishable [39, 60]. The experiments described above add weight to these

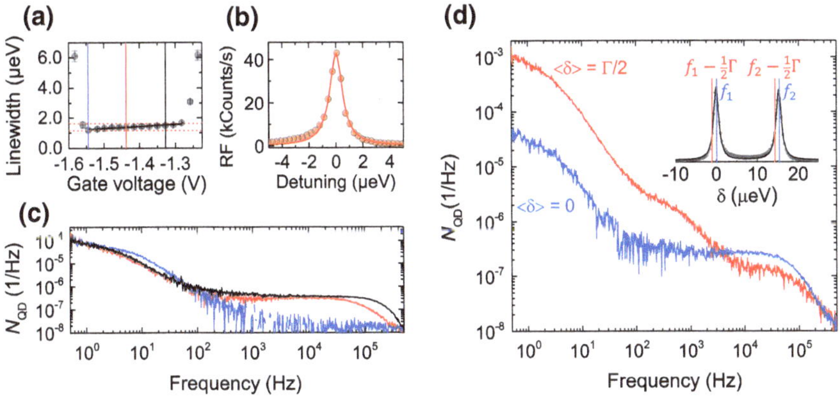

Fig. 9.9 $\mathbf{X^0}$ **spin noise within the Coulomb blockade plateau. a** X^0 optical linewidth measured at Rabi energy $\Omega/\Gamma_0 = 0.50$ for different gate voltages by sweeping the laser frequency through the resonance and integrating 100 ms per point. Γ decreases from 1.66 to 1.19 μeV with decreasing gate voltage. **b** X^0 spectrum with $\Gamma = 1.15$ μeV at $V_g = -1.54$ V. **c** X^0 noise spectra recorded at Rabi energies $\Omega/\Gamma_0 = 0.65$ for different voltages, indicated in **a** by *solid lines*. Maximum/minimum spin noise (*black/blue*) is correlated with the largest/smallest Γ. **d** $N_{QD}(f)$ on X^0 recorded with two lasers of frequencies f_1 and f_2 and a frequency splitting $f_1 - f_2$ equal to the fine structure splitting for $\langle\delta\rangle = 0$ (*blue*) and $\langle\delta\rangle = \Gamma/2$ (*red*). *Inset* shows the laser frequency detuning relative to the optical resonance

assertions: transform-limited linewidths are observed on fast scanning [28, 56]. Charge noise leads to an inhomogeneous broadening of the X^0 as for the X^{1-}, Fig. 9.5. The X^0 is also sensitive to spin noise, i.e. fluctuations in the Overhauser field, but with reduced sensitivity with respect to X^{1-}. For X^0, the sensitivity is second order as the hole "shields" the electron from the spin noise (a consequence of the fine structure); for X^{1-} the sensitivity is first order on account of the unpaired electron in the X^{1-} ground state. Despite the different sensitivity to spin noise the X^0 and X^{1-} linewidths are very similar [23, 24, 28].

A typical X^0 resonance fluorescence spectrum is shown in the ideal case (high quality material at low temperature, resonant excitation on a quantum dot in the Coulomb blockade regime) in Fig. 9.5 with $\Omega/\Gamma_0 = 0.5$ where Ω is the Rabi coupling. The linewidth is a factor of 1.4 larger than the transform-limit (for this particular quantum dot, $\Gamma_0^{X^0} = 0.92 \pm 0.10$ μeV). Figure 9.9a shows Γ versus V_g on X^0, measured below but close to saturation, $\Omega/\Gamma_0 = 0.5$. At the edges of the Coulomb blockade plateau, Γ rises rapidly on account of fast electron spin dephasing via co-tunneling with the Fermi sea [17]. This process slows down as V_g moves away from the plateau edges. The prominent feature is that a "sweet-spot" exists close to the negative V_g-end of the plateau with minimum linewidth 1.19 ± 0.13 μeV, Fig. 9.9a, b. Accounting for the small power broadening, the ideal limit is $\Gamma(\Omega) = \Gamma_0[1 + 2(\Omega/\Gamma_0)^2]^{\frac{1}{2}} = 1.10 \pm 0.10$ μeV. Within the measurement uncertainties of 10%, the transform-limit is therefore achieved. As V_g is raised to the positive side of the "sweet-spot", Γ increases beyond the ideal limit, Fig. 9.9a.

As for the X^{1-}, a diagnostic of the X^0 linewidth is a noise spectrum $N_{QD}(f)$, a Fourier transform of the RF time-trace. The increase in linewidth above the transform-limit represents a sum over all noise sources from the scanning frequency, about 1 Hz, to $\Gamma_0/\hbar$, about 1 GHz. The noise spectra at the low-bias end (the "sweet-spot"), the centre of the plateau and the positive-bias end are shown in Fig. 9.9c. There is a Lorentzian feature with linewidth 30 Hz (noise correlation time 30 ms) and a second Lorentzian feature at higher frequencies with linewidth 200 kHz (correlation time 5 μs). The second feature disappears at the "sweet-spot".

A spectroscopic diagnostic for charge versus spin noise can be established, as for the X^{1-}. Now however two lasers are required: one laser drives one of the X^0 transitions, the other laser drives the other X^0 transition. The scheme exploits the different X^0 response to charge noise and spin noise: charge noise moves both X^0 peaks rigidly together along the detuning axis; spin noise moves them apart or closer together, a "breathing" motion. Specifically, X^0 noise spectra are recorded with two lasers whose frequencies are separated by the fine structure. On detuning both lasers from $\delta = 0$ to $\delta = \Gamma/2$, the sensitivity to charge noise increases (changing from second order to first order) yet the sensitivity to spin noise decreases (remaining second order but with a reduced pre-factor). In the experiment, switching from $\langle\delta\rangle = 0$ to $\langle\delta\rangle = \Gamma/2$ causes the noise power of the low frequency component to increase markedly, Fig. 9.9d, identifying it as charge noise. However, as for X^{1-}, the frequency-sum over the charge noise gives a contribution to Γ of $<0.05\,\mu$eV, a negligible value. (Note that both the dc Stark coefficient and Γ vary from quantum dot to quantum dot yet there is no correlation between the two, pointing also to the unimportance of charge noise in the optical linewidth.) Conversely, the noise power of the high frequency component decreases on detuning both lasers from $\delta = 0$ to $\delta = \Gamma/2$, identifying it as spin noise, Fig. 9.9d. Furthermore, noise spectra measured at $\langle\delta\rangle = 0$ but with a single laser tuned to one of the X^0 transitions show that the low frequency noise, the charge noise, is similar for all three biases yet the high frequency noise, the spin noise, increases with increasing bias, Fig. 9.9c. This confirms that the high frequency noise, the spin noise, is responsible for the inhomogeneous linewidth: the integrated spin noise is vanishingly small at the "sweet-spot", increasing at the centre of the plateau, and increasing further at the positive bias edge.

The "spin noise" of the X^0 is not understood microscopically. If the noise is assigned entirely to an Overhauser field, B_N, very large values are required to account for the experimental results, hundreds of mT, with values increasing with optical Rabi coupling: optical driving apparently agitates the nuclei. Also, the correlation time of the noise is just 5 μs (independent of optical Rabi coupling), considerably smaller than the timescale typical of the nuclear spin-nuclear spin dipole-dipole interaction. An alternative is to parameterize the X^0 spin noise as a B_N-induced fluctuation of the fine structure splitting. This however remains conjecture. The experiment demonstrates only that there is noise in the frequency separation of the two X^0 transitions. This speculation notwithstanding, the "spin noise" disappears at the "sweet spot" and this is a robust phenomenon occurring on all the quantum dots investigated in this sample [56].

9.3.3 Locking the Quantum Dot Optical Resonance to a Frequency Standard

Unlike a real atom, the exact transition wavelength of a quantum dot is not locked to any particular wavelength and varies considerably from quantum dot to quantum dot. However, the host semiconductor can be designed so that considerable possibilities for tuning the emission wavelength exist. Electric field tuning [61, 62] and strain tuning [63, 64] allow the emission wavelength to be tuned over several nanometres. A major problem remains. As shown above, the emission wavelength is not constant: it varies randomly over time, even in very controlled environments at low temperature. As shown in the noise analysis, the culprit at low frequency is electrical noise in the semiconductor which shifts the emission wavelength via the Stark effect [28]. This noise has a $1/f$-like power spectrum resulting in, first, large and uncontrolled drifts at low frequencies and second, an undefined mean value. This noise, while poorly understood, is ubiquitous in semiconductors and makes it very difficult to couple an individual quantum dot to another quantum system, another quantum dot for instance, or an ensemble of cold atoms. Described here is a scheme to lock the quantum dot emission frequency to a frequency standard [65].

The output of the device is a stream of single photons generated by resonance fluorescence (RF) from a single quantum dot. A sketch of the experimental concept is shown in Fig. 9.10a. A linearly-polarized resonant laser is focused onto the sample surface and drives the optical transition. The resonance fluorescence of the quantum dot is collected with the polarization-based dark field technique [23, 27–29, 42], Fig. 9.10b. Simultaneously, the optical resonance is detected in transmission [24] by superimposing a sub-linewidth modulation to the gate, Fig. 9.10c. The transmission signal arises from an interference of quantum dot scattering with the driving laser [6]. The incoherent part, i.e. the spontaneous emission, averages to zero in transmission; what is detected instead is the coherent scattering, i.e. the Rayleigh scattering. In this way, the experiment utilizes both incoherent and coherent parts of the scattered light, for the single photon output and control, respectively. With a small modulation, the transmission signal has a large slope with zero crossing at zero detuning, Fig. 9.10c, and is therefore ideal for the generation of an error signal. $\Delta T/T$, the error signal, is recorded with a lock-in amplifier to reject noise and the lock-in output is fed into a classical feedback scheme. The feedback output is, like the modulation, applied to the gate electrode of the device. The set-point of the control loop is the zero crossing.

The long-term performance of the frequency locking scheme was tested by recording the RF over several hours, Fig. 9.11, without (blue) and with (red) the stabilizing loop. Without feedback, the RF exhibits fluctuations up to a factor of 2 (blue curve). The origin are slow electrical fluctuations in the sample which cause the transition to drift out of resonance with the laser. With feedback, these fluctuations disappear and the RF remains at a constant level (red curve) with noise determined almost entirely from shot noise in the detector, Fig. 9.11b. The average RF signal is a little smaller with feedback because the applied modulation broadens slightly the resonance.

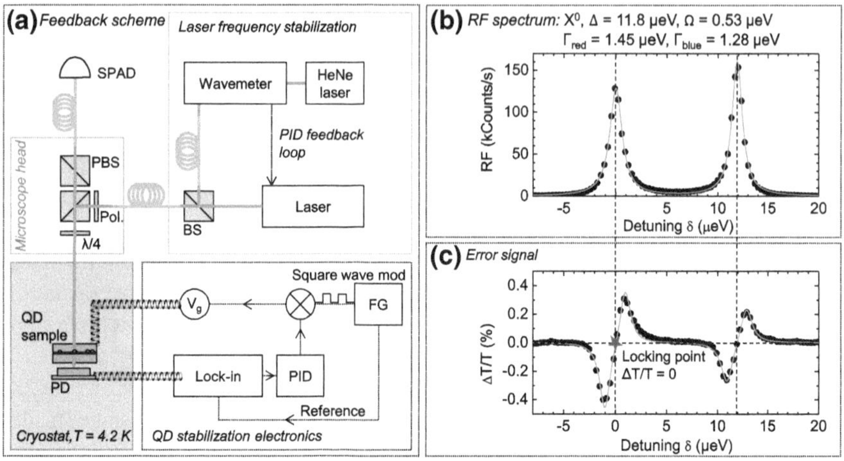

Fig. 9.10 Locking the optical frequency of a single quantum dot to a reference laser: scheme. **a** Schematic view of the experiment. The narrowband laser is stabilized to a fixed frequency by a wavemeter which in turn is stabilized to a HeNe laser. Laser light is guided through optical fibres (*yellow*) and microscope optics before it is focused onto the sample, driving the X^0 transition resonantly (BS = beam-splitter, PBS = polarizing BS, Pol. = linear polarizer). Two simultaneous measurements of X^0 scattering are performed: resonance fluorescence (RF), detected with an avalanche photodiode (APD), and absorption with a photodiode (PD) underneath the sample. The dynamic stabilization is realized with an active PID feedback loop which corrects for fluctuations in the transition energy using the gate voltage V_g and the square wave modulation of a function generator (FG). **b** RF signal of the fine structure split X^0 emission of a single quantum dot at wavelength 936.5 nm, a power corresponding to a Rabi energy Ω of 0.74 μeV and a temperature of 4.2 K. A detuning is achieved by sweeping the gate voltage. The *solid red line* is a Lorentzian fit to the data with linewidth $\Gamma = 1.28$ μeV (309 MHz) and $\Gamma = 1.45$ μeV (350 MHz) and with a fine structure splitting $\Delta = 11.8$ μeV. The linewidths are close to the transform limit of $\Gamma_0 = \hbar/\tau_r = 0.93$ μeV (220 MHz) where τ_r is the radiative lifetime of the exciton transition ($\tau_r = 0.71 \pm 0.01$ ns here). **c** The differential transmission ($\Delta T/T$) signal on the same quantum dot with integration time 100 ms per point using an in situ photodiode. A sub-linewidth square-wave modulation at 527 Hz is applied to the Schottky gate. This broadens both X^0 transitions slightly, here the lower frequency transition from $\Gamma = 1.45$ to $\Gamma = 2.58$ μeV. The *red curve* is a fit to the derivative of the two Lorentzians. The signal around the zero crossing point ($\Delta T/T = 0$) is used to generate an error signal for the feedback scheme. For the PID loop, the proportional factor $P = 0.1$ is chosen with respect to the slope of the error signal, while the integral $I = 0.06$ and the derivative constant $D = 6 \times 10^{-5}$ were obtained by tuning methods (color figure online)

The bandwidth of the feedback can be determined from the RF noise spectrum. $N_{QD}(f)$ corresponding to the time traces of Fig. 9.11a are shown in Fig. 9.11b. Without feedback, $N_{QD}(f)$ has a $1/f$-like dependence on f as a consequence of charge noise in the device, as shown in Fig. 9.7. With feedback, $N_{QD}(f)$ is reduced by up to a factor of 20 at the lowest frequencies, and is constant: the $1/f$-like noise is eliminated. The two curves meet at $f \simeq 130$ Hz once the bandwidth of the PID circuit has been exceeded. At higher frequency the noise spectrum is dominated by spin noise

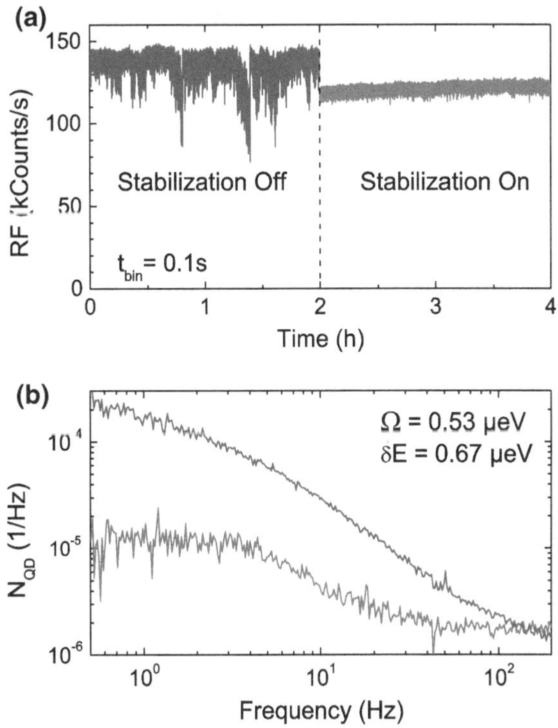

Fig. 9.11 Locking the optical frequency of a single quantum dot to a reference laser: performance. **a** Time trace of the resonance fluorescence (RF) of a single quantum dot (the one from Fig. 9.10) with $\delta = 0\ \mu\text{eV}$ recorded over several hours. The binning time was $t_{\text{bin}} = 100$ ms. The time trace is plotted with (*red*) and without (*blue*) the dynamic stabilization scheme. **b** Noise spectra of the normalized RF signal, $S(t)/\langle S(t)\rangle$, corresponding to the time traces of **a** after correction for external noise sources (color figure online)

[28] and the PID bandwidth is presently too slow to deal with it. This is however conceivable once the extraction efficiency is much improved.

9.4 Electron Spin Dephasing via the Hyperfine Interaction

A key issue for electron spin dephasing of a spin qubit in GaAs is that an electron spin couples to the nuclear spins in the host material via the contact hyperfine interaction [66], Fig. 9.12. The electron states in a self-assembled quantum dot are constructed from atomic s orbitals, Fig. 9.12b. The large amplitude of the s orbital at the location of each nucleus i results in a Fermi contact hyperfine interaction for an electron spin $\mathbf{S}$ with N nuclear spins $\mathbf{I}_i$

$$H_{\text{hf}}^{e} = \Omega \sum_{j=1}^{N} A_e^j |\Psi_e(\mathbf{R}_j)|^2 (I_z^j S_z + I_x^j S_x + I_y^j S_y). \tag{9.3}$$

A_e^j is the coupling coefficient, Ψ_e the electron envelope function and Ω the unit cell volume. The nuclei create an effective magnetic field, the Overhauser field B_N, which

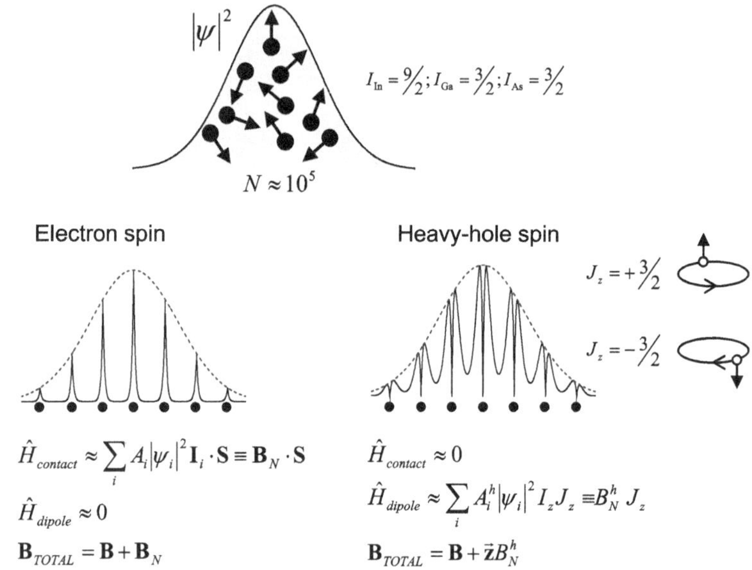

Fig. 9.12 Hyperfine interaction of an electron spin and a hole spin to the nuclear spins in the host material. a The electron or hole spin has a wave function extending over a few nm in all three directions such that it overlaps with about $N \sim 10^5$ atoms in the host material, each atom containing a nucleus with non-zero spin in the case of InGaAs. The nuclear spins are given for the main isotopes. **b** The conduction state is constructed largely from atomic s orbitals, each localized to a unit cell (shown schematically with the *black line* with the *black circles* representing the nuclei), modulated by the envelope function which extends over the quantum dot (shown schematically with the *blue dashed line*). The contact part of the hyperfine interaction dominates. A_i is the coupling coefficient with nuclear spin i, $\mathbf{I}_i$ the nuclear spin, and ψ_i the electron wave function at the location of nuclear spin i. The contact term resembles the interaction of the electron spin $\mathbf{S}$ in a fictitious magnetic field, $\mathbf{B}_N$. The dipole-dipole hyperfine interaction is zero for a pure s orbital, and close to zero in practice. **c** The hole state is constructed largely from atomic p orbitals, each localized to a unit cell, such that the wave function amplitude is small at the location of each nucleus. This suppresses the contact part of the hyperfine interaction. The dipole-dipole part is non-zero however. A heavy hole spin has $J_z = \pm\frac{3}{2}$, corresponding, in a semi-classical interpretation, to a circulating microscopic current clockwise with spin up, or counter-clockwise with spin down. The magnetic dipole moment points therefore either along $+z$ or $-z$ such that the dipole-dipole Hamiltonian has an Ising form, $\propto I_z J_z$. Equivalently, the fictitious magnetic field describing the nuclear spins lies solely along the z-direction (color figure online)

fluctuates in time resulting in spin dephasing [46, 47]. For self-assembled quantum dots, $N \sim 10^5$, $B_{\mathrm{N}} \sim 20\,\mathrm{mT}$ resulting in an energy fluctuation in the electron Zeeman energy of $\delta E_z \simeq 0.6\,\mu\mathrm{eV}$ and $T_2^* \sim 1$ ns.

The first order electron spin-nuclear spin flip-flop processes can be suppressed simply by applying a magnetic field, exploiting the mismatch in electron and nuclear gyromagnetic ratios. The interaction along the magnetic field direction remains and leads to the small electron spin T_2^* times [67]. One mitigating strategy is to prepare the nuclear spins carefully [68–70]. Another is to exploit the low frequency of the nuclear spin dynamics with dynamic decoupling or real-time Hamiltonian estimation methods [52, 70–72], techniques yet to be implemented on self-assembled quantum dots. Optimistically, these first order hyperfine effects can be dealt with. However, the second order processes remain and it has been proposed that they determine the ultimate limit on electron spin coherence [47, 73, 74]. This point is difficult to prove experimentally – many factors (phonons, co-tunneling, charge noise [75]) influence the electron spin decoherence – and the theory is complex [76, 77].

One way to probe the hyperfine interaction between an electron spin and the nuclear spins is to probe the interaction from the point of view of the nuclear spins. In the absence of an electron, the nuclear spins are largely isolated, interacting with each other only by the weak dipole-dipole interaction, leading to long coherence times (milli-second regime) [78, 79]. This is an ideal starting point: the nuclear spin coherence is a sensitive probe of any interaction turned on by the presence of a single electron. Specifically, the second order flip-flop process should lead to a measurable effect on the nuclear spin coherence. The particular process is a spin flip-flop between two nuclear spins. The two nuclear spins are far apart such that they are not coupled by the dipole-dipole interaction. Instead, the coupling is mediated by a common coupling to the delocalized electron. This mechanism couples all nuclear spins together which can conserve energy on undergoing a flip-flop, "shells" of nuclear spins along a contour of constant electron density. In a quantum dot, this electron-mediated nuclear spin-nuclear spin flip-flop couples many nuclear spins together, and the problem has a complex, mesoscopic character.

This approach, addressing the hyperfine-decoherence of the electron spin via the nuclear spin decoherence, hinges on the ability to perform nuclear magnetic resonance (NMR) on the nuclear spins. Self-assembled quantum dots have advantages here. First, the pronounced Coulomb blockade allows the nuclear spin coherence to be measured both with and without an electron. Secondly, the nuclear spins can be both polarized ("dynamic nuclear polarization", DNP) and read-out optically [80]. DNP represents a laser-cooling of the nuclear spins down to mK temperatures [81] creating large population differences amongst the nuclear spin levels, boosting the otherwise very weak NMR signal. A nuclear polarization results in subtle shifts to the optical resonance frequency which can be measured very precisely. Nuclear magnetic resonance sensitive to just $\sim$1,000 nuclear spins has been achieved [79, 81, 82]. Finally, the nuclei in InGaAs quantum dots have large quadrupole shifts [81–83] and, arguably, this represents a simplification: in an NMR experiment on nuclear spins with $I > \frac{1}{2}$, a narrowband NMR drive selects only one transition and the nuclei can be treated as a collection of spin-$\frac{1}{2}$ spins, an ideal test-bed for theory.

The experiment measures the coherence of the nuclear spins associated with an InGaAs quantum dot for different charge states, empty (0), singly-occupied ($1e$) and doubly-occupied ($2e$). A static magnetic field is applied along the growth direction, z; an oscillating magnetic field is applied in-plane, in the x-direction, by sending a current through an on-chip micro-wire, Fig. 9.13d, e. Specifically, the coherence associated with the "central" transition (nuclear spin $I_z = -\frac{1}{2} \leftrightarrow I_z = +\frac{1}{2}$) of the ^{75}As and ^{115}In isotopes is measured. On account of the in-built strain, each nucleus

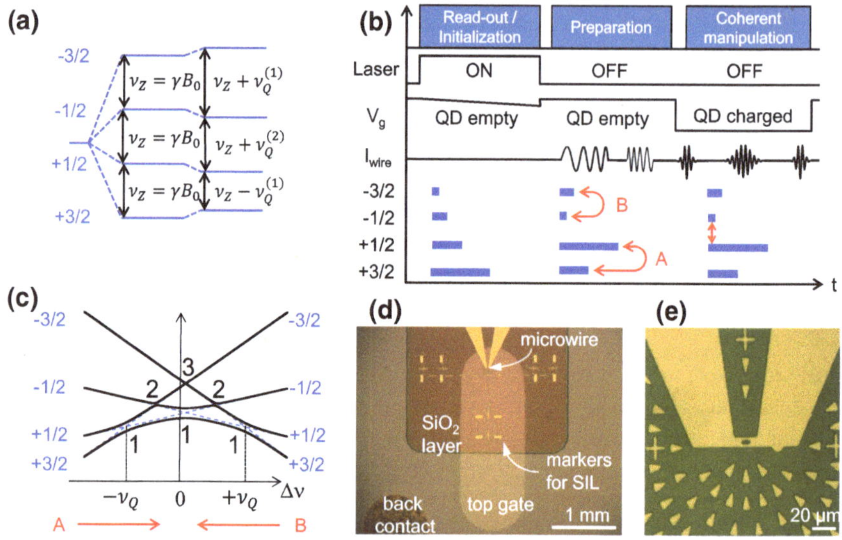

Fig. 9.13 Concepts of the quantum dot NMR experiment. a Energy levels for a spin $I = \frac{3}{2}$ in the laboratory frame. The static field B_z causes the spin states to split into the Zeeman ladder with spacing Larmor frequency ν_Z. Quadrupole effects result in alterations to the level spacing; the central transition frequency is only shifted by second order terms. **b** Measurement cycle: *Read-out/initialization* involves detecting resonance fluorescence from the empty quantum dot (neutral exciton) excited with a narrow-band laser: this both reads the previous nuclear spin polarization and sets a new state. In the *Preparation* part, two chirped pulses are applied, A and B, which swap the populations $+\frac{3}{2} \leftrightarrow +\frac{1}{2}$, $-\frac{3}{2} \leftrightarrow -\frac{1}{2}$ in order to maximize the population difference between the $+\frac{1}{2}$ and $-\frac{1}{2}$ states. In the *Coherent manipulation* part, a particular bias (which controls the quantum dot charge) is applied to the gate of the device, and then a pulse of ac current is applied to the microwire at the radio frequency of the central transition. Finally, the bias is re-set to the starting value. **c** Energy levels for a $\frac{3}{2}$ spin in the rotating frame versus radio frequency detuning in the limit of $\nu_Q \gg \nu_{RF}$ where ν_{RF} is the Rabi coupling. The preparation pulses are indicated by red arrows. **d** Top view of sample showing back contact, top gate, SiO_2 spacer layer, microwire and markers for positioning the solid immersion lens (SIL). **e** Zoom-in of the microwire. A hole in the wire enables optical access to the quantum dot; the *triangular markers* facilitate positioning of the sample in the microscope

experiences an electric field gradient which leads to quadrupole shifts of the bare levels [81–83], as shown in Fig. 9.13a, the eigenenergies in the laboratory frame. The in-built strain is site-dependent resulting in a spread of electric field gradients across the quantum dot, in particular across the electron wave function. For ^{75}As with $I = \frac{3}{2}$, the first-order quadrupole effect shifts the $I_z = -\frac{3}{2} \leftrightarrow I_z = -\frac{1}{2}$ transition to lower frequencies, the $I_z = +\frac{1}{2} \leftrightarrow I_z = +\frac{3}{2}$ transition to higher frequencies, yet the frequency of the central transition is shifted only in second-order. The NMR spectrum consists of a central peak at ν_z, inhomogeneously broadened by the second order quadrupole effects, and well-separated sidebands at $\nu_z \pm \nu_Q^{(1)}$ where ν_z is the Zeeman frequency and $\nu_Q^{(1)}$ the first-order quadrupole shift [82]. For ^{115}In with $I = \frac{9}{2}$, there are 9 NMR transitions but, as for ^{75}As, the central transition is unaffected by the quadrupole interaction to first order. Hence, in the limit $\nu_Q^{(1)} \gg \nu_{RF}$, for each nuclear spin, the $I_z = -\frac{1}{2} \leftrightarrow I_z = +\frac{1}{2}$ NMR transition can be thought of as a quasi two-level system: on driving at frequency ν_z, population is largely confined to the $I_z = \pm\frac{1}{2}$ sub-space. The eigenenergies in the rotating frame are shown in Fig. 9.13b for a realistic quadrupole frequency $\nu_Q^{(1)}$ (2 MHz) and Rabi coupling ν_{RF} (100 kHz). The strongest avoided crossings occur when the bare states have a difference in angular momentum of one quantum unit [81].

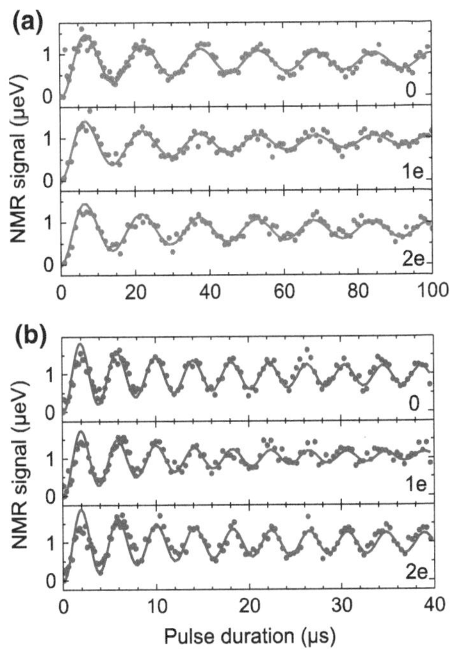

Fig. 9.14 Rabi oscillations of the nuclear spin ensemble. Following nuclear state preparation, the quantum dot is charged (0, 1e or 2e states) and a radio-frequency pulse resonant with the nuclear central transition is applied. The bias is set to the centre of the charging plateau in the case of an occupancy of one electron. Plotted is the NMR signal (the optical shift) versus pulse duration for **a** arsenic, **b** indium, in each case for occupancy zero, one electron and two electrons. The *solid-lines* represent the calculated response of an inhomogeneous distribution (FWHM of central frequency $\delta\nu_Q^{(2)}$) of two-level emitters with Gaussian distribution of resonance frequency. For ^{75}As, the Rabi coupling is fitted to $\nu_{eff} = 64$ kHz, $\delta\nu_Q^{(2)} = 71.7$ kHz, radio frequency magnetic field $B_{RF} = 4.4$ mT from the data at occupancy zero (where T_2 is large, 5 ms). In the 1e state, the fit is excellent with the same ν_{eff} and $\delta\nu_Q^{(2)}$ but with $T_2 = 108$ μs. For In, for occupancy zero, the fit yields $\nu_{eff} = 241$ kHz, $\delta\nu_Q^{(2)} = 146$ kHz and $B_{RF} = 5.2$ mT. For occupancy of one electron, the fit is excellent with the same ν_{eff} and $\delta\nu_Q^{(2)}$ but with $T_2 = 25$ μs. For both ^{75}As and ^{115}In, the fit for occupancy two electrons is excellent with the same parameters as for occupancy zero. The static magnetic field is $B_z = 6.6$ T, temperature 4.2 K

The measurement protocol is shown in Fig. 9.13b. The nuclear spins are polarized and read-out optically (charge state 0); during the NMR protocol the laser is turned off. Figure 9.14 shows the results of a Rabi oscillation experiment. A pulse at the ^{75}As central transition frequency is applied to the microwire. Figure 9.14a shows the NMR signal as a function of pulse duration. A clear oscillation is observed, a Rabi oscillation, as the population is driven coherently between the $I_z = -\frac{1}{2}$ and $+\frac{1}{2}$ states. The period corresponds closely to the expected result, twice the ^{75}As gyromagnetic ratio (the factor of two is the effective coupling and arises on folding the system to an effective spin-$\frac{1}{2}$ system [84]). When the quantum dot is empty, the Rabi oscillations decay with a time constant of 50 μs. Given that the coherence time associated with this transition is in the ms regime [79], this decay corresponds not to a loss of coherence but to an inhomogeneous broadening, the second-order quadrupole shift to the central transition $\nu_Q^{(2)}$. To determine the inhomogeneous broadening $\delta\nu_Q^{(2)}$, the response of an ensemble of coherent two-level systems with a Gaussian distribution in centre frequencies is calculated. Figure 9.14a shows an excellent fit to the data with $\delta\nu_Q^{(2)} = 71.7$ kHz. When the quantum dot is occupied with a single electron, the Rabi oscillations retain the same frequency but they decay sooner (decay time constant 30 μs). When the quantum dot is occupied with two electrons, the Rabi oscillations behave exactly as for an empty quantum dot. Figure 9.14b shows also a Rabi experiment performed at the ^{115}In central transition. The period of the Rabi oscillations, noticeably smaller than that of ^{75}As, reflects both the different gyromagnetic ratio and the increase in the effective coupling (factor 5 for the spin-$\frac{9}{2}$ ^{115}In). The decay of the Rabi oscillations follows the same pattern as for the isotope ^{75}As:

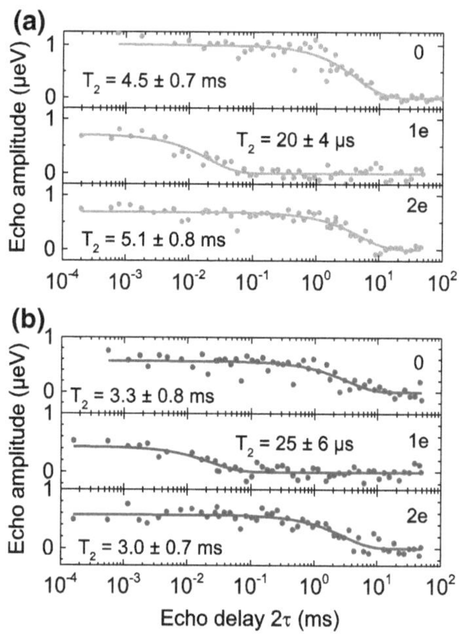

Fig. 9.15 Hahn echo T_2 measurement. NMR signal following a Hahn echo sequence for **a** arsenic, **b** indium, in each case for the three charge states, 0, 1*e*, 2*e*. The Hahn echo consists of the standard $\pi/2 - \tau - \pi - \tau - \pi/2$ sequence. The echo amplitude is plotted against the total delay 2τ. Single exponential fits ($\exp(-2\tau/T_2)$) determine the coherence times. T_2 for the singly charged dot is more than a factor 100 lower than for the empty or doubly charged QD

the decay is the same for the empty and doubly-charged states, more pronounced for the single-charged state.

The Rabi oscillations provide the first indication that the nuclear spin dynamics depend on the quantum dot charge. In fact, they reveal a dependence on electron spin: there is a lone spin in the $1e$ state but the two electrons in the $2e$ state form a spin singlet. However, the faster decay of the Rabi oscillations in the presence of a single electron could signify a decreased nuclear spin coherence or an increased ensemble broadening (for instance through the Knight field). To distinguish between these two cases, it is necessary to measure the coherence associated with the nuclear central transition. The Hahn echo is perfect for this as it removes the dependence on the inhomogeneity in the quadrupole shift. Figure 9.15 presents the Hahn echo amplitude as a function of echo delay for both ^{75}As and ^{115}In, in each case for three charge states. The Hahn echo for the $1e$ state was recorded at the centre of the single-electron charging plateau. For both ^{75}As and ^{115}In, a very pronounced dependence on spin is revealed: the Hahn echo decay time (T_2) *decreases* by more than two orders of magnitude in the presence of a lone electron spin.

For an empty quantum dot, T_2 is a few ms for both ^{75}As and ^{115}In, agreeing with previous experiments [79]. The general timescale points to decoherence via a dipole-dipole interaction. For a singly-occupied quantum dot however the T_2 times fall to just $\sim 20\,\mu s$, a timescale far too short for a dipole-dipole interaction, and an additional decoherence mechanism is clearly turned on. Figure 9.16 shows T_2 versus bias, marking the extents of the $1e$ charging plateau. Far from the charging bias, T_2 is independent of bias for the 0 state, falling monotonically as the charging plateau is crossed. T_2 reaches a minimum at the centre of the $1e$ plateau. In fact T_2 is symmetric about the centre of the $1e$ charging plateau, recovering completely in the $2e$ plateau. This is a striking result: the nuclear spins are least coherent in the $1e$ plateau centre when the electronic degrees of freedom (charge, electron spin, exciton) are most coherent.

In the $1e$ plateau, the electron spin relaxation time T_1^e and the nuclear spin relaxation time T_1 follow exactly the opposite dependence on bias as compared to the nuclear spin T_2. T_1^e is very small close to the edges of the $1e$ plateau edge on account of co-tunneling (the quantum dot electron spin relaxes rapidly by swapping its spin with an electron in the Fermi sea) [17, 19, 85]. In the $1e$ plateau centre, co-tunneling is suppressed at low temperature by the gap between the quantum dot ground state and the Fermi energy of the Fermi sea such that T_1^e increases (by about 4 orders of magnitude with respect to the plateau edge [17]). The electron T_1 process determines the nuclear spin T_1 process: nuclear spin leaks into the Fermi sea [86]. This anti-correlation between electron spin relaxation and the nuclear spin coherence is particularly pronounced at the plateau edge itself. Here electron spins relax extremely rapidly (revealed also in an increase in the optical resonance fluorescence linewidth), and the nuclear spin polarization decays relatively quickly. Nevertheless, this rapid electron relaxation has a relatively benign effect on the nuclear spin decoherence. The recovery of the nuclear spin T_2 in the $2e$ state is also completely consistent with this link of nuclear spin coherence to electron spin: in the $2e$ state, the two electrons form a singlet with zero total spin.

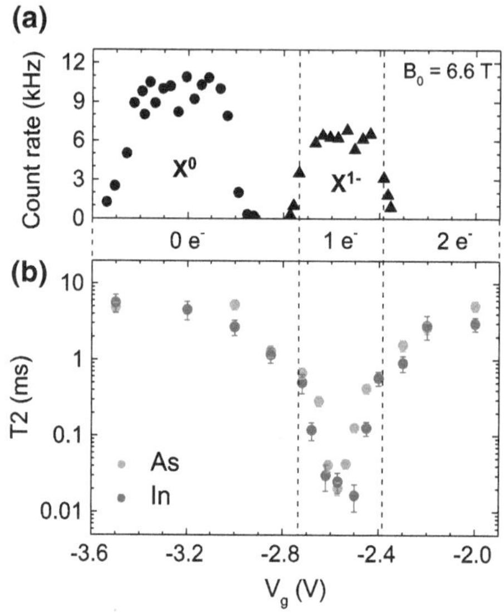

Fig. 9.16 Nuclear spin coherence time as a function of gate voltage. a Peak resonance fluorescence intensity for both neutral (X^0) and charged (X^{1-}) excitons versus bias, V_g, for constant laser intensity. *Dashed lines* correspond to an X^{1-} emission intensity drop of 50% and indicate the boundaries of the charging plateau. At the boundary on the "left", the 0 and e states are degenerate and are thus occupied 50:50; equivalently, at the boundary on the "right", the e and $2e$ states are degenerate. **b** Nuclear spin coherence time, T_2, versus bias, V_g

The experiments demonstrate that slow electron spin relaxation promotes nuclear spin decoherence. This points to a nuclear spin-nuclear spin interaction facilitated by a common interaction with an electron spin. Qualitatively, this interaction accounts for the experimental results. First, although this electron-mediated nuclear spin-nuclear spin interaction arises only in second order, it provides a means for many nuclei of a particular isotope in the quantum dot to couple together such that it has significant consequences. Secondly, the interaction is turned off in the $2e$ ground state, a singlet, accounting for the recovery of the nuclear spin coherence in this regime. Thirdly, electron spin relaxation via co-tunneling is fast relative to the nuclear spin dynamics away from the plateau centre such that the nuclear spin ensemble interacts with a time-averaged electron spin $\langle S_z \rangle$. At the edges of the $1e$ plateau, $\langle S_z \rangle$ is small, suppressing the electron-mediated nuclear spin-nuclear spin interaction: this accounts for the anti-correlation between nuclear spin T_2 and electron spin T_1^e.

A quantitative account of the nuclear spin T_2 in the $1e$ plateau centre has been developed [87]. At the plateau centre, the central transitions of a particular isotope represent a closed system, i.e. coupled quasi-spin-$\frac{1}{2}$ spins, as the average quadrupole splittings (~2 MHz [81]) are larger than the co-tunneling rate (~0.1 MHz). Provided the electron Zeeman energy is larger than the averaged hyperfine coupling, the electron-mediated nuclear spin-nuclear spin interaction results in a Hamiltonian [74],

$$V = \frac{1}{8Z_e}\sum_{j\neq l} A_j A_l \left[\left(\tfrac{1}{2}+S_z\right)\left(I_j^- I_l^+ + I_l^- I_j^+\right) - \left(\tfrac{1}{2}-S_z\right)\left(I_j^+ I_l^- + I_l^+ I_j^-\right)\right]. \tag{9.4}$$

where Z_e is the electron Zeeman energy (in the total magnetic field, external magnetic field plus Overhauser field), A_j is the hyperfine coupling constant of the jth nuclear spin, S_z is the z-component of the electron spin operator, and I_j is the nuclear spin operator of the jth nuclear spin. The terms $I_j^- I_l^+ + I_l^- I_j^+$ and $I_j^+ I_l^- + I_l^+ I_j^-$ represent nuclear spin-nuclear spin flip-flop processes and these terms lead to nuclear spin decoherence. The dynamics of the transverse components of a single nuclear spin are described in the presence of a coupling to all the others using a master equation to second order in V followed by a calculation of the ensemble decoherence rate. Despite the complexity of the problem, an analytical result for the ensemble decoherence rate $\hat{\Gamma}$ for ^{75}As was derived,

$$\hat{\Gamma}_{\mathrm{As}} = \frac{2\sqrt{2}A_{\mathrm{As}}^3}{9\hbar\omega^2 N}, \tag{9.5}$$

where A_{As} is the hyperfine constant of ^{75}As in GaAs. The total number of nuclear spins in the quantum dot N is known from other experiments [58], $N = (8.5 \pm 0.9) \times 10^4$. The hyperfine coupling, $A_{\mathrm{As}} = 86 \pm 10\,\mu$eV, is the standard literature value [57, 88]. The Zeeman energy under these conditions, $Z_e = 246 \pm 30\,\mu$eV, is measured in situ; note that the condition $A < Z_e$ is met in the experiment. The final theoretical result is that $1/\hat{\Gamma}_{\mathrm{As}} = 17 \pm 5\,\mu$s. The error specified represents a random error arising from the uncertainties in the input parameters. The experimental result for ^{75}As in the plateau centre is $T_2 = 20 \pm 4\,\mu$s, Fig. 9.16: the experimental and theoretical results agree well. Away from the plateau centre, the co-tunneling rate increases above $\nu_Q^{(1)}$ eventually becoming larger than the total spread in ν_z (50 MHz in this experiment) such that all transitions of all nuclei can in principle be coupled together via the common interaction with the electron; in practice, the co-tunneling rapidly reduces $\langle S_z \rangle$, shutting off the interaction. A complete theory in the co-tunneling regime is formidably complex. This comment notwithstanding, the agreement with the theoretical result at the plateau centre adds considerable weight to the assertion that an electron-mediated interaction is responsible for the decoherence of the nuclear spins in the presence of a single electron.

The overriding point is that probing the nuclear spins in the quantum dot turns out to be a sensitive probe of the interactions turned on by the presence of a single electron. Here, the loss of nuclear spin coherence can be attributed unambiguously to an electron-mediated nuclear spin-nuclear spin coupling. In turn, the prediction is that this interaction limits the electron spin coherence, to time-scales of tens of microseconds in these quantum dots at these magnetic fields. This mechanism determines a hard limit on the electron spin coherence time. The decoherence time can only be increased by increasing the Zeeman splitting (by applying larger magnetic fields or by engineering the electron g-factor) or by increasing the quantum dot size, both hard to achieve in practice.

9.5 Hole Spin Dephasing

The nuclear spins in GaAs lead to a rapid loss of electron spin coherence (both T_2 and T_2^* processes) [2, 46, 47]. Clearly, an approach which retains the GaAs system while suppressing the interaction of the spin qubit to the nuclear spins is attractive. A hole spin offers an alternative platform.

A hole is the absence of an electron in an otherwise occupied valence level. A hole spin has a fundamentally different hyperfine interaction to the electron spin. The valence states are constructed from atomic p orbitals with zero wavefunction amplitude at the location of the nuclei, Fig. 9.12c. The Fermi contact hyperfine interaction is therefore suppressed [2, 89, 90]. The dipole-dipole part of the hyperfine interaction remains [89–92]. For a *pure* heavy hole (HH) state the hyperfine interaction has an Ising form,

$$H_{\mathrm{hf}}^{\mathrm{HH}} = \Omega \sum_{j=1}^{N} A_{h,z}^{j} |\Psi_h(\mathbf{R}_j)|^2 I_z^j S_z. \tag{9.6}$$

$A_{h,z}^j$ is the coupling coefficient, Ψ_h the hole envelope function, and $S_z = \pm\frac{1}{2}$ represents $J_z = \pm\frac{3}{2}$. The absence of transverse terms means that the heavy hole spin experiences just the z-component of the noisy Overhauser field, Fig. 9.12c. Furthermore, the heavy hole coupling coefficients are reduced with respect to the electron coupling coefficients: $A_{h,z}^j/A_e^j \simeq -10\%$ [90, 92]. The most important consequence of the Ising form is that application of a transverse magnetic field suppresses hole spin dephasing by the nuclear spins [90]. This is so effective that the hyperfine interaction is to all intents and purposes switched off for a *pure* heavy hole spin [93].

A close-to-ideal heavy hole state exists in unstrained, highly confined GaAs quantum wells [94, 95]. Quantum dots however have mixed states. Even for an ideal quantum dot shape, symmetry does not prevent heavy hole-light hole coupling [96, 97], and heavy hole-light hole coupling is an experimental fact [19, 98–100], revealed by deviations in the optical selection rules from the heavy hole limit. For strained InGaAs quantum dots, the light hole accounts for 5–10% of the hole state [19, 98, 100].

The light hole component in the quantum dot hole state has important consequences for the hole spin hyperfine coupling [92, 93, 101, 102]. Additionally, admixture of the conduction s orbitals should be taken into account [93, 101]: while s admixture is small on account of the fundamental energy gap of the semiconductor, it turns on the large Fermi contact part of the Hamiltonian. In a $\mathbf{k}.\mathbf{p}$-description, the band admixtures are described by an 8×8 Hamiltonian (conduction, heavy hole, light hole and spin-orbit split-off states); the hyperfine interaction consists of a Fermi contact term and dipole-dipole-like interactions [90, 92]. For the hole states, provided the admixtures of light hole and conduction states are small, the hyperfine interaction can be folded down to an effective 2×2 Hamiltonian which operates on the mixed hole states. The two mixed hole states are described as a spin-$\frac{1}{2}$ pseudospin, $\mathbf{S}$: $S_z = +\frac{1}{2} \equiv |\Uparrow\rangle$ represents one of the mixed states, $S_z = -\frac{1}{2} \equiv |\Downarrow\rangle$ represents the other. The final result is that the hole hyperfine interaction is no longer purely Ising-like:

$$H_{\mathrm{hf}}^{h} = \Omega \sum_{j=1}^{N} |\Psi_h(\mathbf{R}_j)|^2 (A_{h,z}^{j} I_z^j S_z + A_{h,x}^{j} I_x^j S_x + A_{h,y}^{j} I_y^j S_y). \qquad (9.7)$$

$A_{h,x}^{j}$ and $A_{h,y}^{j}$ are the transverse coupling coefficients and arise from the admixture of both conduction [93, 101] and light hole states [92, 93, 101, 102], both couplings giving terms with the same functional form. In each case, $A_{h,x}^{j}$ and $A_{h,y}^{j}$ depend on $A_{h,z}^{j}$ multiplied by an admixture coefficient. The heavy hole-light hole coupling also introduces non-colinear terms [102]. The transverse coupling makes the hole spin vulnerable to spin dephasing via the in-plane components of the nuclear spins, an interaction which cannot be suppressed in an in-plane magnetic field. In fact the *anisotropy* (rather than the magnitude of $A_{h,z}^{j}$) represents a crucial issue in the development of a hole spin qubit.

Experiments have established long hole spin relaxation times [20, 103–105], coherence times T_2 in the μs range [106, 107], fast spin rotations [107–109] (see Chap. 10) and control of two tunnel-coupled hole spin qubits [108]. The hole spin T_2 falls rapidly above 10 K, a consequence of a spin-orbit mediated phonon interaction [110]. Conveniently, 4 K is cold enough to achieve a highly coherent hole spin. The existence of the longitudinal hole hyperfine interaction has been established [91]. Experimentally, $A_{h,z}^{j}$ averaged over the quantum dot, $\langle A_{h,z} \rangle$, has been measured to be -10% of the average value of A_i^e, $\langle A_i^e \rangle$, on self-assembled quantum dots by dynamically polarizing the nuclear spins along the z-direction and measuring the changes to the electron and hole Zeeman energies [111–113], confirming theoretical expectations [90, 92] albeit with some discussion on the sign [66, 114].

Two difficulties are encountered in probing the hole spin hyperfine interaction optically. First, optical excitation of a hole spin populates an exciton state consisting of two holes in a singlet state but an unpaired electron spin. In this situation it is not trivial to assign any nuclear spin effects unambiguously to the hole spin given the strong hyperfine interaction of the electron spin. Secondly, p-type devices tend to be considerably noisier than n-type devices yet the hole g-factor is very sensitive to an electric field [115–117] such that charge noise results in spin dephasing [107, 116, 117]: in noisy devices this effect completely obscures the hyperfine couplings.

It is important to reduce radically the charge noise in p-type devices. This has been achieved not only by working with ultra-clean material but also by inverting the standard design, switching from the standard p-i-Schottky structure to an n-i-p device. It is also important to carry out an experiment which is sensitive to the transverse terms in the hole hyperfine interaction. This was achieved by polarizing the nuclear spins along a transverse direction, monitoring the polarization via the lone electron spin in the exciton, and measuring the hole Zeeman splitting Z_h ultra-precisely by means of dark state spectroscopy, i.e. coherent population trapping (CPT) [48, 106, 116, 118, 119]. The combination of a coherent hole spin, resonance fluorescence detection (RF) [120] and low-noise samples resulted in a spectral resolution in Z_h of just 10 neV (2.4 MHz).

9.5.1 Coherence Population Trapping on a Single Hole Spin in a Quantum Dot

CPT is a useful spectroscopic tool to measure the hole Zeeman energy with high resolution. CPT is a quantum interference in a Λ-system where two ground states are coupled individually by "pump" and "probe" optical fields to a common upper level, Fig. 9.17. Here, the two ground states correspond to the Zeeman-split hole spin states, described as $|\Uparrow\rangle_x$ and $|\Downarrow\rangle_x$, and the upper level to an exciton, $|\Uparrow\Downarrow,\downarrow\rangle_x$ or $|\Uparrow\Downarrow,\uparrow\rangle_x$, where $|\Uparrow\rangle_x$ and $|\Downarrow\rangle_x$ are the eigenstates of the hole pseudospin in the x-basis and $|\uparrow\rangle_x, |\downarrow\rangle_x$ refer to the electron spin states, also in the x-basis, Fig. 9.17. A transverse magnetic field (in the x-direction) establishes the quantization axis and the Λ-system, Fig. 9.17. This applies to a hole spin provided the in-plane g-factor is non-zero: the interference occurs when the frequency difference of the lasers matches the hole Zeeman splitting, the two-photon resonance. A dark state results, revealed by a dip in the probe spectrum. The spectral position of the dip measures Z_h. Specifically, when $\hbar\Omega_1 \ll \hbar\Omega_2 \ll \hbar\Gamma_r$ ($\hbar\Omega_1$, $\hbar\Omega_2$ are the probe and pump couplings, Γ_r the spontaneous emission rate) the CPT dip has width $\hbar\Omega_2^2/\Gamma_r$. The depth of the dip is sensitive to the hole spin coherence: only for $1/T_2 \ll \Omega_2^2/\Gamma_r$ does the emission in the dip go to zero. Hence, provided the hole spin coherence is high enough, the width of the CPT dip can be much less than the optical linewidth, enabling a highly accurate measurement of Z_h. Furthermore, the location of the CPT dip is determined only by the two-photon resonance. CPT is therefore an ideal technique to extract Z_h. Fluctuations in exciton energy (via charge noise and the Overhauser field acting on the electron spin) modify the emission envelope [106, 118] but not the location of the CPT dip.

CPT on a single quantum dot containing a single hole is shown in Fig. 9.17. The new p-type devices are very important: they remove the charge-noise-induced fluctuations of the CPT dip position which plagued earlier experiments [116]. The occupation of the upper level is monitored with high signal:noise by detecting the resonance fluorescence [28, 120], Fig. 9.17b, c. The resonance fluorescence exhibits a Lorentzian envelope with full-width-at-half-maximum (FWHM) 2.5 μeV and a pronounced dip with FWHM 80 neV (19.3 MHz). A zoom-in of the CPT dip is shown in Fig. 9.17c along with CPT from a quantum dot in sample B with CPT dip width 33 neV (8.0 MHz). These spectra enable the determination of Z_h with a resolution of ~10 neV. The hole g-factor (in-plane magnetic field) is $g_{h,x} = 0.063$ for quantum dot A, $g_{h,x} = 0.035$ for quantum dot B. Averaged over many quantum dots in these samples, $\langle g_{h,x}\rangle = 0.12 \pm 0.10$; in the vertical direction, the hole g-factor is much larger, $\langle g_{h,z}\rangle = 1.22 \pm 0.03$. This is reminiscent of the close-to-ideal heavy hole state in an unstrained quantum well for which $g_{h,x} \ll g_{h,z}$. However, the

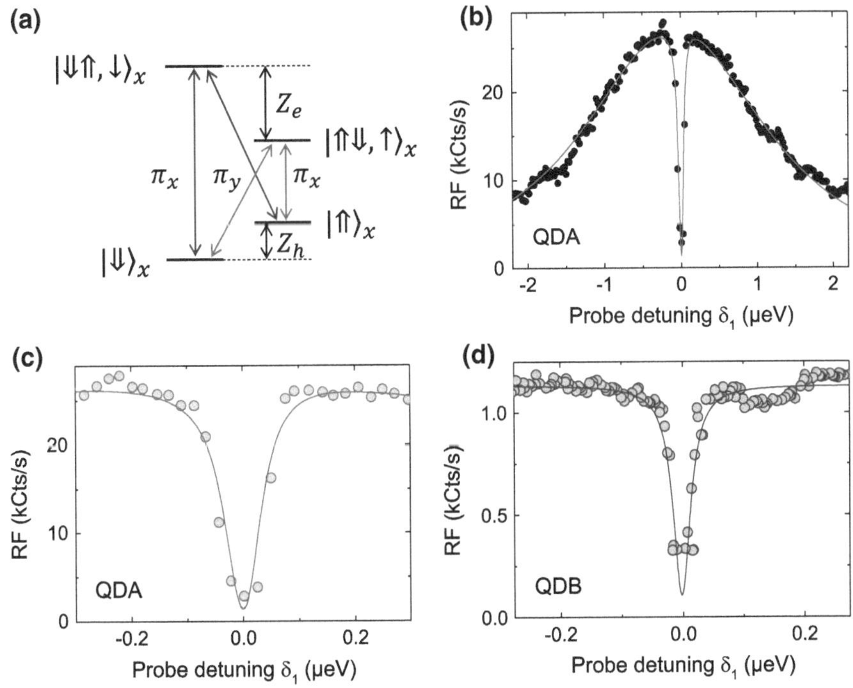

Fig. 9.17 Coherent population trapping on a single hole spin in a quantum dot. a The quantum states. Two optical Λ-systems (*blue* and *red lines*) are established in a magnetic field along the x-direction: $|\Uparrow\rangle_x$ and $|\Downarrow\rangle_x$ are the hole pseudospin eigenstates in the x-basis, split by the Zeeman energy Z_h; the upper levels are the X^{1+} excitons consisting of two holes in a singlet state and an unpaired electron with spin $\pm\frac{1}{2}$, again in the x-basis. The optical transitions are linearly polarized, either π_x or π_y, with equal optical dipoles, at wavelengths close to 950 nm. **b** RF spectrum on a single quantum dot QDA containing a single hole in sample A using the "*blue*" Λ-system (pump on higher energy "vertical" transition). The pronounced dip signifies CPT. The *solid line* shows the result of a 3-level density matrix model (probe coupling $\hbar\Omega_1 = 0.06\,\mu$eV, pump coupling $\hbar\Omega_2 = 0.40\,\mu$eV, spontaneous emission rate $\Gamma_r = 0.68\,\mu$eV, $T_2 > 1\,\mu$s, $T_1 \gg T_2$) convoluted with a Lorentzian with FWHM $\Gamma = 2.5\,\mu$eV to describe slow exciton dephasing, and then with a Lorentzian with FWHM 8.3 neV (2.0 MHz) to account for the mutual coherence of the lasers. The data were recorded with 0.1 s integration time per point at a magnetic field $B_x = 3.00$ T and temperature $T = 4.2$ K. **c, d** Two exemplary CPT dips of QDA and QDB, respectively. The dip from QDA has a FWHM of 80 neV (19.3 MHz) and is modelled with the parameters of (**b**). The limited mutual coherence of the lasers is the main reason that the signal in the dip centre does not go down completely to zero. The dip from QDB has a FWHM of 33 neV (8.0 MHz), 5 s integration per point. The CPT simulation uses $\hbar\Omega_1 = 0.1\,\mu$eV, $\hbar\Omega_2 = 0.49\,\mu$eV, and, as in **b**, $\Gamma_r = 0.68\,\mu$eV, $T_2 > 1\,\mu$s, $T_1 \gg T_2$. In this case, the remaining signal in the dip centre is likely to be a consequence of the small value of Z_h: the dark state can be destroyed by the weak coupling of the pump to the probe transition (color figure online)

magnitude of $g_{h,x}$ is an unreliable measure of the heavy hole-light hole admixture as $g_{h,x}$ is very sensitive to the indium concentration via the strong dependence of the band structure parameters on indium concentration [117]. A small $g_{h,x}$ is encouraging but in itself does not represent a suppressed hole spin hyperfine interaction.

The transverse hole hyperfine interaction is measured by combining CPT and DNP. At zero magnetic field, the resonance fluorescence spectra have a straightforward Lorentzian lineshape, Fig. 9.1. This can change in an applied magnetic field where the resonance has a "top-hat" shape extending over tens of μeV and a strong hysteresis on reversing the scan direction [111, 121, 122]. The interpretation is that as the laser is tuned, the nuclei polarize in such a way that resonance with the laser is maintained. This effect, referred to as "dragging", occurs also here and is used as a tool to create a large DNP. Dragging arises through the hyperfine interaction of the lone electron spin in the exciton. Furthermore, the exact change in electron Zeeman energy under DNP can be probed spectroscopically by measuring a change in transition energy of one of the exciton transitions.

In the experiment, the DNP is controlled via the detuning δ_2 of the pump laser with respect to the bare transition frequency. A strong constant frequency pump laser defines the nuclear spin state and a weak probe laser ($\Omega_1 \ll \Omega_2$) measures both Z_h and the electron Zeeman splitting Z_e. The probe laser is scanned across the vertical and diagonal transitions, Fig. 9.18. A pronounced dip in the spectrum indicates CPT and measures Z_h with ultra-high resolution. For zero pump detuning (zero DNP), the probe response at much lower frequencies determines Z_e: an increase in RF is observed when the probe comes into resonance with the lower energy "vertical" exciton transition, $|\Uparrow\rangle_x \rightarrow |\Uparrow\Downarrow, \uparrow\rangle_x$, Fig. 9.18a, b. As the pump is detuned, dragging causes Z_e to change and the change ΔZ_e can be simply monitored via a shift in the exciton transition, Fig. 9.18a, b. Importantly, the probe coupling is lowered in these experiments until the probe itself is too weak to induce DNP, i.e. the frequency of the low energy resonance does not depend on Ω_1. At each pump detuning, equivalently at each value of ΔZ_e, the hole Zeeman energy Z_h is determined with ultra-high resolution by measuring the exact spectral location of the CPT dip, Fig. 9.18a, b. Figure 9.18c, d plots Z_h versus ΔZ_e. Although Z_e changes by almost 20 μeV, Z_h remains constant to within 20 neV for both quantum dots. This is the main result of this experiment: large values of $\langle I_x \rangle$ do not result in a measurable change in Z_h even when Z_h is measured with high resolution.

Quantitatively, this experiment shows that $|\Delta Z_h/\Delta Z_e| < 0.1\%$ in the presence of a transverse DNP. This result can be interpreted in terms of averaged hyperfine couplings, $\Delta Z_h/\Delta Z_e \simeq \langle A_{h,x}\rangle/\langle A_e\rangle$. Hence, $|\langle A_{h,x}\rangle/\langle A_e\rangle| < 0.1\%$. Furthermore, with $|\langle A_{h,z}\rangle/\langle A_e\rangle| = 10\%$ [111–113], the anisotropy of the hole spin hyperfine interaction can be quantified, $|\langle A_{h,x}\rangle/\langle A_{h,z}\rangle| < 1\%$. This is consistent with generic theoretical estimates [93, 102]; a full calculation specific to an InGaAs quantum dot including all admixtures is presently lacking. In terms of energies, $|\langle A_{h,x}\rangle| < 0.1$ μeV. This implies a very small energy broadening δZ_h in the presence of un-polarized but noisy nuclei ($\delta Z_e = 600$ neV): $\delta Z_h^{\text{spin}} < 0.6$ neV. The energy broadening arising from the longitudinal coupling, i.e. from $\langle A_{h,z}\rangle$, is sub-neV for all transverse fields above about 500 mT.

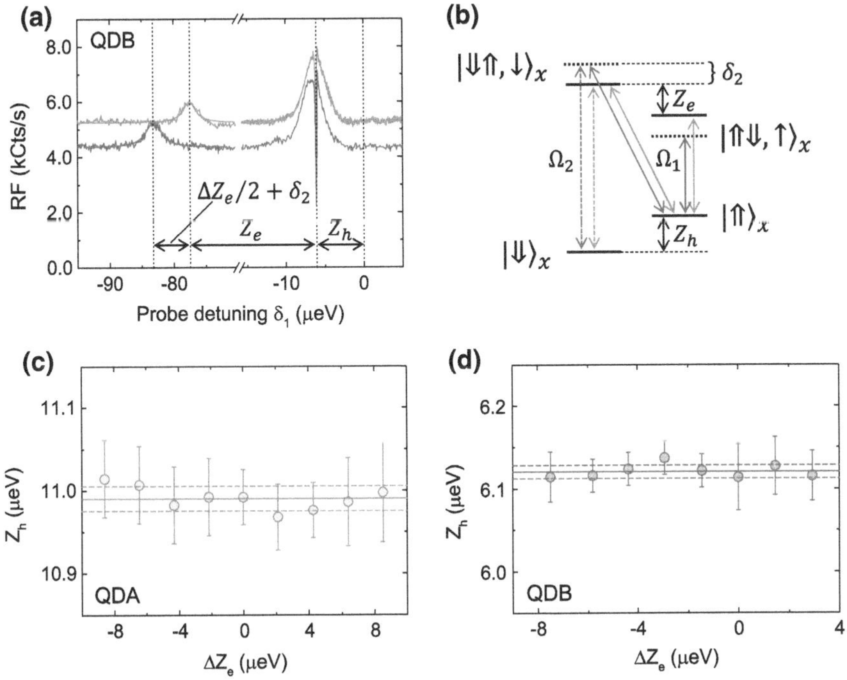

Fig. 9.18 **The transverse hyperfine coupling of a single hole spin.** **a**, Measured probe RF spectrum on QD B in the presence of a much stronger, constant frequency pump laser, *red*: $\delta_2 = 0$; *blue*: $\delta_2 = 3.4$ μeV. The frequency of the probe laser is scanned across the "vertical" and "diagonal" transitions and is plotted with respect to the pump frequency in both cases. The pronounced dip signifies CPT and occurs when $\delta_1 = Z_h$. The peak at large and negative δ_1 arises when the probe is in resonance with the lower energy "vertical" transition. At $\delta_2 = 0$ the separation between this resonance and the CPT dip determines Z_e, the Zeeman energy of the exciton (determined by the lone electron spin). The shift in this resonance signifies a DNP: Z_e changes in response to the change in pump detuning. The measured Rabi energies are $\hbar\Omega_1 = 0.049$ μeV and $\hbar\Omega_2 = 0.49$ μeV; magnetic field 3.00 T; integration time per point 5 s; temperature 4.2 K. **b** The quantum states of the system: the *red arrows* indicate the optical transitions addressed by scanning the probe laser for $\delta_2 = 0$, *blue* for $\delta_2 > 0$. **c**, **d** Z_h versus the change of the electron Zeeman energy ΔZ_e for samples A (g-factor 0.063) and B (g-factor 0.035), respectively. The *solid line* represents the average value, the *dashed lines* represent $\pm\sigma$ where σ is the standard deviation. At the one-σ level, $dZ_h/d\Delta Z_e = 0.1\%$ (color figure online)

9.5.2 *Hole Spin Dephasing*

The CPT experiments can be used to estimate T_2^*. In a CPT experiment, ensemble broadening (described with a T_2^* time, $T_2^* = \hbar/\delta Z_h$) reveals itself by a lifting of the signal in the dip away from zero and an increase in the dip width. An analysis of the CPT spectra of Fig. 9.17, taking into account the mutual coherence of the lasers, results in an energy broadening $\delta Z_h = 3.3 \pm 2.2$ neV. T_2^* is so large that it is very challenging to measure it with small error. To reduce the error, the CPT result can be complemented with all the other spectroscopic results. First, the contribution to δZ_h from charge noise is estimated from the noise analysis. Charge noise is particularly small on sample A contributing 0.10 ± 0.05 μeV to the optical linewidth [56] and results in a Z_h fluctuation of $\delta Z_h^{\text{charge}} = 0.3$ neV at $B_x = 3.00$ T. Secondly, the contribution to δZ_h from spin noise can be determined in situ. The X^{1+} linewidth measures the rms fluctuations in electron Zeeman energy, $\delta Z_e = 1.43 \pm 0.07$ μeV at $B_x = 3.00$ T. This noise arises from a fluctuation in the nuclear spin polarization projected along x, the magnetic field direction, and it corresponds to an Overhauser field of ~40 mT. (Incidentally, this value demonstrates that the nuclear spin distribution is not narrowed in this experiment.) From the upper limit on the coupling coefficient, the corresponding fluctuations in Z_h amount to $<1.43 \pm 0.25$ neV. These results from the linewidths are completely consistent with the CPT result. The final result is that $T_2^* > (460 \pm 80)$ ns.

The long value of hole spin T_2^* arises from the application of an in-plane magnetic field to suppress the longitudinal hyperfine interaction; a very small transverse hyperfine interaction; and low levels of charge noise to reduce charge-noise-induced-dephasing. This T_2^* value is considerably larger than others reported in other experiments [107–109]: it is likely that this is entirely related to the different levels of charge noise in the various experiments. T_2^* is limited by charge and spin noise. In both cases, most of the noise lies at frequencies below 100 kHz [28] such that dynamical decoupling schemes are likely to be highly effective at prolonging the usable coherence beyond T_2^*. Additionally, the nuclear spin coupling can be reduced even further by fabricating flatter quantum dots with circular cross-section to reduce the heavy hole-light hole admixture. A realistic prospect is to push T_2^* into the μs regime. As for a quantum dot electron spin [123], a quantum dot hole spin can be rotated in ~10 ps [107–109]. This combination makes the hole spin in an InGaAs quantum dot an attractive platform. As described in Chap. 12, hole spin-hole spin entanglement has been achieved.

9.6 Conclusions

A self-assembled quantum dot is, under the right conditions (high quality material at low temperature, resonant excitation on a quantum dot in the Coulomb blockade regime), a close-to-ideal emitter of single photons. The noise experiments show that a quantum dot can emit a train of micro-second duration containing photons whose indistinguishability is very high. The train can be prolonged under certain conditions. An electron spin trapped in a self-assembled quantum dot is dephased by the nuclear spins leading to short T_2^* times and T_2 times in the tens of micro-second regime. This interaction can however be very effectively suppressed by applying an in-plane magnetic field to a hole spin. In the presence of noisy nuclear spins, a hole spin is a superior spin qubit using a self-assembled quantum dot as host provided the charge noise is low.

Acknowledgements All this work was carried out by the first generation of researchers in the Nano-Photonics Group, Department of Physics, University of Basel, Switzerland: Lukas Greuter, Julien Houel, Andreas Kuhlmann, Mathieu Munsch, Jonathan Prechtel and Gunter Wüst. The excellent samples were fabricated by Arne Ludwig and Andreas Wieck at Ruhr-University Bochum, Germany. Franziska Maier and Daniel Loss (Department of Physics, University of Basel) carried out the calculations of the nuclear spin coherence. The work on NMR was carried out as part of a fruitful collaboration with Martino Poggio (also Department of Physics, University of Basel).

References

1. J. Fischer, D. Loss, Science **324**, 1277 (2009)
2. R.J. Warburton, Nat. Mater. **12**, 483 (2013)
3. J.M. Garcia, T. Mankad, P.O. Holtz, P.J. Wellman, P.M. Petroff, Appl. Phys. Lett. **72**, 3172 (1998)
4. W. Langbein, P. Borri, U. Woggon, V. Stavarache, D. Reuter, A.D. Wieck, Phys. Rev. B **69**, 161301 (2004)
5. P.A. Dalgarno, J.M. Smith, J. McFarlane, B.D. Gerardot, K. Karrai, A. Badolato, P.M. Petroff, R.J. Warburton, Phys. Rev. B **77**, 245311 (2008)
6. K. Karrai, R.J. Warburton, Superlattices Microstruct. **33**, 311 (2003)
7. J. Claudon, J. Bleuse, N.S. Malik, M. Bazin, P. Jaffrennou, N. Gregersen, C. Sauvan, P. Lalanne, J.M. Gerard, Nat. Photonics **4**, 174 (2010)
8. D. Loss, D.P. DiVincenzo, Phys. Rev. A **57**, 120 (1998)
9. A. Greilich, D.R. Yakovlev, A. Shabaev, A.L. Efros, I.A. Yugova, R. Oulton, V. Stavarache, D. Reuter, A. Wieck, M. Bayer, Science **313**, 341 (2006)
10. S.A. Crooker, J. Brandt, C. Sandfort, A. Greilich, D.R. Yakovlev, D. Reuter, A.D. Wieck, M. Bayer, Phys. Rev. Lett. **104**, 036601 (2010)
11. H. Drexler, D. Leonard, W. Hansen, J.P. Kotthaus, P.M. Petroff, Phys. Rev. Lett. **73**, 2252 (1994)
12. R.J. Warburton, C. Schaflein, D. Haft, F. Bickel, A. Lorke, K. Karrai, J.M. Garcia, W. Schoenfeld, P.M. Petroff, Nature **405**, 926 (2000)
13. J.H. Prechtel, A.V. Kuhlmann, J. Houel, A. Ludwig, S.R. Valentin, A.D. Wieck, R.J. Warburton, Nat. Mater. **15**, 981 (2016)
14. P.A. Dalgarno, M. Ediger, B.D. Gerardot, J.M. Smith, S. Seidl, M. Kroner, K. Karrai, P.M. Petroff, A.O. Govorov, R.J. Warburton, Phys. Rev. Lett. **100**, 176801 (2008)
15. B.D. Gerardot, R.J. Barbour, D. Brunner, P.A. Dalgarno, A. Badolato, N. Stoltz, P.M. Petroff, J. Houel, R.J. Warburton, Appl. Phys. Lett. **99**, 243112 (2011)
16. S. Seidl, M. Kroner, P.A. Dalgarno, A. Högele, J.M. Smith, M. Ediger, B.D. Gerardot, J.M. Garcia, P.M. Petroff, K. Karrai, R.J. Warburton, Phys. Rev. B **72**, 195339 (2005)
17. J.M. Smith, P.A. Dalgarno, R.J. Warburton, A.O. Govorov, K. Karrai, B.D. Gerardot, P.M. Petroff, Phys. Rev. Lett. **94**, 197402 (2005)
18. M. Atatüre, J. Dreiser, A. Badolato, A. Högele, K. Karrai, A. Imamoglu, Science **312**, 551 (2006)
19. J. Dreiser, M. Atatüre, C. Galland, T. Mueller, A. Badolato, A. Imamoglu, Phys. Rev. B **77**, 075317 (2008)
20. B.D. Gerardot, D. Brunner, P.A. Dalgarno, P. Ohberg, S. Seidl, M. Kroner, K. Karrai, N.G. Stoltz, P.M. Petroff, R.J. Warburton, Nature **451**, 441 (2008)
21. J. McFarlane, P.A. Dalgarno, B.D. Gerardot, R.H. Hadfield, R.J. Warburton, K. Karrai, A. Badolato, P.M. Petroff, Appl. Phys. Lett. **94**, 093113 (2009)
22. N.A.J.M. Kleemans, J. van Bree, A.O. Govorov, J.G. Keizer, G.J. Hamhuis, R. Notzel, A.Y. Silov, P.M. Koenraad, Nat. Phys. **6**, 534 (2010)

23. J. Houel, A.V. Kuhlmann, L. Greuter, F. Xue, M. Poggio, B.D. Gerardot, P.A. Dalgarno, A. Badolato, P.M. Petroff, A. Ludwig, D. Reuter, A.D. Wieck, R.J. Warburton, Phys. Rev. Lett. **108**, 107401 (2012)
24. A. Högele, S. Seidl, M. Kroner, K. Karrai, R.J. Warburton, B.D. Gerardot, P.M. Petroff, Phys. Rev. Lett. **93**, 217401 (2004)
25. A. Muller, E.B. Flagg, P. Bianucci, X.Y. Wang, D.G. Deppe, W. Ma, J. Zhang, G.J. Salamo, M. Xiao, C.K. Shih, Phys. Rev. Lett. **99**, 187402 (2007)
26. A.N. Vamivakas, Y. Zhao, C.Y. Lu, M. Atatüre, Nat. Phys. **5**, 198 (2009)
27. S.T. Yilmaz, P. Fallahi, A. Imamoglu, Phys. Rev. Lett. **105**, 033601 (2010)
28. A.V. Kuhlmann, J. Houel, A. Ludwig, L. Greuter, D. Reuter, A.D. Wieck, M. Poggio, R.J. Warburton, Nat. Phys. **9**, 570 (2013)
29. A.V. Kuhlmann, J. Houel, D. Brunner, A. Ludwig, D. Reuter, A.D. Wieck, R.J. Warburton, Rev. Sci. Instr. **84**, 073905 (2013)
30. M. Kroner, S. Remi, A. Hoegele, S. Seidl, A.W. Holleitner, R.J. Warburton, B.D. Gerardot, P.M. Petroff, K. Karrai, Physica E **40**, 1994 (2008)
31. M. Kroner, C. Lux, S. Seidl, A.W. Holleitner, K. Karrai, A. Badolato, P.M. Petroff, R.J. Warburton, Appl. Phys. Lett. **92**, 031108 (2008)
32. A.J. Ramsay, T.M. Godden, S.J. Boyle, E.M. Gauger, A. Nazir, B.W. Lovett, A.M. Fox, M.S. Skolnick, Phys. Rev. Lett. **105**, 177402 (2010)
33. S.M. Ulrich, S. Ates, S. Reitzenstein, A. Löffler, A. Forchel, P. Michler, Phys. Rev. Lett. **106**, 247402 (2011)
34. M. Bayer, A. Forchel, Phys. Rev. B **65**, 041308 (2002)
35. W. Langbein, P. Borri, U. Woggon, V. Stavarache, D. Reuter, A.D. Wieck, Phys. Rev. B **70**, 033301 (2004)
36. P. Borri, W. Langbein, U. Woggon, V. Stavarache, D. Reuter, A.D. Wieck, Phys. Rev. B **71**, 115328 (2005)
37. N. Sangouard, H. Zbinden, J. Mod. Opt. **59**, 1458 (2012)
38. P. Michler, A. Kiraz, C. Becher, W.V. Schoenfeld, P.M. Petroff, L.D. Zhang, E. Hu, A. Imamoglu, Science **290**, 2282 (2000)
39. C. Santori, D. Fattal, J. Vučković, G.S. Solomon, Y. Yamamoto, Nature **419**, 594 (2002)
40. A.J. Shields, Nat. Photonics **1**, 215 (2007)
41. H.S. Nguyen, G. Sallen, C. Voisin, P. Roussignol, C. Diederichs, G. Cassabois, Appl. Phys. Lett. **99**, 261904 (2011)
42. C. Matthiesen, A.N. Vamivakas, M. Atatüre, Phys. Rev. Lett. **108**, 093602 (2012)
43. C. Matthiesen, M. Geller, C.H.H. Schulte, C. Le Gall, J. Hansom, Z. Li, M. Hugues, E. Clarke, M. Atatüre, Nat. Commun. **4**, 1600 (2013)
44. F. Klotz, V. Jovanov, J. Kierig, E.C. Clark, D. Rudolph, D. Heiss, M. Bichler, G. Abstreiter, M.S. Brandt, J.J. Finley, Appl. Phys. Lett. **96**, 053113 (2010)
45. J. Pingenot, C.E. Pryor, M.E. Flatte, Phys. Rev. B **84**, 195403 (2011)
46. I.A. Merkulov, A.L. Efros, M. Rosen, Phys. Rev. B **65**, 205309 (2002)
47. A.V. Khaetskii, D. Loss, L. Glazman, Phys. Rev. Lett. **88**, 186802 (2002)
48. X. Xu, B. Sun, P.R. Berman, D.G. Steel, A.S. Bracker, D. Gammon, L.J. Sham, Nat. Phys. **4**, 692 (2008)
49. D. Press, K. De Greve, P.L. McMahon, T.D. Ladd, B. Friess, C. Schneider, M. Kamp, S. Hoefling, A. Forchel, Y. Yamamoto, Nat. Photonics **4**, 367 (2010)
50. C. Barthel, J. Medford, C.M. Marcus, M.P. Hanson, A.C. Gossard, Phys. Rev. Lett. **105**, 266808 (2010)
51. G. de Lange, Z.H. Wang, D. Riste, V.V. Dobrovitski, R. Hanson, Science **330**, 60 (2010)
52. H. Bluhm, S. Foletti, I. Neder, M. Rudner, D. Mahalu, V. Umansky, A. Yacoby, Nat. Phys. **7**, 109 (2011)
53. B. Alen, F. Bickel, K. Karrai, R.J. Warburton, P.M. Petroff, Appl. Phys. Lett. **83**, 2235 (2003)
54. A.N. Vamivakas, Y. Zhao, S. Fält, A. Badolato, J.M. Taylor, M. Atatüre, Phys. Rev. Lett. **107**, 166802 (2011)

55. C. Schulhauser, D. Haft, R.J. Warburton, K. Karrai, A.O. Govorov, A.V. Kalameitsev, A. Chaplik, W. Schoenfeld, J.M. Garcia, P.M. Petroff, Phys. Rev. B **66**, 193303 (2002)
56. A.V. Kuhlmann, J.H. Prechtel, J. Houel, A. Ludwig, D. Reuter, A.D. Wieck, R.J. Warburton, Nat. Commun. **6**, 8204 (2015)
57. W.A. Coish, J. Baugh, Phys. Stat. Sol. (b) **246**, 2203 (2009)
58. C. Kloeffel, P.A. Dalgarno, B. Urbaszek, B.D. Gerardot, D. Brunner, P.M. Petroff, D. Loss, R.J. Warburton, Phys. Rev. Lett. **106**, 046802 (2011)
59. S. Machlup, J. Appl. Phys. **25**, 341 (1954)
60. O. Gazzano, S. Michaelis de Vasconcellos, C. Arnold, A. Nowak, E. Galopin, I. Sagnes, L. Lanco, A. Lemaître, P. Senellart, Nat. Commun. **4**, 1425 (2013)
61. R.J. Warburton, C. Schulhauser, D. Haft, C. Schäflein, K. Karrai, J.M. Garcia, W. Schoenfeld, P.M. Petroff, Phys. Rev. B **65**, 113303 (2002)
62. A.J. Bennett, M.A. Pooley, R.M. Stevenson, M.B. Ward, R.B. Patel, A.B. de la Giroday, N. Sköld, I. Farrer, C.A. Nicoll, D.A. Ritchie, A.J. Shields, Nat. Phys. **6**, 947 (2010)
63. S. Seidl, M. Kroner, A. Högele, K. Karrai, R.J. Warburton, A. Badolato, P.M. Petroff, Appl. Phys. Lett. **88**, 203113 (2006)
64. K.D. Jöns, R. Hafenbrak, R. Singh, F. Ding, J.D. Plumhof, A. Rastelli, O.G. Schmidt, G. Bester, P. Michler, Phys. Rev. Lett. **107**, 217402 (2011)
65. J.H. Prechtel, A.V. Kuhlmann, J. Houel, L. Greuter, A. Ludwig, D. Reuter, A.D. Wieck, R.J. Warburton, Phys. Rev. X **3**, 041006 (2013)
66. E.A. Chekhovich, M.N. Makhonin, A.I. Tartakovskii, A. Yacoby, H. Bluhm, K.C. Nowack, L.M.K. Vandersypen, Nat. Mater. **12**, 494 (2013)
67. J.R. Petta, A.C. Johnson, J.M. Taylor, E.A. Laird, A. Yacoby, M.D. Lukin, C.M. Marcus, M.P. Hanson, A.C. Gossard, Science **309**, 2180 (2005)
68. D.J. Reilly, J.M. Taylor, J.R. Petta, C.M. Marcus, M.P. Hanson, A.C. Gossard, Science **321**, 817 (2008)
69. X. Xu, W. Yao, B. Sun, D.G. Steel, A.S. Bracker, D. Gammon, L.J. Sham, Nature **459**, 1105 (2009)
70. M.D. Shulman, S.P. Harvey, J.M. Nichol, S.D. Bartlett, A.C. Doherty, V. Umansky, A. Yacoby, Nat. Commun. **5**, 5156 (2014)
71. W.M. Witzel, S.D. Sarma, Phys. Rev. Lett. **98**, 077601 (2007)
72. L. Cywiński, W.M. Witzel, S. Das Sarma, Phys. Rev. Lett. **102**, 057601 (2009)
73. W. Yao, R.B. Liu, L.J. Sham, Phys. Rev. B **74**, 195301 (2006)
74. D. Klauser, W.A. Coish, D. Loss, Phys. Rev. B **78**, 205301 (2008)
75. R. Hanson, L.P. Kouwenhoven, J.R. Petta, S. Tarucha, L.M.K. Vandersypen, Rev. Mod. Phys. **79**, 1217 (2007)
76. W.A. Coish, J. Fischer, D. Loss, Phys. Rev. B **81**, 165315 (2010)
77. A. Faribault, D. Schuricht, Phys. Rev. Lett. **110**, 040405 (2013)
78. Y. Kondo, M. Ono, S. Matsuzaka, K. Morita, H. Sanada, Y. Ohno, H. Ohno, Phys. Rev. Lett. **101**, 207601 (2008)
79. E.A. Chekhovich, M. Hopkinson, M.S. Skolnick, A.I. Tartakovskii, Nat. Commun. **6**, 6348 (2015)
80. B. Urbaszek, X. Marie, T. Amand, O. Krebs, P. Voisin, P. Maletinsky, A. Högele, A. Imamoglu, Rev. Mod. Phys. **85**, 79 (2013)
81. M. Munsch, G. Wüst, A.V. Kuhlmann, F. Xue, A. Ludwig, D. Reuter, A.D. Wieck, M. Poggio, R.J. Warburton, Nat. Nanotechnol. **9**, 671 (2014)
82. E.A. Chekhovich, K.V. Kavokin, J. Puebla, A.B. Krysa, M. Hopkinson, A.D. Andreev, A.M. Sanchez, R. Beanland, M.S. Skolnick, A.I. Tartakovskii, Nat. Nanotechnol. **7**, 646 (2012)
83. C. Bulutay, E.A. Chekhovich, A.I. Tartakovskii, Phys. Rev. B **90**, 205425 (2014)
84. E. Van Veenendaal, B. Meier, A. Kentgens, Mol. Phys. **93**, 195 (1998)
85. M. Kroner, K.M. Weiss, B. Biedermann, S. Seidl, A.W. Holleitner, A. Badolato, P.M. Petroff, P. Öhberg, R.J. Warburton, K. Karrai, Phys. Rev. B **78**, 075429 (2008)
86. P. Maletinsky, A. Badolato, A. Imamoglu, Phys. Rev. Lett. **99**, 056804 (2007)

87. G. Wüst, M. Munsch, F. Maier, A.V. Kuhlmann, A. Ludwig, A.D. Wieck, D. Loss, M. Poggio, R.J. Warburton, Nat. Nanotechnol. **11**, 885 (2016)
88. D. Paget, G. Lampel, B. Sapoval, V.I. Safarov, Phys. Rev. B **15**, 5780 (1977)
89. E. Grncharova, V. Perel, Sov. Phys. Semicond. **11**, 977 (1977)
90. J. Fischer, W.A. Coish, D.V. Bulaev, D. Loss, Phys. Rev. B **78**, 155329 (2008)
91. B. Eble, C. Testelin, P. Desfonds, F. Bernardot, A. Balocchi, T. Amand, A. Miard, A. Lemaître, X. Marie, M. Chamarro, Phys. Rev. Lett. **102**, 146601 (2009)
92. C. Testelin, F. Bernardot, B. Eble, M. Chamarro, Phys. Rev. B **79**, 195440 (2009)
93. J. Fischer, D. Loss, Phys. Rev. Lett. **105**, 266603 (2010)
94. H.W. van Kesteren, E.C. Cosman, W.A.J.A. van der Poel, C.T. Foxon, Phys. Rev. B **41**, 5283 (1990)
95. R.W. Martin, R.J. Nicholas, G.J. Rees, S.K. Haywood, N.J. Mason, P.J. Walker, Phys. Rev. B **42**, 9237 (1990)
96. G. Bester, A. Zunger, Phys. Rev. B **71**, 045318 (2005)
97. J.W. Luo, G. Bester, A. Zunger, Phys. Rev. B **92**, 165301 (2015)
98. D.N. Krizhanovskii, A. Ebbens, A.I. Tartakovskii, F. Pulizzi, T. Wright, M.S. Skolnick, M. Hopkinson, Phys. Rev. B **72**, 161312 (2005)
99. T. Belhadj, T. Amand, A. Kunold, C.M. Simon, T. Kuroda, M. Abbarchi, T. Mano, K. Sakoda, S. Kunz, X. Marie, B. Urbaszek, Appl. Phys. Lett. **97**, 051111 (2010)
100. C.Y. Lu, Y. Zhao, A.N. Vamivakas, C. Matthiesen, S. Fält, A. Badolato, M. Atatüre, Phys. Rev. B **81**, 035332 (2010)
101. F. Maier, D. Loss, Phys. Rev. B **85**, 195323 (2012)
102. H. Ribeiro, F. Maier, D. Loss, Phys. Rev. B **92**, 075421 (2015)
103. D. Heiss, S. Schaeck, H. Huebl, M. Bichler, G. Abstreiter, J.J. Finley, D.V. Bulaev, D. Loss, Phys. Rev. B **76**, 241306 (2007)
104. Y. Li, N. Sinitsyn, D.L. Smith, D. Reuter, A.D. Wieck, D.R. Yakovlev, M. Bayer, S.A. Crooker, Phys. Rev. Lett. **108**, 186603 (2012)
105. R. Dahbashi, J. Hübner, F. Berski, K. Pierz, M. Oestreich, Phys. Rev. Lett. **112**, 156601 (2014)
106. D. Brunner, B.D. Gerardot, P.A. Dalgarno, G. Wüst, K. Karrai, N.G. Stoltz, P.M. Petroff, R.J. Warburton, Science **325**, 70 (2009)
107. K. De Greve, P.L. McMahon, D. Press, T.D. Ladd, D. Bisping, C. Schneider, M. Kamp, L. Worschech, S. Hoefling, A. Forchel, Y. Yamamoto, Nat. Phys. **7**, 872 (2011)
108. A. Greilich, S.G. Carter, D. Kim, A.S. Bracker, D. Gammon, Nat. Photonics **5**, 703 (2011)
109. T.M. Godden, J.H. Quilter, A.J. Ramsay, Y. Wu, P. Brereton, S.J. Boyle, I.J. Luxmoore, J. Puebla-Nunez, A.M. Fox, M.S. Skolnick, Phys. Rev. Lett. **108**, 017402 (2012)
110. S. Varwig, A. René, A. Greilich, D.R. Yakovlev, D. Reuter, A.D. Wieck, M. Bayer, Phys. Rev. B **87**, 115307 (2013)
111. P. Fallahi, S.T. Yilmaz, A. Imamoglu, Phys. Rev. Lett. **105**, 257402 (2010)
112. E.A. Chekhovich, A.B. Krysa, M.S. Skolnick, A.I. Tartakovskii, Phys. Rev. Lett. **106**, 027402 (2011)
113. H. Kurtze, D.R. Yakovlev, D. Reuter, A.D. Wieck, M. Bayer, Phys. Rev. B **85**, 195303 (2012)
114. E.A. Chekhovich, M.M. Glazov, A.B. Krysa, M. Hopkinson, P. Senellart, A. Lemaître, M.S. Skolnick, A.I. Tartakovskii, Nat. Phys. **9**, 74 (2013)
115. A.J. Bennett, M.A. Pooley, Y. Cao, N. Sköld, I. Farrer, D.A. Ritchie, A.J. Shields, Nat. Commun. **4**, 1522 (2013)
116. J. Houel, J.H. Prechtel, A.V. Kuhlmann, D. Brunner, C.E. Kuklewicz, B.D. Gerardot, N.G. Stoltz, P.M. Petroff, R.J. Warburton, Phys. Rev. Lett. **112**, 107401 (2014)
117. J.H. Prechtel, F. Maier, J. Houel, A.V. Kuhlmann, A. Ludwig, A.D. Wieck, D. Loss, R.J. Warburton, Phys. Rev. B **91**, 165304 (2015)
118. M. Fleischhauer, A. Imamoglu, J.P. Marangos, Rev. Mod. Phys. **77**, 633 (2005)
119. J. Hansom, C.H.H. Schulte, C. Le Gall, C. Matthiesen, E. Clarke, M. Hugues, J.M. Taylor, M. Atatüre, Nat. Phys. **10**, 725 (2014)
120. A.V. Kuhlmann, J. Houel, D. Brunner, A. Ludwig, D. Reuter, A.D. Wieck, R.J. Warburton, Rev. Sci. Instrum. **84**, 073905 (2013)

121. C. Latta, A. Hoegele, Y. Zhao, A.N. Vamivakas, P. Maletinsky, M. Kroner, J. Dreiser, I. Carusotto, A. Badolato, D. Schuh, W. Wegscheider, M. Atatüre, A. Imamoglu, Nat. Phys. **5**, 758 (2009)
122. A. Högele, M. Kroner, C. Latta, M. Claassen, I. Carusotto, C. Bulutay, A. Imamoglu, Phys. Rev. Lett. **108**, 197403 (2012)
123. D. Press, T.D. Ladd, B. Zhang, Y. Yamamoto, Nature **456**, 218 (2008)

Chapter 10
Ultrafast Manipulation of Excitons and Spins in Quantum Dots

Alistair J. Brash, Feng Liu and A. Mark Fox

Abstract This chapter reviews the coherent manipulation of excitons and spins in self-assembled InGaAs quantum dots by ultrafast laser pulses. We begin with a review of the basic theory of coherent control of two-level systems, followed by a discussion of the beneficial features of quantum dots. Experiments on ultrafast coherent control of excitons in neutral dots and spins in charged dots are then presented, before concluding with a comparison of the two different approaches in the context of applications in quantum information processing.

10.1 Introduction

The coherent manipulation of quantum systems lies at the heart of most quantum information processing (QIP) schemes. Many of the techniques for coherent control were originally developed in the fields of nuclear magnetic resonance (NMR) and atomic physics, and have only recently been applied to semiconductor systems. The reason for this is the problem of decoherence: the coherence times of optical transitions in most semiconductors are extremely short, making the task of observing coherent phenomena particularly challenging. It is in this context that quantum dots, with their long dephasing times, come into their own. The subject of this chapter is precisely on the coherent control experiments that are facilitated by the long coherence times of quantum dots.

An appreciation of coherent control experiments first requires that the basic concepts should be well understood. The chapter therefore starts in Sect. 10.2 with a summary of the main concepts of two-level systems interacting with resonant laser pulses. The second part of the chapter then focuses on the observation of

A.J. Brash (✉) · F. Liu · A.M. Fox
Department of Physics and Astronomy, University of Sheffield, Sheffield, S3 7RH, UK
e-mail: a.brash@sheffield.ac.uk

F. Liu
e-mail: lf19902001@gmail.com

A.M. Fox
e-mail: mark.fox@sheffield.ac.uk

P. Michler (ed.), *Quantum Dots for Quantum Information Technologies*,
Nano-Optics and Nanophotonics, DOI 10.1007/978-3-319-56378-7_10

coherent phenomena in quantum dots. After discussing the reasons why quantum dots are good for coherent control experiments (Sect. 10.3), two different approaches are considered, namely the control of excitons in neutral dots (Sect. 10.4), and of spins in charged dots (Sect. 10.5). The chapter concludes by comparing the two approaches and giving an outlook for further work.

10.2 Concepts of Coherent Control Experiments

This section gives a tutorial review of the coherent manipulation of two-level systems by resonant laser pulses, starting from first principles. Readers who are familiar with this text-book material may skip ahead to the specific sections on quantum dots, beginning in Sect. 10.3.

10.2.1 The Two-Level Atom Approximation and the Bloch Sphere

The starting point for understanding the coherent interaction between laser light and quantum dots is the *two-level atom approximation*. A real atom has many quantised energy levels, giving rise to a rich spectrum of optical transitions, and the two-level approximation applies when the laser frequency is close to resonance with one of them. As we shall see, the interaction depends very strongly on the detuning Δ of the laser relative to the transition frequency, and becomes negligibly small when Δ is much larger than the line width. We can then neglect the other transitions, and focus exclusively on the one transition that is resonant with the laser.

In applying the two-level approximation to quantum dots, we make use of their discrete energy level spectrum that follows from the three-dimensional confinement of the electrons and holes. The optical transitions consist of sharp lines, corresponding to neutral exciton, biexciton, charged exciton, etc. transitions. The laser can be tuned close to resonance with one of these, while being many line widths away from the others, justifying the use of the two-level approximation. The fact that the model works well is a clear confirmation of the well-used description of quantum dots as "solid-state atoms".

An arbitrary superposition state of a two-level system has a wave function of the form:

$$|\psi\rangle = c_1|1\rangle + c_2|2\rangle \,, \tag{10.1}$$

where c_1 and c_2 are the amplitude coefficients for the two states. The normalization condition requires that:

$$|c_1|^2 + |c_2|^2 = 1 \,, \tag{10.2}$$

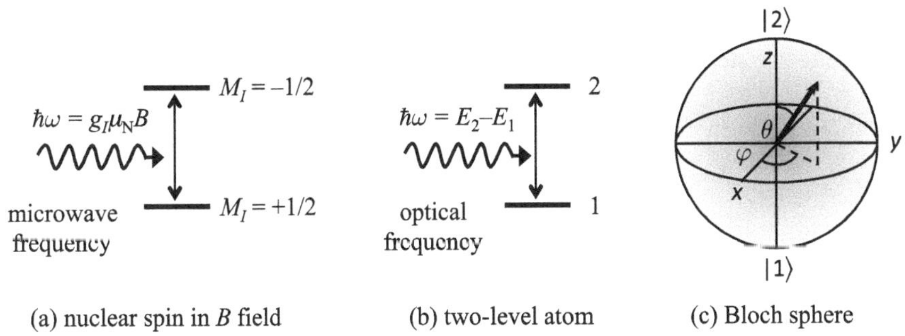

Fig. 10.1 **a** Spin 1/2 system in a magnetic field B interacting resonantly with microwaves. **b** Two-level atom interacting resonantly with a laser. **c** Bloch sphere for representing the state of a two-level system. For historical reasons, we usually consider nuclear spin systems in **a**, in which case the Zeeman splitting is equal to $g_I \mu_N B$, where g_I is the gyromagnetic ratio and μ_N is the nuclear Bohr magneton

which suggests that we can represent the state by a vector of unit length starting at the origin. This geometric interpretation of coherent superposition states is called the *Bloch representation*. The vector that describes the state is called the *Bloch vector*, and the sphere it defines is the *Bloch sphere*.

The Bloch representation was originally developed by Felix Bloch in 1946 to model NMR phenomena, and was adapted to two-level atoms by Feynman, Vernon, and Hellwarth in 1957, where they showed that a two-level atom can be regarded as a pseudo-spin 1/2 system [1]. The corresponding optical Bloch equations were derived by Arecchi and Bonifacio in 1965 [2]. This analogy with a spin 1/2 particle in a magnetic field is shown schematically in Fig. 10.1.

The direction of the Bloch vector $\boldsymbol{s}$ can be specified either in Cartesian co-ordinates (x, y, z) or spherical polar co-ordinates (r, θ, φ). The requirement that the vector has unit length implies that $r^2 = (x^2 + y^2 + z^2) = 1$, so that only two independent variables are required to define an arbitrary state, for example the angles (θ, φ). This allows us to make a unique mapping between the wave function amplitudes (c_1, c_2) and the direction of the Bloch vector.

The connection between the Bloch sphere and a two-level system may be made by defining the poles to correspond to the $|1\rangle$ and $|2\rangle$ states respectively, as shown in Fig. 10.1c. The ground state at the south pole with $|\psi\rangle = |1\rangle$ thus corresponds to $(0, 0, -1)$ in Cartesian co-ordinates or $\theta = \pi$ in polar co-ordinates. Similarly, the pure excited state $|2\rangle$ corresponds to $(0, 0, 1)$ or $\theta = 0$. An arbitrary state is given in Cartesian co-ordinates as:

$$\begin{aligned} x &= 2\,\mathrm{Re}\langle c_1 c_2 \rangle\,, \\ y &= 2\,\mathrm{Im}\langle c_1 c_2 \rangle\,, \\ z &= |c_2|^2 - |c_1|^2\,, \end{aligned} \tag{10.3}$$

where the notation $\langle \cdots \rangle$ indicates that we take the average of repeated measurements on identical systems. In polar co-ordinates this simplifies to:

$$\begin{aligned} c_1 &= \sin(\theta/2)\,, \\ c_2 &= \mathrm{e}^{\mathrm{i}\varphi}\cos(\theta/2)\,. \end{aligned} \tag{10.4}$$

This one-to-one mapping allows us to visualize an arbitrary superposition state of a two-level atom in a geometric way, which is very useful when considering the resonant interaction with an intense optical field.

The coherence of a two-level system relies on having a definite phase relationship between c_1 and c_2, leading to non-zero x and y components of the Bloch vector. By contrast, in a completely incoherent system (i.e. a *statistical mixture*) we only know the probability that the atom is in the upper or lower level, i.e. the z component of the Bloch vector. The phase relationship between c_1 and c_2 is random, so that $\langle c_1 c_2 \rangle = 0$, and the x and y components are zero. Statistical mixtures thus correspond to points *inside* the Bloch sphere with $r < 1$. The system is completely incoherent if both the x and y components are zero, and partially coherent for non-zero x and y but $r < 1$. In the equivalent language of the density matrix $\rho_{ij} = \langle c_i c_j^* \rangle$, the coherence is determined by the off-diagonal elements ρ_{12} and ρ_{21}, while statistical mixtures only contain information about the diagonal elements ρ_{11} and ρ_{22}. Note, however, that there is no difference between superposition states and statistical mixtures for the pure states $|i\rangle$ at the poles of the Bloch sphere with $c_i = 1$.

10.2.2 Rabi Oscillations

The effect of a resonant laser on a two-level atom can understood by solving the time-dependent Schrödinger equation:

$$\hat{H}\Psi = \mathrm{i}\hbar\frac{\partial\Psi}{\partial t}\,. \tag{10.5}$$

We start by splitting the Hamiltonian into a time-independent part $\hat{H}_0$ which describes the atom in the dark, and a perturbation term $\hat{V}(t)$ which accounts for the light-atom interaction:

$$\hat{H} = \hat{H}_0(\boldsymbol{r}) + \hat{V}(t)\,. \tag{10.6}$$

Since we are dealing with a two-level atom, there will be two solutions for the unperturbed system:

$$\hat{H}_0\Psi_i = \mathrm{i}\hbar\frac{\partial\Psi_i}{\partial t}\,, \tag{10.7}$$

with

$$\Psi_i(\boldsymbol{r},t) = \psi_i(\boldsymbol{r})\exp(-\mathrm{i}E_i t/\hbar) \quad \{i = 1, 2\}\,, \tag{10.8}$$

and

$$\hat{H}_0(\boldsymbol{r})\psi_i(\boldsymbol{r}) = E_i\,\psi_i(\boldsymbol{r}) \quad \{i = 1, 2\}\,. \tag{10.9}$$

The general solution for a two-level atom is:

$$\Psi(\boldsymbol{r}, t) = c_1(t)\psi_1(\boldsymbol{r})\mathrm{e}^{-\mathrm{i}E_1 t/\hbar} + c_2(t)\psi_2(\boldsymbol{r})\mathrm{e}^{-\mathrm{i}E_2 t/\hbar}\,. \tag{10.10}$$

On substituting (10.10) into (10.5) with $\hat{H}$ given by (10.6), we obtain:

$$\begin{aligned}&(\hat{H}_0 + \hat{V})\left(c_1\psi_1\mathrm{e}^{-\mathrm{i}E_1 t/\hbar} + c_2\psi_2\mathrm{e}^{-\mathrm{i}E_2 t/\hbar}\right)\\ &= \mathrm{i}\hbar\left((\dot{c}_1 - \mathrm{i}E_1 c_1/\hbar)\psi_1\mathrm{e}^{-\mathrm{i}E_1 t/\hbar} + (\dot{c}_2 - \mathrm{i}E_2 c_2/\hbar)\psi_2\mathrm{e}^{-\mathrm{i}E_2 t/\hbar}\right)\,. \end{aligned} \tag{10.11}$$

On using (10.9), and cancelling several terms, this becomes:

$$c_1\hat{V}\psi_1\mathrm{e}^{-\mathrm{i}E_1 t/\hbar} + c_2\hat{V}\psi_2\mathrm{e}^{-\mathrm{i}E_2 t/\hbar} = \mathrm{i}\hbar\dot{c}_1\psi_1\mathrm{e}^{-\mathrm{i}E_1 t/\hbar} + \mathrm{i}\hbar\dot{c}_2\psi_2\mathrm{e}^{-\mathrm{i}E_2 t/\hbar}\,. \tag{10.12}$$

On multiplying by ψ_i^*, integrating over space, and making use of the orthonormality of the eigenfunctions, we find that:

$$\begin{aligned}\dot{c}_1(t) &= -\frac{\mathrm{i}}{\hbar}\left(c_1(t)V_{11} + c_2(t)V_{12}\mathrm{e}^{-\mathrm{i}\omega_0 t}\right)\,,\\ \dot{c}_2(t) &= -\frac{\mathrm{i}}{\hbar}\left(c_1(t)V_{21}\mathrm{e}^{\mathrm{i}\omega_0 t} + c_2(t)V_{22}\right)\,,\end{aligned} \tag{10.13}$$

where $\omega_0 = (E_2 - E_1)/\hbar$ is the transition frequency, and

$$V_{ij}(t) \equiv \langle i|\hat{V}(t)|j\rangle = \int \psi_i^*\hat{V}(t)\psi_j\,\mathrm{d}^3\boldsymbol{r}\,. \tag{10.14}$$

To proceed further we must consider the explicit form of the perturbation $\hat{V}$. In the semi-classical approach, the light-atom interaction is given by the energy shift of the atomic dipole in the electric field of the light:

$$\hat{V}(t) = e\boldsymbol{r}\cdot\boldsymbol{E}(t)\,. \tag{10.15}$$

We arbitrarily choose the x axis as the direction of the polarization so that we can write $\boldsymbol{E}(t) = (E_0, 0, 0)\cos\omega t$, where E_0 is the amplitude of the light wave, and ω is its angular frequency. The perturbation then simplifies to:

$$\hat{V}(t) = exE_0\cos\omega t = \frac{exE_0}{2}\left(\mathrm{e}^{\mathrm{i}\omega t} + \mathrm{e}^{-\mathrm{i}\omega t}\right)\,, \tag{10.16}$$

and the perturbation matrix elements are given by:

$$V_{ij}(t) = \frac{eE_0}{2}\left(\mathrm{e}^{\mathrm{i}\omega t} + \mathrm{e}^{-\mathrm{i}\omega t}\right)\int \psi_i^* x \psi_j \, \mathrm{d}^3\boldsymbol{r} \,. \tag{10.17}$$

We now introduce the *dipole matrix element* μ_{ij} given by:

$$\mu_{ij} = -e\int \psi_i^* x \psi_j \, \mathrm{d}^3\boldsymbol{r} \equiv -e\langle i|x|j\rangle \,. \tag{10.18}$$

Since x is an odd parity operator and atomic states have well-defined parities, it follows that $\mu_{11} = \mu_{22} = 0$. Moreover, μ_{ij} represents a measurable quantity and must therefore be real, implying $\mu_{21} = \mu_{12}$, because $\mu_{21} = \mu_{12}^*$. With these simplifications, (10.13) reduces to:

$$\begin{aligned}\dot{c}_1(t) &= \mathrm{i}\frac{E_0\mu_{12}}{2\hbar}\left(\mathrm{e}^{\mathrm{i}(\omega-\omega_0)t} + \mathrm{e}^{-\mathrm{i}(\omega+\omega_0)t}\right)c_2(t)\,,\\ \dot{c}_2(t) &= \mathrm{i}\frac{E_0\mu_{12}}{2\hbar}\left(\mathrm{e}^{-\mathrm{i}(\omega-\omega_0)t} + \mathrm{e}^{\mathrm{i}(\omega+\omega_0)t}\right)c_1(t)\,.\end{aligned} \tag{10.19}$$

We now introduce the *Rabi frequency* defined by:

$$\Omega_\mathrm{R} = |\mu_{12}E_0/\hbar| \,. \tag{10.20}$$

We then finally obtain:

$$\begin{aligned}\dot{c}_1(t) &= \frac{\mathrm{i}}{2}\Omega_\mathrm{R}\left(\mathrm{e}^{\mathrm{i}(\omega-\omega_0)t} + \mathrm{e}^{-\mathrm{i}(\omega+\omega_0)t}\right)c_2(t)\,,\\ \dot{c}_2(t) &= \frac{\mathrm{i}}{2}\Omega_\mathrm{R}\left(\mathrm{e}^{-\mathrm{i}(\omega-\omega_0)t} + \mathrm{e}^{\mathrm{i}(\omega+\omega_0)t}\right)c_1(t)\,.\end{aligned} \tag{10.21}$$

These are the equations that must be solved to understand the behaviour of the atom in the light field. The solutions in the weak-field limit (i.e. low light intensity) correspond to the incoherent Einstein B coefficient analysis of transition rates. In what follows, we focus instead on the strong-field limit.

The equations can be simplified by applying the *rotating wave approximation* in which we move to a rotating frame at frequency ω_0. The term at $(\omega + \omega_0)$ oscillates very rapidly in this frame, while the one at $(\omega - \omega_0)$ is nearly stationary. We therefore neglect the former and focus on the latter to obtain:

$$\begin{aligned}\dot{c}_1(t) &= \frac{\mathrm{i}}{2}\Omega_\mathrm{R}\mathrm{e}^{\mathrm{i}\Delta t}c_2(t)\,,\\ \dot{c}_2(t) &= \frac{\mathrm{i}}{2}\Omega_\mathrm{R}\mathrm{e}^{-\mathrm{i}\Delta t}c_1(t)\,,\end{aligned} \tag{10.22}$$

where $\Delta = \omega - \omega_0$ is the detuning. For exact resonance with $\Delta = 0$, we find

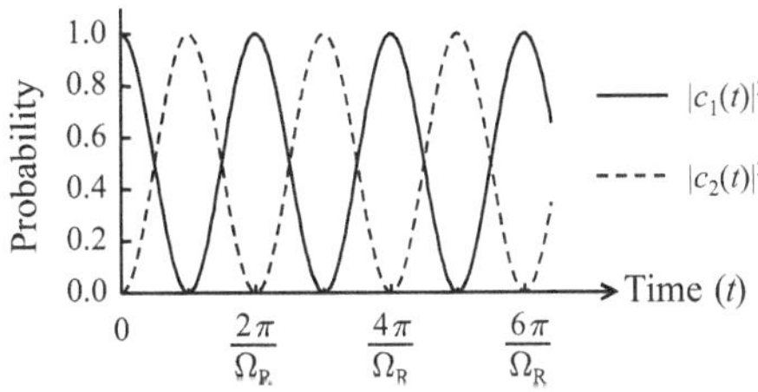

Fig. 10.2 Rabi oscillations for a two-level atom interacting with a resonant laser in the absence of damping. The electron oscillates back and forth between the two levels at the Rabi frequency, Ω_{R}

$$\ddot{c}_1 = \frac{\mathrm{i}}{2}\Omega_{\mathrm{R}}\dot{c}_2 = \left(\frac{\mathrm{i}}{2}\Omega_{\mathrm{R}}\right)^2 c_1 . \tag{10.23}$$

We then obtain the equation of motion:

$$\ddot{c}_1 + \left(\frac{\Omega_{\mathrm{R}}}{2}\right)^2 c_1 = 0 , \tag{10.24}$$

which describes oscillatory motion at angular frequency $\Omega_{\mathrm{R}}/2$. If the particle is in the lower level at $t = 0$ so that $c_1(0) = 1$ and $c_2(0) = 0$, the solution is:

$$\begin{aligned} c_1(t) &= \cos\left(\Omega_{\mathrm{R}} t/2\right) , \\ c_2(t) &= \mathrm{i}\sin\left(\Omega_{\mathrm{R}} t/2\right) . \end{aligned} \tag{10.25}$$

The probabilities for finding the electron in the upper or lower levels are then:

$$\begin{aligned} |c_1(t)|^2 &= \cos^2\left(\Omega_{\mathrm{R}} t/2\right) , \\ |c_2(t)|^2 &= \sin^2\left(\Omega_{\mathrm{R}} t/2\right) . \end{aligned} \tag{10.26}$$

The time dependence of these probabilities is shown in Fig. 10.2. At $t = \pi/\Omega_{\mathrm{R}}$ the electron is in the upper level, whereas at $t = 2\pi/\Omega_{\mathrm{R}}$ it is back in the lower level. The process then repeats itself with a period equal to $2\pi/\Omega_{\mathrm{R}}$. The electron thus oscillates back and forth between the lower and upper levels at a frequency equal to $\Omega_{\mathrm{R}}/2\pi$. This oscillatory behaviour in response to the strong light field is called *Rabi oscillation* or *Rabi flopping*. It has been observed in many systems, and, as we shall see in Sect. 10.2.4, can be given a geometric interpretation in terms of rotations of the Bloch vector. Observations of Rabi flopping in quantum dots will be discussed in Sect. 10.4.2.

For the more general case where $\Delta \neq 0$, it can be shown that:

$$|c_2(t)|^2 = \frac{\Omega_{\mathrm{R}}^2}{\Omega^2}\sin^2\left(\Omega t/2\right) , \tag{10.27}$$

where $\Omega = \sqrt{\Delta^2 + \Omega_R^2}$ is the generalised Rabi frequency. This shows that the frequency of the Rabi oscillations increases but their amplitude decreases as the light is tuned away from resonance, which explains why we can neglect off-resonant transitions.

The observation of Rabi flopping often requires powerful, pulsed lasers, so that the electric field amplitude E_0, and hence Ω_R, varies with time. It is then useful to define the *pulse area* Θ according to:

$$\Theta = \left| \frac{\mu_{12}}{\hbar} \int_{-\infty}^{+\infty} E_0(t)\, \mathrm{d}t \right| . \tag{10.28}$$

This is a dimensionless parameter that serves the same purpose as $\Omega_R t$ in the analysis above. A pulse with $\Theta = \pi$ is called a π-pulse, etc. An atom in the ground state with $c_1 = 1$ at $t = 0$ will thus be promoted to the excited state by a π-pulse, but will end up back in the ground state if it interacts with a 2π-pulse. Note, however, that additional *geometric* phase shifts can be picked up by fermionic particles during Rabi rotations. These are important in the spin control experiments described in Sect. 10.5.

10.2.3 Damping

We have assumed so far that the wave functions remain completely coherent while being driven by the laser. In reality, collisions, or interactions with the environment, randomize the phases, leading to a loss of coherence. The damping mechanisms that cause decoherence are generally determined by two time constants, T_1 and T_2, originally introduced in NMR theory.

The T_1 time characterizes the spontaneous decay rate of the population from the upper state, for example, by a radiative transition:

$$\frac{\mathrm{d}N_2}{\mathrm{d}t} = -\frac{N_2}{T_1} , \tag{10.29}$$

where $N_2 = |c_2(t)|^2 N$ is the population of the upper level in an ensemble of N identical systems. Solution of (10.29) gives exponential decay with $N_2(t) = N_2(0)\exp(-t/T_1)$, which shows that T_1 is the lifetime of the upper level, and characterises the decay of the z component of the Bloch vector, as shown in Fig. 10.3a. Since radiative transitions occur spontaneously (i.e. at random times, triggered by vacuum fluctuations), they destroy all phase information in the system. Non-radiative process can also contribute to T_1. An example of particular relevance to quantum dot physics is the tunnelling of electrons and holes out of a dot in an electric field.

The T_2 time gives the timescale over which coherence is maintained. It is related to T_1 through the following relationship:

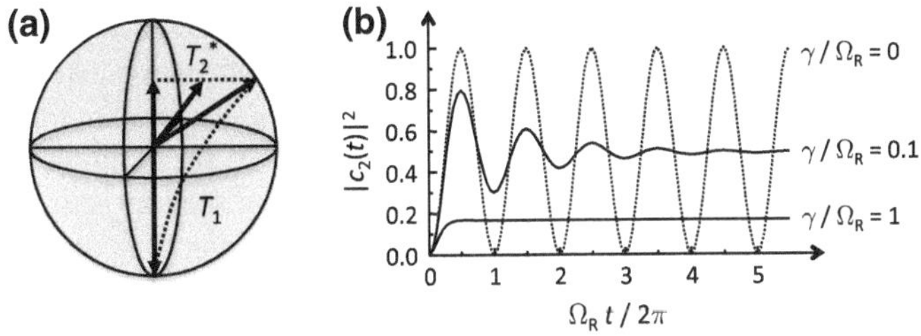

Fig. 10.3 **a** Damping processes in the Bloch representation: T_2^* processes conserve z but T_1 processes do not. **b** Damped Rabi oscillations for two values of the ratio of the damping rate γ to the Rabi oscillation frequency Ω_R. The *dotted curve* shows the oscillations when no damping is present

$$\frac{1}{T_2} = \frac{1}{2T_1} + \frac{1}{T_2^*} \, . \tag{10.30}$$

The first term accounts for the loss of coherence due to population decay. The second is the pure dephasing rate $(T_2^*)^{-1}$. The T_2^* time quantifies pure dephasing processes in which z is unchanged, for example: elastic scattering by phonons or by fluctuating fields from impurities. This difference is illustrated in Fig. 10.3a. The T_2^* processes cause a coherent state on the surface of the Bloch sphere to relax to a mixed state on the z axis, and is therefore called *transverse relaxation* This contrasts with the *longitudinal* T_1 processes that causes changes in z as well as x and y.

The overall coherence time is given by T_2. In a system with negligible pure dephasing (i.e. $T_2^* \gg T_1$), the coherence time is $2T_1$. This means that the ultimate limit on T_2 is set by the lifetime of the upper level. The factor of two between T_2 and T_1 in this limit arises from the fact that coherence is sensitive to the wave function amplitude (i.e. c_2), whereas population depends on $|c_2|^2$. For a system with simple exponential dynamics following (10.29), c_2 decays at half the rate as the population: $|c_2(t)| \propto \exp(-t/2T_1)$.

The probability that a damped two-level system is in the upper level when driven on resonance is given by [3]:

$$|c_2(t)|^2 = \frac{1}{2(1+2\xi^2)}\left[1 - \left(\cos \Omega' t + \frac{3\xi}{(4-\xi^2)^{1/2}} \sin \Omega' t\right) \exp\left(-\frac{3\gamma t}{2}\right)\right] , \tag{10.31}$$

where $\xi = \gamma/\Omega_R$ and $\Omega' = \Omega_R\sqrt{1-\xi^2/4}$. The parameter γ that enters here is the damping rate $1/T_2$. It is easily verified that this formula reduces to the undamped case given in (10.26) when $\gamma = 0$. The effect of damping on Rabi oscillations is illustrated in Fig. 10.3b. The dotted line shows the undamped case with $\gamma = 0$ shown previously in Fig. 10.2. The two other graphs demonstrate the effect of increasing damping. With light damping ($\gamma/\Omega_\mathrm{R} = 0.1$), the system performs a few damped oscillations and then approaches the asymptotic limit with $|c_1|^2 = |c_2|^2 = 1/2$. This asymptotic limit is exactly the behaviour predicted by the incoherent analysis based on the Einstein B coefficients, where the rates of stimulated emission and absorption eventually equal out at high pumping rates, leading to identical upper and lower level populations. With stronger damping ($\gamma/\Omega_R = 1$), no oscillations are observed, and we recover the fully incoherent picture where $|c_2(\infty)|^2$ is proportional to the laser power, independent of time. This asymptotic limit is best seen by setting $\xi \gg 1$, in which case the probability of occupation of the upper level at long times is equal to $\xi^{-2}/4 = \mu_{12}^2 E_0^2/4\hbar^2\gamma^2$, i.e. proportional to the Einstein B coefficient via μ_{12}^2 and the laser power via E_0^2.

The conclusion of this analysis is that Rabi oscillations can only be observed in highly coherent systems where the damping rate is significantly smaller than the Rabi frequency. In atomic gases, the damping rate depends on the collision rate and the radiative lifetime, which gives $\gamma \sim 10^7 - 10^9\,\mathrm{s}^{-1}$. In semiconductors the dephasing times are often much shorter due to phonon scattering, scattering by free charge carriers, or tunnelling. This makes the task of demonstrating Rabi oscillations somewhat difficult, which explains why they are not routinely observed. The art of coherent control experiments thus entails reducing the damping rate as much as possible (e.g. by working with very pure samples at low temperatures) and using lasers with pulses that are shorter than the coherence time.

10.2.4 Coherent Rotations on the Bloch Sphere

The previous sub-sections have established the basic principles for the main focus of this chapter, namely the coherent manipulation of a two-level system by a resonant laser pulse with duration shorter than T_2. This process is best visualised by considering the Bloch vector $\boldsymbol{s}$. If damping is negligible, the system remains coherent, and the pulse only changes the direction of $\boldsymbol{s}$ without altering its length. This means that the pulse acts as a *rotation operator*. We showed previously that a π-pulse can convert a system in the ground state $|1\rangle$ to the excited state $|2\rangle$, and vice versa. This corresponds to a change of the Bloch vector angle θ by π radians, and explains the origin of the name 'π-pulse'. In general, it can be shown that the rotation angle is equal to the pulse area defined in (10.28). Hence a pulse of area Θ causes a rotation of $\boldsymbol{s}$ by Θ radians.

In analysing coherent manipulations of Bloch vectors, we can usually assume that the system is initially in the ground state. The azimuthal angle of the Bloch sphere is then undefined, which means that the azimuthal angle of the rotation axis is also undefined. It is therefore convenient to choose the x and y axis directions so that the first pulse in a sequence produces a rotation about, say, the y axis, leaving the Bloch vector in x–z plane at the end of the pulse. The axis about which subsequent rotations take place is then determined by the phases of the pulses relative to the first one. Combinations of pulses of the appropriate area and phase can then be used to move the Bloch vector to any particular point on the Bloch sphere.

Figure 10.4 illustrates how this works in Ramsey interference experiments. The system is initially in the ground state as shown in Fig. 10.4a. A resonant $\pi/2$ pulse

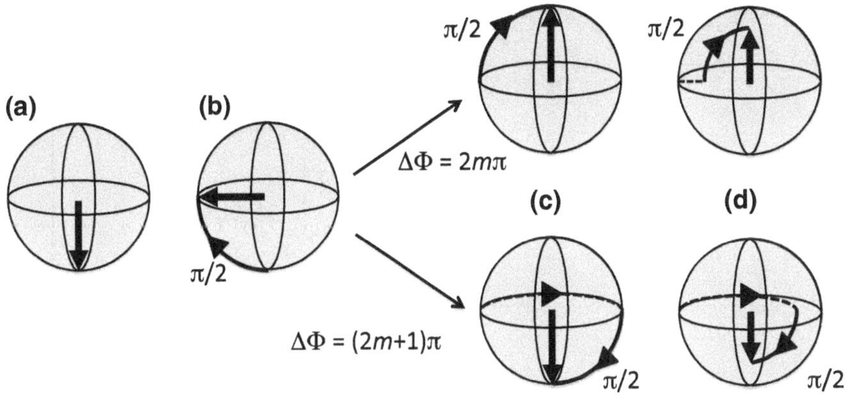

Fig. 10.4 Ramsey interference experiment illustrated on the Bloch sphere. **a** The system is initially in the ground state. **b** A resonant pulse of area $\pi/2$ rotates the system to a point on the equator. **c** A second resonant $\pi/2$ pulse with relative phase $\Delta\Phi$ rotates the system to the north or south pole respectively when $\Delta\Phi = 2m\pi$ or $(2m+1)\pi$, m being an integer. **d** Same as **c**, but with dephasing included

is incident and rotates the Bloch vector by 90° to a point on the equator, as shown in Fig. 10.4b. The system is then interrogated with a second $\pi/2$ pulse with phase $\Delta\Phi$ relative to the first one. If the second pulse is in phase ($\Delta\Phi = 2m\pi$, $m =$ integer, Fig. 10.4c, top half), the second 90° rotation takes place about the same axis as the first, and the system ends up at the north pole. However, if the second pulse is out of phase ($\Delta\Phi = (2m+1)\pi$, Fig. 10.4c, bottom half), the rotation takes place about an axis pointing in the opposite direction to the first, leaving the system at the south pole. Figure 10.4d illustrates the equivalent picture when dephasing is included. The system ends up in a mixed state along the z axis, either in the upper or lower half of the Bloch sphere, depending on the relative phase of the second pulse.

A final mention should be made of the case of off-resonant pulses. Equation (10.27) implies that an off-resonant pulse with detuning Δ induces Rabi oscillations at a slightly different frequency, and with reduced amplitude. This can be visualised as rotations about an axis with polar angle α relative to the equatorial plane, where $\tan\alpha = \Delta/\Omega_R$ [4]. As with resonant pulses, the azimuthal angle is determined by the phase. This off-equatorial rotation for finite Δ accounts for the reduced amplitude of the Rabi flopping. Note that $\alpha = 0$ for $\Delta = 0$, and that $\alpha = \pi/2$ for $\Omega_R = 0$.

10.3 Quantum Dots as Coherent Systems

In the sections that follow, coherent control experiments on quantum dots (QDs) will be reviewed in detail. However, it is first useful to make some general remarks that explain the interest in QDs for coherent state manipulation.

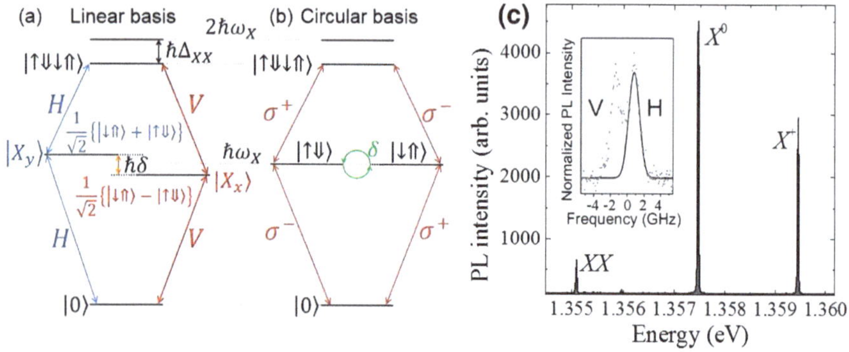

Fig. 10.5 **a** Energy level spectrum of the exciton and biexciton in a neutral dot in the linearly-polarized basis. The two eigenstates $|X_x\rangle$ and $|X_y\rangle$ are split by the fine-structure splitting (FSS) ($\hbar\delta$). **b** Circularly-polarized basis with excitation bandwidth $>\hbar\delta$. The FSS causes precession between the two eigenstates, indicated by the *green arrows*. The exciton and biexciton energies are given respectively by $\hbar\omega_X$ and $\hbar\Delta_{XX}$. For clarity, the splittings are not to scale, as $\hbar\delta \ll \hbar\Delta_{XX} \ll \hbar\omega_X$. **c** Emission spectrum of a typical dot. The neutral exciton and biexciton lines are identified. The X^+ line arise from a positively charged exciton. The *inset* gives a high resolution spectrum of the X^0 line of a similar dot, from which the FSS can be deduced. Data from E.A. Chekhovich and M.N. Makhonin

The observation of intense, discrete lines in the emission spectra of individual dots in 1994 [5] immediately pointed to the 3-D confinement of the electrons and holes. A particularly striking feature was the sharpness of the lines: see, for example, Fig. 10.5c. Since the width ΔE of spectral lines is determined by the dephasing time of the emitters, a sharp line implies a long T_2 time. At the time when single-dot spectra were first observed, the state-of-the-art in semiconductor optics was set by quantum-well samples in GaAs-related materials. The very best samples showed $\Delta E \sim 0.1$ meV, which implied $T_2 \sim 10$ ps at best. While very impressive results were demonstrated (see e.g. [6–8]), the short dephasing time made long-term applications difficult. With InGaAs quantum dots, by contrast, it is relatively straight forward to get $\Delta E \sim 0.01$ meV, with the best samples showing $\Delta E \sim 1\,\mu$eV, limited only by the $\sim$1 ns radiative lifetime [9]. From this we can deduce that, in the right conditions, pure dephasing can effectively be eliminated, leading to $\sim$1 ns coherence times. Such long T_2 times were indeed verified from four-wave mixing experiments on QD ensembles in 2001 [10].

Another very striking feature of QDs is the absence of phonon side-bands. Most localized emitters in solid-state hosts (e.g. NV centres in diamond, Ti^{3+} ions in sapphire) are strongly coupled to phonons, giving rise to vibronic sidebands in their emission and absorption spectra. In the case of NV centres, for example, the coupling is so strong that only a few percent of the emission occurs in the zero-phonon line, with most of the photons emitted from the sidebands [11]. The relatively weak intensity of the zero-phonon line has serious consequences for practical applications in optical QIP.

The reason for the strong phonon sidebands in materials like diamond NV centres and Ti:sapphire is that both the electronic and vibrational modes are localized on length scales similar to the unit cell size. This means that the overlap between the electronic wave functions and the phonon modes is large. By contrast, InGaAs quantum dots (QDs) have envelope wave functions localized on much larger length scales that are determined by the size of the dot, i.e. $\sim$10 nm. This means that the coupling to phonons is relatively weak, with the dominant interaction being to longitudinal-acoustic (LA) phonons via deformational potential scattering. The weak vibronic coupling gives rise to very strong emission in the zero-phonon line, with only $\sim$8% in the sideband at cryogenic temperature [12, 13]. This makes InGaAs QDs excellent single-photon sources [14]. It also gives a small phonon dephasing rate, and hence explains the long T_2 time discussed above.

It should be noted that the actual coherence times measured in QDs are still shorter than those obtained in atomic systems. This follows immediately from the short ($\sim$1 ns) radiative lifetime of InGaAs QDs, compared to atomic transitions (e.g. 16 ns for sodium D lines at $\sim$589 nm). However, this is not necessarily a drawback provided we work in the regime where T_2 is limited by the radiative lifetime. As discussed in Sect. 10.2.3, the key parameter is the ratio of the Rabi frequency to the damping rate. The Rabi frequency is determined by the dipole moment of the transition (see (10.20)), which also determines the radiative lifetime. Hence a system with a large dipole moment automatically has a short coherence time, but the ratio Ω_R/γ is not necessarily adversely affected. In fact, InGaAs QDs typically have dipole moments of $\sim$30 Debye ($\sim 1 \times 10^{-28}$ C m) [15], which is at least an order of magnitude larger than atoms. The larger dipole moment means that the light-matter coupling is stronger, enabling efficient driving at relatively low powers. Overall, InGaAs QDs represent a good compromise between strong light-matter coupling and freedom from dephasing (at least compared to bulk and quantum well semiconductors.) These properties, combined with their compatibility with advanced technological device processing, makes QDs very attractive for QIP applications.

There are several different types of quantum dots, but this chapter focusses on the self-assembled InGaAs QDs grown by epitaxial methods such as molecular beam epitaxy (MBE) or metal-organic chemical vapour deposition (MOCVD). QDs can also be formed during MBE growth of pure GaAs layers at interface islands due to monolayer fluctuations [16]. In fact, the first coherent control experiment on single III-V QDs was performed on these types of dots [17], and this was soon followed by two other key 'firsts': Rabi flopping [18] and a two-qubit quantum gate [19]. However, the electrons and holes in these interface dots are only weakly confined, making them very sensitive to temperature, and also to scattering form free carriers, which can come from either impurities or optical excitation. This means that their T_2 times, although substantially better than bulk or quantum well semiconductors, are shorter than those of InGaAs dots. There are also colloidal quantum dots, but these are hard to incorporate into photonic devices. For these reasons, InGaAs dots are best suited for most applications in QIP.

We should also clarify that everything considered in this section so far refers to the excitons created when electron and hole pairs are excited in QDs by absorption of photons. These excitons have radiative lifetimes of ~1ns, which sets a similar upper limit on T_2. At the same time, they interact very strongly with light, which facilitates their coherent manipulation (see Sect. 10.4 below). An alternative approach involving the coherent control of spins in charged QDs will be discussed in Sect. 10.5. Charged dots contain electrons or holes before the laser is incident, and the goal is to rotate the carrier spins with the laser. Since the resident carriers cannot recombine, T_2 is no longer limited by the radiative lifetime. On the other hand, light does not interact directly with electron or hole spins, which makes the coherent manipulation techniques more challenging. There are therefore advantages and disadvantages of working with spins, as will be discussed in Sect. 10.5.

10.4 Coherent Control of Excitons

The relatively long coherence times of excitons in QDs facilitate their use in ultrafast coherent control experiments. The DiVincenzo checklist for QIP requires that we demonstrate complete control of single excitonic qubits, and establish at least one excitonic two-qubit gate [20]. We therefore begin this section by considering the coherent control of single excitons, which can be visualized as single qubit rotations, and finish with two-exciton systems. We focus exclusively on bright excitons, leaving the discussion of dark excitons to Chap. 4.

In this chapter we restrict our discussion of QD coherence to the time domain. There is an equivalent approach that investigates QD coherence in the frequency domain through high-resolution spectroscopy. Examples include the observation of the Mollow triplet [21], and the Autler–Townes doublet [22, 23]. Phenomena such as these are discussed elsewhere in this book (see e.g. Chaps. 2 and 3).

10.4.1 Level Structure of Excitons in Neutral Quantum Dots

Before considering the principles of coherent control experiments on QDs, we must first review the excitonic level structure of a typical dot. We assume here that the dot is neutral: i.e. that it contains no free electrons or holes before the laser pulse arrives. (Charged dots will be considered in Sect. 10.5.) Furthermore, we neglect the light-hole (LH) bands, since they have significantly larger confinement energies than heavy-hole (HH) bands.

Absorption of a photon creates an electron in the conduction band and a hole in the valence band, which then bind together to form an exciton through their mutual Coulomb attraction. The exciton state in a neutral dot containing one electron and hole, both in their lowest confined energy states (i.e. the *s*-shell), is typically notated as X^0. Since heavy holes have $m_z^{hh} = \pm 3/2$, the possible z-axis spin projections of the exciton are given by:

$$S_z = m_z^{hh} + m_z^e = \pm\frac{3}{2} \pm \frac{1}{2} = -2,\ -1,\ 1,\ 2\,. \tag{10.32}$$

Absorption/emission of a single circularly-polarized photon may only impart a change in angular momentum of ± 1. Hence $S_z = \pm 1$ "bright excitons" with circular polarization $|\uparrow\Downarrow\rangle$ and $|\downarrow\Uparrow\rangle$ are optically allowed whilst $S_z = \pm 2$ "dark excitons" are optically forbidden. Here, and throughout the whole of this chapter, the notation $|\uparrow\rangle$ and $|\Uparrow\rangle$ refers to electron and heavy-hole spin states respectively.

An ideal QD exhibits radial symmetry in the growth (x–y) plane. However, the self-assembly process typically leads to QDs with some degree of asymmetry. In this case the two bright exciton states become coupled by the electron-hole exchange interaction [24], causing a precession between the $S_z = \pm 1$ states with angular frequency δ [25]. As a result, the eigenstates are linearly polarized and split by the fine-structure splitting (FSS) $\hbar\delta$ as shown in Fig. 10.5a, with:

$$\begin{aligned} |X_x\rangle &= \frac{1}{\sqrt{2}}\{|\downarrow\Uparrow\rangle + |\uparrow\Downarrow\rangle\} \\ |X_y\rangle &= \frac{1}{\sqrt{2}}\{|\downarrow\Uparrow\rangle - |\uparrow\Downarrow\rangle\}\,, \end{aligned} \tag{10.33}$$

where the axes x and y are defined by the asymmetry of the dot, often coinciding with the [111] and [$1\bar{1}1$] crystal axes. The magnitude of $\hbar\delta$ varies significantly from dot to dot, as it originates from the randomness of the self-assembly process [26]. Its value is a very important parameter in the coherent control of excitons (see Sect. 10.4.3), and also in the initialisation of spin states in charged dots (see Sect. 10.5.2). Moreover, its minimisation is highly important for the generation of entangled photons [27, 28]. The control of the fine-structure splitting is therefore an important research field (see Chap. 7).

The s-shells of a QD can be occupied by two carriers with opposite spins. It is therefore possible for both of the $S_z = \pm 1$ neutral excitons to exist simultaneously, forming a *biexciton* (generally denoted as either XX or $2X$). The Coulomb interaction between the excitons results in a binding energy ($E_{XX} = \hbar\Delta_{XX}$) that reduces the biexciton energy to less than that of two excitons. The resulting energy level structure is illustrated in Fig. 10.5. Figure 10.5a shows the linearly polarised basis, where the two X^0 eigenstates defined in (10.33) are split by the FSS energy $\hbar\delta$. For circular excitation ($\sigma^\pm = \hat{x} \pm \mathrm{i}\hat{y}$) as shown in Fig. 10.5b, both X^0 spin states are excited and the energy splitting is zero. However, as the eigenstates of the system are linear, the exciton spin precesses with a frequency δ.

Figure 10.5c shows a typical emission spectrum of a single InGaAs quantum dot at 4 K. The neutral exciton and biexciton states are labelled. The energy shift between the X^0 and XX states is equal to 2.37 meV, which equates with the biexciton binding energy $\hbar\Delta_{XX}$ for this dot. The FSS of the X^0 line is clearly shown in the high-resolution spectrum for another dot in the inset. A value of $\hbar\delta = 9.9\,\mu$eV is deduced, together with an exciton linewidth of 4.3 μeV, corresponding to a coherence time $T_2 \sim 150$ ps. This linewidth is smaller than the resolution limit of most spectrometers, and its measurement usually requires the use of a scanning Fabry–Perot interferometer. The lines do, of course, broaden with temperature, implying a reduction in T_2. For this reason, the best coherent control results are generally obtained at 4 K.

A third strong line labelled X^+ is also clearly visible in Fig. 10.5c. This line is due to a positively-charged exciton state and will be discussed in Sect. 10.5. Charged excitons can be observed in "neutral" dots through the capture of free electrons or holes from the wetting layer.

10.4.2 Rabi Flopping

The first observation of Rabi flopping for excitons in a single quantum dot was obtained for GaAs interface dots in 2001 [18]. Similar observations were soon obtained for single InGaAs/AlGaAs dots [29] and for ensembles of self-assembled InGaAs/GaAs dots [30, 31]. In these experiments, the exciton population was measured as a function of increasing pulse area, as in Fig. 10.3b. The Rabi oscillations were subsequently observed directly in the time domain by measuring the second-order correlation of the photons emitted [32] or by performing pump-probe experiments using two pulses [33]. Rabi flopping has also been observed for biexcitons by tuning the laser to the two-photon resonance midway between the X^0 and XX transition frequencies [34].

In 2002, Zrenner et al. established the photocurrent technique that has been used extensively for the experimental work reviewed in this chapter [35]. A schematic of the method is given in Fig. 10.6a. The self-assembled InGaAs quantum dots are embedded within a reverse-biassed Schottky diode, and are excited through a nano-aperture within a metal shadow mask. With nano-apertures in the sub-μm range, the number of dots interrogated by the laser is reduced to $\sim$10 for dot densities of $\sim 10^9\,\text{cm}^{-2}$. The randomness of the self-assembly leads to fluctuations in the size and shape, and hence confinement energies, so that individual QDs can be addressed by tuning the laser. Coherent effects from single dots can then be observed when excited resonantly by a laser pulse with duration shorter than the coherence time.

An attractive feature of the photocurrent technique is that the final state of the dot can be deduced very easily from the photocurrent measured in the external circuit. With negative bias applied to the diode, the dots experience a strong electric field, and tunnel out towards the contacts, as shown schematically in Fig. 10.6a. A π-pulse leaves the dot containing a single exciton, which then generates one electron in the external circuit. For a laser repetition rate of f, the current is equal to fe, where e is the electron charge. This gives a current of around 13 pA for a typical laser repetition rate of $\sim$80 MHz, which is easily measurable with a precision pico-ammeter. The actual current measured is typically lower than this, due to competition with radiative recombination before tunnelling occurs.

Figure 10.6b shows a typical Rabi oscillation measurement on a single self-assembled InGaAs/GaAs quantum dot at 4 K. Six oscillatory periods are clearly

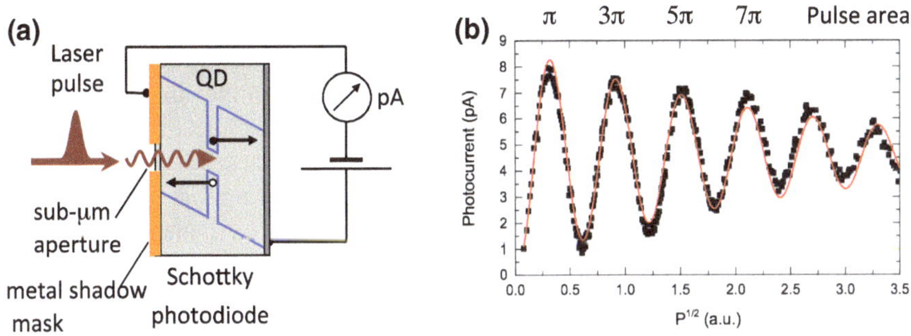

Fig. 10.6 **a** Schematic of experiment to observe Rabi rotations in a quantum dot photodiode. The laser pulse width must be shorter than the exciton coherence time. **b** Rabi rotation of the neutral exciton transition of a single InAs/GaAs quantum dot at 4 K. The photocurrent (final exciton population) oscillates according to the pulse area Θ (which is proportional to the square root of the applied laser power). The *red line* shows a damped sinusoidal fit to the data

visible, but with substantial damping as the pulse area Θ increases. This damping is mainly caused by the way the measurements are made. In NMR experiments, the pulse area defined in (10.28) is varied by changing the pulse duration while keeping its amplitude constant. This is not practical for ultrafast laser experiments, since the pulse length cannot be changed easily. Moreover, since the pulse length changes the laser bandwidth, the excitation conditions are also changed by varying the pulse width. For these reasons, the pulse area is varied by keeping the pulse duration constant, and increasing its amplitude. This means that the laser driving power P is varying as $\sqrt{P}$ along the x axis in Fig. 10.6b, and the damping at high pulse areas is related to this increase in the excitation intensity. The damping is therefore termed Excitation Induced Dephasing (EID). It is important to point out that qubit control experiments on QDs are typically carried out at pulse areas of $\sim\pi$ (see Sect. 10.4.3), where EID is small.

Ramsay et al. performed careful studies of the effect of temperature on Rabi flopping, and demonstrated that EID arises from phonon interactions [36, 37]. The dominant coupling is to longitudinal-acoustic (LA) phonons via deformation potential scattering, which is quantified by the function $J(\omega)$:

$$J(\omega) = \frac{\omega^3}{4\pi^2\rho\hbar v_c^5}\left[D_{\rm e}\,{\rm e}^{\left(-\omega^2 a_{\rm e}^2/4v_c^2\right)} - D_{\rm h}\,{\rm e}^{\left(-\omega^2 a_{\rm h}^2/4v_c^2\right)}\right]^2 . \tag{10.34}$$

The parameters that enter here are the mass density ρ, the sound velocity v_c, the deformation potential constants for electrons and holes $D_{\rm e/h}$, and the electron/hole confinement lengths inside the dot, $a_{\rm e/h}$. The function $J(\omega)$ increases at first with ω due to the ω^3 factor from the LA-phonon density of states. It passes through a peak, and then rolls off rapidly above the "cut-off frequency" due to the exponential form-factors that characterize the physical size of the dot. For typical InGaAs dots, values of $a_{\rm e} = 4.5$ nm and $a_{\rm h} = 1.8$ nm are obtained [38], and $J(\omega)$ peaks at about

1–2 meV. The significance is that the damping rate for Rabi rotations is governed by $J(\Omega_R)$, i.e. the electron-phonon coupling at the Rabi frequency. For the data shown in Fig. 10.6b, the largest Rabi frequency is still smaller than the cut-off frequency, and so the damping gets stronger at higher pulse areas. In principle, the damping rate should weaken for very strong driving when Ω_R exceeds the cut-off frequency [39]. This phenomenon is sometimes called phonon revival, but not yet been observed experimentally.

10.4.3 Manipulation of Exciton States

The demonstration of Rabi flopping confirms the possibility of moving the exciton state around the Bloch sphere in a coherent way. However, a single rotation does not give access to all points on the Bloch sphere, as required for full single-qubit control: two rotations are needed, about different axes. In this sub-section, we shall see how this is done.

The simplest way to achieve full Bloch sphere control is to use two resonant pulses with a well-defined phase difference. This works because the azimuthal angle for the second pulse is determined by its phase relative to the first. A key demonstration of this principle is the observation of Ramsey interference, as explained in Fig. 10.4. Such Ramsey interference was first observed for InGaAs dots within a Schottky photodiode by using two phase-locked excitation pulses of area $\pi/2$ [40, 41]. The time between the pulses had to be less than the exciton coherence time, which was determined by the loss of electrons out of the dot by electric-field-induced tunnelling.

While proving the principle, the Ramsey method suffers from the need to keep the phases of the pulses locked together over long periods, which requires an actively-stabilised interferometer. For this reason, simpler methods have been employed that exploit the fine-structure splitting of the dot. As noted in Sect. 10.4.1, a typical neutral InGaAs dot has two linearly polarized excitons split by the FSS $\hbar\delta$, as shown in Fig. 10.5a. If the dot is excited with a $\boldsymbol{\sigma}^+ \equiv \hat{\boldsymbol{x}} + \mathrm{i}\hat{\boldsymbol{y}}$ pulse with bandwidth $> \delta$ at time $t = 0$, both eigenstates are excited, and the wave function evolves as:

$$\Psi(t) = \frac{1}{\sqrt{2}}\left(|X_x\rangle + \mathrm{i}|X_y\rangle \mathrm{e}^{\mathrm{i}\delta t}\right) , \tag{10.35}$$

where the phase factor accounts for the frequency difference between $|X_x\rangle$ and $|X_y\rangle$. This evolves to a linear state $[(|X_x\rangle - |X_y\rangle)/\sqrt{2}]$ at $t = \pi/2\delta$, through to an opposite circular state $[(|X_x\rangle - \mathrm{i}|X_y\rangle)/\sqrt{2}]$ at $t = \pi/\delta$, to an opposite linear state $[(|X_x\rangle + |X_y\rangle)/\sqrt{2}]$ at $t = 3\pi/2\delta$, and finally back to the original circular state at $t = 2\pi/\delta$. On realizing that the polarization Poincaré sphere maps directly to the exciton spin Bloch sphere, with the $\sigma^{\pm}$ and diagonal linear states at the four cardinal points of the equator, it is apparent that the FSS causes a precession about the $\hat{z}$ axis at a rate δ. This can then be combined with a single x-axis rotation by a resonant

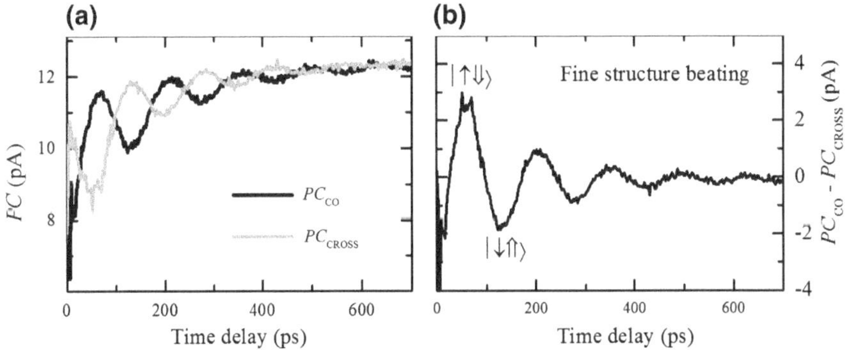

Fig. 10.7 Fine-structure beats from a single InGaAs dot measured by the photocurrent (*PC*) technique. **a** Raw data for co- and cross-polarized probe pulses. **b** Difference of the two signals in **a**. After [46], data from J.H. Quilter

optical pulse to reach arbitrary points on the exciton Bloch sphere [42]. A second "control" pulse can also be added to achieve optical rotation of the exciton spin while it is precessing [43–45].

The fine-structure beats were first observed for single GaAs interface QDs in 1998 [17], and then for an ensemble of InGaAs dots in 2004 [25]. Recent results on fine-structure beating from a single InGaAs dot are shown in Fig. 10.7 [46]. The dot was in a Schottky diode as in Fig. 10.6a, and was excited with a resonant σ^+ pulse of area π at $t = 0$. The dot was then probed by a second π pulse with either σ^+ or σ^- polarization at varying time delay. When the spin of the exciton is co-polarized with the probe, the probe de-excites the exciton and reduces the photocurrent (*PC*) signal. When the spin of the exciton is cross-polarized with the probe, the probe does nothing. Hence the photocurrent signal oscillates at the same rate as the spin precession, as clearly seen in Fig. 10.7a. Note that the signals for opposite polarizations are in anti-phase. The difference between the co- and cross-polarized signals, which is proportional to $\langle S_z \rangle$, is shown in Fig. 10.7b. At least four oscillations can be observed, with the period of 145 ps implying an FSS of 28 μeV. The damping of the oscillations is mainly caused by electron tunnelling out of the dot within the photodiode.

The final state of the exciton qubit in the experiments described above is highly sensitive to changes in the laser pulse area, which means that precise control on the pulse intensity is required. Several authors have explored alternative methods that are less sensitive to the exact pulse area. One such approach is to use adiabatic rapid passage with chirped optical pulses [47, 48]. In these experiments, the frequency of the laser is swept during the pulse, giving a time-varying detuning relative to the transition frequency. In the right conditions, the final state of the dot is relatively insensitive to the exact pulse area. Another method is to pump the dot within the LA-phonon sideband [49]. Pumping via phonon-assisted transitions is, of course, incoherent. However, if the pumping is hard enough, near perfect inversion of the dot is possible, producing a final state very close to the north pole of the Bloch sphere, which is insensitive to phase variations. Such phonon-sideband pumping was recently observed by three groups [38, 50, 51].

10.4.4 Two-Qubit Gates

The DiVincenzo check-list for QIP [20] requires at least one two-qubit gate in addition to the full control of single qubits. One way to achieve this is via coupling between two excitons of opposite spins inside a single dot. As noted in the discussion of Fig. 10.5, the biexciton state does not have twice the energy of the individual excitons, which indicates coupling via the Coulomb interaction. This allows a two-qubit conditional rotation gate (CROT) to be performed by using two laser pulses tuned respectively to the exciton and biexciton transitions of the dot. The method was initially demonstrated for a GaAs interface dot by Li et al. in 2003 [19]. Below, we describe the equivalent experiment on a self-assembled InGaAs dot reported by Boyle et al. in 2008 [52].

A simplified version of the levels used in the CROT gate is given in Fig. 10.8a. The control qubit $|q_1\rangle$ is the $S_z = -1$ exciton, while the target qubit $|q_2\rangle$ corresponds to $S_z = +1$. The combined state of the system $|q_1q_2\rangle$ is denoted by the number of excitons (either 0 or 1) in the respective qubit states. The aim of the experiment is to demonstrate a rotation of the target qubit conditional on the state of $|q_1\rangle$: i.e. we need to show that we rotate $|q_2\rangle$ only when $q_1 = 1$.

The state of the control qubit is determined by a π pulse with σ^- polarization tuned to the X^0 transition. If this pulse is present, we have $q_1 = 1$; if not, $q_1 = 0$. The conditional rotation of $|q_2\rangle$ is performed by a σ^+ pulse of variable area tuned

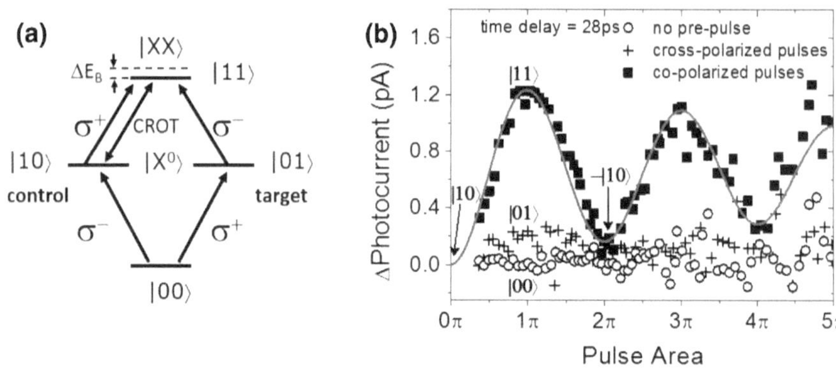

Fig. 10.8 CROT gate in an InGaAs dot using the exciton and biexciton transitions. **a** Level scheme showing the transitions involved. **b** Experimental results. Note that the data were collected in the *linear* polarization basis, where the conditional Rabi rotation is expected for co-polarized pulses, rather than the cross-polarized configuration indicated in **a**. After [52]

to the XX transition. The rotation can only occur if $q_1 = 1$, since a pulse tuned to the biexciton only drives the dot when an exciton of the opposite spin is already present. If the dot is empty, no exciton can be created, as the laser has the wrong frequency. The key result is to therefore to demonstrate Rabi rotation of the biexciton conditional on the state of $|q_1\rangle$.

The results of the experiment are presented in Fig. 10.8b. For technical reasons, the experiment was performed in the linear polarization basis rather than the more natural circular one. In this linear basis, we expect to excite the biexciton when the CROT pulse is co-polarized with the control pulse, and not when it is cross-polarized. This is exactly what is observed: the biexciton Rabi rotation is only observed when the co-polarized control pulse has acted first on the dot. If the control pulse has the wrong polarization, or is not present at all, the initial state of the dot is $|01\rangle$ or $|00\rangle$ respectively, and no Rabi oscillation is observed. On the other hand, when the control pulse has the right polarization, the initial state of the dot is $|10\rangle$. A CROT pulse of area π then drives the system to the $|11\rangle$ state (i.e. the biexciton state), while a 2π pulse leaves the system in the $-|10\rangle$ state, where the $-$ sign originates from the geometric phase of π that is accumulated on completing a 2π rotation of the Bloch sphere. The data in Fig. 10.8b thus establishes the four outcomes of the CROT truth table. The gate fidelity deduced from the data was 0.87 ± 0.04, which was significantly higher than that obtained for interface dots in [19], mainly due to the longer dephasing time of InGaAs dots.

A more scalable QIP architecture would require that the excitons should be localized in separate dots. In the short term, however, the results in [19, 52] prove the principle that QD excitons can be used to demonstrate two-qubit gates, laying the foundations for further work.

10.5 Coherent Control of Spins

Single carriers (electrons or holes) confined within charged QDs may also be manipulated by ultrafast laser pulses. A number of different approaches are used to obtain charged dots. Typically, QDs embedded within diode structures are employed either to ionize a photoexcited exciton [46, 53–55] or to deterministically charge the QDs [56–61]. Alternatively, dopant layers may be added during sample growth; by tuning the doping concentration, it is possible to produce a mean QD carrier occupation of unity at equilibrium [62, 63]. Since resident carriers do not recombine, the T_2 time of their spins is no longer limited by the radiative lifetime. The coherence times can therefore be much longer than for excitons, which motivates their use in coherent control experiments.

10.5.1 Energy Level Structure of Charged Dots

The ground state of a charged dot consists of a dot containing a single electron or hole in their respective s-shells. Absorption of a photon adds an additional electron-hole pair to the system, leading to the formation of a charged exciton called a *trion*. The trion is split from the neutral exciton line by a binding energy typically $\sim$2 meV (see X^+ peak in Fig. 10.5). A single electron leads to a negatively charged trion (e.g. $|X^-\rangle = |\uparrow\Uparrow\downarrow\rangle\rangle$) whilst a single hole leads to a positively charged trion (e.g. $|X^+\rangle = |\Uparrow\uparrow\Downarrow\rangle\rangle$). The trion transitions for opposite spins have orthogonal circular polarizations, which can be exploited for spin-readout, as discussed in Sect. 10.5.5.

The application of a magnetic field (B) splits the electron and hole levels by the *Zeeman* effect into their two m_j states. The splitting is given by:

$$E_Z = g\mu_B B\,, \tag{10.36}$$

where g is the Landé g-factor and μ_B is the Bohr magneton. The hole is regarded as having a *pseudo-spin* of $\pm 1/2$, with the factor of three from the $m_j = \pm 3/2$ states included in its g-factor. The splitting of the ground and trion states of a positive dot in a *Faraday* geometry field (i.e. $\boldsymbol{B}$ parallel to the growth (z) axis) is shown in Fig. 10.9a. Note that the circular polarization selection rules decouple the states of opposite spin.

A *Voigt* geometry field (i.e. $\boldsymbol{B}$ oriented within the sample x-y plane) can mix the bright ($S_z = \pm 1$) and dark ($S_z = \pm 2$) exciton states.. The exciton eigenstates are no longer well defined and it is simpler to consider single carrier states, as shown for a positive dot in Fig. 10.9b. The hole and trion states are defined in terms of their orientation with respect to the magnetic field and are split by the hole and electron Zeeman energies respectively. The hole-trion transitions are linearly polarized, with a

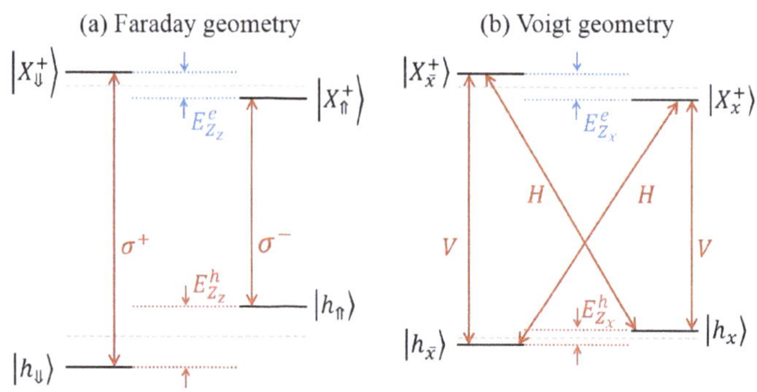

Fig. 10.9 Energy level spectrum of a positively charged dot in a magnetic field, with the Zeeman splittings exaggerated for clarity. **a** Faraday geometry with $\boldsymbol{B}$ directed along the growth (z) axis. **b** Voigt geometry with $\boldsymbol{B}$ directed along the x-axis. $x/\bar{x}$ and $\Downarrow/\Uparrow$ represents opposite hole spin states along the magnetic field. For negative dots, the charges and Zeeman splittings are reversed

pair of cross-polarized diagonal transitions coupling the hole states to the orthogonal trion state. The up and down hole spin states ($|h_{\Uparrow/\Downarrow}\rangle$) are superpositions of x-axis eigenstates, and spins initialised along z therefore precess about the in-plane field at the Larmor frequency $\omega_Z = E_Z/\hbar$. This implements a coherent rotation about the field axis on the Bloch sphere, and is widely used for coherent control of single carrier spins (see Sects. 10.5.3 and 10.5.4).

10.5.2 Spin Initialization

The DiVincenzo checklist for QIP [20] includes the requirement to begin with a well-defined qubit state. To achieve this, single carrier spins are generally *initialised* to either up or down. A widely used method is optical pumping, in which one of the trion transitions is continuously driven by a laser [59, 60]. Over time, the population becomes shelved in the undriven state, as any population that relaxes into the driven state is immediately re-pumped. These methods have reached fidelities as high as 99.8% in Faraday geometry [59], with ~μs initialization times that rely on weak spin-flip processes in the trion state. Faster (ns) initialization times have been observed in Voigt geometry on account of the allowed diagonal transitions (see Fig. 10.9b) to the opposite spin state [60]. The fidelity, however, is slightly lower. Coherent population trapping methods can also be employed [61, 64].

A fault-tolerant QIP implementation [65, 66] requires rapid initialization compared to decoherence, and this prompts research into faster schemes such as exciton ionization in a QD photodiode [53–55, 67]. In this method, an exciton with well defined spin is prepared in a neutral dot with circularly-polarised light. In reverse bias, the electric field causes fast electron tunnelling, leaving a hole behind with its spin determined by the polarization of the laser, as shown schematically in Fig. 10.10a. (Electron initialization is also possible if suitable tunnel barriers are included to reduce the electron tunnelling rate below that of the holes.) The strong circular selection rules for the trion transitions (see right of Fig. 10.10a) underpin methods to measure the fidelity of the initialisation mechanism. The dot is excited with a circularly-polarized π pulse tuned to the neutral exciton and probed by a co- or cross-polarized π pulse of variable frequency at a delay time longer than the electron tunnelling time, as shown in Fig. 10.10b. The negative signal at zero detuning identifies the X^0 transition, since the second π pulse moves the system back to the ground state, leading to a reduction in the photocurrent. The strongly polarization-sensitive signal at the X^+ line confirms the high fidelity of the ionization method.

Exciton ionization schemes can offer both picosecond initialization times and on-demand operation. Unfortunately, the anisotropic exchange interaction [24, 68] typically reduces fidelity by causing spin precession during the exciton lifetime [54, 69, 70]: see Fig. 10.10a and discussion of Fig. 10.7. Speeding up the ionisation process minimizes this effect [67, 71] but also significantly reduces the qubit lifetime due to faster tunnelling rates. The best solution is to select or tune QDs such that

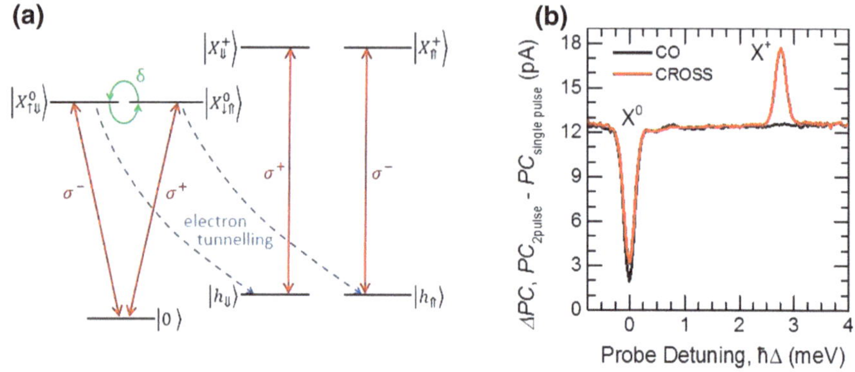

Fig. 10.10 **a** Initialisation of a single hole spin by exciton ionization. On the left are the exciton energy levels of a neutral dot in the circular basis as in Fig. 10.5. After electron tunnelling only a single hole remains. The spin of the hole can be measured by exciting with circularly polarized light to create a positively charged trion as illustrated on the right. **b** Two-color pump-probe photocurrent spectra of QDs exhibiting negligible (2.01 μeV) FSS. The probe is delayed by longer than the electron tunnelling time. *Black (red) line* corresponds to a co- (cross-) polarized probe laser. The figure is adapted from [46]

the exchange interaction becomes negligible [46, 72]. Figure 10.10b illustrates that a fidelity >99% can be obtained in this way for a dot with FSS close to zero. This approach has allowed the demonstration of fast, high-fidelity initialisation with long qubit lifetimes. Modulation of the sample electric field to suppress tunnelling after initialisation [73] has the potential to lead to further increases in lifetime.

10.5.3 Coherent Control of Single Electron Spins

Coherent control of a single electron spin is based on the electron-trion system discussed in Sect. 10.5.1. In Voigt geometry, the energy levels may be considered as a pair of independent Λ systems incorporating the two electron spin states and one of the trion states (see Fig. 10.11a). Under circular excitation, the probability amplitudes from the two systems add. By using a large detuning, unwanted population in excited states is minimised and the upper states may be adiabatically eliminated as shown in Fig. 10.11b. Hence a single broadband circularly-polarised pulse will produce coherent rotations via Stimulated Raman Adiabatic Passage (STIRAP). The rotation angle is given by:

$$\phi_z = \int \lambda_-(t)\,\mathrm{d}t\,, \tag{10.37}$$

where $\lambda_-(t)$ is the eigenenergy of the dressed states given by:

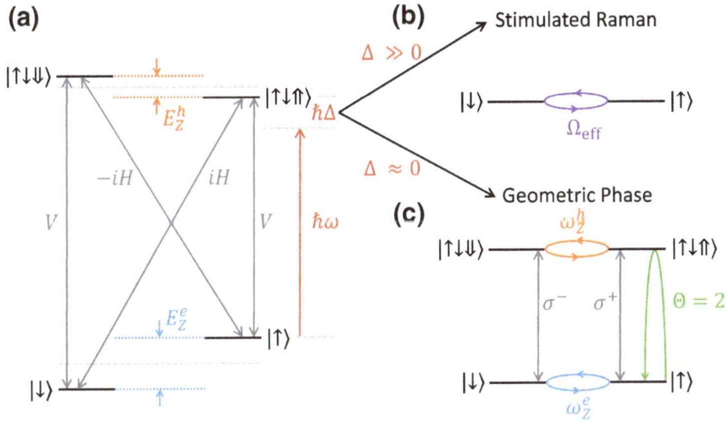

Fig. 10.11 Optical spin rotation methods for electrons in Voigt geometry. The *dot* is excited with circularly polarized laser pulses of energy $\hbar\omega$ and bandwidth broader than the electron/hole Zeeman energies $E_Z^{e/h}$. **a** Energy levels of the electron-trion system in the z basis. Note that this contrasts with Fig. 10.9b, which would correspond to the x basis. The system has four linearly-polarised transitions with the diagonal transitions carrying a $\pi/2$ phase factor. **b** For large laser detuning Δ, the trion levels may be adiabatically eliminated and the laser drives a stimulated Raman transition with Rabi frequency $\Omega_{\mathrm{eff}} \approx \frac{\Omega_H \Omega_V}{\Delta}$ between the two spin states where $\Omega_{\mathrm{H/V}}$ are the Rabi frequencies of the H/V transitions shown in **a**. Changing the laser power allows control of the Rabi rotation angle. **c** For $\Delta = 0$, a 2π laser pulse drives a spin state (selected by the choice of $\sigma^{+/-}$ polarisation) to the trion state and back. The driven spin state acquires a "geometric phase" of π. Control of the geometric phase is achieved by varying Δ

$$\lambda_{\pm} = \frac{1}{2}\sqrt{\Delta^2 \pm |\Omega_R|^2}\,. \tag{10.38}$$

The rotations are about the z axis on the Bloch sphere and may also be interpreted in terms of the AC Stark effect [75]. Full spin control can be achieved by combining with x axis rotations caused by Larmor precession about the $\boldsymbol{B}$ field (see Sect. 10.5.1) [74] optical rotations of electron spins have also been demonstrated by using the geometric phase approach (see Fig. 10.11c) [76, 77]. The principles of this method will be explained in Sect. 10.5.4 below in the context of hole spin control.

A two-qubit spin register with similar possibilities to the biexciton approach discussed in Sect. 10.4.4 may be realized with a QD molecule comprising two vertically stacked QDs [78]. The interaction between the two QDs is facilitated by coherent tunnelling and may be controlled by the applied electric field. Coherent control of an electron spin weakly coupled to an L3 photonic crystal cavity has also been demonstrated [79]. Using a scheme similar to that of [74], the spin rotation is performed by pulses that are detuned from both the trion transitions and the cavity mode. Theoretical work has proposed a more complex scheme whereby two transitions of the electron-trion system are coupled to two non-degenerate cavity modes [80]. This scheme has the potential both to increase initialisation fidelity and reduce the rotation time. Other experiments on electron spin control in nanocavities are discussed in Chap. 11.

10.5.4 Coherent Control of Single Hole Spins

Optical coherent control of a single hole spin may be achieved by similar methods to those for electrons. Schemes using the AC Stark shift for z-axis rotation [81, 82] are the direct equivalent of those for electrons discussed in Sect. 10.5.3, with x-axis rotations again implemented by Larmor precession. In addition, Greilich et al. successfully demonstrated control of a two hole-spin state for two dots interacting through tunnel coupling [81].

An alternative approach was employed by Godden et al. [83] in which a single hole spin was initialised by the exciton ionization method discussed in Sect. 10.5.2, and then z-axis rotations were implemented via the geometric phase previously established for electrons [76, 77]. Excitation by a laser pulse with bandwidth larger than the Zeeman splittings simplifies the hole-trion system to a pair of independent two-level systems that may be selected by orthogonal circular polarisations (see Fig. 10.11c for a schematic of the equivalent level scheme for electrons). A resonant σ^+ pulse of area Θ then drives a Rabi rotation between the spin-down hole and the corresponding trion state, so that an arbitrary initial hole state $|\Psi\rangle = h_{\Uparrow}|\Uparrow\rangle + h_{\Downarrow}|\Downarrow\rangle$ evolves as [84]:

$$|\Psi\rangle \rightarrow |\Psi'\rangle = h_{\Uparrow}|\Uparrow\rangle + h_{\Downarrow}\left[\cos\left(\frac{\Theta}{2}\right)|\Downarrow\rangle + \mathrm{i}\sin\left(\frac{\Theta}{2}\right)|\Downarrow\Uparrow\downarrow\rangle\right]. \qquad (10.39)$$

For $\Theta = 2\pi$, we obtain $|\Psi'\rangle = h_{\Uparrow}|\Uparrow\rangle - h_{\Downarrow}|\Downarrow\rangle$ when trion dephasing is negligible, which is equivalent to a π rotation about z. Arbitrary rotation angles ϕ_z about $\hat{z}$ are implemented by detuning the laser [83, 84]:

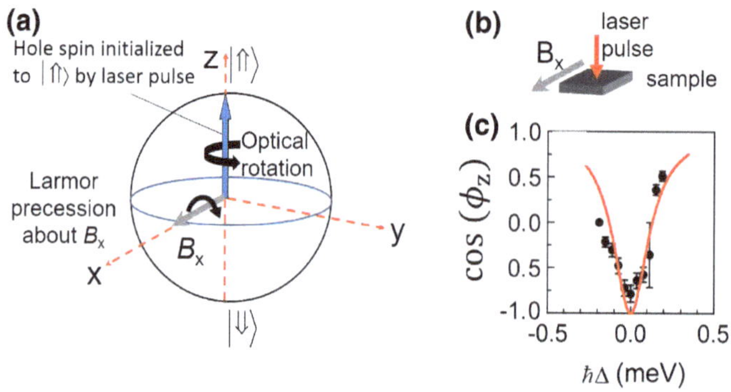

Fig. 10.12 Coherent control of a single hole spin in a Voigt geometry field B_x. **a** Schematic of the method. The hole spin is initialised to $|\Uparrow\rangle$ by the exciton ionization method. Full Bloch sphere control is achieved by combining z axis rotations induced by the geometric phase shift from a detuned 2π pulse as in Fig. 10.11c, with x axis rotations by Larmor precession about the field. **b** Experimental geometry. **c** Optical rotation angle ϕ_z versus laser detuning Δ. The red line is a fit according to (10.40). The figure is adapted from [83]

$$\phi_z = 2\arctan(\beta_z/\Delta)\,, \tag{10.40}$$

where β_z is the pulse bandwidth and Δ the detuning from the hole-trion transition. Figure 10.12 shows a schematic of the method and experimental geometry, together with a fit to the experimental results showing excellent agreement with (10.40). Theoretical proposals indicate that rotations about an arbitrary axis are possible, thereby eliminating the need for Larmor precession, and potentially reducing the gate time [84].

10.5.5 Spin Readout

In the experiments described above, measurement of the spin is performed by averaging over many experimental cycles. For QIP it is desirable to perform "single-shot" readout where the state of the qubit is determined faster than the back-action of the measurement. In Faraday geometry, the trion transitions are spin-selective and have a very weak coupling to the orthogonal spin state. Single-shot spin readout may then be accomplished by driving one transition continuously and collecting the spin-sensitive resonance fluorescence. This approach has been realised experimentally with a fidelity of 82.3% for a measurement time of 800 ns [85].

A drawback to this approach is the Faraday geometry, as coherent spin control requires Voigt geometry. A recent theoretical proposal envisions spin initialisation and manipulation in Voigt geometry, before applying a detuned laser to AC Stark-shift the energy levels into a pseudo-Faraday configuration to perform single-shot readout [86]. Significant potential exists to increase the readout performance by exploiting photonic nanostructures to enhance both the collection efficiency and the emission rate.

10.6 Dephasing: Comparison of Qubits

A critical parameter for any qubit is its coherence time, since this determines how many gate operations can be performed. Table 10.1 compares measurements of T_1, T_2^* and T_2 for exciton and single carrier spin qubits. The basic concepts of these three time constants were outlined in Sect. 10.2.3.

Single exciton qubits have been observed to have lifetime-limited coherence [87, 88], i.e. $T_2 = 2T_1$, where T_1 is typically around 1 ns owing to radiative recombination [88]. Single spins cannot undergo radiative recombination and therefore generally have far longer T_1 times. For InGaAs QDs, electron spin dephasing times (T_2^*) of the order of 1–2 ns are typically measured [89–91] with values of around 10 ns measured for holes [61, 81–83]. The primary source of electron spin dephasing is postulated to be hyperfine interactions with the nuclear spins within the dot [9, 90]. The longer T_2^* time for holes is related to its $\sim 10\times$ smaller hyperfine constant due to

Table 10.1 Table of dephasing timescales for different single qubit implementations in InGaAs self-assembled QDs. The coherence of neutral excitons has been observed to be lifetime-limited and thus the limit $T_2 = 2T_1$ is taken here. In the case of single electrons and holes, the T_2 values reported here are measured by spin echo methods to suppress pure dephasing (T_2^*)

Qubit	T_1	T_2^*	T_2	References
Neutral exciton	~1 ns	n/a	~2 ns	[10, 87, 88]
Electron spin	~20 ms	~2 ns	~2 μs	[53, 89–91]
Hole spin	~270 μs	~10 ns	1.1 μs	[61, 81–83, 92]

the primarily p-type orbital structure [93]. The dominant source of hole dephasing is instead attributed to charge noise [81, 82, 94], most probably originating from charges trapped at the capping layer interface [95, 96]. Further discussion of the exciton and spin dephasing mechanisms may be found in Chap. 9.

The use of spin echo techniques [97] can suppress pure dephasing, allowing longer coherence times (T_2) to be measured. For both electrons [90, 91] and holes [82], this has resulted in T_2 times of the order of μs. It is worth noting that these values are still orders of magnitude below the lifetime (T_1) limit. Recent studies [98] have shown that strain in the sample wafer acts to reduce fluctuations of the nuclear spin bath, potentially offering a route to increasing both T_2 and T_2^* for electron spins. Meanwhile, improvements in sample quality should lead to reduced charge fluctuations for hole spins (see Chap. 9). In addition, a recent theoretical proposal envisions using the AC Stark shift to oppose changes in the Zeeman energy, suppressing charge noise dephasing for holes [99].

10.7 Outlook

The experiments described in Sects. 10.4 and 10.5 show that ultrafast coherent control techniques for single excitons and spins in QDs are now well established. The primary need for future QIP applications is to increase the coherence time. Significant progress has already been made here, with studies clarifying the origins of the dephasing and proposing strategies to reduce it (see Sect. 10.6).

The development of quantum processors based on spin networks will require scaling to multiple qubits. The approaches based on biexcitons (Sect. 10.4.4) or vertically stacked QDs do not scale easily beyond two qubits, and it is therefore necessary to consider spins confined in separate QDs, as discussed in Chap. 12. Entanglement has been observed between QD spins and emitted photons [100–102], illustrating conversion between stationary and flying qubits. The complementary process of transferring a quantum state from a photonic qubit to a QD spin qubit was also demonstrated [103]. Combining these two concepts has recently led to the demonstration of entanglement between two hole spins separated by 5 m [104], a critical development for any spin-based QIP architecture.

The methods described in this chapter are also applicable to other QIP applications. For example, current state-of-the-art single [105, 106] and entangled [107] photon sources are driven by resonant π-pulses acting on the neutral exciton or biexciton. Moving beyond this simple case, coherent control of single spins has been proposed as a means to generate more complex photonic states for QIP such as cluster states [108], and these have recently been observed using dark excitons as the qubits [109] (see Chap. 4). Coherent control methods are also of significant interest for controlling light-matter interactions in cavity-QD systems, as discussed in Chap. 11. Examples include polarization shifts of light that are conditional on the state of a spin [110, 111] and controlling the coherent energy transfer between an emitter and a cavity [112].

Acknowledgments The work reported in this chapter was supported by the Engineering and Physical Sciences Research Council (EPSRC). We are grateful to E.A. Chekhovich and M.N. Makhonin for the data in Fig. 10.5c, and to J.H. Quilter for Fig. 10.10b. We would like to thank our colleagues and collaborators who have contributed to this work, especially: V.M. Axt, A.M. Barth, F. Bello, S.J. Boyle, P. Brereton, R.J. Coles, P.W. Fry, E.M. Gauger, M. Glässl, T.M. Godden, A.V. Gopal, A.P. Heberle, M. Hopkinson, A.F.A. Khatab, R.S. Kolodka, H.Y. Liu, B.W. Lovett, I.J. Luxmoore, L.M.P.P. Martins, A. Nazir, W.K. Ng, J.B.B. Oliveira, J. Puebla-Nunez, J.H. Quilter, A.J. Ramsay, J. Skiba-Szymanska, M.S. Skolnick, A. Tahraoui, D.M. Whittaker, and Y. Wu.

References

1. R.P. Feynman, F.L. Vernon, R.W. Hellwarth, J. Appl. Phys. **28**, 49 (1957)
2. F.T. Arecchi, R. Bonifacio, IEEE J. Quantum Electron. **1**, 169 (1965)
3. R. Loudon, *The Quantum Theory of Light*, 3rd edn. (Oxford University Press, Oxford, 2000)
4. B.W. Shore, *Manipulating Quantum Structures Using Laser Pulses* (Cambridge University Press, Cambridge, 2011)
5. J.Y. Marzin, J.M. Gérard, A. Izraël, D. Barrier, G. Bastard, Phys. Rev. Lett. **73**, 716 (1994)
6. A.P. Heberle, J.J. Baumberg, Phys. Rev. Lett. **75**, 2598 (1995)
7. O. Svelto, S.D. Silvestri, G. Denardo, *Ultrafast Processes in Spectroscopy* (Springer, Berlin, 1996)
8. J. Shah, *Ultrafast Spectroscopy of Semiconductors and Semiconductor Nanostructures*, 2nd edn. (Springer, Berlin, 1999)
9. A.V. Kuhlmann, J.H. Prechtel, J. Houel, A. Ludwig, D. Reuter, A.D. Wieck, R.J. Warburton, Nat. Commun. **6**, 8204 (2015)
10. P. Borri, W. Langbein, S. Schneider, U. Woggon, R.L. Sellin, D. Ouyang, D. Bimberg, Phys. Rev. Lett. **87**, 157401 (2001)
11. R. Schirhagl, K. Chang, M. Loretz, C.L. Degen, Ann. Rev. Phys. Chem. **65**, 83 (2014)
12. I. Favero, G. Cassabois, R. Ferreira, D. Darson, C. Voisin, J. Tignon, C. Delalande, G. Bastard, P. Roussignol, Phys. Rev. B **68**, 233301 (2003)
13. S.L. Portalupi, G. Hornecker, V. Giesz, T. Grange, A. Lemaître, J. Demory, I. Sagnes, N.D. Lanzillotti-Kimura, L. Lanco, A. Auffèves, P. Senellart, Nano Lett. **15**, 6290 (2015)
14. P. Lodahl, S. Mahmoodian, S. Stobbe, Rev. Mod. Phys. **87**, 347 (2015)
15. P.G. Eliseev, H. Li, A. Stintz, G.T. Liu, T.C. Newell, K.J. Malloy, L.F. Lester, Appl. Phys. Lett. **77**, 262 (2000)
16. D. Gammon, E. Snow, B. Shanabrook, D. Katzer, D. Park, Science **273**, 87 (1996)
17. N.H. Bonadeo, J. Erland, D. Gammon, D. Park, D.S. Katzer, D.G. Steel, Science **282**, 1473 (1998)

18. T.H. Stievater, X. Li, D.G. Steel, D. Gammon, D.S. Katzer, D. Park, C. Piermarocchi, L.J. Sham, Phys. Rev. Lett. **87**, 133603 (2001)
19. X. Li, Y. Wu, D. Steel, D. Gammon, T.H. Stievater, D.S. Katzer, D. Park, C. Piermarocchi, L.J. Sham, Science **301**, 809 (2003)
20. D.P. DiVincenzo, in *Mesoscopic Electron Transport*, ed. by L. Sohn, L. Kowenhoven, L. Schön (Kluwer Ac. Publ., Dordrecht, 1997)
21. A.N. Vamivakas, Y. Zhao, C.Y. Lu, M. Atatüre, Nat. Phys. **5**, 198 (2009)
22. X. Xu, B. Sun, P.R. Berman, D.G. Steel, A.S. Bracker, D. Gammon, L.J. Sham, Science **317**, 929 (2007)
23. B.D. Gerardot, D. Brunner, P.A. Dalgarno, K. Karrai, A. Badolato, P.M. Petroff, R.J. Warburton, New J. Phys. **11**, 013028 (2009)
24. D. Gammon, E.S. Snow, B.V. Shanabrook, D.S. Katzer, D. Park, Phys. Rev. Lett. **76**, 3005 (1996)
25. A.I. Tartakovskii, J. Cahill, M.N. Makhonin, D.M. Whittaker, J.P.R. Wells, A.M. Fox, D.J. Mowbray, M.S. Skolnick, K.M. Groom, M.J. Steer, M. Hopkinson, Phys. Rev. Lett. **93**, 057401 (2004)
26. R. Seguin, A. Schliwa, S. Rodt, K. Pötschke, U.W. Pohl, D. Bimberg, Phys. Rev. Lett. **95**, 257402 (2005)
27. R.M. Stevenson, R.J. Young, P. Atkinson, K. Cooper, D.A. Ritchie, A.J. Shields, Nature **439**, 179 (2006)
28. N. Akopian, N.H. Lindner, E. Poem, Y. Berlatzky, J. Avron, D. Gershoni, B.D. Gerardot, P.M. Petroff, Phys. Rev. Lett. **96**, 130501 (2006)
29. H. Kamada, H. Gotoh, J. Temmyo, T. Takagahara, H. Ando, Phys. Rev. Lett. **87**, 246401 (2001)
30. H. Htoon, T. Takagahara, D. Kulik, O. Baklenov, A.L. Holmes, C.K. Shih, Phys. Rev. Lett. **88**, 087401 (2002)
31. P. Borri, W. Langbein, S. Schneider, U. Woggon, R.L. Sellin, D. Ouyang, D. Bimberg, Phys. Rev. B **66**, 081306 (2002)
32. E.B. Flagg, A. Muller, J.W. Robertson, S. Founta, D.G. Deppe, M. Xiao, W. Ma, G.J. Salamo, C.K. Shih, Nat. Phys. **5**, 203 (2009)
33. S.J. Boyle, A.J. Ramsay, A.M. Fox, M.S. Skolnick, A.P. Heberle, M. Hopkinson, Phys. Rev. Lett. **102**, 207401 (2009)
34. S. Stufler, P. Machnikowski, P. Ester, M. Bichler, V.M. Axt, T. Kuhn, A. Zrenner, Phys. Rev. B **73**, 125304 (2006)
35. A. Zrenner, E. Beham, S. Stufler, F. Findeis, M. Bichler, G. Abstreiter, Nature **418**, 612 (2002)
36. A.J. Ramsay, A.V. Gopal, E.M. Gauger, A. Nazir, B.W. Lovett, A.M. Fox, M.S. Skolnick, Phys. Rev. Lett. **104**, 017402 (2010)
37. A.J. Ramsay, T.M. Godden, S.J. Boyle, E.M. Gauger, A. Nazir, B.W. Lovett, A.M. Fox, M.S. Skolnick, Phys. Rev. Lett. **105**, 177402 (2010)
38. J.H. Quilter, A.J. Brash, F. Liu, M. Glässl, A.M. Barth, V.M. Axt, A.J. Ramsay, M.S. Skolnick, A.M. Fox, Phys. Rev. Lett. **114**, 137401 (2015)
39. A. Vagov, M.D. Croitoru, V.M. Axt, T. Kuhn, F.M. Peeters, Phys. Rev. Lett. **98**, 227403 (2007)
40. S. Stufler, P. Ester, A. Zrenner, M. Bichler, Phys. Rev. Lett. **96**, 037402 (2006)
41. R.S. Kolodka, A.J. Ramsay, J. Skiba-Szymanska, P.W. Fry, H.Y. Liu, A.M. Fox, M.S. Skolnick, Phys. Rev. B **75**, 193306 (2007)
42. Y. Benny, S. Khatsevich, Y. Kodriano, E. Poem, R. Presman, D. Galushko, P.M. Petroff, D. Gershoni, Phys. Rev. Lett. **106**, 040504 (2011)
43. Y. Kodriano, I. Schwartz, E. Poem, Y. Benny, R. Presman, T.A. Truong, P.M. Petroff, D. Gershoni, Phys. Rev. B **85**, 241304 (2012)
44. K. Müller, T. Kaldewey, R. Ripszam, J.S. Wildmann, A. Bechtold, M. Bichler, G. Koblmüller, G. Abstreiter, J.J. Finley, Sci. Rep. **3**, 1906 (2013)
45. Y. Kodriano, E.R. Schmidgall, Y. Benny, D. Gershoni, Semicond. Sci. Technol. **29**, 053001 (2014)

46. A.J. Brash, L.M.P.P. Martins, F. Liu, J.H. Quilter, A.J. Ramsay, M.S. Skolnick, A.M. Fox, Phys. Rev. B **92**, 121301(R) (2015)
47. Y. Wu, I.M. Piper, M. Ediger, P. Brereton, E.R. Schmidgall, P.R. Eastham, M. Hugues, M. Hopkinson, R.T. Phillips, Phys. Rev. Lett. **106**, 067401 (2011)
48. C.M. Simon, T. Belhadj, B. Chatel, T. Amand, P. Renucci, A. Lemaitre, O. Krebs, P.A. Dalgarno, R.J. Warburton, X. Marie, B. Urbaszek, Phys. Rev. Lett. **106**, 166801 (2011)
49. M. Glässl, A.M. Barth, V.M. Axt, Phys. Rev. Lett. **110**, 147401 (2013)
50. P.L. Ardelt, L. Hanschke, K.A. Fischer, K. Müller, A. Kleinkauf, M. Koller, A. Bechtold, T. Simmet, J. Wierzbowski, H. Riedl, G. Abstreiter, J.J. Finley, Phys. Rev. B **90**, 241404(R) (2014)
51. S. Bounouar, M. Müller, A.M. Barth, M. Glässl, V.M. Axt, P. Michler, Phys. Rev. B **91**, 161302 (2015)
52. S.J. Boyle, A.J. Ramsay, F. Bello, H.Y. Liu, M. Hopkinson, A.M. Fox, M.S. Skolnick, Phys. Rev. B **78**, 075301 (2008)
53. M. Kroutvar, Y. Ducommun, D. Heiss, M. Bichler, D. Schuh, G. Abstreiter, J.J. Finley, Nature **432**, 81 (2004)
54. A.J. Ramsay, S.J. Boyle, R.S. Kolodka, J.B.B. Oliveira, J. Skiba-Szymanska, H.Y. Liu, M. Hopkinson, A.M. Fox, M.S. Skolnick, Phys. Rev. Lett. **100**, 197401 (2008)
55. D. Heiss, V. Jovanov, M. Bichler, G. Abstreiter, J.J. Finley, Phys. Rev. B **77**, 235442 (2008)
56. R.J. Warburton, C. Schaflein, D. Haft, F. Bickel, A. Lorke, K. Karrai, J.M. Garcia, W. Schoenfeld, P.M. Petroff, Nature **405**, 926 (2000)
57. M. Baier, F. Findeis, A. Zrenner, M. Bichler, G. Abstreiter, Phys. Rev. B **64**, 195326 (2001)
58. D.V. Regelman, E. Dekel, D. Gershoni, E. Ehrenfreund, A.J. Williamson, J. Shumway, A. Zunger, W.V. Schoenfeld, P.M. Petroff, Phys. Rev. B **64**, 165301 (2001)
59. M. Atatüre, J. Dreiser, A. Badolato, A. Högele, K. Karrai, A. Imamoglu, Science **312**, 551 (2006)
60. X. Xu, Y. Wu, B. Sun, Q. Huang, J. Cheng, D.G. Steel, A.S. Bracker, D. Gammon, C. Emary, L.J. Sham, Phys. Rev. Lett. **99**, 097401 (2007)
61. D. Brunner, B.D. Gerardot, P.A. Dalgarno, G. Wüst, K. Karrai, N.G. Stoltz, P.M. Petroff, R.J. Warburton, Science **325**, 70 (2009)
62. S. Cortez, O. Krebs, S. Laurent, M. Senes, X. Marie, P. Voisin, R. Ferreira, G. Bastard, J.M. Gérard, T. Amand, Phys. Rev. Lett. **89**, 207401 (2002)
63. A.B. Henriques, A. Schwan, S. Varwig, A.D.B. Maia, A.A. Quivy, D.R. Yakovlev, M. Bayer, Phys. Rev. B **86**, 115333 (2012)
64. X. Xu, B. Sun, P.R. Berman, D.G. Steel, A.S. Bracker, D. Gammon, L.J. Sham, Nat. Phys. **4**, 692 (2008)
65. D.P. DiVincenzo, Fortschritte der Physik **48**, 771 (2000)
66. J. Preskill, Proc. R. Soc. A **454**, 385 (1998)
67. K. Müller, A. Bechtold, C. Ruppert, C. Hautmann, J.S. Wildmann, T. Kaldewey, M. Bichler, H.J. Krenner, G. Abstreiter, M. Betz, J.J. Finley, Phys. Rev. B **85**, 241306 (2012)
68. M. Bayer, G. Ortner, O. Stern, A. Kuther, A.A. Gorbunov, A. Forchel, P. Hawrylak, S. Fafard, K. Hinzer, T.L. Reinecke, S.N. Walck, J.P. Reithmaier, F. Klopf, F. Schäfer, Phys. Rev. B **65**, 195315 (2002)
69. T.M. Godden, S.J. Boyle, A.J. Ramsay, A.M. Fox, M.S. Skolnick, Appl. Phys. Lett. **97**, 061113 (2010)
70. T.M. Godden, J.H. Quilter, A.J. Ramsay, Y. Wu, P. Brereton, I.J. Luxmoore, J. Puebla, A.M. Fox, M.S. Skolnick, Phys. Rev. B **85**, 155310 (2012)
71. J.D. Mar, J.J. Baumberg, X. Xu, A.C. Irvine, D.A. Williams, Phys. Rev. B **90**, 241303 (2014)
72. J.D. Mar, J.J. Baumberg, X.L. Xu, A.C. Irvine, D.A. Williams, Phys. Rev. B **93**, 045316 (2016)
73. J.H. Quilter, R.J. Coles, A.J. Ramsay, A.M. Fox, M.S. Skolnick, Appl. Phys. Lett. **102**, 181108 (2013)
74. D. Press, T.D. Ladd, B. Zhang, Y. Yamamoto, Nature **456**, 218 (2008)

75. J. Berezovsky, M. Mikkelsen, N.G. Stoltz, L. Coldren, D. Awschalom, Science **320**, 349 (2008)
76. A. Greilich, S.E. Economou, S. Spatzek, D.R. Yakovlev, D. Reuter, A.D. Wieck, T.L. Reinecke, M. Bayer, Nat. Phys. **5**, 262 (2009)
77. E.D. Kim, K. Truex, X. Xu, B. Sun, D.G. Steel, A.S. Bracker, D. Gammon, L.J. Sham, Phys. Rev. Lett. **104**, 167401 (2010)
78. D. Kim, S.G. Carter, A. Greilich, A.S. Bracker, D. Gammon, Nat. Phys. **7**, 223 (2010)
79. S.G. Carter, T.M. Sweeney, M. Kim, C.S. Kim, D. Solenov, S.E. Economou, T.L. Reinecke, L. Yang, A.S. Bracker, D. Gammon, Nat. Photon. **7**, 329 (2013)
80. A. Majumdar, P. Kaer, M. Bajcsy, E.D. Kim, K.G. Lagoudakis, A. Rundquist, J. Vučković, Phys. Rev. Lett. **111**, 027402 (2013)
81. A. Greilich, S.G. Carter, D. Kim, A.S. Bracker, D. Gammon, Nat. Photon. **5**, 702 (2011)
82. K. De Greve, P.L. McMahon, D. Press, T.D. Ladd, D. Bisping, C. Schneider, M. Kamp, L. Worschech, S. Hofling, A. Forchel, Y. Yamamoto, Nat. Phys. **7**, 872 (2011)
83. T.M. Godden, J.H. Quilter, A.J. Ramsay, Y. Wu, P. Brereton, S.J. Boyle, I.J. Luxmoore, J. Puebla-Nunez, A.M. Fox, M.S. Skolnick, Phys. Rev. Lett. **108**, 017402 (2012)
84. S.E. Economou, T.L. Reinecke, Phys. Rev. Lett. **99**, 217401 (2007)
85. A. Delteil, W.b. Gao, P. Fallahi, J. Miguel-Sanchez, A. Imamoğlu, Phys. Rev. Lett. **112**, 116802 (2014)
86. E.B. Flagg, G.S. Solomon, Phys. Rev. B **92**, 245309 (2015)
87. M. Bayer, A. Forchel, Phys. Rev. B **65**, 041308 (2002)
88. W. Langbein, P. Borri, U. Woggon, V. Stavarache, D. Reuter, A.D. Wieck, Phys. Rev. B **70**, 033301 (2004)
89. A. Greilich, D.R. Yakovlev, A. Shabaev, A.L. Efros, I.A. Yugova, R. Oulton, V. Stavarache, D. Reuter, A. Wieck, M. Bayer, Science **313**, 341 (2006)
90. D. Press, K. De Greve, P.L. McMahon, T.D. Ladd, B. Friess, C. Schneider, M. Kamp, S. Hofling, A. Forchel, Y. Yamamoto, Nat. Photon. **4**, 367 (2010)
91. A. Bechtold, D. Rauch, F. Li, T. Simmet, P.L. Ardelt, A. Regler, K. Muller, N.A. Sinitsyn, J.J. Finley, Nat. Phys. **11**, 1005 (2015)
92. D. Heiss, S. Schaeck, H. Huebl, M. Bichler, G. Abstreiter, J.J. Finley, D.V. Bulaev, D. Loss, Phys. Rev. B **76**, 241306 (2007)
93. E.A. Chekhovich, M.M. Glazov, A.B. Krysa, M. Hopkinson, P. Senellart, A. Lemaitre, M.S. Skolnick, A.I. Tartakovskii, Nat. Phys. **9**, 74 (2013)
94. J. Houel, J.H. Prechtel, A.V. Kuhlmann, D. Brunner, C.E. Kuklewicz, B.D. Gerardot, N.G. Stoltz, P.M. Petroff, R.J. Warburton, Phys. Rev. Lett. **112**, 107401 (2014)
95. J. Houel, A.V. Kuhlmann, L. Greuter, F. Xue, M. Poggio, B.D. Gerardot, P.A. Dalgarno, A. Badolato, P.M. Petroff, A. Ludwig, D. Reuter, A.D. Wieck, R.J. Warburton, Phys. Rev. Lett. **108**, 107401 (2012)
96. A.V. Kuhlmann, J. Houel, A. Ludwig, L. Greuter, D. Reuter, A.D. Wieck, M. Poggio, R.J. Warburton, Nat. Phys. **9**, 570 (2013)
97. E.L. Hahn, Phys. Rev. **80**, 580 (1950)
98. A.M. Waeber, M. Hopkinson, I. Farrer, D.A. Ritchie, J. Nilsson, R.M. Stevenson, A.J. Bennett, A.J. Shields, G. Burkard, A.I. Tartakovskii, M.S. Skolnick, E.A. Chekhovich, Nat. Phys. **12**, 688 (2016)
99. A.J. Ramsay, Phys. Rev. B **93**, 075303 (2016)
100. W.B. Gao, P. Fallahi, E. Togan, J. Miguel-Sanchez, A. Imamoglu, Nature **491**, 426 (2012)
101. K. De Greve, L. Yu, P.L. McMahon, J.S. Pelc, C.M. Natarajan, N.Y. Kim, E. Abe, S. Maier, C. Schneider, M. Kamp, S. Hofling, R.H. Hadfield, A. Forchel, M.M. Fejer, Y. Yamamoto, Nature **491**, 421 (2012)
102. J.R. Schaibley, A.P. Burgers, G.A. McCracken, L.M. Duan, P.R. Berman, D.G. Steel, A.S. Bracker, D. Gammon, L.J. Sham, Phys. Rev. Lett. **110**, 167401 (2013)
103. W. Gao, P. Fallahi, E. Togan, A. Delteil, Y. Chin, J. Miguel-Sanchez, A. Imamoğlu, Nat. Commun. **4**, 539 (2013)
104. A. Delteil, Z. Sun, W.B. Gao, E. Togan, S. Faelt, A. Imamoğlu, Nat. Phys. **12**, 218 (2016)

105. X. Ding, Y. He, Z.C. Duan, N. Gregersen, M.C. Chen, S. Unsleber, S. Maier, C. Schneider, M. Kamp, S. Höfling, C.Y. Lu, J.W. Pan, Phys. Rev. Lett. **116**, 020401 (2016)
106. N. Somaschi, V. Giesz, L. De Santis, J.C. Loredo, M.P. Almeida, G. Hornecker, L. Portalupi, T. Grange, C. Antón, J. Demory, C. Gómez, I. Sagnes, N.D. Lanzillotti-Kimura, A. Lemaítre, A. Auffeves, A.G. White, L. Lanco, P. Senellart, Nat. Photon. **10**, 340 (2016)
107. M. Müller, S. Bounouar, K.D. Jons, M. Glassl, P. Michler, Nat. Photon. **8**, 224 (2014)
108. N.H. Lindner, T. Rudolph, Phys. Rev. Lett. **103**, 113602 (2009)
109. I. Schwartz, D. Cogan, E.R. Schmidgall, Y. Don, L. Gantz, O. Kenneth, N.H. Lindner, D. Gershoni, Science **354**, 434 (2016)
110. H. Kim, R. Bose, T.C. Shen, G.S. Solomon, E. Waks, Nat. Photon. **7**, 373 (2013)
111. S. Sun, H. Kim, G.S. Solomon, E. Waks, Nat. Nanotechnol. **11**, 539 (2016)
112. R. Bose, T. Cai, K.R. Choudhury, G.S. Solomon, E. Waks, Nat. Photon. **8**, 858 (2014)

Chapter 11
Interfacing Single Quantum Dot Spins with Photons Using a Nanophotonic Cavity

Shuo Sun and Edo Waks

Abstract The spin of a single electron or hole trapped inside a quantum dot offers a promising quantum memory. These qubits are embedded in a host semiconductor material that can be directly patterned to form compact integrated nanophotonic devices. These devices efficiently interconnect single solid-state qubits with photons, a crucial requirement for quantum networks, quantum repeaters, and photonic quantum computation. This chapter reviews recent experimental progress towards achieving strong spin-photon interactions based on coupled quantum dot and nanophotonic cavity system. Especially we introduce a recent work that reports a coherent spin-photon quantum switch operating at the fundamental quantum limit, where a single photon flips the orientation of a quantum dot spin and the spin flips the polarization of the photon. These strong spin-photon interactions open up a promising direction for solid-state implementations of high-speed quantum networks and on-chip quantum photonic circuits using nanophotonic devices.

11.1 Introduction

Interactions between single spins and photons play a central role in the field of quantum information processing. Spin is a pristine quantum memory while photons are ideal carriers of quantum information. Efficient interfaces between these systems are essential for development of future quantum networks [1, 2] and distributed quantum computers [3]. They also enable critical functionalities such as entanglement distribution [4, 5], non-destructive qubit measurements [6–8], and strong photon-photon interactions for photonic quantum computation [9, 10].

S. Sun (✉) · E. Waks
Department of Electrical and Computer Engineering, Institute for Research in Electronics and Applied Physics (IREAP), and Joint Quantum Institute (JQI), University of Maryland, College Park, MD 20742, USA
e-mail: shuosun@umd.edu

E. Waks
e-mail: edowaks@umd.edu

P. Michler (ed.), *Quantum Dots for Quantum Information Technologies*, Nano-Optics and Nanophotonics, DOI 10.1007/978-3-319-56378-7_11

The spin of a singly charged quantum dot has attracted significant interests for implementing a spin-photon interface. This trapped spin system provides a promising quantum memory with microsecond coherence time [11, 12] and picosecond timescale single-qubit gates [12–17], enabling a large number of quantum operations prior to qubit decoherence. Furthermore, the spin ground states of the charged quantum dot are optically coupled to excited trion states that exhibit nearly radiatively limited emission [18]. These properties have enabled post-selected spin-photon entanglement [19–22], spin-photon teleportation [23], and spin-spin entanglement [24], which are essential capabilities for quantum networks.

Quantum dots are also embedded in a high dielectric solid-state substrate that can be directly patterned to form nanophotonic cavities that enhance light-matter interactions [25–28]. These devices can be integrated on-a-chip to attain a compact architecture for quantum circuits [29, 30]. Tremendous experimental progress has been made in the last decade using a quantum dot strongly coupled to a nanophotonic cavity, including cavity reflectivity control [31], ultrafast optical switch [32–34], single photon level nonlinearities [35–37], non-classical light generation [38, 39], and spin-exciton quantum logic operations [40]. The effort to integrate quantum dot spins with cavities has also experienced rapid progress. Several works demonstrated deterministic loading of a spin in a quantum dot coupled to a nanophotonic cavity [41–43], and more recently coherent control of the loaded spin [44] and spin-dependent Kerr rotation of photons [45, 46]. Very recently, a coherent quantum switch between a quantum dot spin and a photon has also been demonstrated [47].

In this chapter, we review recent experimental progress towards achieving strong spin-photon interactions based on coupled quantum dot and nanophotonic cavity system. This chapter is organized as follows. Section 11.2 provides a theoretical background for interfacing single spins and photons based on a cavity quantum electrodynamics (QED) system. Section 11.3 reviews the experimental efforts to integrate quantum dot spins with different nanophotonic cavities, including micropillar cavities and photonic crystal cavities. In Sect. 11.4, we focus on a recent experimental work that demonstrates a coherent spin-photon quantum switch, where through the mediation of a strongly coupled photonic crystal cavity, a single photon flips the orientation of a quantum dot spin and the spin flips the polarization of the photon. Section 11.5 concludes our discussion and provides outlook for future works.

11.2 Theoretical Background

In this section, we provide theoretical background on using a cavity QED system to interconnect an optical active qubit with photons. We focus our analysis on a generic system consisted of an optical active qubit coupled to an optical cavity, as shown in Fig. 11.1a. We assume that the qubit system has a λ-type energy structure as shown in Fig. 11.1b, with two ground states that form a stable spin qubit, denoted as $|\uparrow\rangle$ and $|\downarrow\rangle$, and one excited state $|e\rangle$ that gives rise to spin-dependent optical transitions

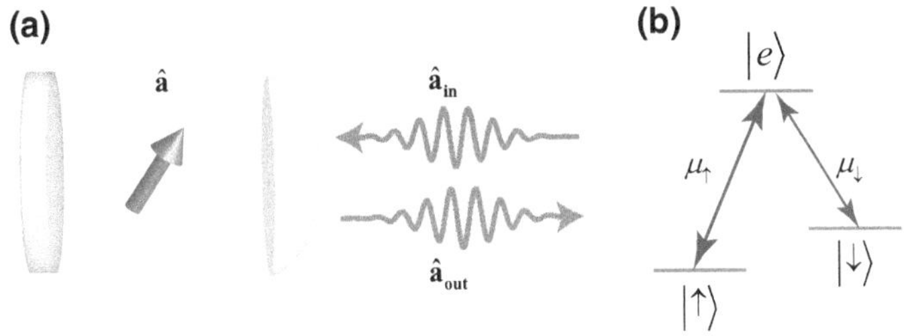

Fig. 11.1 Theoretical model for a generic cavity QED system composed of an optical active qubit coupled to an optical cavity. **a** A schematic cavity QED system. **b** Energy level structure of a generic optical active qubit system

$\mu_\uparrow$ and $\mu_\downarrow$ respectively. The spin-dependent optical transitions provide a mechanism to interconnect the spin qubit with photons. This energy structure exists in many qubit systems that are optically addressable, such as cold atoms, trapped ions, color centers, and charged quantum dots, and therefore represents a generic model.

In order to induce strong spin-photon interactions, we selectively couple the optical transition $\mu_\uparrow$ to a cavity mode, while decouple the other transition $\mu_\downarrow$ to the cavity, either by a large detuning or by selection rules if transition $\mu_\downarrow$ emits a photon with a different polarization than the cavity mode. In this configuration the coupling between the atom and the cavity depends on the spin state. The cavity thus exhibits spin-dependent reflection or transmission coefficients, enabling control of a reflected or transmitted photon by the spin qubit. In our model, we assume that the cavity field only couples to its reflective mode without loss of generality. In this case the spin only modulates the reflection coefficient of the cavity. Double-sided cavities would work similarly with minor modifications.

11.2.1 Calculation of Spin-Dependent Cavity Reflection Coefficients

We calculate the cavity reflection coefficients using cavity input-output formalism [48]. We define $\hat{\mathbf{a}}$ as the bosonic annihilation operator for the cavity field, and $\hat{\mathbf{a}}_{\mathbf{in}}$ and $\hat{\mathbf{a}}_{\mathbf{out}}$ as the operators for the cavity coupled incidence and reflection modes. These operators are related by the cavity input-output relation

$$\hat{\mathbf{a}}_{\mathbf{out}} = \hat{\mathbf{a}}_{\mathbf{in}} - \sqrt{\kappa_{ex}}\hat{\mathbf{a}}, \tag{11.1}$$

where κ_{ex} is the cavity energy decay rate to the reflection mode of the cavity.

In order to calculate the reflection coefficients, we need an expression for the cavity field operator $\hat{\mathbf{a}}$. We assume that the incident photon is quasi-monochromatic with a frequency of ω. We express the Hamiltonian for the coupled atom and cavity system in the rotating reference frame with respect to ω, given by

$$\mathbf{H} = \hbar(\omega_c - \omega)\hat{\mathbf{a}}^\dagger\hat{\mathbf{a}} + \hbar(\omega_x - \omega)\hat{\sigma}_-^\dagger\hat{\sigma}_- + ig\hbar(\hat{\mathbf{a}}\hat{\sigma}_-^\dagger - \hat{\sigma}_-), \tag{11.2}$$

where $\hat{\sigma}_-$ is the lowering operator for transition $\mu_\uparrow$, ω_c and ω_x are the resonance frequencies of the cavity mode and transition $\mu_\uparrow$ respectively, and g is the coupling strength between the cavity mode and transition $\mu_\uparrow$. In the weak excitation limit, the Heisenberg-Langevin equations are [49–51]

$$\frac{d\hat{\mathbf{a}}}{dt} = -\big[i(\omega_c - \omega) + \frac{\kappa}{2}\big]\hat{\mathbf{a}} - ig\hat{\sigma}_- + \sqrt{\kappa_{ex}}\hat{\mathbf{a}}_{\mathbf{in}} \tag{11.3}$$

$$\frac{d\hat{\sigma}_-}{dt} = -\big[i(\omega_x - \omega) + \gamma\big]\hat{\sigma}_- - ig\hat{\mathbf{a}}, \tag{11.4}$$

where κ is the total cavity energy delay rate given by $\kappa = \kappa_{ex} + \kappa_i$, κ_i is the intrinsic loss rate of the cavity due to material absorption and coupling to undesired leaky modes, and γ is the dipole decay rate of for transition $\mu_\uparrow$.

We calculate the cavity field operator by taking the steady solution of (11.3) and (11.4). When the spin is in spin-down state, we have $\langle\hat{\sigma}_-\rangle = 0$, therefore the steady solution for $\hat{\mathbf{a}}$ can be calculated from (11.3) and is given by

$$\langle\hat{\mathbf{a}}\rangle = \frac{\sqrt{\kappa_{ex}}\langle\hat{\mathbf{a}}_{\mathbf{in}}\rangle}{i(\omega_c - \omega) + \frac{\kappa}{2}}. \tag{11.5}$$

When the spin is in spin-up state, the expression for $\langle\hat{\mathbf{a}}\rangle$ is given by

$$\langle\hat{\mathbf{a}}\rangle = \frac{\sqrt{\kappa_{ex}}\big[i(\omega_x - \omega) + \gamma\big]}{\big[i(\omega_c - \omega) + \frac{\kappa}{2}\big]\big[i(\omega_x - \omega) + \gamma\big] + g^2}\langle\hat{\mathbf{a}}_{\mathbf{in}}\rangle. \tag{11.6}$$

We calculate the cavity reflection coefficients $r_\downarrow$ and $r_\uparrow$ for both the spin-down and spin-up cases by combining (11.5) or (11.6) with (11.1). For the spin-down case, we obtain $\langle\hat{\mathbf{a}}_{\mathbf{out}}\rangle = r_\downarrow\langle\hat{\mathbf{a}}_{\mathbf{in}}\rangle$, where $r_\downarrow$ is given by,

$$r_\downarrow = 1 - \frac{\alpha\kappa}{i(\omega_c - \omega) + \frac{\kappa}{2}}, \tag{11.7}$$

where α is the interference contrast given by $\alpha = \kappa_{ex}/\kappa$. For the spin-up case, we obtain $\langle\hat{\mathbf{a}}_{\mathbf{out}}\rangle = r_\uparrow\langle\hat{\mathbf{a}}_{\mathbf{in}}\rangle$, where $r_\uparrow$ is given by

$$r_\uparrow = 1 - \frac{\alpha\kappa\big[i(\omega_x - \omega) + \gamma\big]}{\big[i(\omega_c - \omega) + \frac{\kappa}{2}\big]\big[i(\omega_x - \omega) + \gamma\big] + g^2}. \tag{11.8}$$

11.2.2 Resonance Case: A Spin-Photon Quantum Switch

We focus on the resonance condition where $\omega = \omega_c = \omega_x$. In this case, the cavity reflection coefficients simplify to

$$r_{\downarrow} = 1 - 2\alpha \tag{11.9}$$

$$r_{\uparrow} = 1 - \frac{2\alpha}{1+C}, \tag{11.10}$$

where C is the cooperativity of the system given by $C = 2g^2/\kappa\gamma$.

When $\alpha > 0.5$ and $C > 2\alpha - 1$, $r_{\uparrow}$ and $r_{\downarrow}$ have opposite signs. Thus, the spin state conditionally shifts the phase of a reflected photon by π, implementing a quantum phase operation. An ideal phase switch would be attained in the limit of large cooperativity ($C \gg 1$) and a perfect single sided cavity ($\alpha = 1$) where the reflection coefficient becomes $r_{\downarrow} = -1$ and $r_{\uparrow} = 1$.

The quantum phase switch allows one qubit to conditionally switch the other qubit between its two orthogonal eigenstates. We consider the case where the polarization state of the photon encodes quantum information. We assume that the cavity mode has a well defined polarization direction $\hat{y}$. Therefore only a y-polarized photon experiences spin-dependent phase shift upon reflection, whereas an x-polarized incident photon does not couple to the cavity and gets directly reflected without a phase shift. If the cavity mode does not have a well-defined polarization (i.e. the cavity supports polarization degenerate modes), we could use a simple polarization interferometry setup as illustrated in Fig. 11.2 to implement the similar idea.

We express the state of a photon incident on the cavity in the basis states $|x\rangle$ and $|y\rangle$, which denote the polarization states oriented orthogonal and parallel to the cavity mode respectively. For a right-circularly-polarized incident photon $|x\rangle + i\,|y\rangle$, the reflected state is given by $|x\rangle + ir_{\uparrow(\downarrow)}\,|y\rangle$ (before renormalization). In the limit of large cooperativity and perfect single-sided cavity, the state of the reflected photon remains right-circularly polarized if the atom is in the spin-up state, but becomes left-circularly polarized for spin-down. Similarly, a single control photon can flip the state of the spin. If the spin is prepared in the state $|\uparrow\rangle + |\downarrow\rangle$, then after a

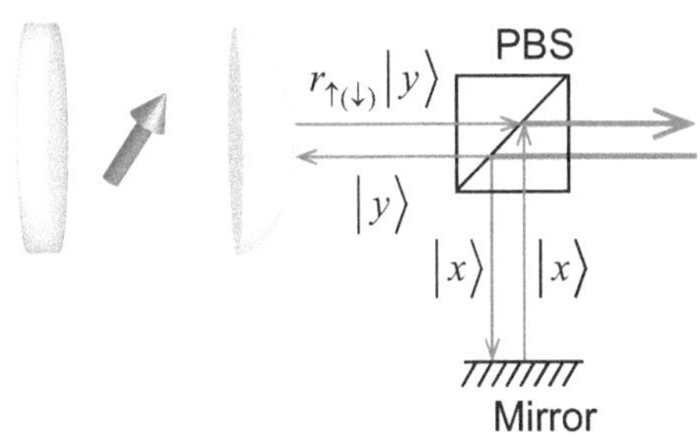

Fig. 11.2 Schematic setup to implement a spin-photon quantum switch where the polarization states of the photon encode quantum information

y-polarized photon reflects from the cavity the spin-state transforms to $|\uparrow\rangle - |\downarrow\rangle$, but an x-polarized photon does not flip the spin.

We note that when $\alpha > 0.5$ but $C < 2\alpha - 1$, $r_\uparrow$ and $r_\downarrow$ have the same sign. Thus the quantum phase operation is not available for a resonant photon if the system cooperativity is too low.

11.2.3 Detuned Case: Spin-Dependent Kerr Rotation

The general expression for $r_\downarrow$ and $r_\uparrow$ are given by (11.7) and (11.8) respectively. We can rewrite $r_{\downarrow(\uparrow)}$ as $r_{\downarrow(\uparrow)} = |r_{\downarrow(\uparrow)}|e^{i\phi_{\downarrow(\uparrow)}}$, where $\phi_{\downarrow(\uparrow)}$ represents the phase of the reflection coefficient $r_{\downarrow(\uparrow)}$. Since in general $\phi_\downarrow \neq \phi_\uparrow$, the spin can still apply spin-dependent phase shift on the photon. The phase difference between the spin-up and spin-down state ϕ is given by $\phi = \phi_\downarrow - \phi_\uparrow$, which is typically nonzero but not necessarily π any more.

The nonzero value of ϕ can be utilized to realize spin-dependent Kerr rotation of a photon, as demonstrated in [45, 46]. Assuming the polarization of the incident photon is in the state $|x\rangle + i\,|y\rangle$, the reflected state becomes $|P_{\uparrow(\downarrow)}\rangle = |x\rangle + i\,|r_{\uparrow(\downarrow)}|e^{i\phi_{\uparrow(\downarrow)}}\,|y\rangle$ (before renormalization). Therefore, the polarizations of the reflected photon are different for the spin-up and spin-down cases, as long as $\langle P_\uparrow|P_\downarrow\rangle \neq 1$, which is equivalent to $\phi \neq 0$.

Similarly, a single detuned photon can also rotate the state of the spin. If the spin is prepared in the state $|\uparrow\rangle + |\downarrow\rangle$, then after a y-polarized photon reflects from the cavity the spin-state transforms to $|r_\uparrow|\,|\uparrow\rangle + |r_\downarrow|e^{i\phi}\,|\downarrow\rangle$ (after taking out an overall phase factor). If $|r_\uparrow| \simeq |r_\uparrow|$, this operation corresponds to the rotation of the spin Bloch vector by an angle ϕ along the equator of the spin Bloch sphere.

For the detuned case, we do not require the cooperativity to be greater than $2\alpha - 1$ in order to induce spin-dependent phase shift. Indeed [45] demonstrated spin-dependent Kerr rotation of a photon with a coopeartivity of $C = 0.2$. However, there is still significant difference between the regime $C > 2\alpha - 1$ and $C < 2\alpha - 1$. Figure 11.3 shows numerically calculated phase shift ϕ as a function of the detuning

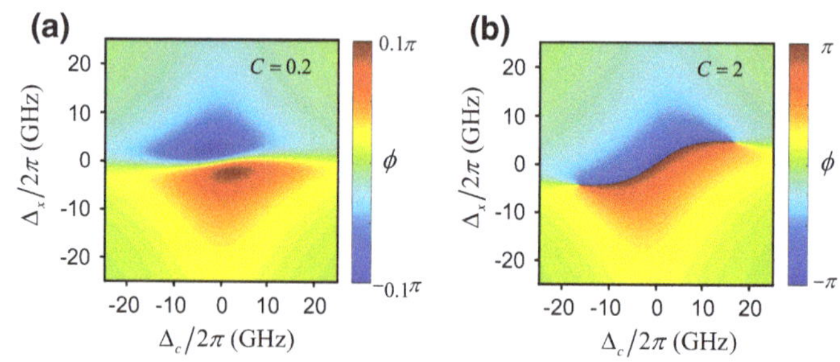

Fig. 11.3 Phase shift ϕ as a function of the detuning Δ_c and Δ_x. **a** $C = 0.2$. **b** $C = 2$

Δ_c and Δ_x, where we define $\Delta_c = \omega - \omega_c$ and $\Delta_x = \omega - \omega_x$. In both Fig. 11.3a, b, we use the following parameters from a realistic quantum dot based cavity QED system [47]: $\alpha = 0.8$, $\kappa/2\pi = 36$ GHz, $\gamma/2\pi = 3$ GHz. Figure 11.3a shows the case where we set $g/2\pi = 10.4$ GHz so that the cooperativity is $C = 2$ and satisfies $C > 2\alpha - 1$. Figure 11.3b shows the case where we set $g/2\pi = 3.3$ GHz so that the cooperativity is $C = 0.2$ and does not meet the condition $C > 2\alpha - 1$. For the case of $C = 2$, we are able to tune the phase shift ϕ to an arbitrary value between $-\pi$ and π, by simply controlling the detuning Δ_c and Δ_x. However, when the cooperativity is small, we can only tune the phase shift ϕ between $-\phi_{max}$ and ϕ_{max}, where $\phi_{max} < \pi$ is an upper limit that is determined by the system cooperativity. As an example, in Fig. 11.3a we have $\phi_{max} = 0.1\pi$.

11.3 Quantum Dot Spins in a Nanophotonic Cavity

Charged quantum dots exhibit spin-dependent optical transitions. As described in Sect. 11.2, by selectively coupling the spin-dependent optical transitions of a charged quantum dot to a cavity mode, one can induce strong interactions between the quantum dot spin and a photon. In this section, we review the experimental efforts to integrate quantum dot spins with different nanophotonic cavities, including micropillar cavities and photonic crystal cavities.

11.3.1 Micropillar Cavities

A micropillar cavity is formed by two Bragg reflectors, which are made of alternative layers GaAs and AlAs. The diameter of the pillar is typically in the order of several micron meters, which lead to a highly localized mode with mode volume in the order of $10(\lambda/n)^3$. One can engineer the cavity transmittance and reflectance by designing the number of layers for the top and bottom Bragg reflector. For example, a single sided cavity can be created by introducing a highly reflective bottom mirror with more layers and an outcoupling top mirror with less layers.

In 2009, Rakher et al. firstly reported integration of a charge tunable quantum dot with a micropillar cavity [41]. The reported device achieves a cooperativity of $C = 2$, which enables significant cavity reflectivity modulation when an extra charge is loaded into the quantum dot. Spin-dependent coupling between a charged quantum dot and a micropillar cavity is reported in [45, 46]. Both works utilized a device with low cooperativity ($C < 0.2$), and observe spin-dependent Kerr rotation of a reflected photon with a polarization rotation degree of $\sim 6°$.

Micropillar cavities enable efficient coupling between an incident field and the cavity mode, because the cavity mode is well matched to a Gaussian mode in the far field. Recent works have demonstrated an input coupling efficiency of exceeding 95% [52], which is a very promising property for achieving deterministic spin-photon interactions. However, coherent optical manipulation of the quantum dot spin in a micropillar cavity is yet to be demonstrated.

11.3.2 Photonic Crystal Cavities

Photonic crystals are periodic nanostructures fabricated on a dielectric material, which lead to periodic modulation of the refractive index in the length scale of an optical wavelength. In a photonic crystal, the motion of a photon obeys optical Bragg scattering, very similarly as the way an electron propagates in ionic lattices. In addition, one can engineer photonic crystals to open a photonic band gap, analogous to an electronic band gap in semiconductors, which prohibits the propagation of photons for certain directions within some frequency range.

Defects in a photonic crystal can support highly localized cavity modes within the photonic band gap, referred as photonic crystal cavities. These cavities support small mode volume (in the order of $(\lambda/n)^3$) and high quality factor, which enable strong light-matter interactions with an embedded quantum emitter [25, 28]. In addition, the properties of a photonic crystal cavity, such as the resonant frequency, mode profile, and polarization can be easily controlled by tailoring the geometry of the photonic crystals or the shape of the defect areas. Photonic crystal cavities can also be easily integrated with other cavities or waveguides using the scalable photonic crystal architecture [33, 53–56]. These assets make the photonic crystal cavity a very attractive platform for realizing integrated photonics.

Several groups have reported deterministic charging of a quantum dot embedded in a photonic crystal cavity [42, 43]. In 2013, Carter et al. firstly demonstrated coherent control of a quantum dot spin embedded in a photonic crystal cavity [44]. This work operated far in the weak coupling regime where the quantum dot produced a spin-dependent cavity reflectivity with contrast of less than 1%. In 2016, Sun et al. firstly demonstrated strong coherent spin-photon interactions based on a strongly coupled charged quantum dot and a photonic crystal cavity [47]. The device has a cooperativity of $C = 2$, enabling a quantum switch between a quantum dot spin and a photon. We will review this work in details in Sect. 11.4.

11.4 Experimental Demonstrations of a Spin-Photon Quantum Switch

In this chapter, we discuss experimental demonstrations of a spin-photon quantum switch using a strongly coupled charged quantum dot and a photonic crystal cavity. We utilized a negatively charged quantum dot containing a single electron. In the presence of a magnetic field applied in the Voigt geometry, the energy structure of the quantum dot is shown in Fig. 11.4. The states of the dot include two ground states,

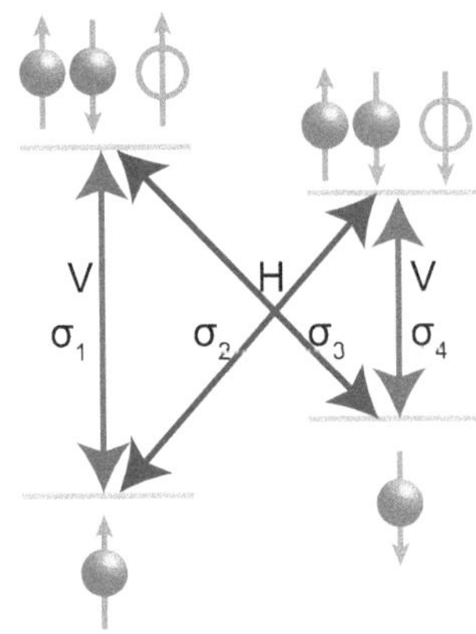

Fig. 11.4 Energy level structure of a charged quantum dot with an external magnetic field applied in the Voigt configuration

corresponding to the two electron spin orientations, and two excited trion states that optically couple to the ground states via four optical transitions denoted as $\sigma_1 - \sigma_4$. The energy level structure consists of two λ-systems and is slightly different from the single λ-system described in Fig. 11.1b. However, we can still induce spin-dependent cavity reflectivity by resonantly coupling only transition σ_1 with the cavity mode, and decoupling all other transitions from the cavity by a large magnetic field induced detuning. Thus, the quantum dot resonantly couples to the cavity only when it is in the spin-up state, inducing a spin-dependent reflection coefficient. As described in Sect. 11.2.2, we are able to realize a quantum switch by utilizing the spin-dependent reflection coefficients.

11.4.1 Device Characterization

To characterize the device, we mount the sample in a closed-cycle liquid-helium cryostat and cool it down to 3.6 K. The sample mount is placed inside the bore of a superconducting magnet that can apply magnetic fields up to 9.2 T. The sample is oriented such that the magnetic field is in the in-plane direction (Voigt configuration), and the cavity axis is approximately 45° with respect to the magnetic field. Sample excitation and collection is performed with a confocal microscope using an objective lens with numerical aperture of 0.68. The coupling efficiency for this configuration is measured to be 1% by measuring the Stark shift of the quantum dot under cavity-resonant excitation [57].

We identify a charged quantum dot coupled to the cavity from the photoluminescence spectrum of the device under a magnetic field applied in the Voigt configuration. Figure 11.5a shows the photoluminescence spectrum from the device used in our measurements when excited using an 860 nm continuous wave laser. At 0 T, the emission spectrum shows a bright peak due to the cavity (labeled as CM) and a second peak due to the quantum dot (labeled as QD), which is red-detuned from the

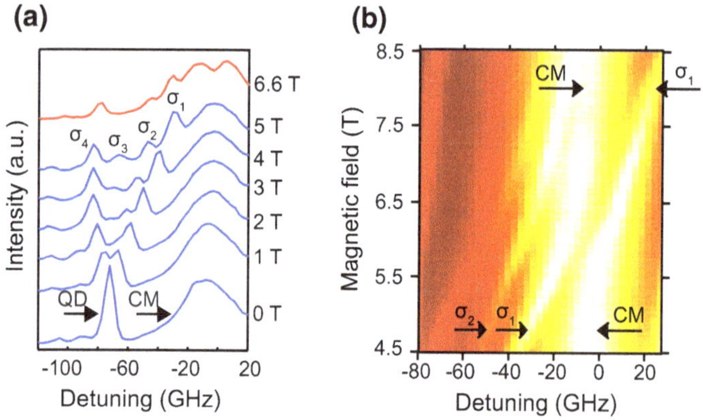

Fig. 11.5 Device characterization with photoluminescence. **a** Photoluminescence spectrum. The *blue lines* show the spectra at various magnetic fields ranging from 0 to 5 T. The *red line* shows the spectrum at 6.6 T, where transition σ_1 is resonant with the cavity. **b** Cavity photoluminescence as a function of the magnetic field

cavity resonance by 0.19 nm (67 GHz). As we increase the magnetic field, the quantum dot splits into four peaks corresponding to the four optical transitions shown in Fig. 11.4.

To demonstrate strong coupling between the quantum dot and the cavity, we finely tune the magnetic field over the range of 4.5–8.5 T and measure the cavity photoluminescence. Figure 11.5b shows the photoluminescence spectrum near the cavity resonance as a function of magnetic field. In this range, transition σ_1 tunes over the cavity resonance and exhibits an anti-crossing, indicating that the system operates in the strong coupling regime.

11.4.2 Spin-Dependent Cavity Reflectivity

To demonstrate that the spin can flip the state of the photon, we use the polarization interferometry set-up shown in Fig. 11.6. We excite the cavity with right-circularly polarized light, and measure the reflected signal along either the left-circularly or right-circularly polarized component. Figure 11.7 shows both the cross-polarized (red diamonds) and co-polarized reflection spectrum (blue circles) when the quantum dot is detuned from the cavity so that the two systems are decoupled. By fitting the reflection spectrum to a Lorentzian lineshape (blue and red solid lines), we determine the cavity energy decay rate to be $\kappa/2\pi = (35.9 \pm 0.7)$ GHz and the interference contrast to be $\alpha/2\pi = 0.81 \pm 0.01$.

We next apply a magnetic field of 6.6 T that tunes transition σ_1 onto cavity resonance via a Zeeman shift. We excite the quantum dot with a narrowband tunable laser to optically pump the spin state [58, 59]. We first tune the optical pumping

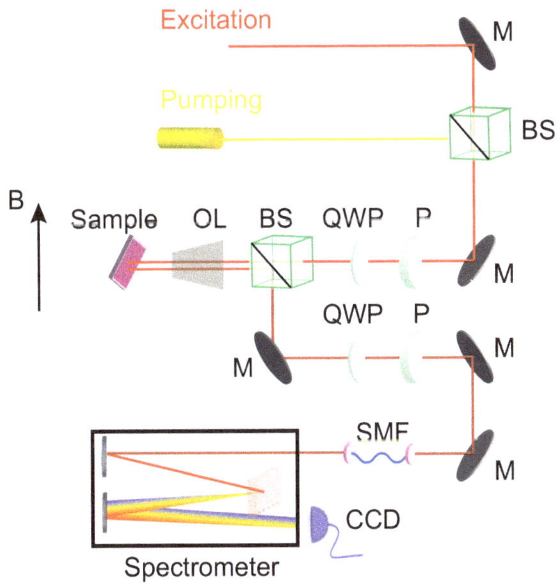

Fig. 11.6 Measurement setup. OL, objective lens; QWP, quarter wave plate; P, polarizer; BS, beam splitter; M, mirror; SMF, single mode fiber; CCD, charged coupled device

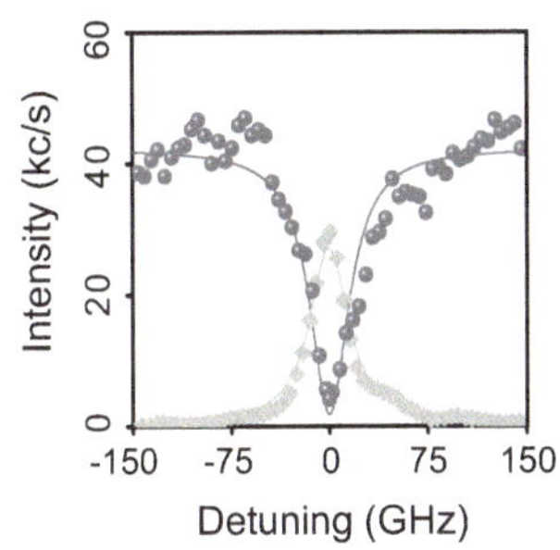

Fig. 11.7 Co-polarized (*blue circles*) and cross-polarized (*red diamonds*) cavity reflection spectrum with no magnetic field. *Blue* and *red solid lines* show the calculated spectrum (color figure online)

laser to transition σ_4 to prepare the quantum dot in the spin-up state. The blue circles in Fig. 11.8a show the cross-polarized reflection spectrum with the optical pumping laser, which exhibits a vacuum Rabi splitting. When we turn off the pumping laser, we observe a reduced contrast due to random spin fluctuations (red diamonds). In contrast, when we optically pump transition σ_2 to initialize the quantum dot to the spin-down state, we observe a spectrum that closely resembles a bare cavity (Fig. 11.8b). This spin-dependent reflection spectrum is one of the essential properties of the phase switch.

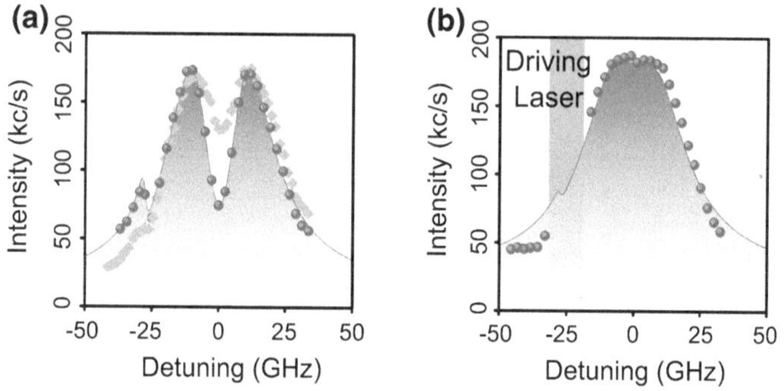

Fig. 11.8 Spin-dependent cavity reflectivity. **a** Cavity reflection spectrum under a 6.6 T magnetic field with (*blue circles*) and without (*red diamonds*) an optical pumping laser resonant with transition σ_4. The *blue solid line* shows the calculated spectrum for the case where the optical pumping laser is turned on. With the pumping laser, we observe a suppression of the cavity response at the σ_1 resonance due to strong coupling. We also observe a Fano-resonant lineshape at 27 GHz detuning, corresponding to the coupling between transition σ_2 and the cavity mode. **b** Cavity reflection spectrum when the pump laser is resonant with transition σ_2. The *blue circles* show the measured spectrum, and the *solid line* shows calculated spectrum. The center wavelength is 927.48 nm for all panels (color figure online)

11.4.3 Coherent Control of Cavity Reflectivity

To demonstrate control of a reflected photon using a coherently prepared spin state, we use all-optical coherent control to manipulate the spin. We fix the magnetic field at 6.6 T. A narrowband continuous-wave laser tuned to transition σ_4 performs spin initialization and circularly polarized picosecond optical pulses generate an effective spin rotation [13, 14]. We perform spin rotations using 6 ps rotation pulses with center frequencies detuned by 520 GHz from the cavity resonance (equal to 15 cavity linewidth). To rotate the spin over the Bloch sphere, we utilize the Ramsey interferometry setup illustrated in Fig. 11.9a, which generates a pair of rotation pulses separated by a time delay τ. A third laser pulse probes the cavity reflectivity a time Δt after the second rotation pulse. We attenuate this laser so that a single pulse contains an average of 0.12 photons coupled to the cavity to ensure a low probability of two-photon events. We set the power of the continuous-wave optical pumping laser to 30 nW. At this power we measure a spin initialization time of (1.27 ± 0.09) ns, which is slow compared with τ and Δt, but fast compared with the repetition time of the experiment (13 ns).

Figure 11.9b shows the reflected probe intensity as a function of rotation pulse power P and delay τ, where we set Δt to 140 ps. We observe Ramsey oscillations in the reflected probe intensity as a function of both P and τ. Figure 11.9c plots the emission intensity of the quantum dot at transition σ_2 for the same measurement, which provides a second readout of the spin state. We observe the same Ramsey oscillation pattern in the quantum dot emission signal, confirming that the reflection modulation shown in Fig. 11.9b is induced by coherent spin manipulation. Figure 11.9d

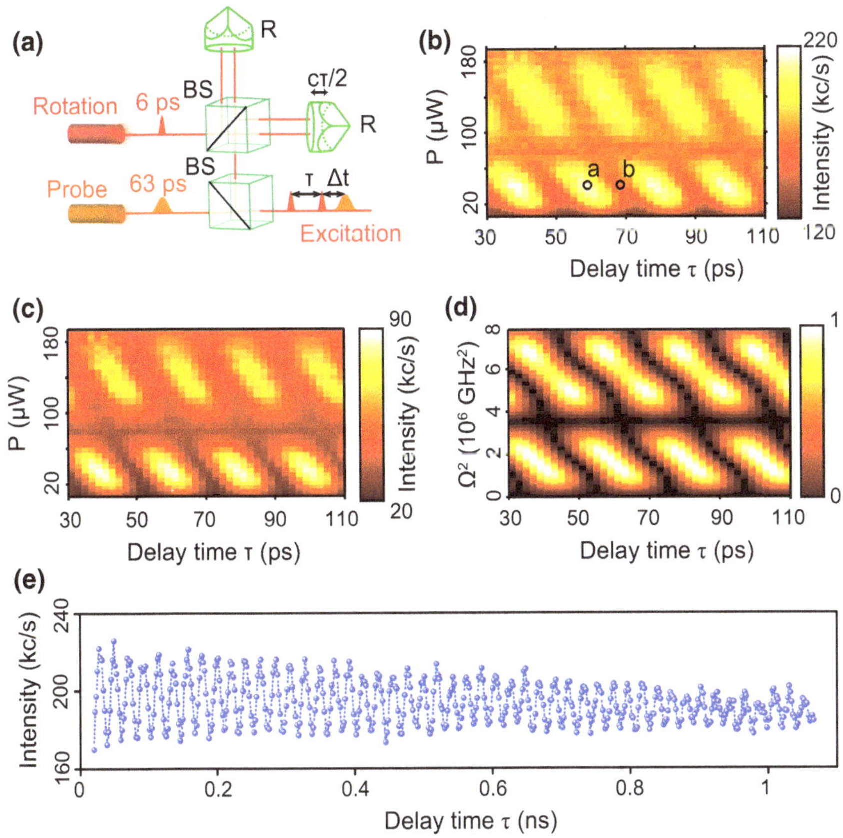

Fig. 11.9 Ramsey interference measurements. **a** Experimental setup for generating the Ramsey pulse sequence. The delay time τ between the two rotation pulses is controlled by a movable retro-reflector mounted on a computer-controlled translation stage. BS, beam splitter; R, retro-reflector. **b** Reflected probe intensity as a function of rotation pulse power P and the delay time τ. **c** Intensity of the quantum dot emission at σ_2 transition frequency as a function of rotation pulse power P and the delay time τ. **d** Calculated spin-down state population as a function of peak rotation pulse power and the delay time τ. We express the rotation pulse as a classical time-varying Rabi frequency with a Gaussian pulse shape and peak power Ω^2. **e** Reflected probe intensity as a function of delay time τ

shows the numerically calculated value for the population of the spin-down state for comparison, which exhibit good agreement with experiments.

In Fig. 11.9e we plot the reflected probe intensity over a larger time range of 1 ns. We fix the power of each rotation pulse to 40 μW, which corresponds to a $\pi/2$-rotation. From the decay of the fringe visibility, we calculate a $T_2^\star$ time of (0.94 ± 0.02) ns. This coherence time is limited by inhomogeneous broadening due to a slowly fluctuating nuclear spin background [11], along with decoherence due to continuous optical pumping during the rotation pulse sequence. We could reduce

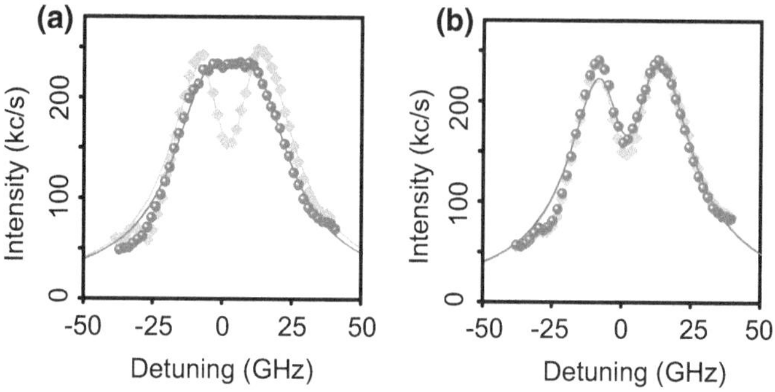

Fig. 11.10 Time-resolved cavity reflection spectrum. **a** and **b**, Reflected probe intensity as a function of probe detuning at the rotation condition indicated by point a and point b in Fig. 11.9b respectively. *Blue circles* are $\Delta t = 140\,\text{ps}$; *red diamonds* are $\Delta t = 13\,\text{ns}$. *Solid lines* are the calculated spectra. The center wavelength is 927.48 nm for both spectra (color figure online)

these effects by turning off the pump laser during the measurement process and using a nuclear field locking [11] or spin echo technique [12], which has been shown to improve the coherence to up to 2.6 μs.

To characterize the fidelity of the spin state preparation, we tune the probe laser across the cavity resonance while setting P and τ to the conditions indicated by the circles in Fig. 11.9b. The resulting cavity spectra are plotted in Fig. 11.10. In Fig. 11.10a the two pulses arrive in-phase with the Larmor precession of the spin, and the quantum dot rotates to the spin-down state. The cavity spectrum (blue circles) is thus similar to the bare cavity Lorentzian lineshape. Figure 11.10b shows the case where the two rotation pulses arrive out-of-phase and the quantum dot rotates back to the spin-up state. The cavity (blue circles) now exhibits a strongly coupled spectrum. We also plot the measured spectrum when $\Delta t = 13\,\text{ns}$ (red diamonds) for comparison. At this condition the spin is re-initialized to the spin-up state in both cases.

11.4.4 Controlling a Spin with a Photon

The previous measurements demonstrate that the spin state of the quantum dot induces a conditional phase shift on the photon. A quantum phase switch would also exhibit the complementary effect, where reflection of a single photon rotates the spin state. To demonstrate this phase shift, we use the experimental configuration shown in Fig. 11.11a. We again perform a Ramsey interference measurement but we inject a weak laser pulse that serves as the control field before the second rotation pulse arrives. We generate the control pulse in the same way as the probe pulse in the previous measurement, with pulse duration of 63 ps. When a control photon couples

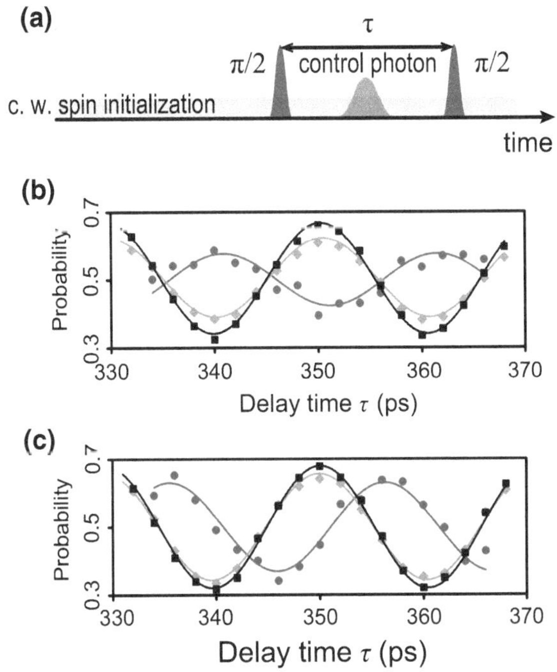

Fig. 11.11 Photon-induced spin phase switch. **a** Pulse timing diagram showing the relative time delays between the rotation pulses and the control field. **b** Occupation probability of the spin-down state as a function of the delay time τ, in the absence of control pulse (*black squares*), conditioned on detecting a reflected control photon polarized along the cavity axis (*blue circles*), and in the presence of the control field but not conditioning on the detection of a control photon (*red diamonds*). The control field is resonant with the σ_1 quantum dot transition. **c** Same as **b**, except that the control field is *blue detuned* from the σ_1 quantum dot transition (color figure online)

to the cavity, it imposes a phase shift on the spin-down state, which shifts the phase of the Ramsey fringes.

We perform statistical spin readout by monitoring the emission at the σ_2 frequency. The blue circles in Fig. 11.11b show the occupation probability of the spin-down state conditioned on the detection of a control photon, as a function of delay between the two rotation pulses. These data are obtained by performing a two photon correlation measurement. The blue solid line is a numerical fit to a sinusoidal function. We compare this curve to the occupation probability of the spin-down state when we block the control field (black squares with black line as a numerical fit). The interference fringe conditioned on detecting a single control photon is shifted in phase by $(1.09 \pm 0.09)\pi$ radians relative to the case where there is no control photon, demonstrating that a single control photon applies a large phase shift to the spin. We attribute the degraded visibility of the Ramsey fringe conditioned on a control photon to finite cooperativity, intrinsic cavity losses and occasional two-photon incidence events.

We can tune the phase shift imparted on the spin by a control photon by introducing a frequency detuning between the control field and transition σ_1, which enables us to apply arbitrary controlled phase shifts. Figure 11.11c shows the same measurement for a blue detuned control field. The conditioned data (blue circles) show a $(0.59 \pm 0.05)\pi$ radian phase shift, which corresponds to a detuning of 7.3 GHz. We also plot the occupation probability of the spin-down state in the presence of the control field but without conditioning on the detection of the control photon (red diamonds in Fig. 11.11b, c). These curves are very similar to the case where the control field is blocked, which indicates that the average number of control photons per pulse coupled to the cavity is much smaller than one.

11.5 Discussions and Outlooks

In this chapter, we reviewed recent experimental progress on interfacing a single quantum dot spin and a single photon using a nanophotonic cavity QED system. Especially, we introduced an experiment that demonstrated a spin-photon quantum phase switch, which achieves strong coherent interactions between a single quantum dot spin and a photon. The strong light-matter coupling strength of quantum dot based cavity QED devices enables a quantum switch operating at unprecedented bandwidths, where the spin can switch photon wavepackets as short as tens of picoseconds [47]. Perhaps the most intriguing aspect of the spin-photon quantum switch is that it monolithically combines spins with strongly interacting nanophotonic structures on a single semiconductor chip, which may have many beneficial properties for future integration and scalability.

The demonstrated spin-photon quantum switch could enable lots of applications in quantum information processing. Recent theory works have showed the potential to deterministically generate spin-photon entanglement based on a similar quantum dot based cavity QED system [60], which is an important step towards solid-state implementations of quantum repeaters and quantum networks. The spin-photon quantum interface might also enable single-shot optical readout of a quantum dot spin in the Voigt configuration [61], an extremely challenging task for quantum dot based quantum information processing. The ultimate direction for this research direction is to construct integrated quantum photonic circuits and on-chip quantum processors using nanophotonic platform with solid-state spins embedded.

Another important aspects of the future works is to improve the performance of the cavity device. Cavity designs with smaller mode-volume could improve the system cooperativity [62, 63], thus enable higher switching fidelity. Using delta-doping layers [14] or active charge stabilization [44] could further improve the spin state preparation fidelity. The spin-photon quantum switch results can also be directly applied in waveguide integrated devices that are more conducive to on-chip integration and can exhibit similar strong light-matter interactions [54]. In such on-chip implementations, waveguide losses create further challenges by degrading the cavity quality factor, which would reduce the cooperativity. Waveguide-coupled devices would therefore require higher bare cavity quality factor to ensure that the light remains on the chip. Past work has demonstrated a quality factor exceeding 50,000 using GaAs photonic crystal cavities operating at near-infrared wavelengths [64], and quality factor exceeding 250,000 in cavities operating at a longer wavelength [65], which could potentially enable both efficient on-chip coupling and high cooperativity. Employing regulated quantum dot growth techniques [66, 67] in conjunction with local frequency tuning [57] could further open up the possibility to integrate multiple quantum dot spins on a single semiconductor chip.

References

1. J.I. Cirac, P. Zoller, H.J. Kimble, H. Mabuchi, Quantum state transfer and entanglement distribution among distant nodes in a quantum network. Phys. Rev. Lett. **78**(16), 3221 (1997)
2. H. Jeff Kimble, The quantum internet. Nature **453**(7198), 1023–1030 (2008)
3. J.I. Cirac, A.K. Ekert, S.F. Huelga, C. Macchiavello, Distributed quantum computation over noisy channels. Phys. Rev. A **59**(6), 4249 (1999)
4. J. Cho, H.-W. Lee, Generation of atomic cluster states through the cavity input-output process. Phys. Rev. Lett. **95**(16), 160501 (2005)
5. S. Ritter, C. Nölleke, C. Hahn, A. Reiserer, A. Neuzner, M. Uphoff, M. Mücke, E. Figueroa, J. Bochmann, G. Rempe, An elementary quantum network of single atoms in optical cavities. Nature **484**(7393), 195–200 (2012)
6. J. Volz, R. Gehr, G. Dubois, J. Estève, J. Reichel, Measurement of the internal state of a single atom without energy exchange. Nature **475**(7355), 210–213 (2011)
7. A. Reiserer, S. Ritter, G. Rempe, Nondestructive detection of an optical photon. Science **342**(6164), 1349–1351 (2013)
8. T.G. Tiecke, J.D. Thompson, N.P. de Leon, L.R. Liu, V. Vuletić, M.D. Lukin, Nanophotonic quantum phase switch with a single atom. Nature **508**(7495), 241–244 (2014)
9. L.-M. Duan, H.J. Kimble, Scalable photonic quantum computation through cavity-assisted interactions. Phys. Rev. Lett. **92**(12), 127902 (2004)
10. A. Reiserer, N. Kalb, G. Rempe, S. Ritter, A quantum gate between a flying optical photon and a single trapped atom. Nature **508**(7495), 237–240 (2014)
11. X. Xiaodong, W. Yao, B. Sun, D.G. Steel, A.S. Bracker, D. Gammon, L.J. Sham, Optically controlled locking of the nuclear field via coherent dark-state spectroscopy. Nature **459**(7250), 1105–1109 (2009)
12. D. Press, K. De Greve, P.L. McMahon, T.D. Ladd, B. Friess, C. Schneider, M. Kamp, S. Höfling, A. Forchel, Y. Yamamoto, Ultrafast optical spin echo in a single quantum dot. Nat. Photonics **4**(6), 367–370 (2010)
13. J. Berezovsky, M.H. Mikkelsen, N.G. Stoltz, L.A. Coldren, D.D. Awschalom, Picosecond coherent optical manipulation of a single electron spin in a quantum dot. Science **320**(5874), 349–352 (2008)
14. D. Press, T.D. Ladd, B. Zhang, Y. Yamamoto, Complete quantum control of a single quantum dot spin using ultrafast optical pulses. Nature **456**(7219), 218–221 (2008)
15. E.D. Kim, K. Truex, X. Xu, B. Sun, D.G. Steel, A.S. Bracker, D. Gammon, L.J. Sham, Fast spin rotations by optically controlled geometric phases in a charge-tunable inas quantum dot. Phys. Rev. Lett. **104**(16), 167401 (2010)
16. K. De Greve, P.L. McMahon, D. Press, T.D. Ladd, D. Bisping, C. Schneider, M. Kamp, L. Worschech, S. Höfling, A. Forchel et al., Ultrafast coherent control and suppressed nuclear feedback of a single quantum dot hole qubit. Nat. Phys. **7**(11), 872–878 (2011)

17. T.M. Godden, J.H. Quilter, A.J. Ramsay, Y. Wu, P. Brereton, S.J. Boyle, I.J. Luxmoore, J. Puebla-Nunez, A.M. Fox, M.S. Skolnick, Coherent optical control of the spin of a single hole in an inas/gaas quantum dot. Phys. Rev. Lett. **108**(1), 017402 (2012)
18. Y. He, Y.-J. Yu-Ming He, X.J. Wei, M.-C. Chen, F.-L. Xiong, Y. Zhao, C. Schneider, M. Kamp, S. Höfling et al., Indistinguishable tunable single photons emitted by spin-flip raman transitions in ingaas quantum dots. Phys. Rev. Lett. **111**(23), 237403 (2013)
19. K. De Greve, Y. Leo, P.L. McMahon, J.S. Pelc, C.M. Natarajan, N.Y. Kim, E. Abe, S. Maier, C. Schneider, M. Kamp et al., Quantum-dot spin-photon entanglement via frequency downconversion to telecom wavelength. Nature **491**(7424), 421–425 (2012)
20. W.B. Gao, P. Fallahi, E. Togan, J. Miguel-Sánchez, A. Imamoglu, Observation of entanglement between a quantum dot spin and a single photon. Nature **491**(7424), 426–430 (2012)
21. J.R. Schaibley, A.P. Burgers, G.A. McCracken, L.-M. Duan, P.R. Berman, D.G. Steel, A.S. Bracker, D. Gammon, L.J. Sham, Demonstration of quantum entanglement between a single electron spin confined to an inas quantum dot and a photon. Phys. Rev. Lett. **110**(16), 167401 (2013)
22. K. De Greve, P.L. McMahon, Yu. Leo, J.S. Pelc, C. Jones, C.M. Natarajan, N.Y. Kim, E. Abe, S. Maier, C. Schneider, et al., Complete tomography of a high-fidelity solid-state entangled spin-photon qubit pair. Nat. Commun. **4**, (2013)
23. W.B. Gao, P. Fallahi, E. Togan, A. Delteil, Y.S. Chin, J. Miguel-Sanchez, A. Imamoğlu, Quantum teleportation from a propagating photon to a solid-state spin qubit. Nat. Commun. **4** (2013)
24. A. Delteil, Z. Sun, W.-b. Gao, E. Togan, S. Faelt, A. Imamoglu, Generation of heralded entanglement between distant hole spins. Nat. Phys. (2015)
25. T. Yoshie, A. Scherer, J. Hendrickson, G. Khitrova, H.M. Gibbs, G. Rupper, C. Ell, O.B. Shchekin, D.G. Deppe, Vacuum rabi splitting with a single quantum dot in a photonic crystal nanocavity. Nature **432**(7014), 200–203 (2004)
26. J.P. Reithmaier, G. Sek, A. Löffler, C. Hofmann, S. Kuhn, S. Reitzenstein, L.V. Keldysh, V.D. Kulakovskii, T.L. Reinecke, A. Forchel, Strong coupling in a single quantum dot-semiconductor microcavity system. Nature **432**(7014), 197–200 (2004)
27. E. Peter, P. Senellart, D. Martrou, A. Lemaître, J. Hours, J.M. Gérard, J. Bloch, Exciton-photon strong-coupling regime for a single quantum dot embedded in a microcavity. Phys. Rev. Lett. **95**(6), 067401 (2005)
28. K. Hennessy, A. Badolato, M. Winger, D. Gerace, M. Atatüre, S. Gulde, S. Fält, E.L. Hu, A. Imamoğlu, Quantum nature of a strongly coupled single quantum dot-cavity system. Nature **445**(7130), 896–899 (2007)
29. D. Englund, A. Faraon, B. Zhang, Y. Yamamoto, J. Vučković, Generation and transfer of single photons on a photonic crystal chip. Opt. Express **15**(9), 5550–5558 (2007)
30. A. Faraon, A. Majumdar, D. Englund, E. Kim, M. Bajcsy, J. Vučković, Integrated quantum optical networks based on quantum dots and photonic crystals. New J. Phys. **13**(5), 055025 (2011)
31. D. Englund, A. Faraon, I. Fushman, N. Stoltz, P. Petroff, J. Vučković, Controlling cavity reflectivity with a single quantum dot. Nature **450**(7171), 857–861 (2007)
32. D. Englund, A. Majumdar, M. Bajcsy, A. Faraon, P. Petroff, J. Vučković, Ultrafast photon-photon interaction in a strongly coupled quantum dot-cavity system. Phys. Rev. Lett. **108**(9), 093604 (2012)
33. R. Bose, D. Sridharan, H. Kim, G.S. Solomon, E. Waks, Low-photon-number optical switching with a single quantum dot coupled to a photonic crystal cavity. Phys. Rev. Lett. **108**(22), 227402 (2012)
34. T. Volz, A. Reinhard, M. Winger, A. Badolato, K.J. Hennessy, E.L. Hu, A. Imamoğlu, Ultrafast all-optical switching by single photons. Nat. Photonics **6**(9), 605–609 (2012)
35. K. Srinivasan, O. Painter, Linear and nonlinear optical spectroscopy of a strongly coupled microdisk-quantum dot system. Nature **450**(7171), 862–865 (2007)
36. I. Fushman, D. Englund, A. Faraon, N. Stoltz, P. Petroff, J. Vučković, Controlled phase shifts with a single quantum dot. Science **320**(5877), 769–772 (2008)

37. D. Gerace, H.E. Türeci, A. Imamoglu, V. Giovannetti, R. Fazio, The quantum-optical josephson interferometer. Nat. Phys. **5**(4), 281–284 (2009)
38. A. Faraon, I. Fushman, D. Englund, N. Stoltz, P. Petroff, J. Vučković, Coherent generation of non-classical light on a chip via photon-induced tunnelling and blockade. Nat. Phys. **4**(11), 859–863 (2008)
39. K. Müller, A. Rundquist, K.A. Fischer, T. Sarmiento, K.G. Lagoudakis, Y.A. Kelaita, C.S. Muñoz, E. del Valle, F.P. Laussy, J. Vučković, Coherent generation of nonclassical light on chip via detuned photon blockade. Phys. Rev. Lett. **114**(23), 233601 (2015)
40. H. Kim, R. Bose, T.C. Shen, G.S. Solomon, E. Waks, A quantum logic gate between a solid-state quantum bit and a photon. Nat. Photonics **7**(5), 373–377 (2013)
41. M.T. Rakher, N.G. Stoltz, L.A. Coldren, P.M. Petroff, D. Bouwmeester, Externally mode-matched cavity quantum electrodynamics with charge-tunable quantum dots. Phys. Rev. Lett. **102**(9), 097403 (2009)
42. D. Pinotsi, J.M. Sanchez, P. Fallahi, A. Badolato, A. Imamoglu, Charge controlled self-assembled quantum dots coupled to photonic crystal nanocavities. Photonics Nanostruct. Fundam. Appl. **10**(3), 256–262 (2012)
43. K.G. Lagoudakis, K. Fischer, T. Sarmiento, A. Majumdar, A. Rundquist, J. Lu, M. Bajcsy, J. Vučković, Deterministically charged quantum dots in photonic crystal nanoresonators for efficient spin-photon interfaces. New J. Phys. **15**(11), 113056 (2013)
44. S.G. Carter, T.M. Sweeney, M. Kim, C.S. Kim, D. Solenov, S.E. Economou, T.L. Reinecke, L. Yang, A.S. Bracker, D. Gammon, Quantum control of a spin qubit coupled to a photonic crystal cavity. Nat. Photonics **7**(4), 329–334 (2013)
45. C. Arnold, J. Demory, V. Loo, A. Lemaître, I. Sagnes, M. Glazov, O. Krebs, P. Voisin, P. Senellart, L. Lanco, Macroscopic rotation of photon polarization induced by a single spin. Nat. Commun. **6** (2015)
46. P. Androvitsaneas, A.B. Young, C. Schneider, S. Maier, M. Kamp, S. Höfling, S. Knauer, E. Harbord, C.Y. Hu, J.G. Rarity, R. Oulton, Charged quantum dot micropillar system for deterministic light-matter interactions. Phys. Rev. B **93**, 241409 (2016). Jun
47. S. Sun, H. Kim, G.S. Solomon, E. Waks, A quantum phase switch between a single solid-state spin and a photon. Nat. Nanotechnol. **11**(6), 539–544 (2016)
48. D.F. Walls, G.J. Milburn, *Quantum Optics* (Springer Science & Business Media, Heidelberg, 2007)
49. S. Hughes, H. Kamada, Single-quantum-dot strong coupling in a semiconductor photonic crystal nanocavity side coupled to a waveguide. Phys. Rev. B **70**(19), 195313 (2004)
50. E. Waks, J. Vuckovic, Dipole induced transparency in drop-filter cavity-waveguide systems. Phys. Rev. Lett. **96**(15), 153601 (2006)
51. J.-T. Shen, S. Fan et al., Theory of single-photon transport in a single-mode waveguide. i. coupling to a cavity containing a two-level atom. Phys. Rev. A **79**(2), 023837 (2009)
52. V. Giesz, N. Somaschi, G. Hornecker, T. Grange, B. Reznychenko, L. De Santis, J. Demory, C. Gomez, I. Sagnes, A. Lemaître, et al., Coherent manipulation of a solid-state artificial atom with few photons. Nat. Commun. **7** (2016)
53. A. Faraon, E. Waks, D. Englund, I. Fushman, J. Vučković, Efficient photonic crystal cavity-waveguide couplers. Appl. Phys. Lett. **90**(7), 073102 (2007)
54. R. Bose, D. Sridharan, G.S. Solomon, E. Waks, Observation of strong coupling through transmission modification of a cavity-coupled photonic crystal waveguide. Opt. Express **19**(6), 5398–5409 (2011)
55. T. Cai, R. Bose, G.S. Solomon, E. Waks, Controlled coupling of photonic crystal cavities using photochromic tuning. Appl. Phys. Lett. **102**(14), 141118 (2013)
56. A.Y. Piggott, K.G. Lagoudakis, T. Sarmiento, M. Bajcsy, G. Shambat, J. Vučković, Photo-oxidative tuning of individual and coupled gaas photonic crystal cavities. Opt. Express **22**(12), 15017–15023 (2014)
57. R. Bose, D. Sridharan, G.S. Solomon, E. Waks, Large optical stark shifts in semiconductor quantum dots coupled to photonic crystal cavities. Appl. Phys. Lett. **98**(12), 121109 (2011)

58. C. Emary, X. Xu, D.G. Steel, S. Saikin, L.J. Sham, Fast initialization of the spin state of an electron in a quantum dot in the voigt configuration. Phys. Rev. Lett. **98**(4), 047401 (2007)
59. X. Xiaodong, W. Yanwen, B. Sun, Q. Huang, J. Cheng, D.G. Steel, A.S. Bracker, D. Gammon, C. Emary, L.J. Sham, Fast spin state initialization in a singly charged inas-gaas quantum dot by optical cooling. Phys. Rev. Lett. **99**(9), 097401 (2007)
60. S. Sun, E. Waks, Deterministic generation of entanglement between a quantum-dot spin and a photon. Phys. Rev. A **90**(4), 042322 (2014)
61. S. Sun, E. Waks, Single-shot optical readout of a quantum bit using cavity quantum electrodynamics. Phys. Rev. A **94**, 012307 (2016). Jul
62. Y. Ota, M. Shirane, M. Nomura, N. Kumagai, S. Ishida, S. Iwamoto, S. Yorozu, Y. Arakawa, Vacuum rabi splitting with a single quantum dot embedded in a h1 photonic crystal nanocavity. Appl. Phys. Lett. **94**(3), 3102 (2009)
63. R. Ohta, Y. Ota, M. Nomura, N. Kumagai, S. Ishida, S. Iwamoto, Y. Arakawa, Strong coupling between a photonic crystal nanobeam cavity and a single quantum dot. Appl. Phys. Lett. **98**(17), 3104 (2011)
64. E. Weidner, S. Combrié, A. De Rossi, J. Nagle, S. Cassette, A. Talneau, H. Benisty et al., Achievement of ultrahigh quality factors in gaas photonic crystal membrane nanocavity. Appl. Phys. Lett. **89**(22), 221104 (2006)
65. S. Combrié, A. De Rossi, Q.V. Tran, H. Benisty, Gaas photonic crystal cavity with ultrahigh q: microwatt nonlinearity at 1.55 μm. Opt. Lett. **33**(16), 1908–1910 (2008)
66. C. Schneider, M. Strauß, T. Sünner, A. Huggenberger, D. Wiener, S. Reitzenstein, M. Kamp, S. Höfling, A. Forchel, Lithographic alignment to site-controlled quantum dots for device integration. Appl. Phys. Lett. **92**(18), 183101 (2008)
67. M.K. Yakes, L. Yang, A.S. Bracker, T.M. Sweeney, P.G. Brereton, M. Kim, C.S. Kim, P.M. Vora, D. Park, S.G. Carter et al., Leveraging crystal anisotropy for deterministic growth of inas quantum dots with narrow optical linewidths. Nano Lett. **13**(10), 4870–4875 (2013)

Chapter 12
Entanglement Generation Based on Quantum Dot Spins

Aymeric Delteil, Wei-bo Gao, Zhe Sun and Ataç Imamoğlu

Abstract Quantum correlations between a confined spin and a propagating single photon can be used to entangle distant spins. In this chapter, we review recent progress in the field culminating in the demonstration of spin-photon entanglement, teleportation of quantum information from a photonic qubit to a quantum dot spin and heralded entanglement of distant hole spins. These results constitute important milestones towards the realization of quantum repeaters and on-chip quantum networks.

12.1 Introduction

12.1.1 Motivation

Quantum networks made of quantum memories interconnected by photons could find applications in long-distance quantum communication based on quantum repeaters as well as in distributed quantum computation [1–4]. Realization of such networks requires a quantum interface between stationary (matter) qubits and flying (photonic) qubits. Self-assembled quantum dots (QDs) feature unique properties that allow for the realization of such an interface:

- they can be deterministically charged with a single electron or hole, the spin of which constitutes a quantum memory of coherence time up to several microseconds, limited by hyperfine interaction [5];

A. Delteil · Z. Sun · A. Imamoğlu (✉)
Institute of Quantum Electronics, ETH Zurich, 8093 Zurich, Switzerland
e-mail: imamoglu@phys.ethz.ch

A. Delteil
e-mail: delteila@phys.ethz.ch

W.-b. Gao
Division of Physics and Applied Physics, Nanyang Technological University, Singapore 637371, Singapore

P. Michler (ed.), *Quantum Dots for Quantum Information Technologies*, Nano-Optics and Nanophotonics, DOI 10.1007/978-3-319-56378-7_12

- they exhibit optical transition to excited states (trions) with a short lifetime of about 1 ns. Strong spin-orbit interaction yields spin-dependent selection rules [6, 7].

As a consequence, QDs allow for fast initialization [8], manipulation [9] and measurement of the spins using laser excitation. In addition, single photon emission is typically much faster than optical transitions in most other matter qubit systems. Moreover, they benefit from the semiconductor technology which allows to design integrated optical and electrical functionalities – for instance optical cavities for efficient light extraction [10, 11], and diodes for electric field control [11, 12] – and open prospects for on-chip integration [13].

In this chapter, we review recent experimental realization of elementary protocols based on self-assembled QDs. We first present the recent demonstration of spin-photon entanglement with QDs (Sect. 12.2), which constitutes the building block for realizing quantum-dot-based quantum networks. We then discuss the indistinguishability of photonic qubits emitted by remote QDs, essential for interconnecting distant nodes (Sect. 12.3). Based on these demonstrations, we review the experimental realization of quantum teleportation from a photonic qubit to a QD spin qubit (Sect. 12.4). We then focus on the recent demonstration of heralded entanglement generation. After having introduced the entanglement generation protocol (Sect. 12.5), we expose a novel method to measure the spin coherence based on interference of inelastic light scattering (Sect. 12.6), followed by the implementation and characterisation of a single qubit phase gate, useful for entanglement verification (Sect. 12.7). In the last part, we discuss the experimental verification of distant entanglement between QD hole spins (Sect. 12.8).

12.1.2 Quantum Dot Structures

The experiments discussed in this chapter are based on InAs/GaAs self-assembled QD samples grown by molecular beam epitaxy (MBE). A single QD layer is embedded in a lossy cavity of $Q \sim 20$, ensuring that a sizeable part of the emitted light will escape the sample in a single free space mode that can be coupled to a fibre. A solid immersion lens (SIL) increases further the collection efficiency, up to a total amount of $\sim$10–20% depending on the particular structure design. A Schottky (Sects. 12.2–12.4) or *p-i-n* (Sects. 12.5–12.8) diode structure allows to control the charge state of the QD as well as fine tuning of the optical transition energy using quantum confined Stark effect.

12.2 Quantum Dot Spin-Photon Interface

12.2.1 Spin-Photon Entanglement Generation Scheme

All demonstrations of entanglement between a photonic qubit and a matter qubit so far are based on excitation to a state external to the qubit subspace, from which radiative decay channels to the qubit states coexist with equal oscillator strength [14–16] (lambda system). Other entanglement generation schemes are in principle possible, for instance using giant Faraday rotation induced by a single spin in a cavity [17], however their experimental realization still remains elusive.

In single QD systems, a lambda scheme is formed when applying in-plane magnetic field (Voigt geometry), leading to Zeeman splitting of the spin ground states, to which any of the two trion excited states can decay at the same rate. The corresponding energy diagram is shown in Fig. 12.1, as well as the allowed optical transitions from one of the excited states (the red trion). All four transitions between the two ground states and the two excited states have identical oscillator strengths and are linearly polarized [18]. The ground states of the QD are identified by the orientation of the electron spin, with $|\uparrow\rangle$ and $|\downarrow\rangle$ respectively denoting spins parallel and antiparallel to the magnetic field direction. Spontaneous emission of an horizontally (H) or vertically (V) polarized photon at frequency ω_r or, respectively, ω_b from the trion state $|T_r\rangle$ at rate $\Gamma/2$ brings the QD back into the state $|\downarrow\rangle$ or, respectively, $|\uparrow\rangle$.

The basic principle behind the deterministic generation of a spin-photon entangled state is straightforward: following the excitation of the QD into one of the trion states (here we consider excitation to the red trion), radiative recombination projects the coupled QD/photonic mode system into the entangled state

$$|\Psi\rangle = \frac{1}{\sqrt{2}}(|\downarrow\rangle|\omega_r; H\rangle + i|\uparrow\rangle|\omega_b; V\rangle) \qquad (12.1)$$

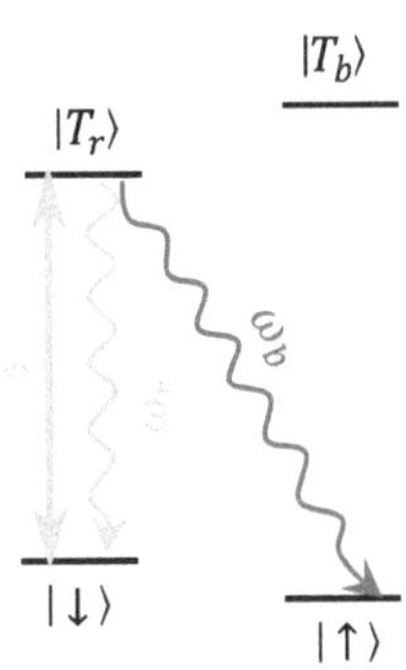

Fig. 12.1 Energy-level diagram of a single-electron-charged InGaAs QD in Voigt geometry. Spontaneous emission following resonant optical excitation of the trion state, $|T_r\rangle$, with Rabi frequency Ω_{res}, leads to the generation of an entangled spin-photon state

written in the computational basis, where $|\omega_r; H\rangle$ and $|\omega_b; V\rangle$ denote propagating single-photon-pulse state vectors with respective centre frequencies ω_r and ω_b and respective polarizations H and V. The electronic Zeeman energy satisfies $\omega_Z = \omega_b - \omega_r$.

As can be seen on (12.1), the spin is entangled with two photonic degrees of freedom: polarization and centre frequency. In order to avoid which-path information leakage, one of these two photonic degrees of freedom has to be erased to end with a Bell state. Erasure of the centre frequency degree of freedom can be obtained by either frequency down-conversion using short pulses [19] or by time filtering/post-selection using a window shorter than the beatnote period [20]. One obtains a state where the spin is entangled with the photon polarization. On the other hand, it is also possible to erase the polarization information using a polarizer at 45°, projecting the photon state on $|H + V\rangle$ or $|H - V\rangle$. This last approach is well suited for fibre-based applications, where the colour degree of freedom is robust to propagation in fibres thanks to a small typical wavelength difference between the two photonic components. In the following we will focus on the latter approach, developed in [21], in which the verification of spin-photon entanglement relies on the possibility to perform high-resolution time-resolved resonance fluorescence [22, 23].

In [21], the spin-photon entangled pair is generated from a single-electron-charged, self-assembled InGaAs QD, with an external magnetic field of 0.7 T applied perpendicular to the growth direction.

12.2.2 Entanglement Verification

Verification of entanglement and estimation of the fidelity to a maximally entangled state need not necessarily a full tomography of the generated state, although the latter can participate to understand fidelity limitations [24]; non-classical correlations can be evidenced by measuring conditional probabilities in two orthogonal bases [25].

In the case of spin-photon frequency entanglement, the two sets of correlations to be measured are:

- (classical) correlations between the spin in the computational basis and the photon colour
- (quantum) correlations between the spin in a superposition state ("rotated basis") and the photon in a superposition of the two frequencies.

In addition to resonant laser excitation of the trion states, the scheme uses non-resonant, right-hand-circularly (σ^+)-polarized, 4-ps-long mode-locked laser pulses that are red-detuned from the trion transitions by ~210 GHz: these pulses induce a coherent rotation of the electron spin [9], with minimal trion excitation (see Sect. 10.5.3 for a detailed description of this coherent control). The set-up, that integrates two-laser excitation together with time- and frequency-resolved resonance fluorescence, is depicted in Fig. 12.2. Suppression of the σ^+-polarized background laser light is ensured by a polarizer that projects the polarization of the resonance

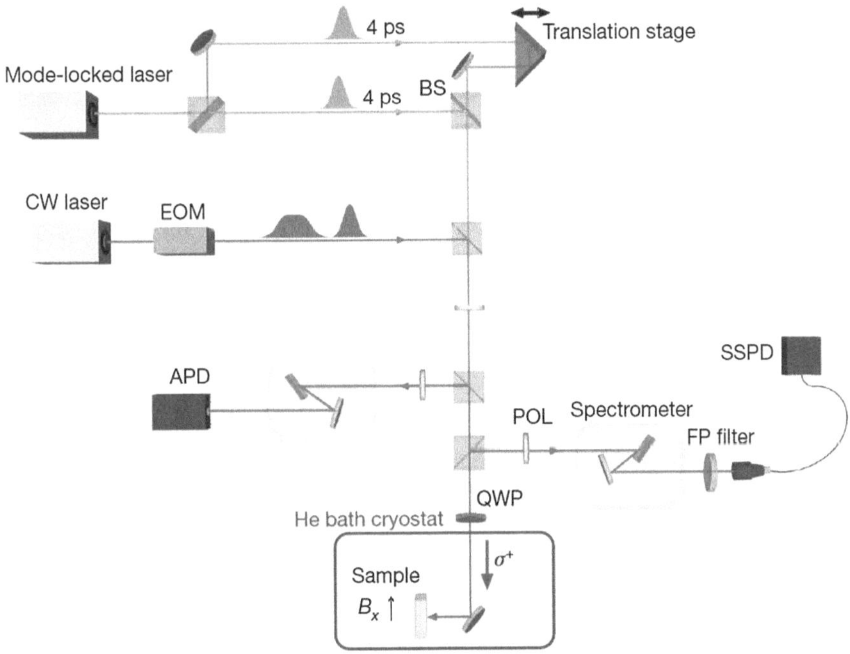

Fig. 12.2 Schematic of the experimental set-up highlighting the important optical elements. BS, beam splitter; CW, continuous wave; EOM, electro-optic modulator, FP, Fabry-Pérot filter; POL, polarizer QWP, quarter-wave plate

fluorescence photons to σ^-. As mentioned in Sect. 12.2.1, it also erases correlations between the electron spin and the polarization of emitted photons; photonic states in the output mode are then fully characterized by their centre frequency, and entanglement is then demonstrated between the electron spin orientation and the centre frequency of the single-photon pulse.

The pulse sequence we used is depicted on Fig. 12.3a. It consists of:

- a 5-ns resonant laser pulse tuned to the $|\downarrow\rangle - |T_r\rangle$ transition, to prepare the QD in state $|\uparrow\rangle$
- a 4-ps π-pulse that transfers the QD to the $|\downarrow\rangle$ state
- a 1.2-ns resonant laser pulse that generates the spin-photon entangled pair
- an additional 4-ps-long spin rotation pulse used to rotate the spin-measurement basis.

The whole pulse sequence is repeated after 13 ns, with the 5-ns resonant laser implementing spin measurement for the preceding pulse sequence as well as ensuring spin pumping/preparation in the $|\uparrow\rangle$ state for the subsequent cycle. Indeed the detection of a photon at the avalanche photodiode (APD) during this pulse tells with a high confidence level that the spin is in the state $|\downarrow\rangle$. Note that, as the average number of photons emitted during this spin measurement/pumping time window is

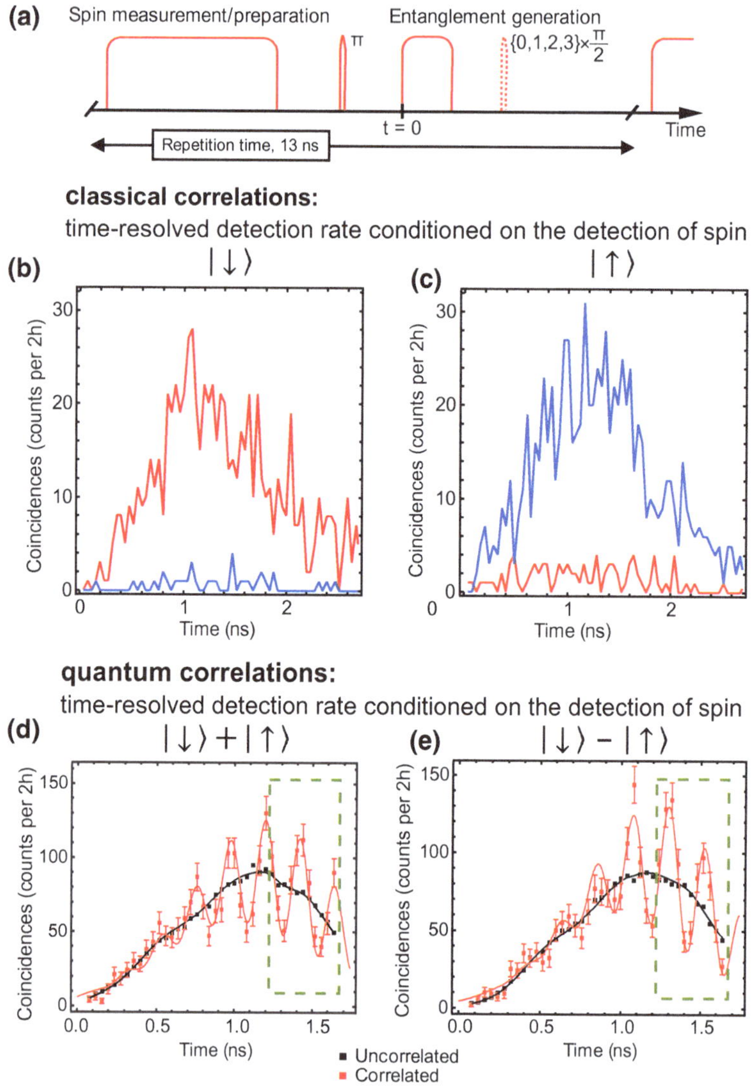

Fig. 12.3 **a** Pulse sequence used to measure the spin-photon correlations. **b** Time-dependent resonance fluorescence at $|\omega_r\rangle$ (*red*) and $|\omega_b\rangle$ (*blue*), conditioned on detection of a photon during the subsequent measurement/preparation pulse. The strong suppression of conditional photon detection events at $|\omega_b\rangle$ demonstrates the strong classical correlation between spin measurement yielding $|\downarrow\rangle$ and a photon detection at $|\omega_r\rangle$. **c** Same as in **d** but now with a π-pulse applied before the measurement/preparation pulse, showing strong correlations between spin detection in $|\uparrow\rangle$ and photon detection at $|\omega_b\rangle$. **d** Time-resolved coincidence events between the single-photon detection following the entanglement pulse and the detection of a photon during the first measurement/preparation pulse following a $\pi/2$-pulse (*red squares*). For comparison, an average of coincidence events between the spin and photon detection events taking place in 25 different excitation/preparation cycles is also shown (*black squares*). The *red curve* is a fit to the experimental data. **e** Same as in **d** but now with a $3\pi/2$-pulse

$\sim$1 and our combined collection and detection efficiency is $\sim$0.1%, the absence of a detection events brings no information, therefore it implies that we extract heralded spin information [7].

Classical Correlations

Demonstration of spin-photon correlations in the computational basis requires the measurement of classical correlation between the spin direction and the photon centre frequency. This is achieved by carrying out a coincidence measurement between resonance fluorescence photons at either $|\omega_b\rangle$ or $|\omega_r\rangle$ generated immediately after the entanglement pulse and those generated during the measurement/preparation pulse. Conditional on the detection of the spin state $|\downarrow\rangle$, we find that the probability of detecting a blue (ω_b) photon at the SSPD is drastically suppressed compared with the probability of detecting a red (ω_r) photon (Fig. 12.3b). With the additional 4-ps π-pulse introduced at $t = 3$ ns, it is possible to condition the single-photon colour measurements on detection of the spin in $|\uparrow\rangle$. In this case, we find that the red-photon detection events are strongly suppressed (Fig. 12.3c). From these measurements, we determine the fidelity of the classical correlations to be $F_1 = 0.87 \pm 0.09$.

Quantum Correlations

To verify that the generated spin-photon state is entangled, correlations between spin orientation and photon colour have to be measured in a rotated basis. This verification makes use of the relative phase between the two components of the entangled state in (12.1). After generation at time t_g, the two components acquire a time-dependent phase that stems from the different propagation phase factors of the blue and red frequency components of the photonic mode. In our experiments, we infer the generation time from the detection time, t_d, of the single photon at the SSPD, through the relation $t_g = t_d - L/c$, where L denotes the distance from the QD to the detector and c is the speed of light. Hence, given a photon detection event at t_d, we can a-posteriori write the time dependence of the spin-photon entangled state of (12.1) as

$$|\Psi(t)\rangle = \frac{1}{\sqrt{2}}\left(|\downarrow\rangle|\omega_r; H\rangle e^{-i\omega_Z(t-t_g)} + i|\uparrow\rangle|\omega_b; V\rangle\right) \tag{12.2}$$

The requisite spin measurement is implemented by rotating the QD electron spin with either a $\pi/2$-pulse or a $3\pi/2$-pulse at $t = t_1 = 1.7$ ns. A photon detection event at the APD during the following measurement/preparation pulse then projects the electron spin into $(|\uparrow\rangle - i|\downarrow\rangle)/\sqrt{2}$ ($\pi/2$-pulse) or $(|\uparrow\rangle + i|\downarrow\rangle)/\sqrt{2}$ ($3\pi/2$-pulse). The photonic wavefunction consistent with the spin measurement after the $\pi/2$-pulse can be written as

$$\left(|\omega_r; H\rangle e^{-i\omega_Z(t-t_g)} - |\omega_b; V\rangle\right)/\sqrt{2} \tag{12.3}$$

After passing through the polarizer, which fixes the polarization of the photon to be $(|H\rangle - i|V\rangle)/\sqrt{2}$, this wavefunction is

$$|\Phi(t)\rangle = \frac{1}{\sqrt{2}}\left(|\omega_r\rangle e^{-i\omega_Z(t-t_g)} - i\,|\omega_b\rangle\right) \tag{12.4}$$

The resulting coincidence probability is $P = \left(1 + \sin\left(\omega_Z\left(t_1 - t_g\right)\right)\right)$. As a consequence of the time-dependent relative phase, two single-photon states generated at different times, t_g^a and t_g^b, such that $\omega_Z\left(t_g^a - t_g^b\right) = \pi$, allow us to infer the overlap of the photonic state with two orthogonal states. Random photon detection times together with the fact that the single-photon pulse width, τ, satisfies $\omega_Z\tau > 1$, effectively implement projective measurements into orthogonal states such as $|\omega_r\rangle \pm |\omega_b\rangle$. The single-photon pulses are measured by a fast superconducting single photon detector (SSPD), which has a jitter of $\Delta\tau = 60$ ps, short enough to resolve the time dependence in (12.4).

Figure 12.3b shows the coincidence between the single-photon detection events induced by the entanglement pulse and the detection of a photon during the measurement/preparation pulse, following a $\pi/2$-pulse: the oscillations with a period given by $2\pi/\omega_Z = 238$ ps in the conditioned single-photon detection events (Fig. 12.3d, red squares) stem from the conditioned photonic wavefunction given in (12.4) and constitute a remarkable manifestation of the quantum coherence of the entangled spin-photon system. For comparison, we also show an average of coincidence events between the spin and photon detection events taking place in different excitation/preparation cycles (Fig. 12.3d, black squares), which do not show any oscillations. Figure 12.3e shows the time-dependent coincidence measurements obtained by applying a $3\pi/2$-pulse before the measurement/preparation pulse. Note that the oscillations of visibility $47 \pm 4\%$ for spin detection along $(|\downarrow\rangle + i|\uparrow\rangle)/\sqrt{2}$ are π out of phase relative to those along $(|\downarrow\rangle - i|\uparrow\rangle)/\sqrt{2}$. Using the data from Fig. 12.3d, e, we determine the fidelity of the quantum correlations in this rotated basis to be the average of the two visibilities, yielding $F_2 = 0.46 \pm 0.04$. The overall measured entanglement fidelity is then $F \geq (F_1 + F_2)/2 = 0.67 \pm 0.05$, limited predominantly by the finite jitter of the SSPD.

12.2.3 Coherence of the Entangled Pair and Spin-Photon Entanglement with Spin-Echo Sequence

A major limitation in the demonstration of quantum correlations between the electron spin and the photon frequency stems from the relatively short ground-state spin decoherence time, T_2^*, of the electron spin, originating from the slowly fluctuating nuclear spin environment: owing to the hyperfine interaction between the electron spin and QD nuclear spin ensemble, the electron spin is subject to an effective magnetic field determined by a combination of the fixed externally applied field and the quasi-static random Overhauser field. For different repetitions of the experiment the nuclear spins will be in different states, yielding a corresponding slowly fluctuating Overhauser field and thereby decreasing of the timescale that electron spin coherence is observable. This timescale is commonly referred to as spin T_2^* dephasing time. We refer the reader to Sect. 9.4 for an extensive description and characterisation of nuclear-spin-induced electron spin dephasing.

The decay of the oscillations as a function of $t_1 - t_g$ in conditioned single-photon detection events (Fig. 12.3b, d) constitutes a measurement of the electron spin decoherence that was previously carried out using Ramsey interferometry [5]. By fitting the decay of coincidence measurements in the time window [0 ns, 1.64 ns] we find that $T_2^* = 1.1 \pm 0.2$ ns for the data in Fig. 12.3b and that $T_2^* = 0.9 \pm 0.2$ ns for the

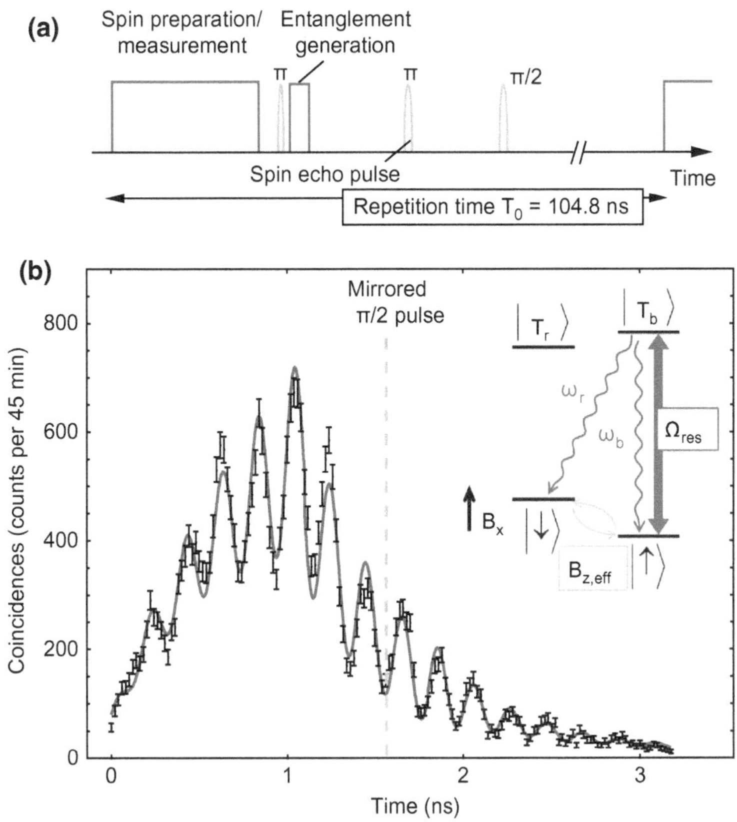

Fig. 12.4 **a** Laser pulse sequence used for the generation of spin-photon entangled state and measurement of quantum correlations after a spin-echo sequence. **b** Spin-photon entanglement in the rotated basis and the relevant energy-level diagram (*right inset*). The main figure shows the photon detection events as a function of time, during and after the entanglement generation pulse, conditioned upon a spin-echo-delayed spin measurement in the $(|\downarrow\rangle + |\uparrow\rangle)/\sqrt{2}$ state. The entanglement generation pulse is turned on at time $t = 0$ and is 1.2 ns long. The oscillations at $\omega_b - \omega_r = 4.9$ GHz are due to the beating between the two frequency components of the projected single-photon superposition state. The vertical *dashed line* indicates the entangled spin-photon generation time for which the spin-echo sequence exactly cancels the random phase accumulated due to hyperfine interactions

data in Fig. 12.3d, for this QD. The coherence time can be prolonged beyond T_2^* thanks to a spin-echo sequence, allowing to utilize the spin-photon entangled pair in quantum communication protocols where the network nodes are separated by macroscopic distances. This has been demonstrated in [26], where a spin-echo sequence is applied to a singly charged QD as illustrated in Fig. 12.4a, together with the relevant energy-level diagram and the allowed optical transitions under an external magnetic field $B_x = 0.7$ T that is applied perpendicular to the growth direction (Fig. 12.4b, inset). The QD is prepared in $|\downarrow\rangle$ state by applying a 10 ns-long resonant laser pulse tuned to the $|\uparrow\rangle - |T_b\rangle$ transition. A subsequent 4 ps-long π-pulse ("rotation pulse") transfers the spin onto $|\uparrow\rangle$. An entangled spin-photon pair is then generated by resonant excitation of the $|T_b\rangle$ with a 1.2 ns resonant laser pulse ("entanglement pulse"). The whole pulse sequence is repeated after 104.8 ns, with the subsequent 10-ns long resonant laser implementing spin measurement for the preceding pulse sequence, as well as ensuring spin pumping/preparation in the $|\downarrow\rangle$ state for the next cycle. For an echo time $T_{echo} = 13$ ns, a measurement of the spin-photon correlations in a rotated basis is shown in Fig. 12.4b. Combined with measurements that project the photon and the spin in different states (similarly to described in Sect. 12.2.2), we can calculate the spin-photon entanglement fidelity to be $F > 0.63 \pm 0.02$. This bound is once again limited by the detection jitter (64 ps). If the detector jitter were absent, we would obtain $F > 0.71 \pm 0.02$ ($F > 0.82 \pm 0.02$) as the lower bound for entanglement fidelity with (without) spin-echo.

An alternative way to retain coherence of a spin-photon entangled pair during a longer time is to use hole spins, which exhibit a similar energy level diagram but have a T_2^* one to two orders of magnitude longer. This is the route taken by the experiments described in Sects. 12.5–12.8.

Another QD based system that exhibits a lambda scheme which is in principle well suited for generated spin-photon entanglement is a QD molecule in the doubly charged regime, at the so-called sweet spot [27]. Although spin-photon entanglement has not been demonstrated so far, the lambda system formed by the singlet and triplet ground states optically coupled to a common trion state has been used to demonstrate coherent population trapping, showing promising spin coherence of tens to hundreds of nanoseconds.

12.3 Indistinguishability of Photonic Qubits Emitted by Different Dots

The implementations of quantum communication protocols that we will be discussing in the Sects. 12.4–12.8 rely on the indistinguishability of single photons emitted by remote QDs. In this section we show how to generate photonic qubits (that are not entangled with any external degree of freedom) and characterize the indistinguishability of such qubits emitted by two different QDs [26].

12.3.1 Generation of Photonic Frequency Qubits

In a neutral self-assembled InGaAs QD, the elementary optical excitations from the unique ground state $|0\rangle$ are the two fundamental exciton states $|X_r\rangle$ and $|X_b\rangle$ that are split in energy by the anisotropic electron-hole exchange interaction [28]. It is therefore possible to generate a photonic qubit (see Sect. 7.1) by resonant pulsed excitations of these excitonic transitions. The exciton state $|X_r\rangle$ ($|X_b\rangle$) decays at a rate Γ_1 by spontaneous emission of a photon at frequency ω_r (ω_b) back into $|0\rangle$. A laser pulse resonant with either of these two exciton states results in the generation of single-colour single photon states denoted by $|\omega_r\rangle$, $|\omega_b\rangle$. Alternatively, applying a two-colour laser pulse that is simultaneously resonant with both $|X_r\rangle$ and $|X_b\rangle$ results in the superposition state $\alpha|\omega_r\rangle + \beta|\omega_b\rangle$ (left inset of Fig. 12.5).

A two-colour laser pulse that excites the QD in such a superposition of exciton states can be obtained by tuning the frequency of a diode laser exactly in the middle of the two neutral exciton transitions and sending the beam to an amplitude electro-optic modulator (EOM) that generates pulses from the continuous-wave laser field. A phase EOM is used to generate laser sidebands that are on resonance with the exciton transitions, and enables us to resonantly drive $|X_r\rangle$ and $|X_b\rangle$ simultaneously. The amplitude of the modulation signal is set such that the central (carrier) peak at the laser frequency is completely suppressed. In this way, most of the laser power is carried by the frequencies ω_r and ω_b. Two orthogonal polarizers on the excitation and collection paths of the QD, set at $|H+V\rangle$ and $|H-V\rangle$, respectively, are used

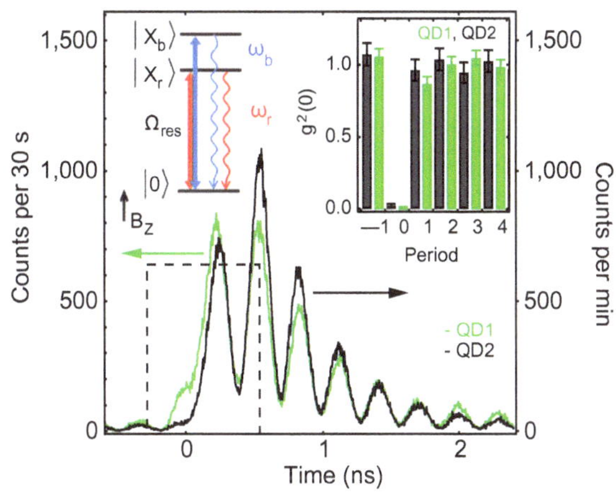

Fig. 12.5 Time resolved resonance fluorescence counts for QD1 (*green*) and QD2 (*black*). A 0.8 ns two-colour laser pulse (*dashed line*) that is resonant with both excitonic transitions excites the QD (*left inset*). Subsequent spontaneous emission generates a single photon in a superposition of ω_r and ω_b. The matched oscillations of the emission from the two QDs at $\omega_b - \omega_r = 3.45$ GHz indicate that the generated photonic qubit states are nearly identical. The *right inset* shows photon correlation ($g^{(2)}$) using the photons emitted after the excitation pulse is turned off, such that the likelihood of two or more photon detection events is very small

to suppress the laser background. Upon spontaneous emission at t_e, the generated photon is in the state $|\psi_p\rangle = \left(e^{i\Delta t_e/2}e^{i\phi_L}|\omega_b\rangle + e^{-i\Delta t_e/2}e^{-i\phi_L}|\omega_r\rangle\right)/\sqrt{2}$, where $\Delta = \omega_b - \omega_r$ and ϕ_L is the tunable relative phase imprinted by phase modulation.

The detection of the photonic frequency qubit can be implemented with a detector having time jitter smaller than $1/\Delta$. Upon absorption of a photon, the qubit is projected onto the state $|\psi_m\rangle = (|\omega_b\rangle + |\omega_r\rangle)/\sqrt{2}$; the probability that a photon is registered is then $|\langle\psi_m|\psi_p\rangle|^2 = (1 + \cos(2\phi_L + \Delta t_e))/2$. With ϕ_L and Δ fixed, the counts show an oscillation as a function of the photon emission time t_e, which can be deduced from the corresponding photon detection time t_d. The result is a temporal beat signal stemming from an interference of the two partial waves of the frequency qubit on the single-photon detector. The measurement of a photonic frequency qubit is depicted in Fig. 12.5 for two QDs (QD1 and QD2): the beats of ω_r and ω_b demonstrate that the single photon is in a coherent superposition of two frequencies. The finite visibility of the interference of the two frequency components stems predominantly from the jitter of the detector. The single photon character of the photonic qubit is verified by the vanishing $g^{(2)}$ (see Sect. 1.2) as shown in the inset of Fig. 12.5.

12.3.2 *Indistinguishability of the Photonic Qubits*

While indistinguishability of consecutive photons emitted by a single QD has been extensively investigated, reaching high values above 99% visibility [11, 29, 30], what is of interest for the implementation of protocols interconnecting remote nodes is the indistinguishability of photons emitted by different emitters, that are subject to independent fluctuations and possibly slightly different characteristics (e.g. lifetime or centre frequency). In the following we present a characterization of the indistinguishability of two photonic qubits emitted by two different QDs using two-photon interference in a Hong-Ou-Mandel (HOM) set-up [31]. The photonic frequency qubits are generated by two QDs (QD1 and QD2) placed in separate cryostats. Obtaining pairs of QDs having identical emission wavelength is challenging since self-assembled QD wavelength distribution typically spreads over a few tens of nanometers, which is about four orders of magnitude larger than the natural linewidth. Although the quantum confined Stark effect allows to tune the resonance to some extend (typically 10–30 GHz) it is necessary to have at one's disposal a large number of dots in order to spectrally select a matching pair from photoluminescence measurements. Once a pair of QDs with nearly identical PL emission wavelengths is found, local gate voltages applied separately to the two QDs are used to fine tune their transition frequencies into resonance. To ensure that both ω_r and ω_b of the two QDs are identical, we additionally apply local magnetic fields.

Figure 12.6 shows the time-resolved coincidences between the two output ports of the HOM interferometer as a function of the time delay between the photon detection times, when the two input photons have either identical of orthogonal polarizations. Our measurements reveal that the visibility is $V = (C_\perp - C_{//})/C_\perp = 80.2 \pm 2.9\%$, when we consider the photon emission after the laser pulse in the time window [0.8 ns,

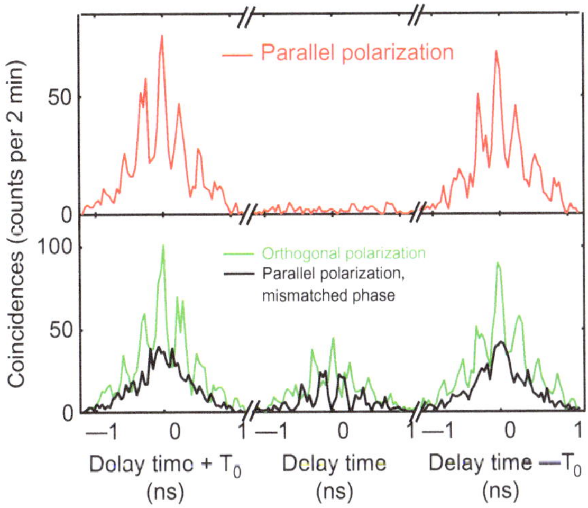

Fig. 12.6 The photonics qubits depicted in Sect. 12.3.1 are incident on a beam splitter. Coincidence counts on the two arms of the beam splitter are plotted as a function of the delay between the recorded photon arrival times. T_0 is the pulse repetition time of 13.1 ns

2 ns]. Here, $C_\perp$ ($C_{//}$) is the total counts in the central peak with orthogonal (parallel) polarization for the input modes of the beam splitter. When the two input photons are in different periods or if they have orthogonal polarizations, then two-photon interference is absent: the observed beat signal in those cases originates exclusively from single-photon interference. With a retro-reflecting prism in one of the arms of the interferometer, we introduce a half-period time delay ($t = \pi/\Delta$) for one of the pulses, which also renders the two photonic qubits distinguishable (Fig. 12.6b). For any given arrival time at the beam splitter, such a delay ensures that the relative phase between the two frequency components of the two single photons differ by π. In this case, the observed beat notes around the center period near zero delay time stem from an interference between the two frequency qubits $\left(|\omega_b\rangle + e^{i\theta}|\omega_r\rangle\right)/\sqrt{2}$ and $\left(|\omega_b\rangle - e^{i\theta}|\omega_r\rangle\right)/\sqrt{2}$. The suppression of the beat signal in different periods stems from a superposition of two π-phase-shifted single-photon interference patterns.

12.4 Photon to Spin Teleportation

The generation of spin-photon entangled pairs (Sect. 12.2) together with the indistinguishability of single photons emitted by remote QDs (Sect. 12.3) opens the way to experimental realization of long distance quantum communication protocols. Of particular interest is the teleportation from a propagating qubit to a stationary qubit [32, 33], that not only constitutes a versatile quantum-state-transfer method but can also serve as a quantum computational primitive [34–36]. In this section we focus on the recent demonstration of teleportation from a photonic frequency qubit to a QD spin qubit [26].

12.4.1 Set-Up and Protocol

In this experiment, the photonic frequency qubit to be teleported is brought into one arm of a Hong-Ou-Mandel (HOM) interferometer, while the other arm is fed with the photonic part of an entangled spin-photon pair of the form $|\Psi\rangle = (|\downarrow\rangle|\omega_b\rangle - |\uparrow\rangle|\omega_r\rangle)/\sqrt{2}$, generated as described in Sect. 12.2. The state of the coupled system consisting of two photons in modes A and B and the QD spin prior to the beam splitter is

$$|\Psi\rangle = \frac{1}{\sqrt{2}}(\alpha|\omega_b\rangle_A + \beta|\omega_r\rangle_A) \otimes (|\downarrow\rangle|\omega_r\rangle_B - |\uparrow\rangle|\omega_b\rangle_B) \tag{12.5}$$

Provided that the single-photon pulses have identical spatiotemporal profiles, a coincidence detection at the output of the interferometer heralds successful teleportation. Indeed if the photons in modes A and B are indistinguishable in every aspect but their internal (frequency/colour) state, the only possibility for a simultaneous coincidence detection at the output of the HOM interferometer is to have the input two-photon state in $|\varphi_S\rangle = (|\omega_b\rangle_A|\omega_r\rangle_B - |\omega_r\rangle_A|\omega_b\rangle_B)/\sqrt{2}$. Therefore, detection of a coincidence projects the input photonic state (in modes A and B) to $|\varphi_S\rangle$.

As mentioned in Sect. 12.2, the entangled state is affected by decoherence of the QD spin at a timescale $T_2^* \sim 1$ ns. To ensure that the electron spin coherence is intact for a longer time period, we introduce an optical spin-echo sequence that removes the effect of static, but random, Overhauser field, as described in Sect. 12.2.3. The electron spin is still subject to decoherence but now at a timescale that is commonly referred to as spin-echo coherence time T_2, such that $T_2 \gg T_2^*$.

The photonic qubit that is coupled to the mode A is generated by the neutral QD3, whose exciton transition energy is nearly identical to that of the QD2 trion. Local electric and magnetic fields ensure that both ω_r and ω_b are identical for both dots. We focus on photons emitted in a 800 ps long time interval, slightly longer than the 650 ps QD lifetime, and observe coincidences at the outputs of the HOM interferometer. The spin state corresponding to this measurement outcome is $\langle\varphi_S|\Psi\rangle = \alpha|\uparrow\rangle + \beta|\downarrow\rangle$, as it can be verified from (12.5).

12.4.2 Classical Correlations

Experimental verification of teleportation is based on three-fold coincidence detection of photons at the two output modes of the HOM interferometer, together with a photon detection during the spin-measurement/preparation pulse (Fig. 12.7a). The spin population in the $|\uparrow\rangle$ state before the start of the spin-measurement pulse determines the probability that resonantly excited QD scatters one or more photons. After a few optical cycles, the spin is pumped to the $|\downarrow\rangle$ state and the QD resonance fluorescence ceases.

In order to demonstrate classical correlations between the colour of the photon to be teleported and the final spin state, we use an input photon that is prepared either in $|\omega_r\rangle_A$ or $|\omega_b\rangle_A$ and measure the spin state projected to either $|\uparrow\rangle$ or $|\downarrow\rangle$. For a mode A photon in $|\omega_r\rangle_A$, a three-fold coincidence projects the photon in mode B onto $|\omega_b\rangle_B$ and the spin onto $|\uparrow\rangle$. Figure 12.7b shows that the same period (Period 0) three-fold coincidences corresponding to a spin measurement in $|\uparrow\rangle$ are a factor ~4 larger than those corresponding to $|\downarrow\rangle$. By using an input photon in state $|\omega_b\rangle_A$, the three-fold coincidences in Period 0 show that detecting the spin in $|\downarrow\rangle$ is now ~4 times more likely than detecting it in $|\uparrow\rangle$ (Fig. 12.7c), in accordance with the predictions

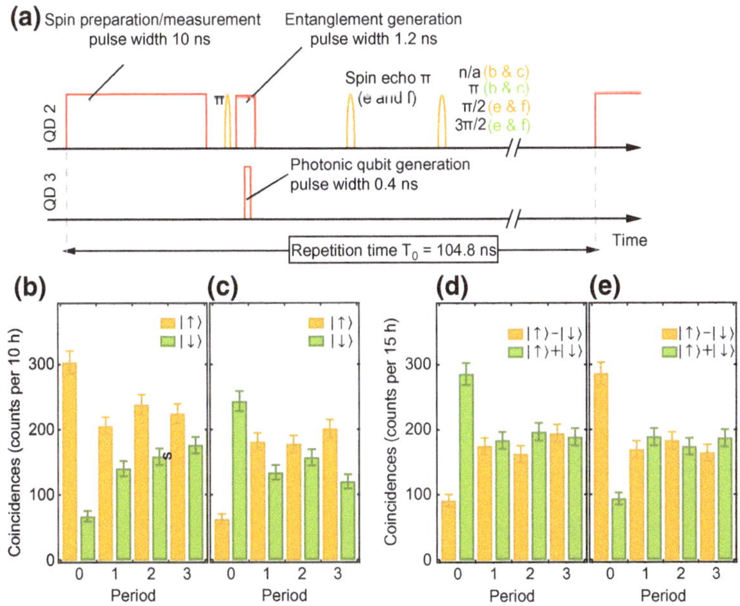

Fig. 12.7 Demonstration of quantum teleportation. **a** Experimental teleportation pulse sequence. The pulses applied to QD2 for entanglement generation are the same as in Fig. 12.3 with an additional spin echo π pulse. In the spin measurement stage, different combinations of spin rotation pulses are used. For QD3, 400 ps excitation pulses are used for generating the input photonic qubit. For teleportation, we use the coincidences at the two outputs of the HOM interferometer in a 800 ps long time-interval (labelled with the *grey box*). **b**, **c** Teleportation with input qubit $|\omega_r\rangle$ (**b**) or $|\omega_b\rangle$ (**c**). The plots show three-fold coincidence counts between the two output arms of the beam splitter and a photon detection during the following spin measurement pulse (Period = 0) or a later pulse period (Period > 0). The *green* and *yellow* columns represent different spin measurement basis. An enhanced probability for the spin state $|\uparrow\rangle$ ($|\downarrow\rangle$) as well as a decreased probability for $|\downarrow\rangle$ ($|\uparrow\rangle$) is observed when the photonic qubit was initially prepared in $|\omega_r\rangle$ ($|\omega_b\rangle$). **d**, **e** Same as **b**, **c** but with different input qubit $|\omega_r\rangle + |\omega_b\rangle$ (**d**) or $|\omega_r\rangle - |\omega_b\rangle$ (**e**). A spin-echo π-pulse is applied to prolong the spin coherence time

of the teleportation protocol. From these measurements, we obtain the teleported state fidelities 0.79 ± 0.1 (0.82 ± 0.09) for $|\omega_r\rangle_A$ ($|\omega_b\rangle_A$), where the fidelity for each teleported state is calculated as the overlap of the ideal electron spin state after the completion of the protocol with the corresponding experimental outcome for the spin state.

The different period three-fold coincidences are obtained by correlating the coincidence events for photons at the output of the HOM interferometer generated within the same period, with a spin measurement in the subsequent period. The corresponding coincidences also show a difference between the spin-up and spin-down populations, but this difference is due to a steady-state population difference between the two spin states and is not affected by the photonic qubit state.

12.4.3 *Quantum Correlations*

Demonstration of quantum teleportation requires that coherences in the photonic superposition state at the input mode A are faithfully transferred onto the spin state. To verify this, we prepare the single photon in mode A in either $|\omega_r\rangle_A + |\omega_b\rangle_A$ or $|\omega_r\rangle_A - |\omega_b\rangle_A$. As the propagation time of the photons onto the superconducting-single-photon detector (SSPD) is about 11 ns, we introduce a spin-echo pulse sequence to ensure that the spin measurement is carried out only after the coincidence detection at the output of the HOM interferometer. The three-fold coincidences now indicate an enhanced probability for detection of the spin in state $|\uparrow\rangle + |\downarrow\rangle$ for an input photon in $|\omega_r\rangle_A + |\omega_b\rangle_A$ (Fig. 12.7d) and $|\uparrow\rangle - |\downarrow\rangle$ for an input photon in $|\omega_r\rangle_A - |\omega_b\rangle_A$ (Fig. 12.7e). From these measurements, we obtain the teleported state fidelities 0.76 ± 0.03 (0.75 ± 0.03) for $|\omega_r\rangle_A + |\omega_b\rangle_A$ ($|\omega_r\rangle_A - |\omega_b\rangle_A$).

The measured teleported state fidelities are primarily limited by the small mismatch between the temporal pulse shapes and the spatial overlap profiles of the two interfering photons, as well as the finite spin-photon entanglement fidelity stemming from hyperfine-interaction-mediated electron spin decoherence. Unlike measurements of spin-photon entanglement fidelity (Sect. 12.2), the experimentally determined teleportation fidelity is independent of the detector jitter. Prolongation of the spin-echo time beyond 25 ns in our experiments is limited by the linearly coupled Overhauser field components along the electron spin quantisation direction [37]. The relatively short timescales at which teleportation can be observed for QD2 limits possible applications of teleportation where longer memory times are needed. We note that at higher magnetic fields spin T_2 coherence times exceeding 1 μs have been reported for single InAs QDs [5, 37].

12.5 Distant Entanglement Generation Protocol

Generation of distant entanglement between two distant QD spins can be thought of as a natural extension of the teleportation experiment, where the photonic qubit is replaced by a photon that is entangled with another QD spin, thus realizing the so-called Simon-Irvine protocol [38]. A two-fold coincidence detection at the output of the HOM interferometer would herald successful generation of entanglement between the two distant spins. Entanglement generation between two distant quantum memories based on such scheme has been demonstrated in several physical systems [39, 40].

The major obstacle to its realization as such with self-assembled QDs in Voigt geometry is the lack of efficient spin measurement: to verify entanglement one would need to measure both QD spins in addition to the two-fold coincidence heralding entanglement, leading to four-fold coincidence whose rate with typical collection efficiencies would be too low for a realistic implementation.

It is however possible to herald entanglement with a single detection event of spin-flip Raman scattering, provided that it is impossible to determine the source of the detected photon [41]. Such a scheme has been previously realized with trapped ions [42] and more recently with QD hole spins [43]. In the following we will focus on the latter realization. Unlike the previously described experiments (Sects. 12.2–12.4), heralded entanglement generation has been realized with hole spin qubits, that have a longer T_2^* coherence time in comparison with the electron spin (see [44–47] and Sect. 12.6). In the present case the holes are optically injected by resonant driving of the neutral exciton followed by tunelling of the electron to the back contact, leaving behind a single hole (see Sect. 10.5.2) as illustrated in Fig. 12.8, which depicts the relevant energy-level diagram as well as the allowed optical transitions for single-hole charged QDs in Voigt geometry. The initial states of the optical transitions in

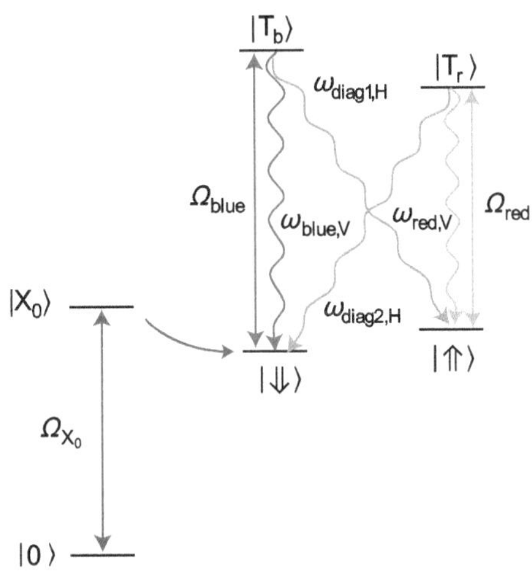

Fig. 12.8 Energy level diagram of a single QD. Upon excitation of the neutral exciton ($|X^0\rangle$) state, the electron can tunnel out, leaving behind a single hole. Application of a finite magnetic field gives rise to spin-dependent optical selection rules with four allowed transitions of identical oscillator strength

the single-hole charged regime are metastable states identified by the orientation of the heavy-hole pseudo-spin, with $|\Uparrow\rangle$ ($|\Downarrow\rangle$) denoting $+3/2$ ($-3/2$) hole angular momentum projection. Presence of $B_x \neq 0$ yields a finite splitting of the pseudo-spin states due to heavy-light hole mixing [28]. As in the previously described case of a negatively charged QD, spontaneous emission of a V (H) polarized photon at frequency ω_{blue} (ω_{diag1}) from the trion state $|T_b\rangle$ at rate $\Gamma/2$ brings the QD back into the $|\Downarrow\rangle$ ($|\Uparrow\rangle$) state. Due to these selection rules, addressing any of the four allowed transitions with a single laser will efficiently transfer the spin population into the opposite ground state within $\sim$10 ns.

Figure 12.9 depicts the experimental set-up where two QDs separated by 5 m are resonantly driven by weak 3.2 ns long pulses from a Ti:Sapphire laser, termed the entanglement laser. Additional diode laser pulses ensure that each QD is optically charged with a single excess heavy-hole and that the hole pseudo-spin is prepared in the requisite state. Since the intensity of entanglement laser is chosen to be well below saturation, the ensuing optical transitions lead to either V-polarized Rayleigh scattering or H-polarized Raman scattering (see Sects. 3.2 and 3.4.3 for an extensive description of light scattering in the low power regime).

The light propagation time from the first beam splitter (BS1) to both dots, as well as from the dots to the second beam splitter (BS2) are rendered nearly identical, such that the photons scattered by the two dots during a single entanglement laser pulse reach the second beam splitter at the same time. When both QDs are initially prepared in the $|\Downarrow\rangle$ state, simultaneous weak excitation of the blue transitions will lead to either a Raman or Rayleigh scattering event with a probability $\varepsilon^2 \ll 1$, leaving the system in the state

$$\begin{aligned} |\Psi\rangle_{12} = \tfrac{1}{\sqrt{2}}[|\Downarrow, 0\rangle + \varepsilon e^{-i\theta_1}(|\Uparrow, 1_{d1,H}\rangle + |\Downarrow, 1_{b,V}\rangle)]_{QD1} \\ \otimes[|\Downarrow, 0\rangle + \varepsilon e^{-i\theta_2}(|\Uparrow, 1_{d1,H}\rangle + |\Downarrow, 1_{b,V}\rangle)]_{QD2} \end{aligned} \qquad (12.6)$$

where $|1_{b,H}\rangle$ refers to a single H-polarized photon with centre frequency ω_{blue} and $|1_{d1,V}\rangle$ refers to a single V-polarized photon with centre frequency ω_{diag1}. To ensure that a click in one of the single-photon detectors stems from Raman scattering we use polarizers, transmission gratings and Fabry-Pérot filters. In this case, detection of a single (Raman) photon projects the composite system wave-function onto the maximally entangled state

$$|\Psi\rangle_{12} = \frac{1}{\sqrt{2}}[|\Uparrow, \Downarrow\rangle + e^{-i\theta}|\Downarrow, \Uparrow\rangle] \qquad (12.7)$$

in the limit where two photon scattering probability ε^4 is vanishingly small. Provided that the Zeeman splitting in the two QDs are rendered identical, the relative phase $\theta = \theta_2 - \theta_1$ is time-independent and is primarily determined by the optical path length difference between the two arms from BS1 to BS2 (Fig. 12.9).

The entanglement generation scheme we use relies crucially on the indistinguishability of the photons emitted by two remote QDs (QD1 and QD2) such that

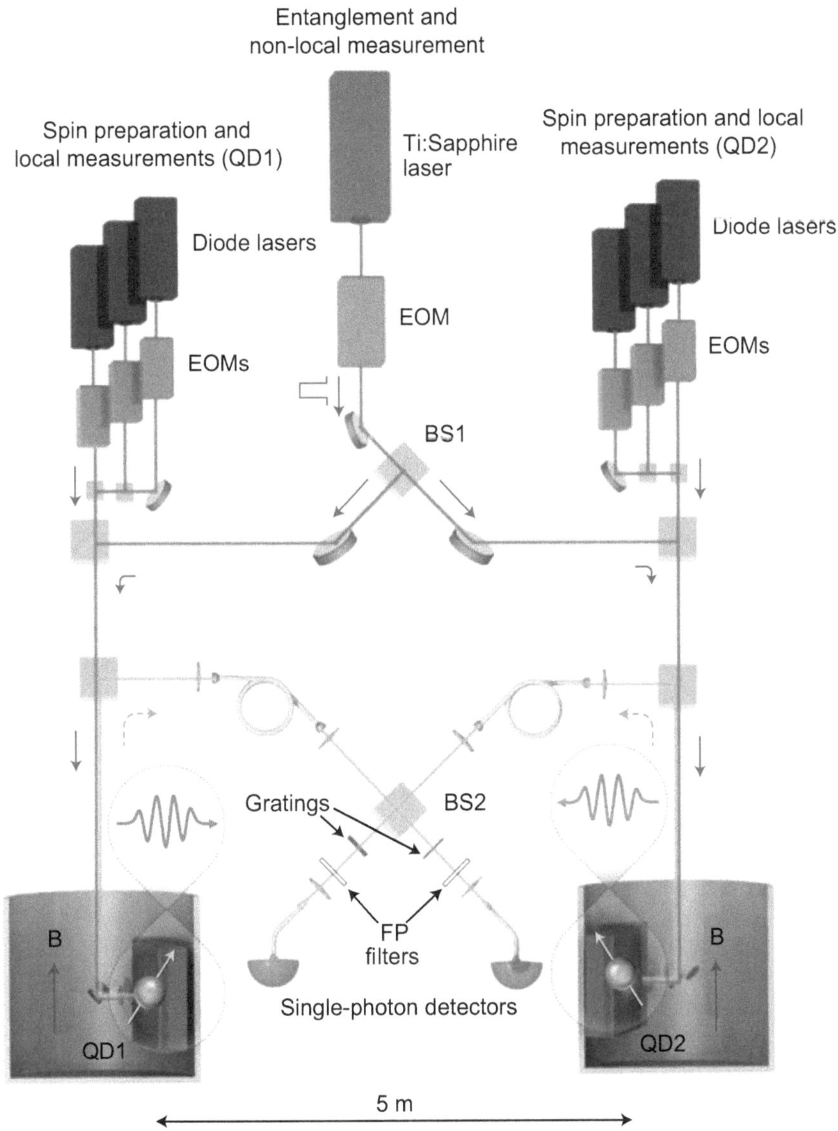

Fig. 12.9 Two bath cryostats separated by 5 m host QD samples in Voigt geometry. The QDs can be addressed by diode lasers (in *black*) for local state preparation and readout, and by a Ti:Sapphire laser (in *blue*) for entanglement generation and non-local measurement. EOM stands for electro-optic modulator (color figure online)

"which-path" information is not available in the single-photon interferometer depicted in Fig. 12.9. The indistinguishability of the Raman scattered photons is characterized by a Hong-Ou-Mandel experiment (see Sect. 12.3). The associated interference visibility deduced from this measurement is $91 \pm 6\%$ guaranteeing that no substantial "which-path" information is conveyed by Raman photons.

For the protocol we implement, it is essential that the QD spins remain coherent during the time it takes for the heralding process to be completed. In our scheme the latter is determined predominantly by the 21.7 ns propagation time from the QDs to the single-photon detectors. In the next section, we show how Raman scattering can be used to demonstrate that the hole pseudo-spin retains its coherence on this timescale.

12.6 Interference of Raman Scattering for Hole Coherence Measurement

12.6.1 Optically Injected Holes as Coherent Spin Qubits

The T_2^* coherence time of the electron spin in self-assembled QDs is short, typically about 1 ns. Such short value originates in hyperfine interaction with the fluctuating bath of nuclear spins, as discussed in Sect. 12.2.3 and in Sect. 9.4. On the other hand, the coherence time of the QD hole pseudo-spin in Voigt geometry has been found to exceed this value by one to two orders of magnitude, thanks to the p-type Bloch part of the hole wavefunction that leads to a vanishing Fermi contact term of the hyperfine interaction. The coherence time is then limited by charge fluctuations, hence being inversely proportional to the external magnetic field [48] and susceptible to vary from sample to sample. We refer the reader to Sect. 9.5 for an extensive characterisation of the hole dephasing mechanisms.

Ramsey interferometry with hole spins has been used by several groups to demonstrate values of hole T_2^* between 2.3 ns [45] and 20 ns [46, 47]. Coherent population trapping experiments have suggested even longer coherence times at low magnetic field, up to a few hundreds of ns [44]. Ramsey interferometry performed on the sample used in [21, 26, 49] confirms such high values, indicating a coherence time of ~250 ns for the optically injected holes at $B = 1$ T as it can be seen on Fig. 12.10.

12.6.2 Measurement of Spin Coherence Time Using Raman Scattering

Ramsey interferometry discussed in the previous subsection is the most usual technique to characterize the T_2^* coherence time of electrons and hole spins in QDs [45, 50]. It is however strongly affected by dynamical nuclear spin polarization effects [47, 51, 52] and requires strong detuned σ-polarized laser pulses to implement coherent rotation of the spin to be measured. An alternative of Ramsey interferometry for spin coherence measurement is single photon interference of spin-flip Raman scattering [53].

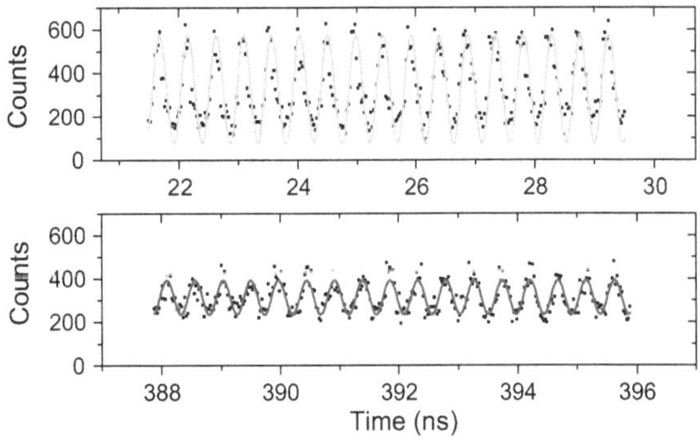

Fig. 12.10 *Black dots* Ramsey interferometry performed on holes spins, revealing oscillations up to ~400 ns delay. *Plain curves* sine fitting of the data

This technique is based on the fact that first-order coherence (as defined in the Sect. 1.2) of spin-flip Raman scattering (inelastic scattering resulting a change of the spin state – see Sect. 3.4.3) is determined by the coherence properties of the excitation laser field and the spin coherence [54, 55]. Therefore, measuring the coherence time of Raman scattered photons upon excitation with a monochromatic laser field is equivalent to a measurement of the spin dephasing time. It is essential to carry out Raman coherence measurements at low excitation limit well below the saturation intensity in order to ensure that spin dephasing induced by Rayleigh scattering remains weak as compared to the inherent T_2^* time. Moreover, dynamical nuclear spin polarization is strongly suppressed in this regime, allowing to observe the expected Gaussian decay of the interference signal.

In [53], this technique is demonstrated with both an electron and a hole spin and at different time delays covering the relevant timescales to observe the Gaussian decay of the interference signal. If the decay to be observed is of order nanosecond or lower, as it is the case for the electron spin, Raman scattering can be performed using a single weak laser pulse (Fig. 12.11a). The loss of signal amplitude due to laser-induced decay of the spin population can be compensated by properly attenuating the transmission of the longer arm of the interferometer. The scattered photons are sent into a stabilized Mach-Zehnder (MZ) interferometer where the visibility of the interference fringes $V_{Ram}(\Delta t)$ can be observed as a function of the time delay Δt. Rayleigh scattering interference visibility $V_{Ray}(\Delta t)$ is measured in the same conditions as a reference. Their ratio $R(\Delta t) = V_{Ram}(\Delta t)/V_{Ray}(\Delta t)$ – expressing the reduction of visibility associated with the spin decoherence – exhibits a Gaussian decay at a timescale that agrees very well with Ramsey interferometry measurements of the same QD (Fig. 12.11b).

Measuring the hole coherence time cannot be done with a single laser pulse since the signal would vanish before the spin has decohered due to spin pumping. In this case, a sequence of two pulses separated by a time delay matching the optical path length difference in the two arms of the interferometer is used (Fig. 12.11c,d), such that spin pumping is interrupted between the two pulses, allowing to observe the decay of hole spin coherence.

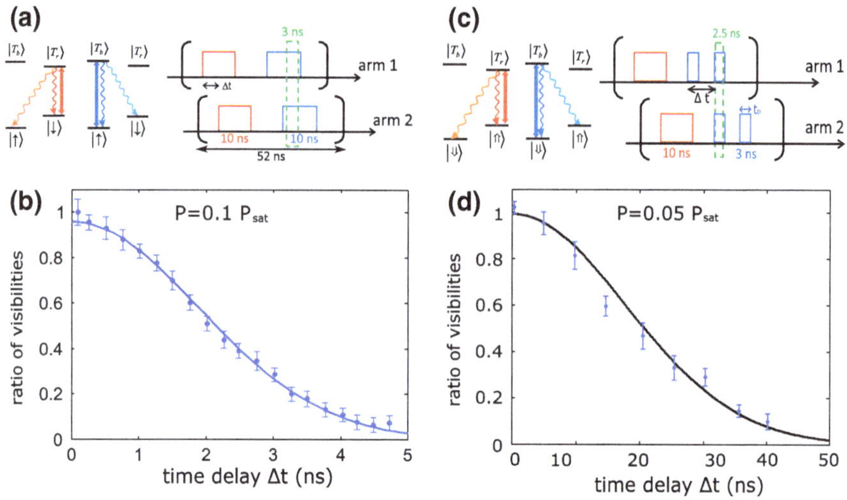

Fig. 12.11 **a** Pulse sequence used for the first-order coherence measurement of an electron spin and relevant transitions. *Red square frame* 10 ns preparation pulse; *Blue square frame* 10 ns excitation pulse; *Green dashed box* 3 ns post-selected time-window. The overall repetition rate is 52 ns. **b** Electron spin: Ratio between the visibility of ω_{blue} photons and ω_{diag1} photons as a function of the time delay Δt for the excitation laser power $P = 0.1\,P_{sat}$. The *solid curve* is a Gaussian fitting of the data. **c** Pulse sequence used for the first-order coherence measurement of a hole spin and relevant transitions. *Red square frame* 10 ns preparation pulse; *blue square frames* 3 ns excitation pulses; *green dashed box* 2.5 ns post-selected time-window. **d** Hole spin: Ratio between the visibility of ω_{blue} photons and ω_{diag1} photons as a function of the time delay Δt for the excitation laser power $P = 0.05\,P_{sat}$. The *solid curve* is a Gaussian fitting of the data

The main conclusion of this study is that in the weak excitation regime, the ratio of Raman and Rayleigh scattering interference visibilities – after a delay Δt – directly gives the spin degree of coherence after this delay. As in the distant entanglement generation experiment, the spin has to remain coherent during the time it takes for the photon to travel from the QD to the detector (22 ns in the present case), the associated constraint on the dot pair is that both of them should display a value of $R(\Delta t = 22\text{ ns}) > 0.5$, signifying that most of their coherence is preserved after this propagation time. In [43], the dots used in the experiment verify $R_1 = 58.3 \pm 3.2\%$ and $R_2 = 53.8 \pm 6.3\%$ at the relevant electric and magnetic field settings, which demonstrates that their coherence time is longer than what is needed for the entanglement generation protocol to take place.

12.7 Implementation and Characterisation of z-Rotation (Phase) Gate

Determination of quantum correlations relies on the ability to control the relative phase θ in the (12.7). The latter can be adjusted using a V-polarized off-resonant laser field that induces different phases on the two spin states due to different magnitude of the ac-Stark effect. In the experiment of [43], the phase shift is performed on QD1 by applying a laser that is red-detuned by ~20 GHz from the red and ~50 GHz from the blue vertical transition (bottom right diagram of Fig. 12.2d). The difference in the ac-Stark shift experienced by the two transitions allows the state $|\Downarrow\rangle$ to accumulate a phase $\varphi = \Omega^2\tau\delta/4\Delta(\Delta+\delta)$, relative to $|\Uparrow\rangle$. Here, Ω is the Rabi frequency

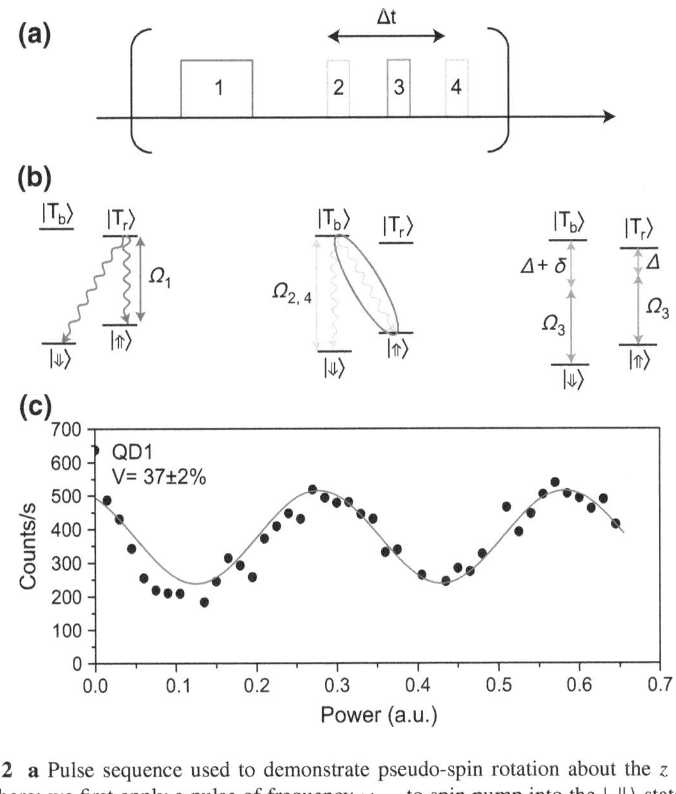

Fig. 12.12 **a** Pulse sequence used to demonstrate pseudo-spin rotation about the z axis of the Bloch sphere: we first apply a pulse of frequency ω_{red} to spin pump into the $|\Downarrow\rangle$ state (pulse 1). We then apply two successive weak pulses (2 and 4) at frequency ω_{blue}. The time offset of the two pulses approximately matches the path length difference of the MZ interferometer. An additional detuned laser pulse of 4 ns (pulse 3) is inserted between the two pulses at ω_{blue} (2 and 4). Light scattered during the pulse 2 and 4 interfere at the second beamsplitter. **b** Corresponding energy diagrams. **c** *black dots* count rate of the output detector, as a function of the detuned laser power, demonstrating control of the pseudo-spin phase. *Red curve* fit to the data. The error bars of the visibilities correspond to one standard deviation (color figure online)

of the laser, τ the pulse duration, Δ the detuning from the red transition and $\delta = \omega_{blue} - \omega_{red}$ the energy difference between the two transitions.

In order to characterize the effect of spin-phase rotation, we performed interferometric measurements similar to those described Sect. 12.6, with the optical path-length difference kept constant. A 4 ns-long detuned laser pulse is applied in between the two weak excitation pulses (Fig. 12.12a). By varying the laser power from 0 to $\sim 2\,\mu$W, we change the relative phase of the two spin states and thus the relative phase of the Raman scattering amplitude before and after the pulse that induces the spin-state dependent ac-Stark shift. The oscillations in the count rate as a function of the laser power (Fig. 12.12c, black dots) unequivocally demonstrate single pseudo-spin rotation about the z axis of the Bloch sphere. The red curve in Fig. 12.12c is a sinusoidal fit to the data, showing that no sizeable loss of visibility is observed for spin rotation up to 4π.

This single qubit phase gate, together with previously demonstrated spin x-rotation using picosecond pulses [9], provides a full control of the spin state over the whole Bloch sphere in the rotating frame. Moreover the single-qubit-phase characterization scheme provides a general way to measure phase modifications of a single qubit.

12.8 Experimental Verification of Entanglement

Our entanglement verification scheme is based on the fact that only certain components of the generated state density matrix enter into the estimation of the state fidelity to a maximally entangled state $|\psi^+\rangle$, defined as $F = \langle\psi^+|\rho|\psi^+\rangle = \frac{1}{2}\left(\rho_{\Uparrow\Downarrow,\Downarrow\Uparrow} + \rho_{\Downarrow\Uparrow,\Uparrow\Downarrow} + \rho_{\Uparrow\Downarrow,\Uparrow\Downarrow} + \rho_{\Downarrow\Uparrow,\Downarrow\Uparrow}\right)$. The diagonal components $\rho_{\Uparrow\Downarrow,\Uparrow\Downarrow} + \rho_{\Downarrow\Uparrow,\Downarrow\Uparrow}$ are estimated by the measure of conditional probabilities of spin population in the computational basis (classical correlations, Sect. 12.8.1) and the sum of the off-diagonal elements $\rho_{\Uparrow\Downarrow,\Downarrow\Uparrow} + \rho_{\Downarrow\Uparrow,\Uparrow\Downarrow}$ is measured using a novel scheme based on a spin-phase dependent non-local measurement (quantum correlations, Sect. 12.8.2).

12.8.1 Classical Correlations

To demonstrate classical correlations between the distant spins, we carry out local single-spin measurement in the computational basis, conditioned upon the detection of a Raman photon during the entanglement pulse. Contrarily to the experiments described in Sects. 12.2 and 12.4 where we address a fixed spin-to-trion transition and the presence or absence of a π-pulse determines the spin state we measure, here we take benefit from the fact that each spin state can be excited to a corresponding trion state with the same oscillator strength and the same laser polarization but using a different resonant laser wavelength. The detection of a photon during a blue (red) laser pulse thus tells with a high confidence level that the state of the

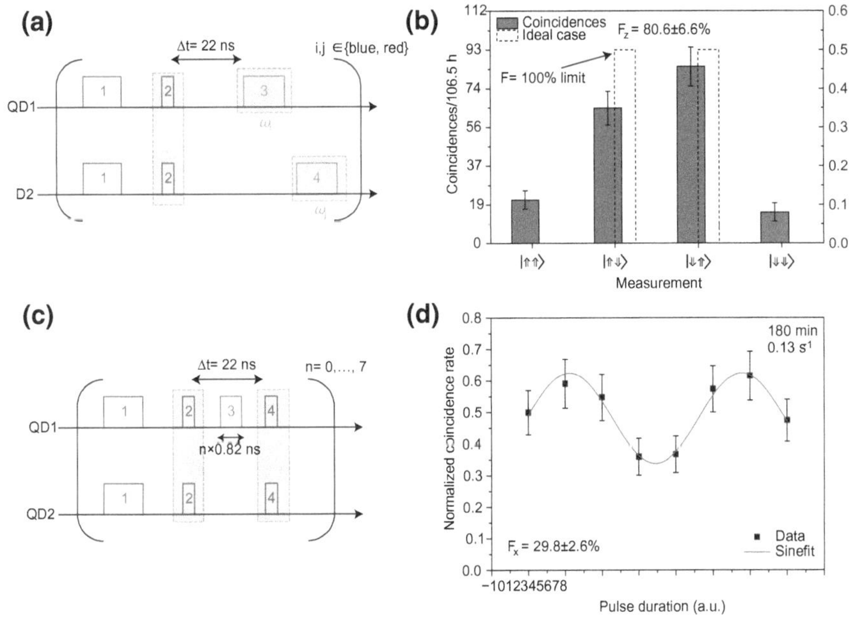

Fig. 12.13 **a** Pulse sequence used for the measurement of classical correlations between the distant spins. After spin pumping into the $|\Downarrow, \Downarrow\rangle$ state (pulse 1), a weak entanglement pulse (pulse 2) is sent simultaneously to both quantum dots (QD1 and QD2). After 22 ns, pulse 3 measures the spin-state of QD1 and then pulse 4 measures the spin-state of QD2. The four measurement combinations are alternated. **b** *Red bars* results of three-fold coincidences between a photon emitted during the entanglement pulse and a photon in each of the two measurement pulses (orange shading in Fig. 12.3a) obtained during a total measurement time span of 106.5 h. The *dashed bars* represent the ideal limit of vanishing even parity spin state detection. The error bars represent one standard deviation deduced from poissonian statistics of the raw detection events. The measured fidelity is $F_z = 80.6 \pm 6.6\%$. **c** Pulse sequence used to measure quantum correlations between the distant spins. After spin pumping into the $|\Downarrow, \Downarrow\rangle$ state, a weak entanglement pulse (pulse 2) is used to drive both QDs. A detuned laser pulse (pulse 3) modifies the phase of the QD1 hole spin phase. After 22 ns, a non-local measurement pulse is applied to both QDs. The pulse sequence is repeated for different values of the duration of the pulse 3 ranging from 0 to 7×0.82 ns, corresponding to a laser-induced QD1 spin phase rotation ranging from 0 to 3π. d *Black dots* two-fold coincidence rate between a photon detected during the entanglement pulse and a second photon detected during the measurement pulse (*orange shading* in Fig. 12.3c), normalized by the average detection rate between photons emitted during different periods, as a function of the pulse-length of pulse 3. The error bars represent one standard deviation deduced from poissonian statistics of the raw detection events. The *red curve* is a sinusoidal fit to the data, yielding a visibility of $V = 29.8 \pm 2.6\%$. The deduced overall fidelity is $F = (F_z + V)/2 = 55.2 \pm 3.5\%$ (color figure online)

spin prior to the measurement pulse was $|\Downarrow\rangle$ ($|\Uparrow\rangle$). In order to measure the four different spin combinations under the same experimental conditions, we alternate in a single experiment four pulses sequences, each performing one of the four requisite measurement combinations. The full pulse sequence is described in Fig. 12.13a: we first prepare the state $|\Downarrow, \Downarrow\rangle$ by spin pumping, then apply the weak entanglement laser pulse. The power used is ~2% of the saturation power and the Raman photon scattering probability is $\varepsilon^2 \sim 7\%$. The detection of a Raman photon during this pulse heralds successful entanglement generation. We then successively measure the state of the two dots. The measurement pulses of the two dots are offset in time, allowing us to extract which-path information. These two measurements are performed close to saturation and the detection efficiencies are rendered similar. The duration of the full sequence is 4 × 104 ns. Figure 12.13b shows the results of the 3-fold coincidences detected during 106.5 h of measurement. As expected, the odd parity events, where the spins of the two dots are opposite, are much more likely than the even parity events, where the two spins are found in the same state. The associated fidelity is $F_z = 80.6 \pm 6.6\%$.

12.8.2 Quantum Correlations

To demonstrate quantum correlations between the two distant spins, we implement a delayed two-photon interference experiment. The key element of this approach for verifying quantum correlations is the possibility to rotate one of the spins along the z-axis after heralded spin entanglement is generated. Application of a detuned laser pulse on QD1, as described in Sect. 12.7, results in rotating the phase of the entangled state by $\alpha(\tau)$ so that the the entangled state becomes $(|\Uparrow, \Downarrow\rangle + e^{-i\theta - i\alpha(\tau)}|\Downarrow, \Uparrow\rangle)/\sqrt{2}$. Subsequent application of a second weak (measurement) pulse, that is identical in intensity and duration to the entanglement pulse, on both QDs simultaneously leads to

$$\begin{aligned} |\Psi\rangle_{12} = & \frac{1}{\sqrt{2}}[|\Uparrow, \Downarrow, 0\rangle + \varepsilon e^{-i\theta_2}|\Uparrow, \Uparrow, 1_{d1,H}\rangle \\ & + e^{-i\theta - i\alpha(\tau)}(\varepsilon e^{-i\theta_1}|\Uparrow, \Uparrow, 1_{d1,H}\rangle + |\Downarrow, \Uparrow, 0\rangle)] \end{aligned} \quad (12.8)$$

$$\begin{aligned} = & \frac{\varepsilon}{\sqrt{2}} e^{-i\theta_2}(1 + e^{-i\alpha(\tau)})|\Uparrow, \Uparrow, 1_{d1,H}\rangle \\ & + \frac{1}{\sqrt{2}}\left(|\Uparrow, \Downarrow, 0\rangle + e^{-i\theta - i\alpha(\tau)}|\Downarrow, \Uparrow, 0\rangle\right). \end{aligned} \quad (12.9)$$

Therefore, conditioned on an initial Raman photon detection event that heralded spin-spin entanglement, the detection of a second time-delayed Raman photon detection probability scales as $\varepsilon^2|1 + e^{-i\alpha(\tau)}|^2$. The expectation value of Raman photon detection can be shown to be

$$\langle E^{(-)}E^{(+)}\rangle \propto 1 + 0.5\langle\sigma_z^1 + \sigma_z^2\rangle - \langle\sigma^2_{\Downarrow\Uparrow}\sigma^1_{\Uparrow\Downarrow} + \sigma^1_{\Downarrow\Uparrow}\sigma^2_{\Uparrow\Downarrow}\rangle. \quad (12.10)$$

The peak-to-peak contrast in $\langle E^{(-)}E^{(+)}\rangle$ obtained by varying $\alpha(\tau)$ therefore gives us the magnitude of non-local quantum correlations between the two spins.

To verify the presence of quantum correlations using such a delayed two-photon interference experiment, we use the pulse sequence described in Fig. 12.13c. We once again prepare the spins in the $|\Downarrow, \Downarrow\rangle$ state by spin pumping and then apply the weak entanglement generation pulse in the same way as for the classical correlation measurement. The phase of the state is then modified by the detuned laser pulse, whose duration is changed within the pulse sequence by alternating eight patterns that differ only by the duration of this particular pulse. Eight evenly distributed durations are chosen to cover more than one full revolution. Finally the measurement pulse is simultaneously sent to both dots. The duration of the full sequence is 8×52 ns. The two-fold coincidences measured for each value of the pulse length are normalized by the uncorrelated coincidence rate obtained by measuring two photons emitted in different periods. Figure 12.13d presents data obtained during 180 min of measurement. The obtained ratio exhibits clear oscillations of visibility $29.6 \pm 2.8\%$.

12.8.3 Discussion

Combining the results depicted in Fig. 12.13b, d, we deduce an overall fidelity of the generated entangled state of $F = 55.2 \pm 3.5\%$. Although this number is relatively modest compared to previous work based on other physical systems [39, 40, 42, 56, 57], it is predominantly limited by the T_2^* coherence time of the hole spins and hence could be substantially increased by either a decrease in the magnetic field [48] or introduction of dynamical decoupling [45]. The detection rate of single photons emitted during the entanglement pulse is 2300 photons per second; the latter directly yields the heralded entanglement generation rate. Such high value has been made possible by the use of a high repetition rate ($\Gamma_{rep} = 1.9 \times 10^7$ s^{-1}) allowed by the fast spin initialization (10 ns) together with a relatively high collection efficiency of $\sim$20% to the objective leading to an overall collection efficiency of $\sim$0.2%, and could be further increased using cavity QED [10]. On the other hand, the lack of efficient spin measurement leads to a low three-fold coincidence rate of $\sim$2/h. The scheme we presented in Sect. 12.8.2 allows verification of non-local quantum coherence using two-fold coincidence, and therefore can be performed much faster than a full state tomography.

12.9 Conclusion and Outlook

In this chapter we have reviewed recent demonstrations of elementary quantum-dot-based quantum communication protocols, namely entanglement between a QD spin and a propagating photon, teleportation from a propagating photon to a QD spin and generation of heralded entanglement between distant QD hole spins.

When it comes to comparing these results with those previously obtained with other physical systems, one can identify a major advantage of QD-based realizations which is the possibility to obtain much higher success rates. This comes as a result of the advantages offered by semiconductor technological possibilities that allow easy integration into semiconductor structures which in turn enables both electrical and photonic control of the environment for optimized photon collection efficiency [11]. Moreover thanks to the short lifetime (< 1 ns) of the excited states it is possible to perform state initialization within 10 ns and therefore operate at high repetition rates.

On the other hand, the main drawback in QD-based implementations of quantum communication protocols such as those presented in this chapter is the fact that the coherence time of QD spins is relatively short as compared to other commonly used physical system, such as atoms or NV centres. The coherence times involved in the presented experiments (both for electrons with echo and holes without echo) are of order of a few tens of nanoseconds, and the best reported spin-echo T_2 coherence times are at most of order microsecond. This limits the use of QDs for quantum communication or distributed quantum computation to about a kilometre, but should however not be a limitation for on-chip applications.

Another major difficulty comes from the inefficient spin measurement for single QDs in Voigt geometry. The use of singlet-triplet qubits in QD molecules could allow to circumvent this obstacle since it can not only prolong the coherence times [27] but also simultaneously allow for single-shot readout [58].

Acknowledgements The authors would like to acknowledge their coworkers Emre Togan, Parisa Fallahi, Martin Kroner, Yves Delley, Javier Miguel-Sanchez and Stefan Fält for their extensive participation in the presented work.

References

1. J.I. Cirac, P. Zoller, H.J. Kimble, H. Mabuchi, Phys. Rev. Lett. **78**, 3221 (1997)
2. L.M. Duan, M.D. Lukin, J.I. Cirac, P. Zoller, Nature **414**, 413 (2001)
3. H.J. Kimble, Nature **453**, 1023 (2008)
4. S. Ritter et al., Nature **484**, 195 (2012)
5. D. Press et al., Nature Photon. **4**, 367 (2010)
6. A. Imamoğlu, D.D. Awschalom, G. Burkard, D.P. DiVincenzo, D. Loss, M. Sherwin, A. Small, Phys. Rev. Lett. **83**, 4204 (1999)
7. S.T. Yilmaz, P. Fallahi, A. Imamoğlu, Phys. Rev. Lett. **105**, 033601 (2010)
8. M. Atatüre, J. Dreiser, A. Badolato, A. Högele, K. Karrai, A. Imamoğlu, Science **312**, 551 (2006)
9. D. Press, T.D. Ladd, B. Zhang, Y. Yamamoto, Nature **456**, 218 (2008)
10. O. Gazzano et al., Nature Commun. **4**, 1425 (2013)
11. N. Somaschi et al., Nature Photon. **10**, 340 (2016)
12. A. Högele et al., Phys. Rev. Lett. **93**, 217401 (2004)
13. G. Reithmaier, S. Lichtmannecker, T. Reichert, P. Hasch, K. Müller, M. Bichler, R. Gross, J.J. Finley, Sci. Rep. **3**, 1901 (2013)
14. B.B. Blinov, D.L. Mohering, L.M. Duan, C. Monroe, Nature **428**, 153 (2004)
15. T. Wilk, S.C. Webster, A. Kuhn, G. Rempe, Science **317**, 488 (2007)
16. E. Togan et al., Nature **466**, 730 (2010)
17. C. Arnold et al., Nature Commun. **6**, 6236 (2015)
18. X. Xu et al., Phys. Rev. Lett. **99**, 097401 (2007)

19. K. De Greve et al., Nature **491**, 421 (2012)
20. J.R. Schaibley et al., Phys. Rev. Lett. **110**, 167401 (2013)
21. W.B. Gao, P. Fallahi, E. Togan, J. Miguel-Sanchez, A. Imamoğlu, Nature **491**, 426 (2012)
22. A. Muller et al., Phys. Rev. Lett. **99**, 187402 (2007)
23. A.N. Vamivakas, Y. Zhao, C.Y. Lu, M. Atatüre, Nature Phys. **5**, 198 (2009)
24. K. De Greve et al., Nature Commun. **4**, 2228 (2013)
25. P.L. McMahon, K. De Greve, in *Engineering the Atom-Photon Interaction*, ed. By A. Predojević, M.W. Mitchell (Springer, 2015), pp. 365–402
26. W.B. Gao, P. Fallahi, E. Togan, A. Delteil, Y.S. Chin, J. Miguel-Sanchez, A. Imamoğlu, Nature Commun. **4**, 2744 (2013)
27. K.M. Weiss, J.M. Elzerman, Y.L. Delley, J. Miguel-Sanchez, A. Imamoğlu, Phys. Rev. Lett. **109**, 107401 (2012)
28. M. Bayer et al., Phys. Rev. B **65**, 195315 (2002)
29. C. Santori, D. Fattal, J. Vučković, G.S. Solomon, Y. Yamamoto, Nature **419**, 594 (2002)
30. S. Ates, S.M. Ulrich, S. Reitzenstein, A. Löffler, A. Forchel, P. Michler, Phys. Rev. Lett. **103**, 167402 (2009)
31. C.K. Hong, Z.Y. Ou, L. Mandel, Phys. Rev. Lett. **59**, 2044–2046 (1987)
32. J.F. Sherson et al., Nature **443**, 557 (2006)
33. Y.A. Chen et al., Nature Phys. **4**, 103 (2008)
34. D. Gottesman, I.L. Chuang, Nature **402**, 390 (1999)
35. E. Knill, R. Laflamme, G.J. Milburn, Nature **409**, 46 (2001)
36. W.B. Gao et al., Proc. Natl. Acad. Sci. **107**, 20869 (2010)
37. R. Stockill, C. Le Gall, C. Matthiesen, L. Huthmacher, E. Clarke, M. Hugues, M. Atatüre, Nature Comm. **7**, 12745 (2016)
38. C. Simon, W. Irvine, Phys. Rev. Lett. **91**, 110405 (2003)
39. D.L. Moehring et al., Nature **449**, 68 (2007)
40. H. Bernien et al., Nature **497**, 86 (2013)
41. C. Cabrillo, J.I. Cirac, P. García-Fernández, P. Zoller, Phys. Rev. A **59**, 1025 (1999)
42. L. Slodička, G. Hétet, N. Röck, P. Schindler, M. Hennrich, R. Blatt, Phys. Rev. Lett. **110**, 083603 (2013)
43. A. Delteil, Z. Sun, W. Gao, E. Togan, S. Faelt, A. Imamoğlu, Nature Phys. **12**, 218 (2016)
44. D. Brunner et al., Science **325**, 70 (2009)
45. K. De Greve et al., Nature Phys. **7**, 872 (2011)
46. K. Greilich, S.G. Carter, D. Kim, A.S. Bracker, D. Gammon, Nature Photon. **5**, 702 (2011)
47. S.G. Carter et al., Phys. Rev. B **89**, 075316 (2014)
48. J.H. Pretchel et al., Phys. Rev. B **91**, 165304 (2015)
49. A. Delteil, W.B. Gao, P. Fallahi, J. Miguel-Sanchez, A. Imamoğlu, Phys. Rev. Lett. **112**, 116802 (2014)
50. N.F. Ramsey, Phys. Rev. **78**, 695 (1950)
51. A. Högele, M. Kroner, C. Latta, M. Claassen, I. Carusotto, C. Bulutay, A. Imamoğlu, Phys. Rev. Lett. **108**, 197403 (2012)
52. T.D. Ladd, D. Press, K. De Greve, P.L. McMahon, B. Friess, C. Schneider, M. Kamp, S. Höfling, A. Forchel, Y. Yamamoto, Phys. Rev. Lett. **105**, 107401 (2010)
53. Z. Sun, A. Delteil, S. Faelt, A. Imamoğlu, Phys. Rev. B **93**, 241302(R) (2016)
54. G. Fernandez, T. Volz, R. Desbuquois, A. Badolato, A. Imamoğlu, Phys. Rev. Lett. **103**, 087406 (2009)
55. T.M. Sweeney et al., Nature Photon. **8**, 442 (2014)
56. D. Hucul et al., Nature Phys. **11**, 37–42 (2015)
57. J. Hofmann et al., Science **337**, 72–75 (2012)
58. Y.L. Delley, M. Kroner, S. Faelt, W. Wegscheider, A. Imamoğlu, arXiv:1509.04171

Chapter 13
Photonic Integrated Circuits with Quantum Dots

Ulrich Rengstl, Michael Jetter and Peter Michler

Abstract The usage of linear optics for quantum computation is a fast evolving field in the quantum optics community. This attention is mainly driven by the conceivable integration of whole quantum processors on a single semiconductor chip, as it is common for classical electrical computers. Although silicon is the most commonly used platform for on-chip photonics due to its high degree of technical development, the usage of III-V semiconductors is a promising candidate for fully integrated quantum optical networks due to the straightforward implementation of semiconductor quantum dots as on-demand single-photon sources. This chapter gives an overview over the current advance on GaAs-based photonic circuits with integrated quantum dots. We will focus on the basic design of integrated photonics and the realizations of the on-chip equivalents of elementary building elements which are needed for quantum computation. This includes the performance analysis of fabricated devices as beam splitters and their compatibility with quantum dots as integrated single-photon sources.

13.1 Introduction

The realization of quantum computational schemes is currently a topic under broad interest. This interest is driven by the development of several algorithms which allow the computation of specific problems with polynomial computational costs on a quantum computer, which would lead to an exponential computational cost on classical computers.

The potential of algorithms using quantum mechanical principals to outperform classical algorithms was first shown by David Deutsch by solving the famous question if a coin is fair or unfair [1]. A fair coin has a head on the one side and a number on the other side whereas an unfair coin has heads or numbers on both sides. Classically it is necessary to look on both sides to decide if the coin is fair or unfair. The Deutsch

U. Rengstl · M. Jetter · P. Michler (✉)
Research Center SCoPE and IQST, Institut für Halbleiteroptik und Funktionelle Grenzflächen (IHFG), University of Stuttgart, Allmandring 3, 70569 Stuttgart, Germany
e-mail: p.michler@ihfg.uni-stuttgart.de

P. Michler (ed.), *Quantum Dots for Quantum Information Technologies*, Nano-Optics and Nanophotonics, DOI 10.1007/978-3-319-56378-7_13

algorithm on the other hand prepares the input state in a superposition state to 'look on both sides simultaneously' [2].

The two algorithms which are nowadays mainly mentioned when talking about quantum algorithms in the public are Grover's algorithm for the efficient search in large unsorted databases [3, 4] and Shor's algorithm for the prime factorization of large integers [5]. Especially the future realization of Shor's algorithm on quantum computers may have a huge impact on the current data communication, as current cryptography systems, like RSA, are based on the non-polynomial computing time for the prime factorization. However, the realization of Shor's algorithm will shift the prime factorization to a problem with polynomial computing time, which may render current cryptography systems useless. On the other hand, a fully functional quantum computer will give researchers a new tool to perform quantum simulations on a native system [6].

Two main steps to the realization of quantum computers are needed. First, the implementation of single quantum bits, qubits, which cannot only represent the states 0 and 1 but also every superposition between these states. And secondly, the execution of operations on single qubits and operations between multiple qubits via gates. Current realization methods for qubits contain e.g. single atoms or superconducting qubits. Another realization possibility was shown by Knill et al. in [7] by exploiting linear optics. In their publication, they proposed the realization of qubits using path-entangled or polarization-entangled photons and the implementation of gates via beam splitters and phase shifters. One appealing facet of this scheme is the possibility of the implementation of the necessary elements on single semiconductor chips, opening a perspective of easy to handle end user applications like it is common for classical computers.

In this contribution, we want to give an introduction into possible realizations of basic elements for quantum computers based on photonic integrated circuits with integrated quantum dots as single-photon sources on the basis of III-V semiconductors. Section 13.2 gives a short overview about the principle of linear optics quantum computation including basic one-qubit and two-qubit operations. In Sect. 13.3, different possible designs of chip-integrated waveguides are discussed as fundamental basis for the design of integrated circuits. This forms the basis for the following Sect. 13.4 about integrated circuits, which covers the state of the art results on integrated beam splitters on GaAs with integrated semiconductor quantum dots. Finally, Sect. 13.5 gives a perspective for a gate operating on multiple qubits from different quantum dots as well as a perspective for universal quantum gates by extending the view to current advances in the integrated silicon photonics community. The last part of Sect. 13.5 takes a look on the advances towards the integration of the fourth part of photonic integrated circuits, the detectors. Here, we focus on the implementation of superconducting single-photon detectors onto GaAs waveguide structures, which leads to the perspective of a fully integrated linear optics quantum circuit with source, gate and detector on one single semiconductor chip.

13.2 Quantum Computing

13.2.1 Universal Set of Gates

After the first proofs of the usefulness of quantum computing the development was at first mainly theoretical and mathematical as the realization of a computer needs the realization of a universal set of gates. So a universal set of adequate operations had to be found first. In irreversible classical computation, one possible set is well known to just be composed of the NAND gate, so the NAND gate itself is universal.

In 1995, skipping several development steps, it was shown that a universal quantum gate can be built by a two-qubit gate of the form

$$\mathsf{A}(\phi,\alpha,\theta)=\begin{pmatrix}1&0&0&0\\0&1&0&0\\0&0&\mathrm{e}^{\mathrm{i}\alpha}\cos\theta&-\mathrm{i}\,\mathrm{e}^{\mathrm{i}(\alpha-\phi)}\sin\theta\\0&0&-\mathrm{i}\,\mathrm{e}^{\mathrm{i}(\alpha+\phi)}\sin\theta&\mathrm{e}^{\mathrm{i}\alpha}\cos\theta\end{pmatrix} \tag{13.1}$$

[8], operating on a two-qubit state, which can be represented by

$$\begin{pmatrix}c_{00}\\c_{01}\\c_{10}\\c_{11}\end{pmatrix}=c_{00}\,|00\rangle+c_{01}\,|01\rangle+c_{10}\,|10\rangle+c_{11}\,|11\rangle\ . \tag{13.2}$$

This rather complex operation can be further decomposed into several one-qubit gates, which allow to reach any point in the Hilbert-space and one single two-bit gate. The first concepts needed at least one single qubit operation with a phase shift of a non-rational multiple of π [8–10]. Later it was shown, that a universal set of quantum gates can also be obtained by two elementary single-qubit gates, the Hadamard gate (H) and a phase shifter of $\pi/4$, the $\pi/8$-gate (T), and one elementary two-qubit gate, the controlled-NOT gate (CNOT) [11].

$$\mathsf{H}=\frac{1}{\sqrt{2}}\begin{pmatrix}1&1\\1&-1\end{pmatrix}\ ,\ \mathsf{T}=\mathrm{e}^{\mathrm{i}\frac{\pi}{8}}\mathrm{e}^{-\mathrm{i}\frac{\pi}{8}\sigma_{\mathrm{z}}}=\begin{pmatrix}1&0\\0&\mathrm{e}^{\mathrm{i}\frac{\pi}{4}}\end{pmatrix}\ ,\ \mathsf{CNOT}=\begin{pmatrix}1&0&0&0\\0&1&0&0\\0&0&0&1\\0&0&1&0\end{pmatrix}, \tag{13.3}$$

where σ_{z} is the third Pauli matrix. A single qubit can be represented by

$$\begin{pmatrix}c_0\\c_1\end{pmatrix}=c_0\,|0\rangle+c_1\,|1\rangle\ . \tag{13.4}$$

While the phase gate T creates a phase shift between the two base states by transforming $c_0\,|0\rangle + c_1\,|1\rangle$ into $c_0\,|0\rangle + \mathrm{e}^{\mathrm{i}\frac{\pi}{4}}c_1\,|1\rangle$, the Hadamard gate H can be used to convert a pure state into a mixed state:

$$\mathsf{H}\,c_0\,|0\rangle = \frac{c_0}{\sqrt{2}}\left(|0\rangle + |1\rangle\right)\ , \qquad \mathsf{H}\,c_1\,|1\rangle = \frac{c_1}{\sqrt{2}}\left(|0\rangle - |1\rangle\right)\ . \tag{13.5}$$

The CNOT gate operates on two qubits, the control and the target qubit. It flips the state of the target qubit if a control qubit is $|1\rangle$. So it is equivalent to writing the XOR of the control and the target qubit into the target qubit while preserving the state of the control qubit.

13.2.2 Linear Optics Quantum Computation

Knill et al. [7] where the first to show that the realization of quantum computing using solely linear optical elements is possible. Therefore, their scheme of linear optics quantum computation (LOQC) is often referred to as the Knill, Laflamme and Milburn (KLM) scheme. It outlines an implementation possibility for quantum computations using only beam splitters, phase shifters, sources of indistinguishable single photons and single-photon detectors.

The basic concept of the KLM-scheme is to realize a qubit with a single photon in two optical modes. This can be realized via polarization-entanglement, so the two states $|0\rangle$ and $|1\rangle$ are encoded using horizontal $|H\rangle$ and vertical $|V\rangle$ polarization, or path-entanglement, where the states are spatial distinguishable. We will focus on the second approach as we will use later separate waveguides for the $|0\rangle \rightarrow |10\rangle$ and $|1\rangle \rightarrow |01\rangle$ states.

So the H, T and CNOT gates need to be implemented using optical elements. The phase shift T is easily implemented using a variable (optical) path on one beam path. For the realization of a Hadamard gate H the interaction between two modes is necessary. This can be realized using a beam splitter. The transmission matrix of a beam splitter cube, where only the reflection on one side obtains an additional π-phase shift, can be written as

$$\mathsf{B}_\eta = \begin{pmatrix} \sqrt{\eta} & \sqrt{1-\eta} \\ \sqrt{1-\eta} & -\sqrt{\eta} \end{pmatrix}\ , \tag{13.6}$$

where η is the reflectivity of the beam splitter. By setting $\eta = 0.5$ we describe a 50:50 beam splitter, and obtain the same transmission matrix as required for the Hadamard gate.

The implementation of the CNOT with linear optics is more challenging as it requires a two photon interaction. This has been proposed by using a non-deterministic CNOT gate [7]. The functionality of this gate is based on the Hong–Ou–Mandel effect [13]. This effect is observed if two indistinguishable photons impinge from different sides on a beam splitter.In contrast to two distinguishable photons, these indistinguishable photons form a Fock state of the form $|11\rangle = \mathsf{a}^\dagger \mathsf{b}^\dagger |00\rangle$, with the two creation operators $\mathsf{a}^\dagger$ and $\mathsf{b}^\dagger$, representing the two photons, acting on the vacuum state $|00\rangle$. If a 50:50 beam splitter (a Hadamard operation like in (13.3)) operates on this state, we obtain an entangled state.

$$\mathsf{H}\,\mathsf{a}^\dagger \mathsf{b}^\dagger |00\rangle \rightarrow \frac{1}{2}\left(\mathsf{a}^{\dagger 2} - \mathsf{b}^{\dagger 2}\right)|00\rangle = \frac{1}{\sqrt{2}}\left(|20\rangle - |02\rangle\right) . \tag{13.7}$$

This is a so-called 2002 state. The surprising result is the cancellation of the possibilities that one photon leaves at each output port of the beam splitter. Using this effect a CNOT gate can be build up by using an array of optical beam splitters with indistinguishable photons as input.

We focus here on the realization possibility shown in Fig. 13.1 [12]. The success probability of this gate is $1/9$ and the correct results have to be isolated using post-selection. So it is not sufficient for scalable quantum computation, but it can be upgraded to a heralded gate, where the success of its operation can be detected by measuring ancilla photons, which is beyond the scope of this work [14]. The transmission matrix of this structure can be obtained using the definition of beam splitters in (13.6). The whole device can be divided into three layers (marked in Fig. 13.1) with the transmission matrices:

$$\mathsf{A}_{1,3}^{\mathrm{CNOT}} = \begin{pmatrix} 1 & & & & \\ & 1 & & & \\ & & 1 & & \\ & & & \frac{1}{\sqrt{2}}\begin{pmatrix} 1 & 1 \\ 1 & -1 \end{pmatrix} & \\ & & & & 1 \end{pmatrix} , \quad \mathsf{A}_{2}^{\mathrm{CNOT}} = \frac{1}{\sqrt{3}} \begin{pmatrix} \begin{pmatrix} -1 & \sqrt{2} \\ \sqrt{2} & 1 \end{pmatrix} & & \\ & \begin{pmatrix} -1 & \sqrt{2} \\ \sqrt{2} & 1 \end{pmatrix} & \\ & & \begin{pmatrix} 1 & \sqrt{2} \\ \sqrt{2} & -1 \end{pmatrix} \end{pmatrix} . \tag{13.8}$$

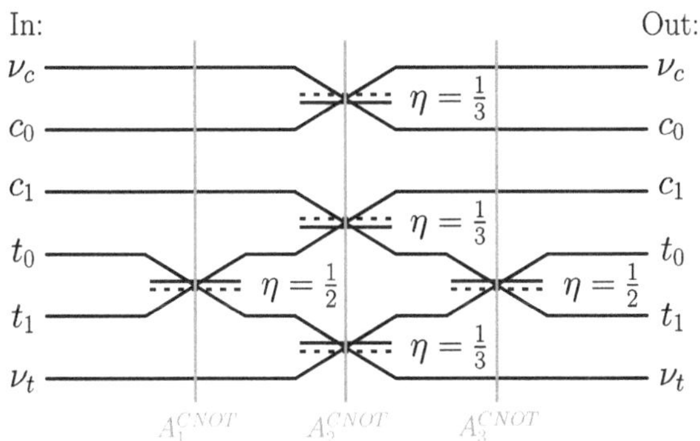

Fig. 13.1 Implementation of a probabilistic CNOT gate based on linear optics as proposed in [12]. The reflectivities of the 5 beam splitters are given by η. The side which introduces a phase shift of π upon reflection is *dashed*. c is the control qubit, t the target qubit and ν are ancilla modes. $\mathsf{A}_1^{\mathrm{CNOT}}$ to $\mathsf{A}_3^{\mathrm{CNOT}}$ mark the operations found in (13.8)

The final transmission matrix can be obtained by composing these matrices:

$$\mathsf{CNOT} = \mathsf{A}_3^{\mathrm{CNOT}}\mathsf{A}_2^{\mathrm{CNOT}}\mathsf{A}_1^{\mathrm{CNOT}} = \frac{1}{\sqrt{3}}\begin{pmatrix} -1 & \sqrt{2} & & & & \\ \sqrt{2} & 1 & & & & \\ & & -1 & 1 & 1 & \\ & & 1 & 1 & & 1 \\ & & 1 & & 1 & -1 \\ & & & 1 & -1 & -1 \end{pmatrix}, \tag{13.9}$$

with the input modes $(\nu_c, c_0, c_1, t_0, t_1, \nu_t)$ composed of the control qubit c, the target qubit t and the two ancilla modes ν_c and ν_t. It can be seen that there is more mode intermixing present than allowed for a pure CNOT gate. This can be quantified by calculating the response of this gate for the essential input modes

$$\begin{aligned} & (\ \nu_c,\ c_0,\ c_1,\ t_0,\ t_1,\ \nu_t\ ) && (\ \nu_c,\ c_0,\ c_1,\ t_0,\ t_1,\ \nu_t\ ) \\ |00\rangle_{\mathrm{In}} &= (\ 0,\ 1,\ 0,\ 1,\ 0,\ 0\ )\ , & |10\rangle_{\mathrm{In}} &= (\ 0,\ 0,\ 1,\ 1,\ 0,\ 0\ )\ , \\ |01\rangle_{\mathrm{In}} &= (\ 0,\ 1,\ 0,\ 0,\ 1,\ 0\ )\ , & |11\rangle_{\mathrm{In}} &= (\ 0,\ 0,\ 1,\ 0,\ 1,\ 0\ )\ , \end{aligned}$$

where the ancilla modes are always unoccupied.

With this method it can be verified that the CNOT gate performs the correct operation if, and only if, there is a coincidence between the control and the target qubit, meaning that no photon leaves the gate in an ancilla mode and both qubits contain one photon each. This leads to the necessity of the mentioned post-selection and therefore of a measurement of the output states. The operation of this gate can be summarized by

$$\begin{aligned} |00\rangle_{\mathrm{In}} &\rightarrow \tfrac{1}{3}\left(|00\rangle_{\mathrm{Out}} + ...\right)\ , & |10\rangle_{\mathrm{In}} &\rightarrow \tfrac{1}{3}\left(|11\rangle_{\mathrm{Out}} + ...\right)\ , \\ |01\rangle_{\mathrm{In}} &\rightarrow \tfrac{1}{3}\left(|01\rangle_{\mathrm{Out}} + ...\right)\ , & |11\rangle_{\mathrm{In}} &\rightarrow \tfrac{1}{3}\left(|10\rangle_{\mathrm{Out}} + ...\right)\ , \end{aligned}$$

where the undesired results '...' are already filtered. These states '...' contain all results where the two output qubits do not contain exactly one photon each and are therefore discarded during the post selection. The success probability of this operation is therefore limited to 1/9 [12]. Despite this limitation and the missing scalability this implementation and its on-chip analog have been established as a test-bed for linear optical quantum computation [15–17].

13.3 Photonic Waveguides with Integrated Quantum Emitters

The first implementation of a CNOT gate in the KLM scheme was done with bulk optics using the scheme presented in Fig. 13.1 [16]. Also the heralded gate mentioned earlier was realized with bulk optics [18]. But to allow the scalability of the LOQC schemes the field has to move from space-filling optical setups over the integration of the linear optical elements towards the full integration of all needed elements on one single chip. The first step in this direction was achieved with silicon photonics and off-chip single-photon sources and detectors [17, 19, 20]. A further step is the additional implementation of single-photon sources into the LOQC-chip.

In this section we cover the basics for the on-chip generation and guiding of single photons. Therefore, we will discuss the principles of waveguides and the integration of QDs as promising candidates for on-demand single-photon sources.

13.3.1 Types of Photonic Waveguides

When changing from free-space optics to integrated photonics the most fundamental change is the principle of light propagation. While in free-space applications the signal is carried in collimated beams, the guiding on a chip is done by waveguides (WGs). These guide the light by confining it to only a few allowed propagating modes defined by a contrast in the refractive index. They typically consist of a core material which is surrounded by a cladding material with lower refractive index, while the highest index step, and therefore the highest mode confinement, would be achieved by a vacuum cladding.

Figure 13.2 gives an overview over some of the most common waveguide types. Type (a) reassembles a typical optical fiber structure with a radial symmetric composition. The light propagates mainly in the waveguide core and is confined by the cladding. This principle can be easily transferred to integrated waveguides using lithographic methods. These involves typically the non-uniform structuring process of the core material using a lithographically defined hard mask. This leads to a more favorable rectangular profile of the waveguide as seen in Fig. 13.2b. The channel waveguide in (b) consists, like the optical fiber (a), of a structured core material with a surrounding cladding material. A special variation of this concept is the free-standing waveguide, which uses vacuum as cladding material.

Another possibility to achieve a confinement is the usage of different cladding materials on the top than on the bottom of the waveguide core (Fig. 13.2c). This is especially interesting if the lower cladding material is an epitaxial grown layer, which was deposited before the core material; while the upper cladding material is vacuum as a result of the structuring process of the core material. These structures are called ridge waveguides. If a slab of the core material is left behind on the whole sample the structure is often referred as rib waveguide as depicted in Fig. 13.2d.

Figure 13.2e shows a waveguide design which is called strip-loaded waveguide and is used by some laser devices. Here the lateral confinement is entirely achieved by the change in the effective refractive index introduced by the low refractive index slab on top of a two-dimensional waveguide layer. Especially this concept demonstrates the necessity to consider the evanescent field of the waveguide modes and their interaction with the surrounding.

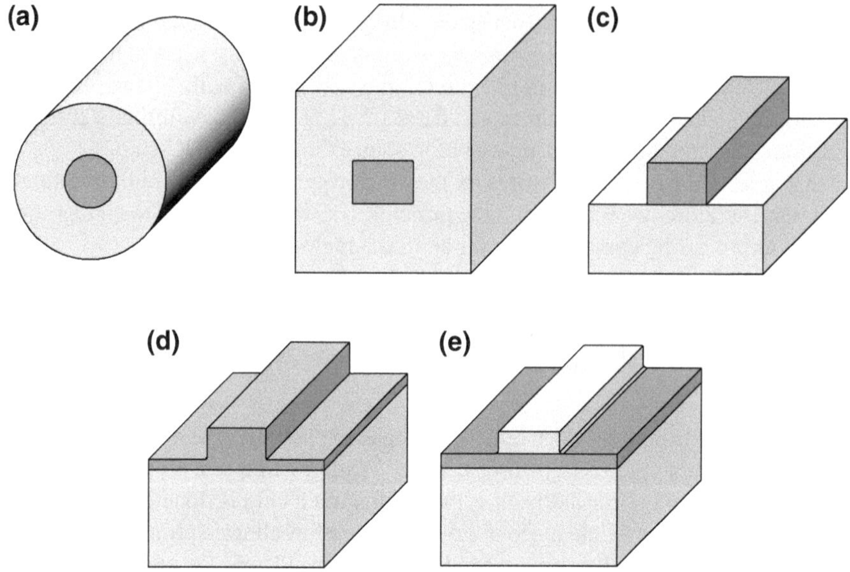

Fig. 13.2 Variety of possible waveguide designs. The refractive index of the core (*dark*) is higher than the one of the cladding (*light*). **a** A fiber and its on-chip equivalent: a channel waveguide (**b**). **c** A ridge waveguide with different cladding material on *top* (commonly vacuum) than on the *bottom*. **d** A rib waveguide, like (**c**) but with an additional slab on the side. **e** A strip-loaded waveguide

A popular method to calculate the field profile of propagating modes inside waveguide structures is the calculation of the eigenmodes of the specific waveguide cross-section in the frequency-domain. This can be done by solving the eigenvalue problem

$$\nabla \times \nabla \times \boldsymbol{E} = n^2 (\boldsymbol{r}) k^2 \boldsymbol{E} , \tag{13.10}$$

where $\boldsymbol{E}$ is the electric field, $k = \omega/c$ the wavenumber and $n(\boldsymbol{r})$ the refractive index profile of the waveguide [21].

Figure 13.3 shows the mode profiles of the fundamental mode of the waveguides depicted in Fig. 13.2. The modes were obtained by finding the fully-vectorial eigenmodes of Maxwell's equations in a plane-wave basis using the freely available MIT Photonic-Bands (MPB) package [22]. This software package was used for all frequency-domain simulations in this chapter. It is worth to note, that the modes in an optical waveguide are never complete transverse electromagnetic (TEM) modes, as the strong interaction of the field with the index step of the waveguide leads to a coupling between all three field components. Nevertheless the modes with a major electric component in the horizontal transverse or vertical transverse direction are called TE or TM modes, respectively.

This non vanishing electric field component in propagation direction leads to the remarkable property, that the guided modes exhibit a chirality. This can be exploited

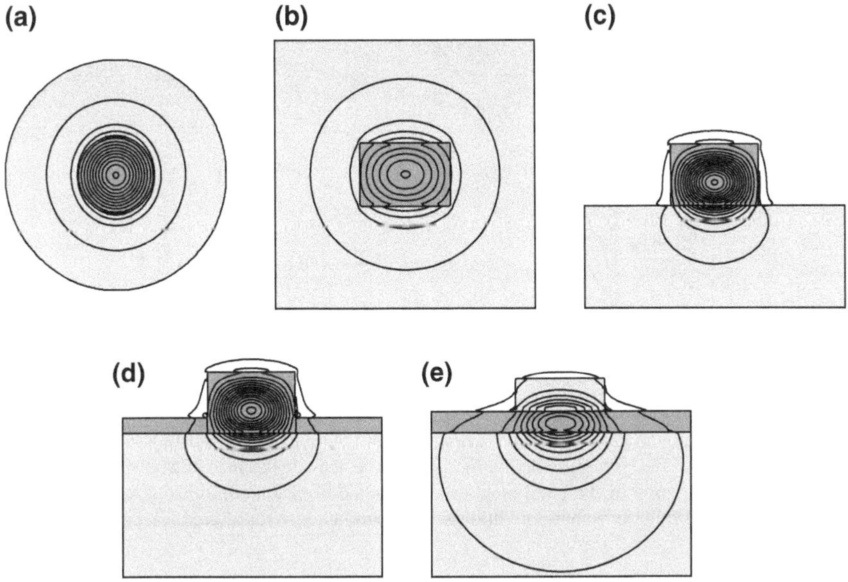

Fig. 13.3 Mode profiles of the fundamental TE-modes inside the waveguide structures illustrated in Fig. 13.2. Depicted is $|\boldsymbol{E}|^2$ in a linear scale. A reduction in the symmetry of the mode can be observed by reducing the symmetry of the waveguide (**a–c**). Furthermore, the expansion of the mode into the cladding layers is more pronounced the lower the effective refractive index of the mode is (**c–e**). These profiles were calculated using MPB and single TE mode GaAs/$Al_{0.42}Ga_{0.58}As$ waveguides

to obtain a unidirectional emission of in-plane circular polarized emitters, dependent on their lateral position inside the waveguide [23].

For the applicability in quantum information processes it is necessary to achieve a single-mode operation of the waveguides, as the excitation of a undefined number of modes would lead to undefined input states of the information processing circuits and therefore to unpredictable results. This can be also understood as the fail of the Hong–Ou–Mandel effect due to the propagation of the photons in distinguishable modes. Therefore, the dimensions of the waveguides have to be chosen to allow only the propagation of the fundamental mode. This is fulfilled if at a given photon energy only the propagation constant k_z of the fundamental mode exists in the light cone between unbound light in the cladding material and unbound light in the core material. In this case, all other modes reach propagation constants of the unbound waves with $k_{z,\text{cut-off}} = \omega/(n_{\text{clad}}c)$, where n_{clad} is the refractive index of the cladding material. So theses modes are not guided in the waveguide, so-called radiative modes, and the guiding is solely enabled by the fundamental mode. Figure 13.4 shows the typical band structure of a channel waveguide, where the region of pure single-mode operation is marked. The usage of the scale invariant variables h/λ_0 over $k_zh/2\pi$ allows the instant dimensioning of the simulated waveguide type with the height h for arbitrary operation wavelengths λ_0.

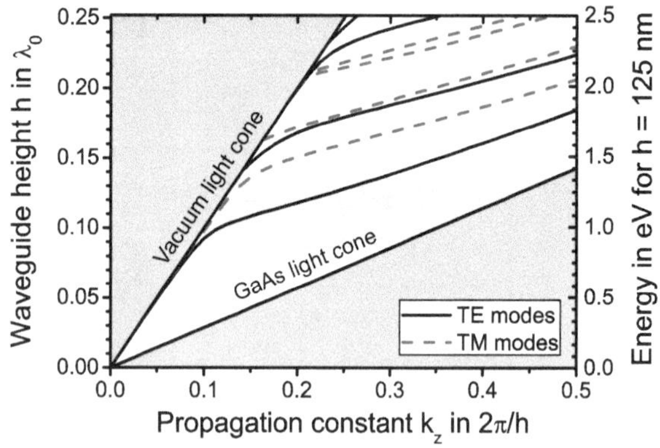

Fig. 13.4 Band structure of a channel waveguide consisting of GaAs in vacuum. The waveguide width w is twice the waveguide height h. The region of single-mode TE operation, with a size of $0.085\lambda_0 < h < 0.135\lambda_0$, is highlighted in *light gray*. The scale invariance of Maxwell's equations allows the representation of the band diagram in the scale invariant variables h/λ_0 over $k_z h/2\pi$. As an example, the right axis shows additionally the band energy for a fixed waveguide height of $h = 125$ nm

13.3.2 Integrated Single-Photon Sources

Keeping the goal of scalable single-chip quantum circuits with implemented sources and detectors in mind, the coupling of on-chip quantum emitters to the propagating modes of photonic waveguides is a major prerequisite of fully integrated waveguide circuits. This is a major drawback of the traditional silicon platform, where the direct implementation of on-demand single-photon sources is rare [24, 25] and their implementation into waveguide circuits have not been shown up to now.

There are several attempts to overcome this issue, like the integration of probabilistic sources using on-chip spontaneous four-wave-mixing. Here the χ^3-nonlinearity of silicon is used to transform a strong pump laser with frequency ν_1 into two additional frequencies with equal distance to the pump frequency $\nu_3 = \nu_1 + \delta$ and $\nu_4 = \nu_1 - \delta$. As the probability of this process scales with the interaction length and the power of the incident beam, the pump power has to be chosen, that the probability to create exactly one photon pair is maximal [26, 27]. By using two pump lasers with the frequencies ν_1 and ν_2 it is even possible to create frequency-degenerated photon pairs with frequency $\nu_3 = (\nu_1 + \nu_2)/2$, which can be used as indistinguishable photons in LOQC. This is equivalent to the reverse of the above process. But still the main drawback of this process is the probabilistic character of the photon creation, which is incompatible with a triggered state preparation. An additional drawback are the high laser powers inside the waveguides, which lead to the necessity of on-chip filters to avoid a blinding of chip integrated detectors due to the laser light.

Another possibility to realize integrated single-photon sources for silicon photonics is the use of hybrid structures with photon emitters based on different optical active material and a subsequent interface to the silicon platform [29, 30].

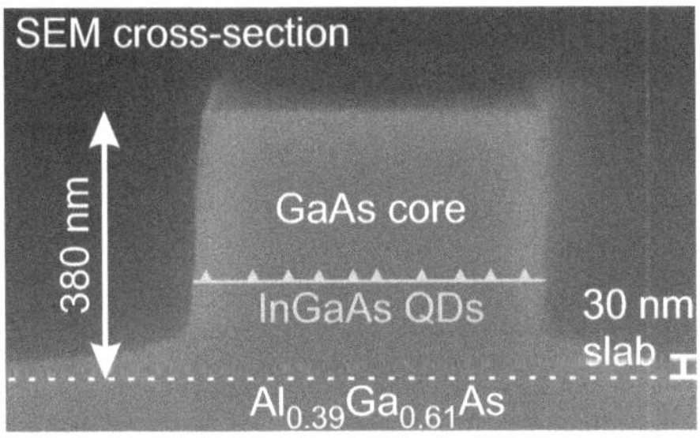

Fig. 13.5 Scanning electron microscopy image of the cross-section of a single TE/TM-mode rib waveguide composed of GaAs as core and AlGaAs as cladding layer. Monolithic integrated InGaAs QDs serve as sources of single photons. Reprinted from [28] with the permission of AIP Publishing

The possibility on which we will have a closer look in this chapter is the usage of an optical active semiconductor as material platform for the whole waveguide circuits. This allows the monolithic integration of semiconductor quantum dots (QDs) as single-photon sources. As pointed out in Chap. 3 by *Portalupi* and *Michler*, QDs exhibit excellent single-photon characteristics, especially when driven resonantly, and are therefore promising candidates for the state preparation of quantum information processes. Furthermore, they can be easily integrated into III-V semiconductors like GaAs, by their direct deposition during the layer growth of the future WG. One minor drawback of semiconductor QDs is the necessity for a cryogenic environment ($\sim$5 K) to obtain a high purity single-photon emission. Figure 13.5 shows the profile of a GaAs/$Al_{0.39}Ga_{0.61}As$ rib waveguide, where the integrated InGaAs QDs are schematically delineated.

13.3.3 Coupling Between Quantum Dot Emission and Waveguides

To achieve an efficient state preparation it is not only necessary to implement bright sources with a triggered single-photon emission, but also a high coupling efficiency of the emitted photons to the fundamental mode of the waveguides is required. There are two main characteristics which are relevant to describe the efficiency of the dipole coupling. First the Purcell-factor

$$F_\mathrm{P} = \frac{\Gamma_\mathrm{total}}{\Gamma_\mathrm{bulk}}, \tag{13.11}$$

which describes the enhancement of the spontaneous emission rate Γ_total relative to the emission rate of the same emitter inside a bulk material Γ_bulk. This enhancement arises from an increase in the local density of states (LDOS) seen by the quantum dot, due to the interaction with the environment leading to a faster decay channel.

The second important factor is the β-factor, giving the fraction of photons Γ_{bound} which are funneled into the guided modes

$$\beta = \frac{\Gamma_{\text{bound}}}{\Gamma_{\text{total}}} . \qquad (13.12)$$

As the dipole moment of the heavy-hole exciton transition inside a QD lies primarily perpendicular to its growth direction, it is sufficient for most approaches to approximate a QD as a single in-plane point dipole coupling to the waveguide. The theoretical study of this process can be done in a straight forward process by simulating the emission characteristics of a point source using finite-difference time-domain (FDTD) simulations. FDTD simulations are based on the numerical integration of Maxwell's equations in the time domain. So the temporal evolution of the light field can be directly observed and the transmitted power through the waveguide can be measured directly by integrating over the Poynting vector over a monitor plane. Typically this is done with the Fourier transformed fields to allow the frequency resolved analysis of the coupling efficiency. The FDTD-simulations shown in this chapter were all performed using the freely available software package Meep [31].

As the waveguides of interest for quantum information processing have to be used in single-mode operation, and therefore are typical designed to only support one or two modes (e.g. the fundamental TE and fundamental TM) for the specified operation wavelength, it is also common to use frequency-domain simulations. This allows the calculation of the coupling between a dipole and a considered mode, normalized to the dipole radiation in bulk material [32]. As this result can be seen as a product between the Purcell-factor and the β-factor of a specific mode, an additional FDTD simulation can be performed to obtain the Purcell-factor to normalize the results of the frequency-domain simulations. A way to directly obtain the true β-factor from frequency-domain simulations is to calculate the coupling of the dipole to all modes. Then the coupling to the guided modes contribute to Γ_{bound}, while the coupling to the unguided modes Γ_{unbound} give rise to radiative losses. So the normalization via the total coupling strength $\Gamma_{\text{total}} = \Gamma_{\text{bound}} + \Gamma_{\text{unbound}}$ can be directly obtained. This has been done for photonic crystal waveguides by calculating not only the Bloch-modes but also the radiative Quasi-Bloch-modes using perfectly matched layers [33].

Figure 13.6a visualizes the emission coupling of a single dipole oriented perpendicular to a rib waveguide depending on its lateral position inside the waveguide. Figure 13.6b shows the measured flux through the waveguide depending on the distance to the source, obtained by a FDTD simulation. The formation of a low-loss propagating mode can be observed. However, as the refractive index between waveguide core (GaAs, $n = 3.52$) and underlying cladding ($Al_{0.42}Ga_{0.56}As$, $n = 3.24$) is very small, most of the emission is lost into the substrate. So the overall coupling efficiency between dipole and waveguide is in the order of 8% to 9% per direction [28]. This can be overcome by using free-standing waveguides where a high refractive index contrast is realized into all directions, allowing for high coupling efficiencies of up to ~48% per direction [34]. A further increase in the β-factor can be achieved by using

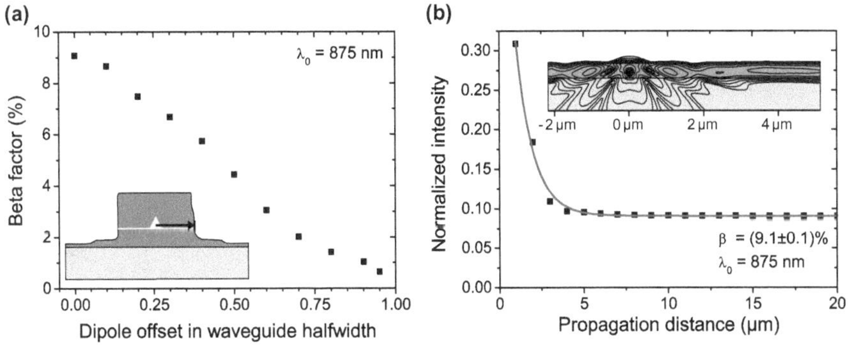

Fig. 13.6 Coupling efficiencies (β-factors) of a single dipole into the propagation modes of a waveguide. **a** Dependency of the horizontal offset of the dipole from its center position. As the overlap with the mode-field drops at the waveguide edges, the dipole coupling is maximal in the center of the waveguide. The inset shows the used waveguide profile as obtained from Fig. 13.5. **b** Coupling of a centered dipole to waveguide modes. The formation of a lossless guided mode can be observed. The inset shows a vertical cross-section of the formed radiation coil. Most of the emission is lost into the substrate. Both results were obtained using FDTD simulations

photonic crystal cavities and waveguides, where over 98% coupling efficiency has been reported [35]. This photonic crystals can then be coupled to ridge or channel waveguides for their usage as single-photon sources in photonic circuits [36, 37].

13.4 Photonic Waveguide Circuits

Integrated waveguide structures allow the realization of flying photonic qubits on a chip. The next step would be the realization of an on-chip interaction with and between these qubits. As we want to realize qubits by the path entanglement of photons between two waveguides, the interaction with a single qubit can be achieved by coupling the propagation modes of two waveguides.

A typical coupler configuration is the 3 dB-coupler where one or two input channels are equally split to two output channels. Such couplers are very common in the classical telecommunication industry were light signals need to be divided to multiple receivers. A common realization of such couplers are directional couplers (DCs) in which two waveguides are brought into close vicinity which allows the interaction of the guided modes of both waveguides via their evanescent fields [17, 28, 34]. If only the propagation mode of one input arm (Port 1) of the coupler is excited an oscillation of the light between both waveguide arms can be observed inside the coupler (Fig. 13.7). This process can be thought of as two coupled harmonic oscillators where the energy is periodically transferred between both oscillators. If both waveguides are separated after the correct interaction length both output ports 3 and 4 will receive 50% of the incident power, forming a 3 dB, or 50:50, coupler.

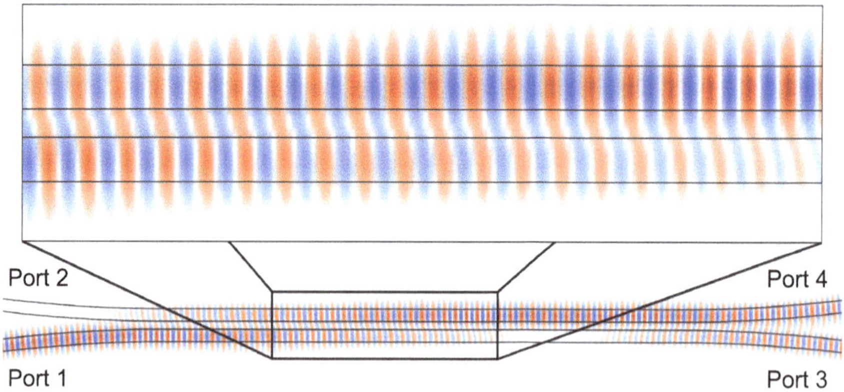

Fig. 13.7 Principle of a directional coupler. Two waveguide arms are brought into close vicinity which leads to a power exchange. The positive (negative) real value of the transverse electric field is depicted in *red* (*blue*). The inset shows a magnification of the coupler region. The $\pi/2$ phase shift of the transmitted power can be observed as described in (13.13). The coupler shown here is a 50:50 coupler with a length of $3L_{3\text{dB}}$

The transmission matrix of directional couplers varies slightly from those of beam splitter cubes as discussed above. Directional couplers are fully symmetrical, therefore both reflected modes receive a phase shift of $\pi/2$. Neglecting a global phase factor, the transmission matrix of a DC for a wave of the form $\exp\{\mathrm{i}\,(\omega t - k_z z)\}$ can be written as

$$\mathsf{A}_{\mathrm{DC}} = \begin{pmatrix} \mathrm{i}\sqrt{\eta} & \sqrt{1-\eta} \\ \sqrt{1-\eta} & \mathrm{i}\sqrt{\eta} \end{pmatrix} . \tag{13.13}$$

A common way to dimension DCs uses again the single-mode characteristic of the input waveguides. As long as both waveguides do not touch each other, the composed coupler supports two supermodes: a symmetric and an asymmetric mode which reassemble the guided part of the orthogonal basis of eigenmodes of the coupler (Fig. 13.8a). The incoming light mode can now be easily expanded in these eigenmodes, leading to 50% of the incident power in each coupler mode (Fig. 13.8b). As these two supermodes have a different field profile and therefore feel a slightly different effective refractive index, their propagation constants $k_{z_{\mathrm{sym}}}$ and $k_{z_{\mathrm{asym}}}$ vary. This leads to a beating between these two modes and therefore an energy transfer between both waveguide arms (Fig. 13.8c). When separating the two waveguides again, the resulting field can be again expanded into the eigenmodes of both waveguides, giving the splitting ratio. As the power is periodically transferred between both waveguide arms, the splitting ratio can be expressed by the beating between the two supermodes. Therefore the power which is coupled between both waveguides T can be expressed as

$$T = \sin^2\left(\frac{L}{2}\left(k_{z_{\mathrm{sym}}} - k_{z_{\mathrm{asym}}}\right)\right) , \tag{13.14}$$

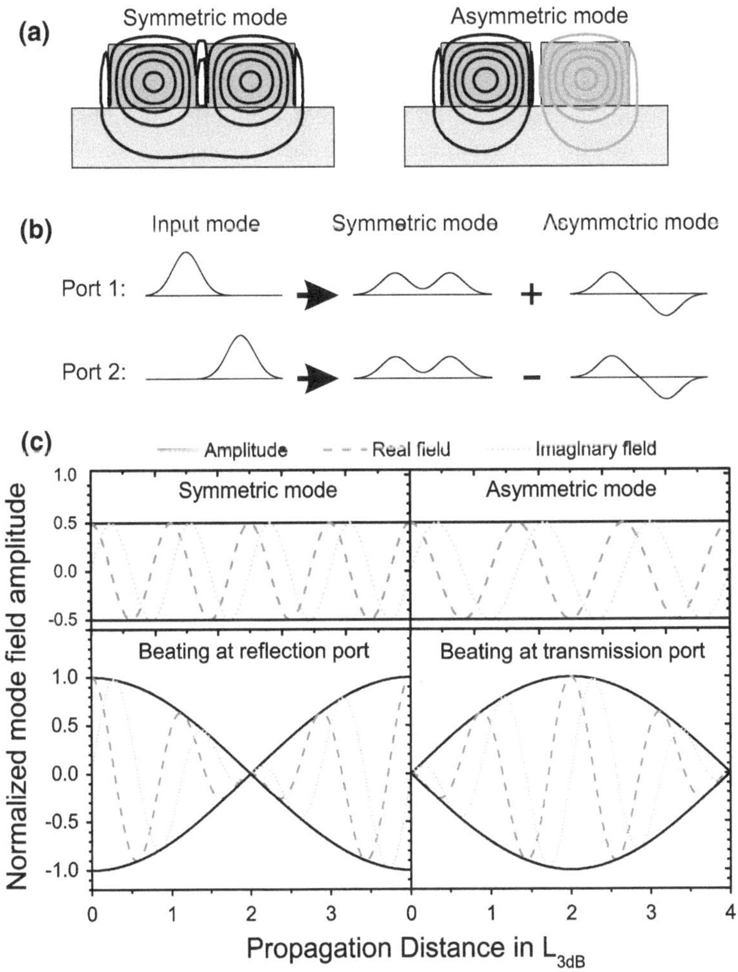

Fig. 13.8 Simplified visualization of the supermode method for a directional coupler. **a** The DC of two single-mode waveguides supports a symmetric and a asymmetric supermode, displayed by the positive (negative) electric field in *black* (*gray*). **b** An incoming single-waveguide mode can be expanded into these supermodes. **c** The lower propagation constant of the asymmetric mode (here: $k_{z_{asym}} = 0.75k_{z_{sym}}$) leads to a beating of the supermodes and therefore to a power exchange between the waveguides. The depicted graphs show the field amplitude, squaring them leads to the expected $\sin^2$-behavior of the transmitted/reflected power shown in (13.14)

where L is the interaction length inside the coupler. $R = 1 - T$ would be the power which is not coupled to the other waveguide. With this, we can define the necessary interaction length for a 3 dB splitter as

$$L_{3dB} = \frac{\pi}{2\left(k_{z_{sym}} - k_{z_{asym}}\right)}. \tag{13.15}$$

The propagation constants $k_{z_{\rm sym}}$ and $k_{z_{\rm asym}}$ can be obtained via frequency domain simulations, analog to isolated WGs (Sect. 13.3.1).

These calculations neglect the coupling which takes already place while the waveguides approach each other. Furthermore it should be noted, that this kind of couplers are very prone to small changes in their design as a small variation in the waveguide separation, size or coupler length directly affect the splitting ratio. So the formula above is sufficient to obtain starting parameters for a planned coupler design. To obtain exactly the desired splitting ratio, several optimization iterations should be taken into account, or, as we will show later, the operation wavelength of the emitter has to be fitted to the 50:50 operation region of the directional coupler.

13.4.1 Basic Performance Analysis

One possible basic test circuit for on-chip photonics with integrated single-photon sources consists of a single beam splitter where both output arms are coupled to a free-space detection setup, while at least one input arm exhibits embedded single-photon sources. We will focus again on the usage of $Al_xGa_{1-x}As$ as main material platform for the waveguide circuit and embedded InGaAs QDs as single-photon emitters. The realization of such devices has been shown both with channel [34] and rib waveguides [28, 38]. While the channel waveguides used in [34] rely on free-standing structures which are typically produced by underetching the defined GaAs waveguides using an AlGaAs sacrificial layer (see Fig. 13.9a), rib waveguides are

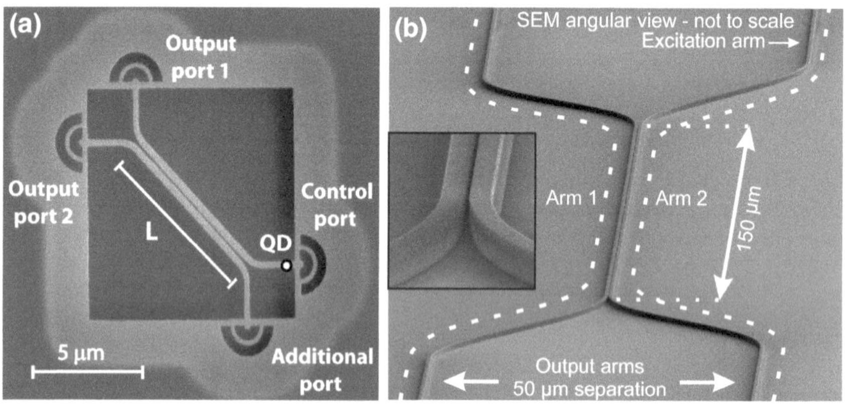

Fig. 13.9 Scanning electron images of two possible waveguide designs with directional couplers using III-V semiconductors with integrated QDs. **a** Free-standing GaAs waveguides. The underlying sacrificial layer was removed to create a membrane. The optical control is achieved via bulls-eye couplers at the ends of the waveguides. Reprinted from [34] with the permission of AIP Publishing. **b** GaAs rib waveguides on an AlGaAs cladding. The optical control of the output arms is obtained by cleaving the chip perpendicular to the waveguides and collecting the transmitted light directly from the output port of the cleaved edge. Reprinted from [28] with the permission of AIP Publishing

directly defined on an AlGaAs cladding layer (see Fig. 13.5). This leads to a weaker confinement and therefore a lower dipole coupling as mentioned before as well as a need for larger structures (Fig. 13.9b). On the other hand, rib waveguides show an excellent mechanical stability even on the cm scale [28, 38]. A possibility to combine both advantages was presented in [39] where GaAs waveguides were placed on a SiO_2/Si_3N_4 substrate which leads to a high refractive index contrast while keeping the mechanical stability of rib waveguides. All three implementations [28, 34, 39] exhibit an epitaxial grown layer of self-assembled quantum dots throughout the whole waveguide structure enabling the creation of single photons at the desired position by direct laser excitation from the top.

Propagation Losses

One main benchmark of optical waveguides are their propagation losses. Only the warranty of low propagation losses enables the future scalability of photonic integrated circuits to devices with multiple qubits and large logic structures. These can be measured statistically by moving the excitation spot over the waveguide and therefore exciting QDs at different positions inside the photonic circuit. Optical losses of rib waveguides are shown in Fig. 13.10. The optical losses range from (1.20 ± 0.33) to (2.61 ± 0.57) dB/mm, which is comparable with other measurements on similar structures [40], but about one order of magnitude higher than comparable silicon technology where propagation losses between 1 and 4 dB/cm are possible for single-mode ridge type waveguides [41, 42].

It can be observed, that the bends seem to add no additional losses to the structure as the losses stay the same over the whole structures. This may be unexpected as bends

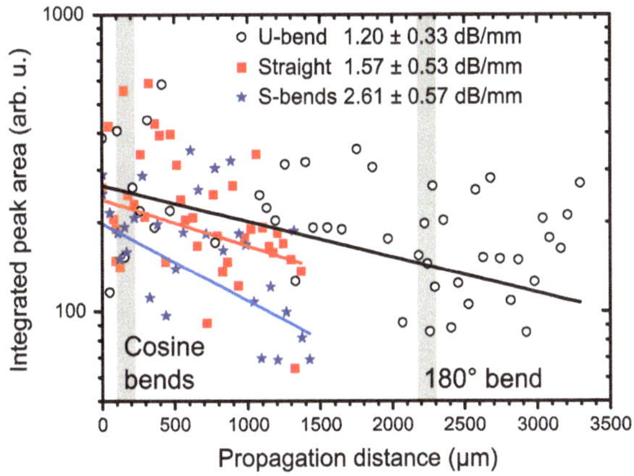

Fig. 13.10 Measured propagation losses of rib waveguide structures as shown in Fig. 13.9b. The implemented structures (s-bends and u-bends) do not introduce observable additional losses. Each dot represents a measurement of the integrated photoluminescence of a QD at a different distance to the output port at the cleaved facet of the chip. The fluctuations originate from varying QD brightnesses

are normally supposed to introduce additional losses due to the mode deformation in bend waveguide sections which lead to coupling losses between bend and straight sections and therefore the loss of power into radiative modes. Additionally this mode deformation leads to a stronger interaction between the photons and the waveguide sidewalls inside the bend region which can result in scattering losses. However, as the bending radii with $\sim 20\,\mu\text{m}$ for the cosine bends and $25\,\mu\text{m}$ for the $180°$ bends are very large, no effects on the propagation losses are notable. So the observed losses and their variation are supposed to stem mainly from imperfections and roughnesses on the waveguide sidewalls introduced during the fabrication process. For bends with smaller curvature radius the mode shift to the outer facet should be taken into account. This can be done by shifting the bend regions relative to the straight ones, which may be interesting for commercial applications where high integration densities are favorable [43–46].

Splitting Ratio

For the realization of quantum photonic circuits the exact adjustment of the splitting ratio of integrated beam splitters is essential to obtain predictable results with high visibility. So it is important to check the splitting ratio of the fabricated beam splitter. The splitting ratio of a directional coupler is not fixed for all guided wavelengths, instead it depends on the propagation constants of the supermodes (13.14). Similar to the propagation constants of the band structures of single waveguides like shown in Fig. 13.4, these propagation constants $k_{z_{\text{sym}}}$ and $k_{z_{\text{asym}}}$ depend on the free space wavevector k. Therefore, the splitting ratio is wavelength dependent.

Figure 13.11 shows the splitting ratio of several QDs of the beam splitter used in [28]. The wavelength dependency is clearly visible and needs to be considered for the design of photonic integrated circuits. In this case, the circuit could be operated around 877 nm, so the emission lines of the excited QDs have to match this operation wavelength.

Single-Mode Operation

As different guided modes experience different coupling strengths inside a directional coupler, it is also necessary to ensure that only one mode is excited to obtain predictable results from a beam splitter. For pure single-mode waveguides this is naturally given, but for rib waveguides like the one used in [28], which support both the fundamental TE and the fundamental TM mode it is necessary to show that the waveguides are only operated in the single-mode regime. This can be done by photoluminescence measurements analyzing the polarization of the light coupled from the output ports to free-space. As mentioned above, the propagating modes in dielectric waveguides are no pure TE or TM modes, therefore we expect no perfect degree of polarization (*DOP*) of 100%. The far-field polarization of guided modes coupled to free-space can be calculated using FDTD simulations and a near- to far-field transformation as described in [47] for silicon on silica waveguides.

Figure 13.12 shows the *DOP* for photons from a QD integrated into a rib waveguides measured after the passing through a beam splitter [28]. The measured value of 96% for a reflected photon is near the expected value of about $DOP = 99.2\%$

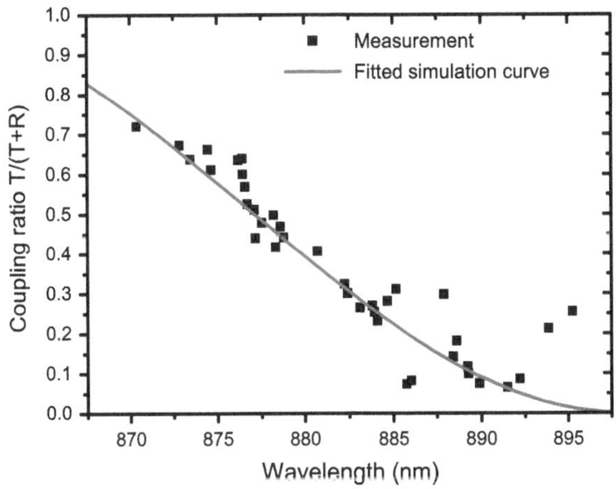

Fig. 13.11 Measured and simulated coupling ratio of the directional coupler depicted in Fig. 13.9b. For the operation of large circuits this wavelength dependent splitting ratio has to be taken into account. The simulated curve was obtained by calculating the supermodes of the fabricated coupler yielding their propagation constants. An additional scaling factor 1.072 ± 0.003 of the coupling constant is used to fit the simulation curve to the data. This accounts for the not perfectly known coupler profile and numerical inaccuracies. The near unity scaling factor shows the high consistency between simulation and measurements

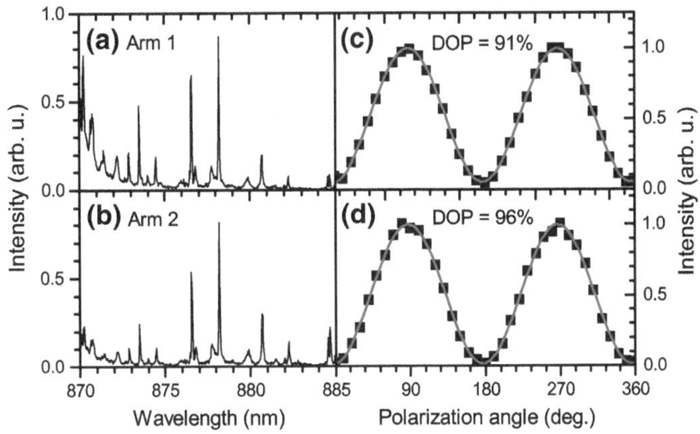

Fig. 13.12 **a+b** Ensemble QD spectra of an excitation location in arm 2 behind the beam splitter measured from both output arms 1 (**a**) and 2 (**b**) consecutively (see Fig. 13.9b). Except for the variation in the splitting ratio, both spectra are similar. **c+d** The high degree of polarization of the observed QD emission is a clear hint for single-mode operation. Reprinted from [28] with the permission of AIP Publishing

for pure TE single-mode operation of a comparable Si waveguide [47] and is therefore a clear hint for the single-mode operation. This single-mode operation is not surprising as the heavy-hole dipole moment of InGaAs QDs lies in the growth plane (horizontal), therefore the QD can only couple weakly to the TM-mode with its main field component perpendicular to the growth plane. The slightly lower *DOP* for a photon coupled to the other arm of the DC may be explained with spurious TE-/TM-mode coupling inside the DC [28].

Single-Photon Operation

After the verification of the single-mode operation it is also necessary to ensure the single-photon operation, as the beam splitter is supposed to work on single qubits and thus with single photons in the LOQC scheme. Therefore, the emission of a single QD behind a beam splitter is detected on both output arms simultaneously and the number of coincidences between the two output ports is measured for varying delay times. The obtained correlation function $g^{(2)}(\tau)$ gives a quantity of the antibunching behavior of a single-photon source, where the possibility to detect two photons at the same time vanishes in the ideal case, corresponding to $g^{(2)}(0) = 0$. These measurements have been done for both free-standing and rib beam splitters and show the single-photon operations with $g^{(2)}(0) = 0.31 \pm 0.03$ [34] and $g^{(2)}(0) = 0.06 \pm 0.14$, respectively [28].

Figure 13.13 shows a typical $g^{(2)}(\tau)$ measurement for a rib waveguide beam splitter. The used QD shown in Fig. 13.13 was excited quasi-resonantly so a nearly background-free QD emission was obtained. The splitting ratio of the QD emission was $(48.7/51.3 \pm 0.9)\%$, corresponding to its wavelength of 876.6 nm. To obtain a good signal-to-noise ratio, as necessary for correlation measurements, it is necessary to filter out the background from the excitation laser. For the early usage of photonic integrated circuits with off-chip detection, as presented here, this can be

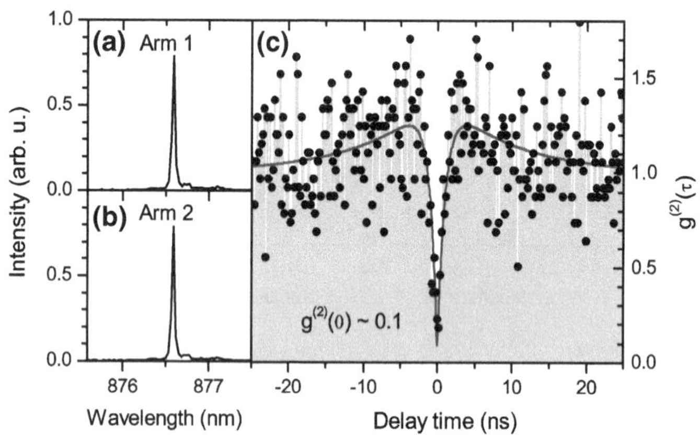

Fig. 13.13 **a+b** QD spectra of a quasi-resonantly excited single quantum dot in arm 2 behind the beam splitter measured from both output arms 1 (**a**) and 2 (**b**) consecutively (see Fig. 13.9b). A near background-free emission is observable. **c** Simultaneous measurement of this line from both output ports allows a cross-correlation measurement of the two output arms, reassembling an auto-correlation measurement of the QD with an on-chip beam splitter. Reprinted from [28] with the permission of AIP Publishing

done by sending each detected signal through a monochromator. Another method is presented in the following section.

With the verification of the low-loss single-mode operation of free-standing and rib waveguides, as well as the examination of the splitting ratio during single-photon operation, the basic operability of the beam splitter is checked. This allows the advancing to more sophisticated devices and methods as outlined in the next sections.

13.4.2 Excitation Methods

For quantum information processes one main goal is the deterministic creation of the desired input state which makes it necessary to create single, background-free photons; so only one single QD transition should be driven. A standard excitation method for semiconductor QDs is the excitation above the band gap of the GaAs barrier or via the wetting layer. This leads to an excitation of all QDs within the excitation spot of typical around 2 μm. Furthermore, the excitation laser also scatters into guided modes of the waveguide. So a broad ensemble of lines can be observed at the output port of the waveguides which makes spectral filtering of the signals necessary. But for a scaling to multi-qubit devices with several detection channels, it will be necessary to detect the light on-chip so the necessity for a spectral filtering of the signal is undesired.

A way to overcome this limitation is the quasi-resonant or resonant excitation of a single QD transition. During the quasi-resonant operation, the excitation laser is tuned near the emission line of a QD. The QD is, for example, then excited into its p-shell, while the emission takes place from the s-shell following a fast non-radiative decay. The benefit is that only one single QD is excited and the filtering has to be done only for the excitation laser. The next step is the tuning of the laser to resonant excitation. This allows the generation of single photons with superior properties as explained in Chap. 3 by *Portalupi* and *Michler*. The disadvantage is the impossibility to spectrally distinguish between laser light and quantum dot emission.

Furthermore, we want a deterministic creation of a single photon. This can be achieved by using pulsed excitation instead of continuous-wave excitation. The probability that the QD is now excited during one pulse depends on the pulse energy of the excitation laser. This leads to the observation of Rabi oscillations when varying the excitation power [48, 49]. When adjusting the laser excitation power to the π-pulse, where the state population is inverted – meaning in our case the certain creation of one exciton – we achieve a deterministic creation of a single photon. In reality, the excitation fidelity is of course limited due to dephasing processes, furthermore the coupling probability between dipole and guided mode reduces the single-photon rate (see Sect. 13.3.3) and laser stray light leads typically to high backgrounds levels.

While the coupling efficiency can be effected by the choice of different waveguide structures or even by using waveguide coupled microresonators like photonic crystal cavities [35–37], a promising way to reduce the laser stray light is the shaping of

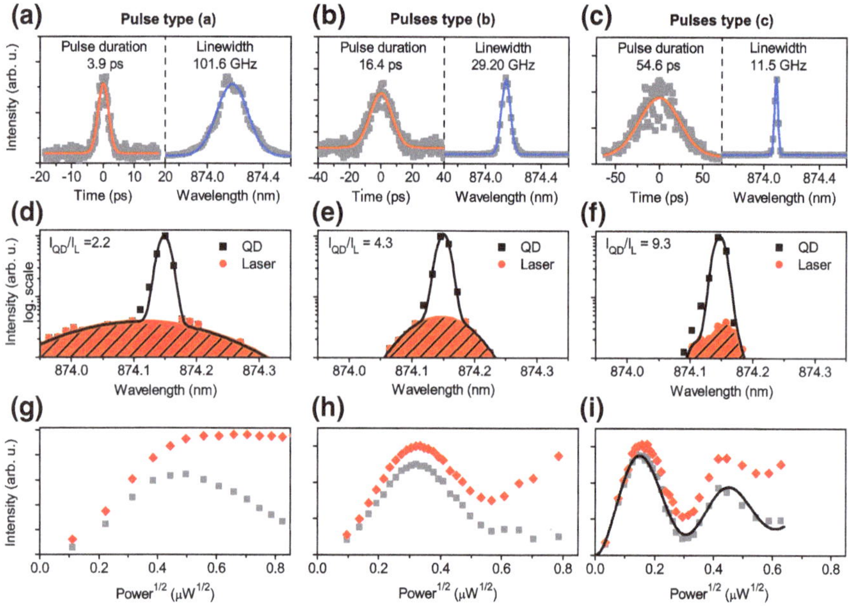

Fig. 13.14 **a–c** Temporal and spectral width of three excitation laser shapes with decreasing spectral width. **d–f** This allows the spectral matching of the laser to the QD emission for resonant excitation with enhanced signal-to-noise ratio. **g–i** Higher excitation efficiencies lead to the observation of Rabi-oscillations for varying laser excitations. Data with (without) laser background correction is plotted in *squares* (*diamonds*). The fitted curve in (**i**) corresponds to a high excitation fidelity of $(72.4 \pm 3.3)\%$. Reprinted from [49] with the permission of OSA Publishing

the excitation laser [49]. This can be done by sending the excitation laser through a monochromator with exchangeable grating. Due to the diffraction of the laser, a spectral filtering can be applied by using a spatial filter, like the coupling to a single-mode fiber. A spectrally shaped laser pulse can be seen in Fig. 13.14a–c for different gratings. It can be also observed that, as the spectral filtering removes parts of the pulses in the frequency domain, the temporal length of the pulses increase. Figure 13.14d–f demonstrate how the spectral width of the laser is shaped to nearly match the emission line of the QD. These spectra where taken from an output arm of a beam splitter. The higher overlap between laser and QD enhances the portion of laser light, which can be effectively absorbed by the QD. Therefore, the laser power can be reduced at smaller excitation linewidths. Figure 13.14g–i show the observed QD emission for varying excitation powers. The discussed Rabi-oscillations are clearly visible, also the shift of the π-pulse excitation to lower laser powers. This leads to a rise in the QD-signal to laser background ratio from 2.2 to 9.3, indicating the possibility of laser background reduction by excitation laser shaping.

13.5 Perspective of Fully Integrated Photonic Quantum Logic

The previous section focused on the realization of basic circuits using networks of photonic waveguides and the integration of single-photon sources. But there are still two main elements missing: the integration of phase-shifters and detectors. These will be covered in this section. Furthermore, a small overview on electric field tuning of semiconductor QDs will be given as this is a promising possibility to tune two different QDs to the same emission wavelength, as necessary for multi-qubit applications.

13.5.1 Phase Shifters

To complete the set of quantum gates the implementation of phase shifters, operating on single arms of the path-encoded qubits, is necessary. This allows the direct realization of phase gates (13.3) with variable phase shifts and the upgrade of directional couplers to Hadamard gates as shown below.

Another exciting possibility is the realization of programmable gate arrays which allow the implementation of arbitrary single qubit operations by adjusting integrated phase shifters. This is equivalent to the realization of arrays of arbitrary transformation in U(2), which can be expressed as

$$\begin{pmatrix} c_0 \\ c_1 \end{pmatrix}_{\text{Out}} = \begin{pmatrix} \mathrm{e}^{\mathrm{i}\gamma} \sin\omega & \mathrm{e}^{\mathrm{i}\gamma} \cos\omega \\ \cos\omega & -\sin\omega \end{pmatrix} \begin{pmatrix} c_0 \mathrm{e}^{\mathrm{i}\phi_0} \\ c_1 \mathrm{e}^{\mathrm{i}\phi_1} \end{pmatrix}_{\text{In}}, \tag{13.16}$$

with the variable reflectivity η and $\sqrt{\eta} = \sin\omega$ and adjustable input phases [50]. This corresponds to a variable beam splitter cube followed by a phase shifter on one output arm. As the absolute phase of the input state has no influence on the performance of the transformation, one of the phase factors ϕ_0, ϕ_1 can be neglected and the relative phase between both input ports can be adjusted using an additional phase shifter ϕ in front of the variable beam splitter.

At a first glance, the beam splitter with variable reflectivity is not realizable using on-chip directional couplers, where the reflectivity depends only on the coupling strength and coupling length between the waveguide arms – both parameters which are defined by the shape of the DC and the operating wavelength. Indeed, it is possible to realize integrated variable beam splitters using Mach–Zehnder interferometers (MZIs) [50]. Here a phase shifter $\exp(\mathrm{i}\theta)$ on one arm between two 50:50 directional couplers can be used to adjust a transmission matrix of the form

$$\begin{pmatrix} c_0 \\ c_1 \end{pmatrix}_{\text{Out}} = \frac{1}{2} \begin{pmatrix} \left(1 - \mathrm{e}^{\mathrm{i}\theta}\right) & \mathrm{i}\left(1 + \mathrm{e}^{\mathrm{i}\theta}\right) \\ \mathrm{i}\left(1 + \mathrm{e}^{\mathrm{i}\theta}\right) & -\left(1 - \mathrm{e}^{\mathrm{i}\theta}\right) \end{pmatrix} \begin{pmatrix} c_0 \\ c_1 \end{pmatrix}_{\text{In}}. \tag{13.17}$$

By carefully adjusting the phase shifter to generate a phase shift of $\theta = 2\arcsin\left(\sqrt{\eta}\right)$ the transmission matrix converts into a variable beam splitter cube (13.6) with an additional phase shift between reflection and transmission. By choosing a 50:50 splitting ratio, the MZI resembles a Hadamard gate.

Incorporating the MZI (13.17) into the universal quantum gate (13.16) yields a device composed of two 50:50 beam splitters and three phase shifters with a transmission matrix

$$\begin{aligned}\begin{pmatrix} c_0 \\ c_1 \end{pmatrix}_{\text{Out}} &= \frac{1}{2}\begin{pmatrix} \mathrm{e}^{\mathrm{i}\gamma} & 0 \\ 0 & 1 \end{pmatrix}\begin{pmatrix} \mathrm{i} & 1 \\ 1 & \mathrm{i} \end{pmatrix}\begin{pmatrix} \mathrm{e}^{\mathrm{i}\theta} & 0 \\ 0 & 1 \end{pmatrix}\begin{pmatrix} \mathrm{i} & 1 \\ 1 & \mathrm{i} \end{pmatrix}\begin{pmatrix} \mathrm{e}^{\mathrm{i}\phi} & 0 \\ 0 & 1 \end{pmatrix}\begin{pmatrix} c_0 \\ c_1 \end{pmatrix}_{\text{In}} \\ &= \frac{1}{2}\begin{pmatrix} \mathrm{e}^{\mathrm{i}\phi}\mathrm{e}^{\mathrm{i}\gamma}\left(1-\mathrm{e}^{\mathrm{i}\theta}\right) & \mathrm{i}\,\mathrm{e}^{\mathrm{i}\gamma}\left(1+\mathrm{e}^{\mathrm{i}\theta}\right) \\ \mathrm{i}\,\mathrm{e}^{\mathrm{i}\phi}\left(1+\mathrm{e}^{\mathrm{i}\theta}\right) & -\left(1-\mathrm{e}^{\mathrm{i}\theta}\right) \end{pmatrix}\begin{pmatrix} c_0 \\ c_1 \end{pmatrix}_{\text{In}}, \end{aligned} \tag{13.18}$$

which allows to perform any qubit rotation. As the phase difference between the output ports is insignificant for the operation of the gate, the third phase shifter $\exp(\mathrm{i}\gamma)$ can be omitted.

An integrated device composed of 15 such universal quantum gates has been shown on a silica platform and therefore with external single-photon sources [15]. This allows to perform any U(6) operation by adjusting the phase shifters. This includes, but is not limited to, a probabilistic CNOT gate (see (13.9)) operating with two qubits and two ancilla modes.

The phase shifters in [15] have been realized using thermo-optic heaters. These use the temperature dependency of the refractive index of silica to change locally the propagation constant of the guided wave, and therefore change the relative phase between the two arms inside or before the MZIs.

For the realization of phase shifters on GaAs we have to keep in mind that the integrated QDs are operated at cryogenic temperatures and the wavelength and especially the half width of their emission depends on the temperature. Therefore the introduction of heat into the waveguide structure is problematic. Fortunately, GaAs is a non-centrosymmetric material with a non-zero linear electro-optic coefficient r_{41} of around $-1.6 \times 10^{-12}\,\text{m/V}$ [51] which allows the usage of the Pockels effect to adjust the refractive index by local electric fields. An advantage of this method is the high modulation frequency of electric fields compared to thermo-optical phase shifters, where the modulation frequency is limited by the maximum heat dissipation. Instead of electrical resistors which introduce locally heat into the sample, two electrodes can be defined near a waveguide arm to introduce an electric field. Here, it is convenient to use a doped substrate as one electrode and a lithographically defined metal stripe as top electrode like schematically shown in Fig. 13.15. With this method a GaAs based MZI with integrated phase shifters was shown using external sources of single photons [52].

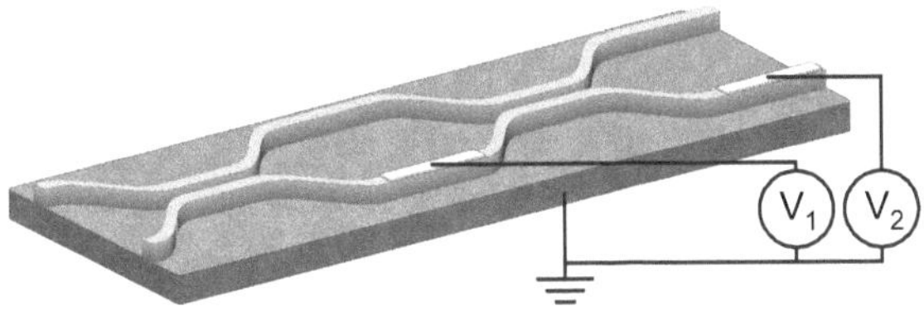

Fig. 13.15 Principle of a universal gate in the U(2) realized with a Mach–Zehnder interferometer. One phase shifter before and one phase shifter in between the directional couplers ensure a full control of the gate operation. These phase shifters are realized via metal contacts for electro-optical modulation of the refractive index

13.5.2 Electric Field Tuning

As we have seen before, it is possible to realize integrated quantum photonic circuits on GaAs. We have also seen that the amount of stray light coupled into the photonic circuits can be reduced drastically by using convenient excitation methods. But for the implementation of more complex computational schemes we need more than one single qubit and therefore several indistinguishable photons. On-chip multi-qubit operations have been mainly done using off-chip sources of single photons as the off-chip creation of indistinguishable single photons can be done by well-developed processes like parametric down-conversion [15, 17].

To use instead QDs as scalable on-chip sources for multiple qubits, the emission of remote QDs has to be matched in frequency to obtain indistinguishable photons. This is difficult to achieve by the growth process of the QDs itself, as the QDs are normally grown in a self-assembled method leading to a broad size- and therefore wavelength-distribution.

A method to tune the QD emission after their growth is the usage of temperature, strain or electric fields. This already allowed the observation of the interference between photons from remote QDs in off-chip setups [53, 54]. As mentioned above, an elevated temperature is undesired as it increases the emission linewidth. Strain fields on the other hand are difficult to introduce locally [55, 56]. So the tuning with electric fields is nowadays a common method [57]. The process requirements are rather the same as for the creation of electro-optic phase shifters. The electric field can be both applied in lateral or vertical direction [58–60].

Figure 13.16 shows possible electrode configurations for electric field tuning in waveguides. In Fig. 13.16a the electrodes can be deposited and lithographically structured nearby the waveguides to create a lateral electric field. A tuning range of about 0.2 nm by applying fields with 13 kV/cm was shown for CdSe/ZnSe QDs embedded in mesa structures [61]. Similar results were achieved on planar samples with a tuning of the excitonic line of InGaAs QDs by about 0.2 nm with up to 10 kV/cm [59].

However, the benefit of vertical electric fields is the possibility to epitaxial grow the electrodes by using doped materials (Fig. 13.16b). Such reverse biased pin-structures allow high electric fields with low voltages due to the small distance between the electrodes. The usage of such layer structures on planar samples allowed the creation

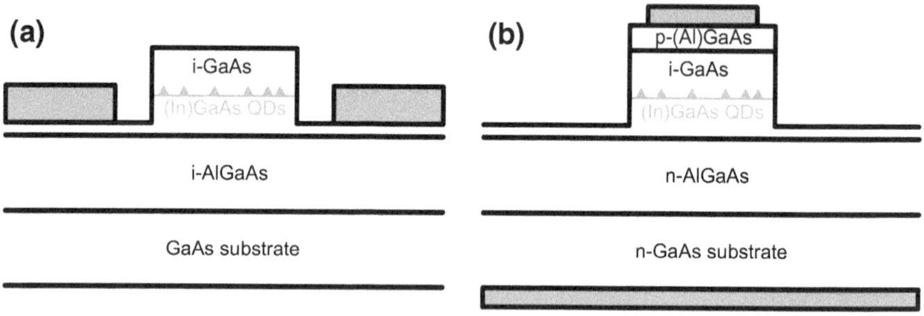

Fig. 13.16 Two possible electrode (*gray*) configurations for electric field tuning of waveguide embedded QDs. **a** Electrodes for lateral fields with two metal electrodes. **b** Electrodes for vertical fields using a pin-structure

of indistinguishable photons from remote QDs [54]. Here the indistinguishably was shown using an off-chip photonic circuits with fiber couplers. The tuning of QDs in photonic integrated circuits may be the next step to allow two-photon interference of the emission of QDs integrated in waveguide circuits.

A drawback of the structure presented in Fig. 13.16b is the doping of the waveguide structure. By doping the layers during the growth process the waveguides of the whole photonic circuit will be doped, which leads to enhanced absorption losses. A way to reduce these losses could be to reduce the overlap of the propagating mode with doped structures, by moving them further away, or replacing the top electrode by a metal electrode with sufficient distance to the waveguide. This will on the other hand reduce the electric field seen by the QDs and therefore will require higher applied voltages. So the perfect structure for electric field tuning is not found yet and it will be interesting which structure will prevail.

13.5.3 Integrated Detectors

The last part for integrated quantum photonic circuits is the detection of the photons. For a scalable device this should also be done on the chip itself. A very promising technology for integrated single-photon detectors on integrated circuits, which we want to have a short look at, are superconducting single-photon detectors (SSPDs) [62]. The principle of an SSPD is briefly summarized in Fig. 13.17a–d. It is composed of thin film superconducting nanowires which are biased close to the critical current, where the superconductivity collapses (a). The absorption of an incident photon heats the superconductor locally and creates a hot-spot where the superconductivity collapses (b). So the bias current is redirected around this hot-spot and exceeds the critical current density around the hot-spot (c). This leads to a collapse of the superconductivity over a whole width of the nanowire and therefore results in a resistance and a measurable voltage peak over the superconductor. The detector

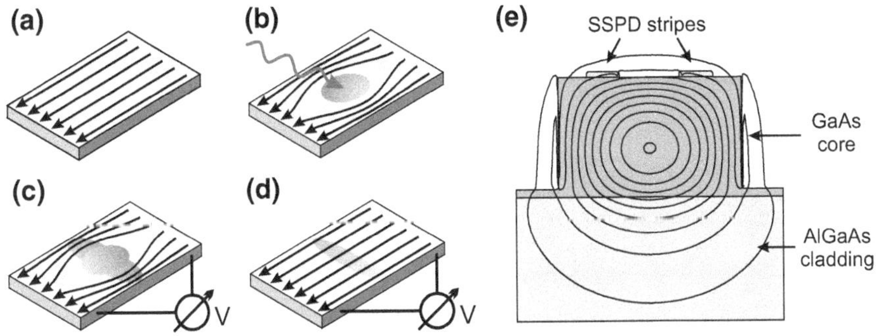

Fig. 13.17 **a–d** Principle of an SSPD: **a** The superconducting film is biased close to its critical current. **b** An absorbed photon generates heat and redirects the current flow. **c** The critical current density is exceeded near the hot-spot leading to a measurable resistivity. **d** The SSPD resets itself by heat dissipation. **e** Overlap of a propagating mode inside a waveguide with two superconducting stripes on its top. This leads to a non-vanishing absorption possibility

resets itself automatically after a short delay time by dissipating the introduced heat (d) [63, 64].

A typical superconducting material for these detectors is niobium nitride (NbN) with critical temperatures on GaAs between 6 and 12 K depending on the film thickness and quality [65]. As thinner film have higher detection efficiencies because of the higher possibility that a hot-spot creates a resistance barrier, it is necessary to reach cryogenic temperatures to achieve the transition of the NbN films to the superconducting state. Also the efficiency and the dark count rates of finished structures are directly improved with lower operation temperatures. As the QDs are also operated at cryogenic temperature, this is no drawback for SSPDs in quantum photonic circuits with integrated InGaAs QDs.

A commonly used structure for SSPDs is a meander type arrangement of the nanowires as seen in Fig. 13.18a. These structures are optimized for a high filling factor and therefore high absorption efficiency for vertical incident photons. In contrast to this, the photons from a photonic integrated circuit will not impinge from the top but are guided inside a waveguide. Here the absorption takes place by an evanescent coupling of the guided photons to the NbN film as seen in Fig. 13.17e. So the absorption probability is not proportional to any filling factor but to the overlap between mode field and superconducting film and the length of the detector. The NbN film is therefore directly deposited onto the GaAs core of the waveguides. A possible design of such a detector is depicted in Fig. 13.18b. The absorption efficiency of the detector can be in principle >99% by just increasing the length of the detector.

The possibility to detect QD fluorescence from InGaAs QDs embedded into GaAs waveguides with integrated SSPDs has already been shown [40]. Figure 13.19 depicts the demonstrated combination of a multimode waveguide with a meander type SSPD. A further improvement is the realization of photon number resolving SSPDs which has also been demonstrated [66]. But there are still struggles with stray light, as

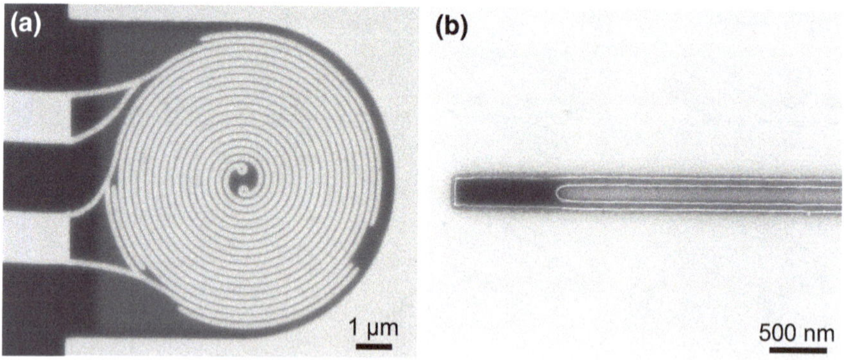

Fig. 13.18 Scanning electron microscopy images of SSPD designs. The niobium nitride is displayed in *black*. **a** Spiral design optimized for vertical incident light with maximized filling factor. **b** Dual stripe design optimized for waveguide integration [68]

Fig. 13.19 Combination of waveguide embedded QDs with integrated SSPDs. The InGaAs QDs are excited from *top* while their emission can be detected off-chip using a confocal photoluminescence geometry and on-chip by the evanescent coupling of the guided mode to a meander type NbN SSPD. Reprinted by permission from Macmillan Publishers Ltd: Scientific Reports [40], copyright 2013

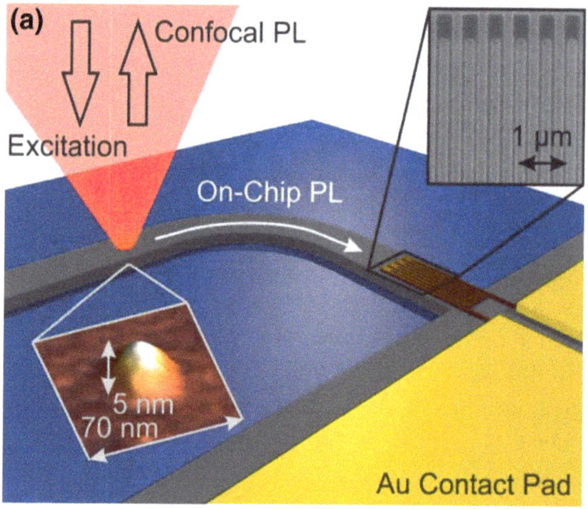

the detector detects any photon impinging on it. To suppress the laser stray light an absorbing backside coating of the substrate and the covering of the detectors has been used [67]. Together with a shaped resonant excitation scheme as presented in Sect. 13.4.2 this may lead to a sufficient laser suppression to allow the usage of integrated SSPDs for future quantum photonic circuits.

13.6 Summary

The integration of quantum photonic circuits on single semiconductor chips is currently a fast evolving field. Semiconductor QDs have proven to be suitable candidates as single-photon emitters for such chips. Several components for the realization of basic fully integrated quantum photonic circuits have been realized in the past years. Nowadays InGaAs QDs integrated in waveguide structures with beam splitters as photonic circuits are an established test bed for the development of more advanced integrated circuits. Further steps will include the manipulation of the emission of the QDs and the state manipulation by tunable phase shifters as they could be implemented by locally adjustable electric fields. This will open the door to multi-photon interference and therefore multi-qubit gates with remote QDs as single-photon sources.

On the other end of the device a lot of work is done on implementing single-photon detectors with high detection efficiencies and shielding them from stray light leading to high background levels. But the current development is not only driven by the technological development of new chips with more complex elements and the merging of developed elements on one single chip, but also by the development of new optical methods to help with the avoidance of stray light.

The future will show how all these and maybe more parts will be joined together to create single chip quantum integrated photonic circuits and if their realization with III-V semiconductors is able to prevail in the race for fully integrated quantum information processors.

Acknowledgements The authors would like to thank Nikola Prtljaga of the group of Maurice S. Skolnick of the University of Sheffield for providing a SEM image of their free-standing waveguide structures and Jonathan Finley of the Technical University of Munich for the sketch of their combination of a waveguide with embedded QDs and a SSPD. We also thank Ekkehart Schmidt of the group of Michael Siegel of the Karlsruhe Institute of Technology for the SEM images of SSPDs.

We would also like to acknowledge the people who made the presented work possible: therefore we would like to thank Matthias Paul for the epitaxial fabrication of the layer structures, Thomas Reindl of the group of Jürgen Weis of the Max-Planck-Institute for Solid State Research Stuttgart for the electron-beam lithography and Mario Schwartz for the optical characterization of the waveguide samples. We also thank Hendrik Niederbracht, Thomas Herzog, Fabian Hargart, Jan Kettler and Simone L. Portalupi for their support with the fabrication techniques and during the optical measurements.

This work was financial supported by the Deutsche Forschungsgemeinschaft DFG via the project MI 500/29-1.

References

1. D. Deutsch, Quantum theory, the church-turing principle and the universal quantum computer. Proc. R. Soc. Lond. A Math. Phys. Eng. Sci. **400**(1818), 97–117 (1985)
2. M. Fox, *Quantum Optics: An Introduction* (Oxford University Press, Oxford, 2006)

3. L.K. Grover, A fast quantum mechanical algorithm for database search, in *Proceedings of the Twenty-eighth Annual ACM Symposium on Theory of Computing, STOC '96* (ACM, New York, 1996), pp. 212–219
4. L.K. Grover. A fast quantum mechanical algorithm for database search (1996). arXiv:quant-ph/9605043
5. P.W. Shor, Algorithms for quantum computation: discrete logarithms and factoring. in *35th Annual Symposium on Foundations of Computer Science, 1994 Proceedings* (1994), pp. 124–134
6. R.P. Feynman, Simulating physics with computers. Int. J. Theor. Phys. **21**(6), 467–488 (1982)
7. E. Knill, R. Laflamme, G.J. Milburn, A scheme for efficient quantum computation with linear optics. Nature **409**, 46–52 (2001)
8. A. Barenco, A universal two-bit gate for quantum computation. Proc. R. Soc. Lond. A Math. Phys. Eng. Sci. **449**(1937), 679–683 (1995)
9. D. Deutsch, Quantum computational networks. Proc. R. Soc. Lond. Ser. A Math. Phys. Sci. **425**(1868), 73–90 (1989)
10. D.P. DiVincenzo, Two-bit gates are universal for quantum computation. Phys. Rev. A **51**, 1015–1022 (1995)
11. P.O. Boykin, T. Mor, M. Pulver, V. Roychowdhury, F. Vatan, On universal and fault-tolerant quantum computing: a novel basis and a new constructive proof of universality for Shor's basis, in *40th Annual Symposium on Foundations of Computer Science, 1999* (1999), pp. 486–494
12. T.C. Ralph, N.K. Langford, T.B. Bell, A.G. White, Linear optical controlled-not gate in the coincidence basis. Phys. Rev. A **65**, 062324 (2002)
13. C.K. Hong, Z.Y. Ou, L. Mandel, Measurement of subpicosecond time intervals between two photons by interference. Phys. Rev. Lett. **59**, 2044–2046 (1987)
14. T.B. Pittman, B.C. Jacobs, J.D. Franson, Probabilistic quantum logic operations using polarizing beam splitters. Phys. Rev. A **64**, 062311 (2001)
15. J. Carolan, C. Harrold, C. Sparrow, E. Martín-López, N.J. Russell, J.W. Silverstone, P.J. Shadbolt, N. Matsuda, M. Oguma, M. Itoh, G.D. Marshall, M.G. Thompson, J.C.F. Matthews, T. Hashimoto, J.L. O'Brien, A. Laing, Universal linear optics. Science **349**(6249), 711–716 (2015)
16. J.L. O'Brien, G.J. Pryde, A.G. White, T.C. Ralph, D. Branning, Demonstration of an all-optical quantum controlled-not gate. Nature **426**(6964), 264–267 (2003)
17. A. Politi, M.J. Cryan, J.G. Rarity, S. Yu, J.L. O'Brien, Silica-on-silicon waveguide quantum circuits. Science **320**(5876), 646–649 (2008)
18. S. Gasparoni, J.-W. Pan, P. Walther, T. Rudolph, A. Zeilinger, Realization of a photonic controlled-NOT gate sufficient for quantum computation. Phys. Rev. Lett. **93**, 020504 (2004)
19. A. Peruzzo, P. Shadbolt, N. Brunner, S. Popescu, J.L. O'Brien, A quantum delayed-choice experiment. Science **338**(6107), 634–637 (2012)
20. J.B. Spring, B.J. Metcalf, P.C. Humphreys, W. Steven Kolthammer, X.-M. Jin, M. Barbieri, A. Datta, N. Thomas-Peter, N.K. Langford, D. Kundys, J.C. Gates, B.J. Smith, P.G.R. Smith, I.A. Walmsley, Boson sampling on a photonic chip. Science **339**(6121), 798–801 (2013)
21. C.L. Xu, W.P. Huang, M.S. Stern, S.K. Chaudhuri, Full-vectorial mode calculations by finite difference method. IEE Proc. Optoelectron. **141**(5), 281–286 (1994)
22. S.G. Johnson, J.D. Joannopoulos, Block-iterative frequency-domain methods for Maxwell's equations in a planewave basis. Opt. Express **8**(3), 173–190 (2001)
23. R.J. Coles, D.M. Price, J.E. Dixon, B. Royall, E. Clarke, P. Kok, M.S. Skolnick, A.M. Fox, M.N. Makhonin, Chirality of nanophotonic waveguide with embedded quantum emitter for unidirectional spin transfer. Nat. Commun. **7**(11183) (2016)
24. M. Benyoucef, H.S. Lee, J. Gabel, T.W. Kim, H.L. Park, A. Rastelli, O.G. Schmidt, Wavelength tunable triggered single-photon source from a single CdTe quantum dot on silicon substrate. Nano Lett. **9**(1), 304–307 (2009)
25. M. Wiesner, W.-M. Schulz, C. Kessler, M. Reischle, S. Metzner, F. Bertram, J. Christen, R. Roßbach, M. Jetter, P. Michler, Single-photon emission from electrically driven InP quantum dots epitaxially grown on CMOS-compatible Si(001). Nanotechnology **23**(33), 335201 (2012)

26. H. Fukuda, K. Yamada, T. Shoji, M. Takahashi, T. Tsuchizawa, T. Watanabe, J.-i. Takahashi, S.-i. Itabashi, Four-wave mixing in silicon wire waveguides. Opt. Express **13**(12), 4629–4637 (2005)
27. J.W. Silverstone, D. Bonneau, K. Ohira, N. Suzuki, H. Yoshida, N. Iizuka, M. Ezaki, C.M. Natarajan, M.G. Tanner, R.H. Hadfield, V. Zwiller, G.D. Marshall, J.G. Rarity, J.L. O'Brien, M.G. Thompson, On-chip quantum interference between silicon photon-pair sources. Nat. Photon **8**(2), 104–108 (2014)
28. U. Rengstl, M. Schwartz, T. Herzog, F. Hargart, M. Paul, S.L. Portalupi, M. Jetter, P. Michler, On-chip beamsplitter operation on single photons from quasi-resonantly excited quantum dots embedded in GaAs rib waveguides. Appl. Phys. Lett. **107**(2) (2015)
29. E. Murray, D.J.P. Ellis, T. Meany, F.F. Floether, J.P. Lee, J.P. Griffiths, G.A.C. Jones, I. Farrer, D.A. Ritchie, A.J. Bennett, A.J. Shields, Quantum photonics hybrid integration platform. Appl. Phys. Lett. **107**(17) (2015)
30. I.E. Zadeh, A.W. Elshaari, K.D. Jns, A. Fognini, D. Dalacu, P.J. Poole, M.E. Reimer, V. Zwiller, Deterministic integration of single photon sources in silicon based photonic circuits. Nano Lett. **16**(4), 2289–2294 (2016)
31. A.F. Oskooi, D. Roundy, M. Ibanescu, P. Bermel, J.D. Joannopoulos, S.G. Johnson, MEEP: a flexible free-software package for electromagnetic simulations by the FDTD method. Comput. Phys. Commun. **181**, 687–702 (2010)
32. G. Angelatos, S. Hughes, Theory and design of quantum light sources from quantum dots embedded in semiconductor-nanowire photonic-crystal systems. Phys. Rev. B **90**, 205406 (2014)
33. G. Lecamp, P. Lalanne, J.P. Hugonin, Very large spontaneous-emission β factors in photonic-crystal waveguides. Phys. Rev. Lett. **99**, 023902 (2007)
34. N. Prtljaga, R.J. Coles, J. O'Hara, B. Royall, E. Clarke, A.M. Fox, M.S. Skolnick, Monolithic integration of a quantum emitter with a compact on-chip beam-splitter. Appl. Phys. Lett. **104**(23) (2014)
35. M. Arcari, I. Söllner, A. Javadi, S. Lindskov Hansen, S. Mahmoodian, J. Liu, H. Thyrrestrup, E.H. Lee, J.D. Song, S. Stobbe, P. Lodahl, Near-unity coupling efficiency of a quantum emitter to a photonic crystal waveguide. Phys. Rev. Lett. **113**, 093603 (2014)
36. S. Fattah poor, T.B. Hoang, L. Midolo, C.P. Dietrich, L.H. Li, E.H. Linfield, J.F.P. Schouwenberg, T. Xia, F.M. Pagliano, F.W.M. van Otten, A. Fiore, Efficient coupling of single photons to ridge-waveguide photonic integrated circuits. Appl. Phys. Lett. **102**(13) (2013)
37. P. Lodahl, S. Mahmoodian, S. Stobbe, Interfacing single photons and single quantum dots with photonic nanostructures. Rev. Mod. Phys. **87**, 347–400 (2015)
38. K.D. Jns, U. Rengstl, M. Oster, F. Hargart, M. Heldmaier, S. Bounouar, S.M. Ulrich, M. Jetter, P. Michler, Monolithic on-chip integration of semiconductor waveguides, beamsplitters and single-photon sources. J. Phys. D Appl. Phys. **48**(8), 085101 (2015)
39. P. Stepanov, A. Delga, X. Zang, J. Bleuse, E. Dupuy, E. Peinke, P. Lalanne, J.-M. Grard, J. Claudon, Quantum dot spontaneous emission control in a ridge waveguide. Appl. Phys. Lett. **106**(4) (2015)
40. G. Reithmaier, S. Lichtmannecker, T. Reichert, P. Hasch, K. Müller, M. Bichler, R. Gross, J.J. Finley, On-chip time resolved detection of quantum dot emission using integrated superconducting single photon detectors. Sci. Rep. **3**, 1901 (2013)
41. M. Gnan, S. Thoms, D.S. Macintyre, R.M.D. La Rue, M. Sorel, Fabrication of low-loss photonic wires in silicon-on-insulator using hydrogen silsesquioxane electron-beam resist. Electron. Lett. **44**(2), 115–116 (2008)
42. Y.A. Vlasov, S.J. McNab, Losses in single-mode silicon-on-insulator strip waveguides and bends. Opt. Express **12**(8), 1622–1631 (2004)
43. P. Bienstman, E. Six, A. Roelens, M. Vanwolleghem, R. Baets, Calculation of bending losses in dielectric waveguides using eigenmode expansion and perfectly matched layers. IEEE Photonics Technol. Lett. **14**(2), 164–166 (2002)
44. E.A.J. Marcatili, Bends in optical dielectric guides. Bell Syst. Tech. J. **48**(7), 2103–2132 (1969)

45. A. Melloni, P. Monguzzi, R. Costa, M. Martinelli, Design of curved waveguides: the matched bend. J. Opt. Soc. Am. A **20**(1), 130–137 (2003)
46. M.K. Smit, E.C.M. Pennings, H. Blok, A normalized approach to the design of low-loss optical waveguide bends. J. Lightwave Technol. **11**(11), 1737–1742 (1993)
47. J. Wang, J.C. Wirth, Y. Xuan, D.E. Leaird, A.M. Weiner, M. Qi, Far-field polarization characterization of the fundamental modes of a strip silicon waveguide. Opt. Lett. **38**(22), 4785–4788 (2013)
48. M.N. Makhonin, J.E. Dixon, R.J. Coles, B. Royall, I.J. Luxmoore, E. Clarke, M. Hugues, M.S. Skolnick, A. Mark Fox, Waveguide coupled resonance fluorescence from on-chip quantum emitter. Nano Lett. **14**(12), 6997–7002 (2014)
49. M. Schwartz, U. Rengstl, T. Herzog, M. Paul, J. Kettler, S.L. Portalupi, M. Jetter, P. Michler, Generation, guiding and splitting of triggered single photons from a resonantly excited quantum dot in a photonic circuit. Opt. Express **24**(3), 3089–3094 (2016)
50. M. Reck, A. Zeilinger, H.J. Bernstein, P. Bertani, Experimental realization of any discrete unitary operator. Phys. Rev. Lett. **73**, 58–61 (1994)
51. Gallium arsenide (GaAs), electrooptic constants, second and third order nonlinear susceptibilities: Datasheet from Landolt-Börnstein - Group III Condensed Matter, Volume 41A1β: Group IV Elements, IV-IV and III-V Compounds. Part b - Electronic, Transport, Optical and Other Properties in SpringerMaterials (doi:10.1007/10832182_216). Copyright 2002 Springer, Berlin
52. J. Wang, A. Santamato, P. Jiang, D. Bonneau, E. Engin, J.W. Silverstone, M. Lermer, J. Beetz, M. Kamp, S. Hfling, M.G. Tanner, C.M. Natarajan, R.H. Hadfield, S.N. Dorenbos, V. Zwiller, J.L. O'Brien, M.G. Thompson, Gallium arsenide (GaAs) quantum photonic waveguide circuits. Opt. Commun. **327**, 49–55 (2014). Special Issue on Nonlinear Quantum Photonics
53. E.B. Flagg, A. Muller, S.V. Polyakov, A. Ling, A. Migdall, G.S. Solomon, Interference of single photons from two separate semiconductor quantum dots. Phys. Rev. Lett. **104**, 137401 (2010)
54. R.B. Patel, A.J. Bennett, I. Farrer, C.A. Nicoll, D.A. Ritchie, A.J. Shields, Two-photon interference of the emission from electrically tunable remote quantum dots. Nat. Photon **4**(9), 632–635 (2010)
55. Y. Chen, J. Zhang, M. Zopf, K. Jung, Y. Zhang, R. Keil, F. Ding, O.G. Schmidt, Wavelength-tunable entangled photons from silicon-integrated III-V quantum dots. Nat. Commun. **7**(10387) (2016)
56. R. Trotta, J. Martin-Sanchez, J.S. Wildmann, G. Piredda, M. Reindl, C. Schimpf, E. Zallo, S. Stroj, J. Edlinger, A. Rastelli, Wavelength-tunable sources of entangled photons interfaced with atomic vapours. Nat. Commun. **7**(10375) (2016)
57. T.B. Hoang, J. Beetz, M. Lermer, L. Midolo, M. Kamp, S. Höfling, A. Fiore, Widely tunable, efficient on-chip single photon sources at telecommunication wavelengths. Opt. Express **20**(19), 21758–21765 (2012)
58. M.E. Reimer, M. Korkusiński, D. Dalacu, J. Lefebvre, J. Lapointe, P.J. Poole, G.C. Aers, W.R. McKinnon, P. Hawrylak, R.L. Williams, Prepositioned single quantum dot in a lateral electric field. Phys. Rev. B **78**, 195301 (2008)
59. M.M. Vogel, S.M. Ulrich, R. Hafenbrak, P. Michler, L. Wang, A. Rastelli, O.G. Schmidt, Influence of lateral electric fields on multiexcitonic transitions and fine structure of single quantum dots. Appl. Phys. Lett. **91**(5) (2007)
60. O. Wolst, M. Schardt, M. Kahl, S. Malzer, G.H. Dhler, A combined investigation of lateral and vertical Stark effect in InAs self-assembled quantum dots in waveguide structures. Phys. E Low-dimensional Syst. Nanostructures **13**(2–4), 283–288 (2002)
61. J. Seufert, M. Obert, M. Scheibner, N.A. Gippius, G. Bacher, A. Forchel, T. Passow, K. Leonardi, D. Hommel, Stark effect and polarizability in a single CdSe/ZnSe quantum dot. Appl. Phys. Lett. **79**(7), 1033–1035 (2001)
62. F. Flassig, M. Kaniber, G. Reithmaier, K. Mller, A. Andrejew, R. Gross, J. Vukovic, J.J. Finley, Towards on-chip generation, routing and detection of non-classical light, *Quantum Dots and Nanostructures: Synthesis, Characterization, and Modeling XII*, vol. 9373 (2015), pp. 937305–937305–9

63. G.N. Gol'tsman, O. Okunev, G. Chulkova, A. Lipatov, A. Semenov, K. Smirnov, B. Voronov, A. Dzardanov, C. Williams, R. Sobolewski, Picosecond superconducting single-photon optical detector. Appl. Phys. Lett. **79**(6), 705–707 (2001)
64. A.D. Semenov, G.N. Gol'tsman, A.A. Korneev, Quantum detection by current carrying superconducting film. Phys. C Supercond. **351**(4), 349–356 (2001)
65. G. Reithmaier, J. Senf, S. Lichtmannecker, T. Reichert, F. Flassig, A. Voss, R. Gross, J.J. Finley, Optimisation of NbN thin films on GaAs substrates for in-situ single photon detection in structured photonic devices. J. Appl. Phys. **113**(14) (2013)
66. D. Sahin, A. Gaggero, Z. Zhou, S. Jahanmirinejad, F. Mattioli, R. Leoni, J. Beetz, M. Lermer, M. Kamp, S. Hfling, A. Fiore, Waveguide photon-number-resolving detectors for quantum photonic integrated circuits. Appl. Phys. Lett. **103**(11) (2013)
67. G. Reithmaier, M. Kaniber, F. Flassig, S. Lichtmannecker, K. Mller, A. Andrejew, J. Vuckovic, R. Gross, J.J. Finley, On-chip generation, routing, and detection of resonance fluorescence. Nano Lett. **15**(8), 5208–5213 (2015). PMID: 26102603
68. SEM images of SSPDs were provided by the Institute of Micro- und Nanoelectronic Systems (IMS) of the Karlsruhe Institute of Technology (KIT), Karlsruhe, Germany

Index

P. Michler (ed.), *Quantum Dots for Quantum Information Technologies*,
Nano-Optics and Nanophotonics, DOI 10.1007/978-3-319-56378-7

FSC
www.fsc.org
MIX
Papier | Fördert gute Waldnutzung
FSC® C083411

Kolibri 360 GmbH
Ferdinand-Jühlke-Straße 7
99095 Erfurt, Deutschland
produktsicherheit@kolibri360.de